T0293064

Differential Equations with MATLAB®

Exploration, Applications, and Theory

TEXTBOOKS in MATHEMATICS

Series Editors: Ken Rosen and Al Boggess

PUBLISHED TITLES

RISK ANALYSIS IN ENGINEERING AND ECONOMICS, SECOND EDITION
Bilal M. Ayyub

INTRODUCTION TO THE CALCULUS OF VARIATIONS AND CONTROL WITH MODERN APPLICATIONS
John T. Burns

MIMETIC DISCRETIZATION METHODS
Jose E. Castillo

AN INTRODUCTION TO PARTIAL DIFFERENTIAL EQUATIONS WITH MATLAB®, SECOND EDITION
Mathew Coleman

RISK MANAGEMENT AND SIMULATION
Aparna Gupta

ABSTRACT ALGEBRA: AN INQUIRY-BASED APPROACH
Jonathan K. Hodge, Steven Schlicker, and Ted Sundstrom

QUADRACTIC IRRATIONALS: AN INTRODUCTION TO CLASSICAL NUMBER THEORY
Franz Holter-Koch

GROUP INVERSES OF M-MATRICES AND THEIR APPLICATIONS
Stephen J. Kirkland

AN INTRODUCTION TO NUMBER THEORY WITH CRYPTOGRAPHY
James Kraft and Larry Washington

REAL ANALYSIS AND FOUNDATIONS, THIRD EDITION
Steven G. Krantz

ELEMENTS OF ADVANCED MATHEMATICS, THIRD EDITION
Steven G. Krantz

APPLYING ANALYTICS: A PRACTICAL APPROACH
Evan S. Levine

ADVANCED LINEAR ALGEBRA
Nicholas Loehr

DIFFERENTIAL EQUATIONS WITH MATLAB®: EXPLORATION, APPLICATIONS, AND THEORY
Mark A. McKibben and Micah D. Webster

PUBLISHED TITLES CONTINUED

APPLICATIONS OF COMBINATORIAL MATRIX THEORY TO LAPLACIAN MATRICES OF GRAPHS
Jason J. Molitierno

ABSTRACT ALGEBRA: AN INTERACTIVE APPROACH
William Paulsen

ADVANCED CALCULUS: THEORY AND PRACTICE
John Srdjan Petrovic

COMPUTATIONS OF IMPROPER REIMANN INTEGRALS
Ioannis Roussos

TEXTBOOKS in MATHEMATICS

Differential Equations with MATLAB®

Exploration, Applications, and Theory

Mark A. McKibben
West Chester University
West Chester, Pennsylvania, USA

Micah D. Webster
Goucher College
Baltimore, Maryland, USA

CRC Press is an imprint of the
Taylor & Francis Group, an **informa** business
A CHAPMAN & HALL BOOK

CRC Press
Taylor & Francis Group
6000 Broken Sound Parkway NW, Suite 300
Boca Raton, FL 33487-2742

ISBN 13: 978-1-4665-5707-9 (hbk)

Visit the Taylor & Francis Web site at
http://www.taylorandfrancis.com

and the CRC Press Web site at
http://www.crcpress.com

To our mentors...
Sergiu Aizicovici, Sam “Cal” Calavitta, Patrick Guidotti, and David Keck.

Contents

List of Figures

List of Tables

Preface

The mathematical modeling of complex phenomena that evolve over time relies heavily on the analysis of a variety of systems of ordinary and partial differential equations. Such models are developed in very disparate areas of study, ranging from the physical and biological sciences, engineering, and population ecology to economics and financial markets. The fact that the use of differential equations is so widespread ensures that students, scholars, and professionals across disciplines, at one time or another, will encounter a differential equation that describes a scenario of particular interest to them. Yet, a traditional course in differential equations focuses on basic techniques of solving certain types of equations and analyzing some very basic mathematical models. Even more innovative differential equations courses only scratch the surface of analyzing differential equations from a qualitative viewpoint. Such first courses simply do not provide the tools necessary to analyze differential equations and their utility in mathematical modeling is not imparted onto the student.

The mathematical community has responded to this apparent gap by developing mathematical modeling courses and second courses in differential equations. There are vast differences in the approaches and topic choice used in these courses. Many such books tend to focus primarily on discrete dynamical systems, while others provide a vast collection of topics (including discrete dynamical systems, optimization methods, probabilistic models, and graph theory) that cover different types of models. The few books that fall into this latter category are thorough, but tend to be encyclopedic compendia of ideas designed for those already familiar with the subject. They would be daunting to a beginning-level student who had only completed a course in multivariable calculus.

This book is an attempt to fill the apparent gap in the existing literature on books related to mathematical modeling and differential equations, and is unique in many ways. Our primary goal is to instill in the reader a sense of intuition and practical and theoretical "know-how" as it pertains to mathematical models involving ordinary and partial differential equations. We provide a bigger, unifying picture inherent to the study and analysis of more than 20 distinct models spanning disciplines, and make this material accessible to a student who has only completed a course in multivariable calculus by conveying the thought process of building the surrounding theory from the ground up, and actively involving them in this mathematical enterprise throughout the text. Pedagogically, the goals of the project are achieved by engaging the reader by posing questions of all types (from verifying details and illustrating theorems with examples to posing (and proving) conjectures of actual results and analyzing broad strokes that occur within the development of the theory itself, and applying it to specific models) throughout the development of the material. Heuristic commentary and motivation are also included to provide an intuitive flow through the chapters. As such, the exposition in the text, at times, may lack the "polished style" of a mathematical monograph, and the language used will be colloquial English rather than the standard mathematical language. But, this style has the benefit of encouraging the reader to not simply passively read the text, but rather work through it, which is essential to obtaining a meaningful grasp of the material.

The book is divided into two main parts. The first is devoted to the study of systems of linear ordinary differential equations. We introduce and develop, to various degrees, mathematical models from ten disparate fields of study, including pharmacokinetics, chemistry, classical mechanics, neural networks, physiology, and electrical circuits. These models play crucial roles throughout the chapter. For one, their analysis motivates different aspects of the theory being developed. We use interactive activities extensively to guide the readers through different processes and sequences of observations that lead to the development of theoretical results. Once the theory is established, the models then serve a secondary role of illustrating the applicability of the abstract theory. The theoretical results in Part I are developed in three distinct cases, each one of which builds upon the previous. These cases are the homogeneous case, the nonhomogeneous case (with a time-dependent forcing term), and the semilinear case (where the forcing term depends on time and state). The topics of existence and uniqueness, continuous dependence on parameters and initial conditions, long-term behavior, and convergence schemes permeate all theoretical chapters.

Part II mimics Part I in structure almost identically, but this time focuses on linear partial differential equations. The guiding principle of the approach in Part II is to express the partial differential equations that arise in the mathematical modeling of phenomena in ten different fields of study, including heat conduction, wave propagation, fluid flow through fissured rocks, pattern formation, and financial mathematics, as abstract version of the linear ordinary differential equations that were studied in Part I. Once this connection is made, the theoretical results from Part I can be subsumed as a special case of those in Part II.

The structure of this book is likely to be somewhat different than others through which you have worked. For one, the text is NOT simply intended to be read passively. Rather, it is chock full of questions posed to the reader! A passive reading of this text will prove to be ineffective, whereas carefully working through it will result in a solid understanding of the material. Everything from computations to underlying assumptions must be carefully examined in order to fully understand the material. An integral part of this text is the collective of questions posed throughout. We consciously indicate when the discussion is intentionally terse by including parenthetical questions and instructions, such as **(Why?)**, **(Do so!)**, and **(Explain how.)**. Sometimes, we even bring the discussion to a screeching halt with a **STOP!** It is imperative that you attempt to answer each of these questions to ensure you develop a solid background as you work through the text.

While the primary goal of the text is to develop an overarching mathematical theory, we are doing so under the assumption that your prior exposure to theoretical mathematics is minimal. We intend to build your intuition and involve you in the enterprise of building the theory from the ground up. That said, we need to arm you with the essential tools of analysis, but in a manner that balances intuition, practicality, and mathematical rigor. This is accomplished with a preliminary chapter in each main part of the book which provides the essential tools to be used in that part, along with some hands-on **EXPLORE!** activities. Certainly, the completion of these portions of the text is by no means a substitute for a formal study of analysis. But, they do provide you with just enough knowledge to proceed through the next step of the development of the theory.

Applying the theory is, to many readers, arguably the most important and satisfying part of the journey. Our dual emphasis on the analysis of mathematical models is not typical in textbooks. We feel that this connection to concrete applications helps build intuition, provides motivation for why someone would want to develop such a theory, and post fortiori arms having developed the theory by examining the plentiful collection of applications. There are **APPLY IT!** activities sprinkled throughout the text, asking you to reexamine models previously mentioned in light of newly-formed theoretical results.

Finally, the glue that binds all aspects of our journey is our extensive use of MATLAB® Graphical User Interfaces (GUIs). As you will see, this feature of the text will enable you to discover patterns and make conjectures that otherwise would be solely reachable using real analysis and, for all practical purposes, be out of reach. Although not a goal of this text, the numerical algorithms used in GUIs to construct solutions to differential equations mimic the theoretical approach outlined in the text. For more details, an interested reader is directed to [44, 35, 8, 23, 29, 28].

It is our hope that this book will be useful not only in the classroom, but also as a book that anyone with an interest can pick up and develop a sense of mastery of the subject. We welcome you and hope that you find your journey through this text as enjoyable and fulfilling as it was for us to write it.

The MATLAB GUIs for use with this textbook can be downloaded from the following webpage. http://www.wcupa.edu/_ACADEMICS/SCH_CAS.MAT/mmckibben/DEwithMATLAB.asp

Please direct any inquiries via email to either Dr. Micah Webster (micah.webster@goucher.edu) or Dr. Mark McKibben (mmckibben@wcupa.edu).

Acknowledgments

The two-plus years that it took to co-write this text was one of the most rewarding, fulfilling, and dare we say, fun, experiences in each of our professional careers. Various stages of writing this book and developing the MATLAB GUIs were truly energizing and uplifting, while others required us to plumb the very depths of our patience and perseverance; it was all we could do to restrain each other from hurling the computer off the desk during momentary lapses of sanity. That said, this book is truly a labor of love, and we hope that the joy we experienced writing it (evident perhaps most outwardly by our occasional playfulness in the writing) and our love of the subject are infectious to our readers.

There are many people without whom the project would likely not have come to fruition. Our respective spouses, Jodi and Noelle, have been indelible sources of encouragement, energy, and support throughout this project.

Mark's mother, Pat, has been supportive of all his academic endeavors for decades and Micah's young son, Zane, taught him to focus, adapt, and to live in the moment. We both would like to thank Dr. Richard Craster and Dr. Roberto Sassi for their support and permission to alter some of their MATLAB codes for use in several Part II GUIs.

Our respective mentors, Sergui Aizicovici, Sam Calavitta, Patrick Guidotti, and David Keck were instrumental in the development of our academic and professional careers. Their unbridled enthusiasm and love of mathematics continue to be a significant driving force.

Cynthia Young, Dean of the College of Arts and Sciences at UCF; Patrick Guidotti, Professor of Mathematics at UCI; and Geoff Cox, Assistant Professor of Mathematics at VMI all gave valuable feedback on initial stages of the proposal.

Our students have been very helpful throughout the process. At Goucher, we have Alexa Gaines, who beta tested Part I of the text, including all the exercises, **EXPLORE!** Projects, and GUIs. And, the MA 347 class we co-taught effectively alpha-tested early versions of Part I of the text. More recently two MAT 343 classes at West Chester University beta-tested certain EXPLORE! Projects and GUIs from Part I.

From West Chester University of Pennsylvania, graduate students Gavin Hobbs and Andrew Stump put forth a concerted effort beta-testing the MATLAB GUIs and writing up careful solutions of the corresponding exercises. This is tedious work that would not have been completed were it not for their assistance. These solutions will be included on the webpage for the book.

We would also like to thank Goucher College for supporting this endeavor through a Junior Faculty Development Grant and a Faculty Development Grant.

We would like to thank the entire Taylor and Francis team. To our editors Bob Stern and Bob Ross, thank you for patiently guiding us through the publication process from beginning to end. To our project coordinator Jessica Vakili, who handled our production, stylistic, and marketing questions in a very helpful and timely manner. And to Pawan Sharma, who professionally produced various diagrams in the text. And finally, to Shashi Kumar, for his efforts in typesetting the final manuscript.

And, last but not least, we would like to thank you, the reader, for embarking on this journey with us through an amazing field of mathematics. We hope your study of it is as fulfilling as ours has been thus far.

MATLAB is a registered trademark of The MathWorks, Inc. For product information, please contact:

The MathWorks, Inc.
3 Apple Hill Drive
Natick, MA 01760-2098 USA
Tel: 508 647 7000
Fax: 508-647-7001
E-mail: info@mathworks.com
Web: www.mathworks.com

Author Bios

Mark A. McKibben earned his Ph.D. in mathematics from Ohio University in 1999 in the area of differential equations under the direction of Sergiu Aizicovici. He joined the faculty of the Mathematics and Computer Science Department at Goucher College in 1999. He earned tenure and achieved the rank of full professor during his time at that institution, and in 2013, he joined the Mathematics Department at West Chester University of Pennsylvania as an associate professor. His research interests include abstract evolution equations, stochastic analysis, and control theory. He has published more than two dozen articles on these subjects in peer-reviewed journals, has delivered national and international talks on the results, and has published two graduate-level books on abstract evolution equations. He uses a discovery learning approach in the classroom that prompts the students to engage in the learning process at all stages of the curriculum. He was recognized by the Mathematical Association of America in 2012 for excellence in teaching and received the *John M. Smith Teaching Award.*

Micah Webster graduated from the University of California, Irvine in 2007 with a Ph.D. in mathematics. There, under the supervision of Patrick Guidotti, his research focused on both theoretical properties of and numerical approximations to solutions of nonlinear diffusion equations. He joined the faculty of the Mathematics and Computer Science Department at Goucher College in 2008. His research interests include abstract evolution equations, numerical analysis, and image processing. His pedagogical interests focus on advancing student engagement and retention by utilizing the inverted classroom, blended learning, and project-based learning. In 2012, he was the first recipient of the *Excellence in Teaching Award* for a non-tenured faculty member at Goucher College.

Part I

Ordinary Differential Equations

Chapter 1

Welcome!

Welcome to the rich world that is differential equations. What prompted you to embark on this journey? Are you hunting for answers to questions about the Universe? Are you, as the result of allergies, contemplating taking two tablets instead of one to alleviate symptoms, but seek mathematical affirmation of the effectiveness of this decision? Have you wondered why it is that while you are standing atop the Empire State Building - swaying back and forth, back and forth, ... - the building doesn't simply concede defeat and snap in half? Are you curious as to how quickly the avian flu virus can spread throughout the world, infecting unsuspecting passerbyes? Or, how about those airplane wings - why don't they simply break apart in the middle of flight?

Whatever your reason for opening this book, your study of this material can shed some light on the answers to such questions. Many interesting, often surprising and sometimes downright unbelievable adventures await you. Again, welcome, hang on, and enjoy the ride.

1.1 Introduction

What are differential equations? No succinct definition truly captures the essence of this sprawling area of study. Researchers study phenomena occurring in nature guided by the goal of predicting the behavior of a solution of a system of equations governed by a set of discipline-specific laws. Relying on experimental observation at the onset of an investigation can be costly, subject to error, and can prove to be futile if the imposed assumptions turn out to be incorrect or infeasible. An often more pragmatic first step is to formulate a mathematical model of the situation and study its properties. Assuming success in this regard, a natural second step is to test the validity of the model using actual experimentation. The equations and/or assumptions are reexamined and tweaked, the abstract theory is modified, and then the system is again tested experimentally.

The phenomena with which we shall concern ourselves in Part I can all be characterized by quantities that change with time and can be formally described using a system of ordinary differential equations. The birth of differential equations as an area of study goes back to 1687 when they first appeared in Sir Isaac Newton's book *Philosophiae Naturalis Principia Mathematics.* The equations he posed were consequences of his observation that one should study forces to which a system responds in order to describe the laws of nature. Since then, many researchers have made significant contributions, creating a massive continually-growing literature.

Many situations in physics, chemistry, economics, etc. that you have likely encountered are described by differential equations. Among the common examples of these phenomena are bacteria and population growth and decay, elementary financial models, and predator-prey systems. As you will discover along your journey through this text, differential equations play a crucial role in the modeling and analysis of numerous complicated scenarios.

1.2 This Book Is a Field Guide. What Does That Mean For YOU?

A perusal of the Table of Contents will reveal that your journey through differential equations will be split into two main parts. The first part is devoted to a discussion of certain classes of linear *ordinary* differential equations. Various theoretical concepts necessary to understand mathematical models whose mathematical description falls into this category will be introduced and developed, and subsequently addressed and generalized several times as the complexity of the models increases. All told, eleven distinct models spanning disciplines will be investigated in Part I, and an encompassing abstract theory will be developed.

Part II focuses on linear partial differential equations. Here, we will study another ten distinct mathematical models, and develop an encompassing abstract theory analogous to the one formulated in Part I. The theoretical results established in Part I can effectively be interpreted as special cases of the theory developed here.

The structure of this book may be somewhat different than others through which you have worked. For one, the text is NOT simply intended to be read passively. Rather, it is chock full of questions posed to you. If this journey were envisioned as a trek through a densely-vegetated rain forest, then our approach is akin to arming each and every reader with a machete intended to be used to help forge the path to an elusive Mayan ruins that legend suggests is home to riches beyond belief. Hyperbole aside, YOU are asked to work through many of the details. A passive reading of this text will prove to be ineffective, whereas carefully working through it will result in a solid understanding of the material. Everything from computations to assessing the validity of underlying assumptions must be carefully examined in order to fully understand the material. We consciously indicate when the discussion is intentionally terse by including parenthetical questions and instructions, such as **(Why?)**, **(Do so!)**, and **(Explain how.)**. Sometimes, we even bring the discussion to a screeching halt with a **STOP!** It is imperative that you not leave any stone unturned and attempt to answer each of these questions. Failing to do so could start as a seemingly minor drop of confusion and then, snowball into a thunderous sea of misunderstanding as you proceed through the book.

While the primary goal of the text is to develop an overarching mathematical *theory*, we are doing so under the assumption that your prior exposure to theoretical mathematics is minimal. We intend to build your intution and involve you in the enterprise of building the theory from the ground up. That said, we need to arm you with the essential tools of analysis, but in a manner that balances intuition, practicality, and mathematical rigor. This is accomplished with a preliminary chapter in both main parts of the book which provides the essential tools to be used in that part, along with the inclusion of hands-on **EXPLORE!** activities. The purpose of the **EXPLORE!** activities is just that, to explore.

We provide questions to guide you and point you in some direction. All we ask is that you walk, or even run, in some direction, even if you do not know the final destination. Please, do not feel that you should know the *correct* answer in the traditional math book sense. The correct answer to an **EXPLORE!** question is more of a thought process followed by articulating what you have considered and your conclusion. Certainly, the completion of these portions of the text is by no means a substitute for a formal study of analysis. But, they do provide you with just enough knowledge to proceed through the next step of the development of the theory.

Applying the theory is, to many readers, arguably analogous to finding the lost treasure as the culmination of a long, tiresome journey. Our dual emphasis on the analysis of mathematical models is not typical in textbooks. We feel that this connection to concrete applications helps build intuition, provides motivation for why such a theory should be developed, and *post fortiori* affirms having developed the theory by examining the plentiful collection of applications. There are **APPLY IT!** activities sprinkled throughout the text, asking you to reexamine models in light of new theoretical results.

Finally, the glue that binds all aspects of our journey is our extensive use of MATLAB Graphical User Interfaces (GUIs). As you will see, this feature of the text enables you to discover patterns and make conjectures that otherwise would be solely reachable using real analysis making them intractible.

1.3 Mired in Jargon—A Quick Language Lesson!

Superficially, a *differential equation* (DE) is an equation involving derivatives. If an equation contains derivatives of functions dependent only on a single variable, then it is called an *ordinary differential equation* (ODE), while an equation containing partial derivatives (which arise when the functions of interest depend on more than one variable) is appropriately called a *partial differential equation* (PDE). Some examples of such equations are

$$u'(x) = f(x, u(x)) \tag{1.1}$$

$$\frac{d^2u(x)}{dx^2} + 3\frac{du(x)}{dx} = \sin x \tag{1.2}$$

$$\frac{\partial^2 u(t,x)}{\partial t^2} + c^2 \frac{\partial^2 u(t,x)}{\partial x^2} = 0. \tag{1.3}$$

Equations (1.1) and (1.2) are ODEs while (1.3) is a PDE. The variables upon which the functions depend, namely x (and also t in the case of (1.3)) are called *independent variables.* Often, the independent variables in an equation are clearly understood and in such cases we suppress the dependence on them and write a cleaner, more succinct equation. For instance, the following equations are often written in place of (1.1) - (1.3), respectively:

$$u' = f(x, u) \tag{1.4}$$

$$\frac{d^2u}{dx^2} + 3\frac{du}{dx} = \sin x \tag{1.5}$$

$$\frac{\partial^2 u}{\partial t^2} + c^2 \frac{\partial^2 u}{\partial x^2} = 0. \tag{1.6}$$

The *order* of a differential equation is the order of the highest derivative appearing in the equation. The order of (1.1) is 1, while the order of both (1.2) and (1.3) is 2.

With the basic terminology in place, we loosely define what is meant by solving a differential equation. The notion is intuitive enough. Indeed, when asked to "solve (1.1)," you are to determine a real-valued *function* $u = u(x)$ that *satisfies* the equation (1.1). Unlike algebraic equations, the unknown quantity of interest is an actual function, rather than a real-number. Defining what precisely is meant by saying "a function satisfies a differential equation" can be very tricky, and we will need to pay more careful attention to various nuances as we proceed through the text. Momentarily, though, it is reasonable to think of this obtaining a true statement as the result of plugging the function into the equation. Consider the following example.

Example 1.3.1. Verify that the function $u(x) = e^{-x}$ is a solution of the ODE $u' = -u$.
<u>*Solution*</u>: Observe that $u'(x) = -e^{-x}$ and that $-u(x) = -e^{-x}$, for any real number x. Substituting these into the given ODE results in a true statement, for any real number x. So, $u(x)$ is a solution to the given ODE.

We make two remarks regarding Example 1.3.1. One, let C be an arbitrary real number and consider the function u_C defined by $u_C(x) = Ce^{-x}$. Clearly, u_C is also a solution of the given ODE. **(Why?)** The presence of the arbitrary constant C prompts us to refer to this as a *general solution* of the DE. This demonstrates that without additional information, a DE can have more than one solution; in fact, it can have infinitely many. Two, the solution of the DE is a function and as such, comes equipped with its own domain which indicates when the DE is solvable. A complete understanding of the behavior of a solution of a DE requires that we know its domain. It is customary to include this domain when describing the solution of a DE.

Exercise 1.3.1. Verify that for any real constant C, $u_C(t) = Ce^{-3t}$ is a general solution of the ODE $u''(t) + 2u'(t) - 3u(t) = 0$. What is the domain of the solution?

Exercise 1.3.2. Verify that $z(x) = C_1 \cos(\ln x) + C_2 \sin(\ln x)$ is a general solution of the Cauchy-Euler ODE $x^2 z''(x) + xz'(x) + z(x) = 0$, where C_1 and C_2 are arbitrary real constants. What is the domain of the solution?

1.4 Introducing MATLAB

MATLAB® is a software package produced by Mathworks, Inc. (mathworks.com) and is available on a systems ranging from personal computers to supercomputers. The name MATLAB stems from "Matrix Laboratory," which alludes to the software package's foundation in linear algebra. MATLAB has evolved into an interactive development tool for scientific and engineering programs, or in fact any setting for which numeric computation is required.

This textbook takes advantage of MATLAB's Graphical User Interface (GUI) environment to minimize the use of MATLAB syntax. To obtain all the GUIs for the textbook go to http://www.wcupa.edu/_academics/sch_cas.mat/mmcKibben/book3.asp and download the files to your computer. Place the files in a folder and record the name of the folder for later use. In order to execute the text's GUIs one only needs to identify the *current folder* and the *command window* in MATLAB.

If necessary, change the current folder to the one that contains this text's GUI files. This can be done by manually typing the location of the folder or by using the *browse for folder* button (the button with three dots located in the top right of screen in Figure 1.1).

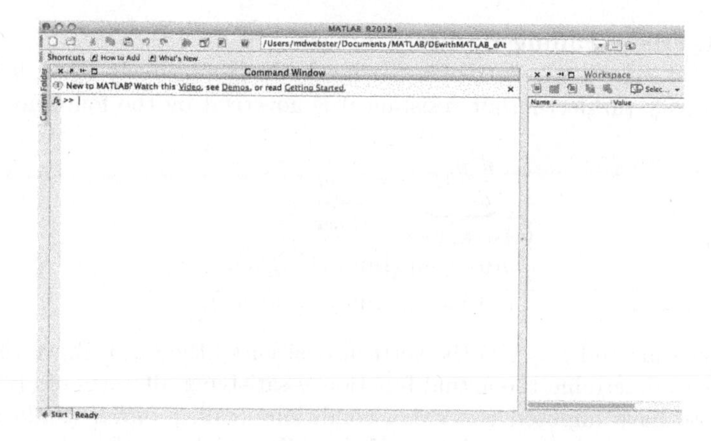

FIGURE 1.1: MATLAB Command Window

For example, in Figure 1.1 the current folder is titled *DEwithMATLAB_eAt*, so in order for the GUIs to work all the files need to be placed in that folder. When prompted by an exercise to use a specific GUI you will be directed to the command window. The command window is the large panel in the middle of the screen in Figure 1.1. A defining feature of the command window is the prompt `>>`. To open a GUI, type the name as provided in the exercise instructions after the `>>` prompt and hit return. For example,

```
>> MATLAB_Elementary_Population_Growth
```

will open the MATLAB GUI of the same name and output an interface for which one can visualize solutions, learn new concepts, and answer questions posed in exercises, MATLAB-Exercises, **EXPLORE!**s, and **APPLY IT!**s. The vast majority of GUIs that accompany this text are self-contained in the sense that all the information you need can be found within the panel of the interface. However, in a few cases additional information can be found in the command window. In these situations the GUI's *Instructions* and *Help* button will direct you to the command window and explain how to retrieve any additional information.

The GUIs in this book have been tested under MATLAB release, version 7.14.0.739 (R2012a). For a detailed description of MATLAB, its features, and syntax refer to the MATLAB manual along with the Mathworks website,

http://www.mathworks.com/help/matlab/index.html.

1.5 A First Look at Some Elementary Mathematical Models

A differential equation has an entire family of solutions associated with it until additional restrictions are imposed on the behavior of the solution. A natural question is, "How do we isolate a particular solution of a DE?" To explore this issue, we shall examine some familiar elementary models.

Model I.1. Freely-Falling Bodies

The height, at any time $t \geq 0$, of a moving body falling freely along a linear vertical path due to gravity (neglecting air resistance) is governed by the following mathematical model :

$$\begin{cases} \underbrace{\frac{d^2y}{dt^2}(t)}_{\text{Acceleration}} = \underbrace{-g}_{\text{Gravity}}, \\ y(0) = s_0 \text{ (Initial Height)}, \\ y'(0) = v_0 \text{ (Initial Velocity)}, \end{cases} \tag{1.7}$$

where t denotes time and $y = y(t)$ the vertical position of the body above the ground.

Our goal is to determine the actual function y satisfying all parts of (1.7). In addition to the ODE, we have imposed two extra conditions (called *initial conditions*) that will enable us to isolate a particular solution of the differential equation that we seek and, as such, further enable us to provide a more precise description of the quantity of interest. Whenever we consider a DE, together with conditions of this type, we refer to the entire problem as an *initial-value problem* (IVP) .

Let us momentarily disregard the initial conditions and attempt to solve the ODE. Upon inspection, it seems reasonable that solving this differential equation for y simply requires us to consecutively integrate both sides with respect to t twice. **(Why is this justified?)** Doing so yields

$$\begin{aligned} \frac{d^2y}{dt^2}(t) = -g \Longrightarrow & \int \frac{d^2y}{dt^2}(t)dt = \int -g\, dt \\ & \Longrightarrow \frac{dy}{dt}(t) + C_1^{\star} = -\, gt + C_1^{\star\star}, \end{aligned}$$

where $C_1^{\star}$ and $C_1^{\star\star}$ are arbitrary real constants (called *parameters*). We can combine these constants into a single arbitrary constant C_1 **(Why?)** to obtain the cleaner equation

$$\frac{dy}{dt}(t) = -gt + C_1.$$

This equation is meaningful for all $t \geq 0$ and so, we can integrate once again to obtain

$$y(t) = -\frac{g}{2}t^2 + C_1 t + C_2, \tag{1.8}$$

where C_1 and C_2 are arbitrary real constants. Equation (1.8) is the general solution of IVP (1.7) and is sometimes referred to as a *two-parameter family of solutions* of (1.7). If we stop here, we have infinitely many solutions to the differential equation portion of IVP (1.7). However, demanding that this solution satisfy the two initial conditions enables us to isolate a unique function that satisfies all three conditions. Using these two conditions enables us to uniquely determine C_1 and C_2 as follows:

$$\begin{aligned} y(0) = & s_0 \Longrightarrow s_0 = 0 + 0 + C_2 = C_2 \\ y'(0) = & v_0 \Longrightarrow v_0 = 0 + C_1 = C_1 \end{aligned}$$

Hence, the *particular solution* of (1.7) is given by $y(t) = -\frac{g}{2}t^2 + v_0 t + s_0$, which is the familiar formula developed in an elementary physics course.

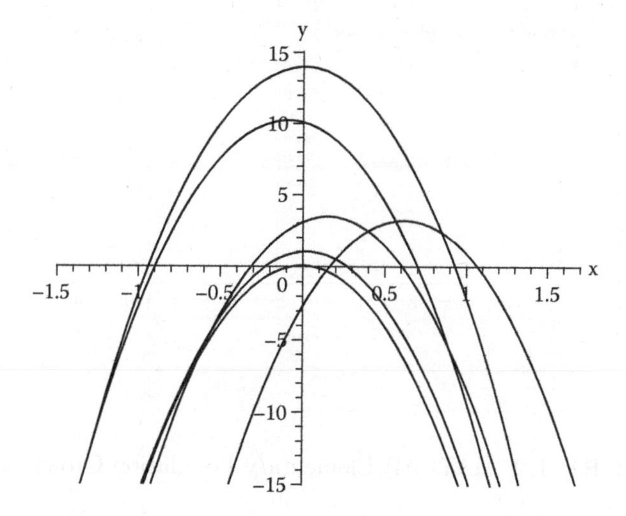

Model I.2 Elementary Population Growth

A very elementary model of population growth that ignores limitation of resources, among other factors, can be formed under the assumption that the substance/population grows simply at a rate that is proportional to its size. This immediately yields the IVP

$$\begin{cases} \frac{dx(t)}{dt} & = kx(t), \ t > t_0, \\ x(t_0) & = x_0, \end{cases} \tag{1.9}$$

where k is the constant of proportionality, $x(t)$ is the size of the population at time t, $\frac{dx}{dt}$ is the rate of growth or decay, and x_0 is the initial size of the population.

It is easy to see that the solution of (1.9) is $x(t) = e^{kt}x_0$. If $k > 0$, then the population is growing exponentially, while if $k < 0$, it decreases exponentially **(Why?)**. Such a model is useful when determining the half-life of a radioactive substance.

Exercise 1.5.1. The population of bacteria in a culture grows at a rate proportional to the number of bacteria present at any time. After 3 hours, it is observed that there are 400 bacteria present. After 10 hours, there are 2,000 bacteria present. What was the initial number of bacteria present?

MATLAB-Exercise 1.5.1. We will use **MATLAB** to investigate the population model given by (1.9), the population model posed in Exercise 1.5.1. Open **MATLAB** and in the command line, type:

MATLAB_Elementary_Population_Growth

Use the GUI to answer the following questions. Make certain to click on the 'Description of System' button for a description of the GUI and instructions on how to use the GUI. A typical screenshot can be found in Figure 1.2.

i.) Use your answers in Exercise 1.5.1 as parameters for the GUI to find the population after 5 hours.

ii.) Suppose there were measurement errors caused by misaligned lab equipment and, in actuality, there were 405 bacteria after 3 hours and 2100 bacteria after 10 hours. Compute the new initial population and proportionality constant.

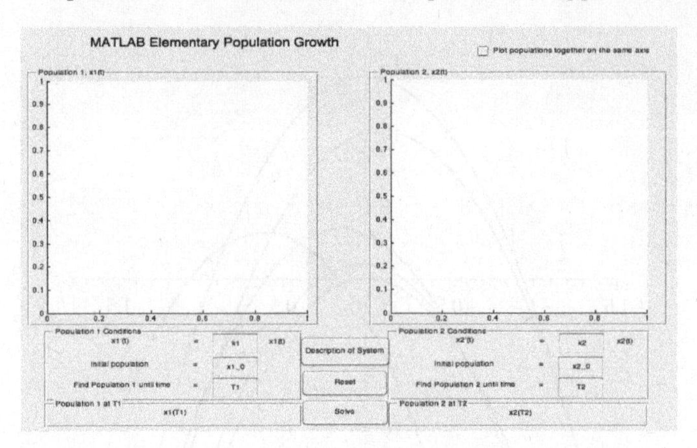

FIGURE 1.2: MATLAB Elementary Population Growth GUI

iii.) Let us refer to the situation described in Exercise 1.5.1 as *Population 1*, and the one in ii) as *Population 2*. Use the GUI to plot both populations on the same axis over a 5 hour time span. Comment on the shapes of the graphs. For example, do they look the same? Are they getting further apart? Closer together?

■

Model I.3 Newton's Law of Heating and Cooling

Did you ever wonder why it seems that within a <u>very</u> short time of taking an ice-cold soda out of the refrigerator to drink outside on a scorching summer day, the soda becomes warm or, why does a Thanksgiving turkey cool so quickly once it is taken out of an oven and set onto the dining room table? These phenomena can be modeled using *Newton's Law of Heating and Cooling* , which says that the rate at which a body cools is proportional to the difference between the temperature of the body and the temperature of the surrounding medium (called the *ambient temperature*).

Let $T(t)$ represent the temperature of the body at time t, T_m the constant temperature of the surrounding medium, and $\frac{dT}{dt}$ the rate at which the body cools. Then, Newton's Law of Heating and Cooling can be expressed mathematically as the following IVP:

$$\begin{cases} \frac{dT(t)}{dt} &= k\left(T(t) - T_m\right),\ t > t_0, \\ T\left(t_0\right) &= T_0, \end{cases} \tag{1.10}$$

where k is the constant of proportionality and T_0 is the temperature of the body at time $t = t_0$. If the body is cooling, then it makes sense that $T(t) \geq T_m$. Likewise, if the body is warming, the reverse inequality is true. **(Why?)** The solution of IVP (1.10) is given by

$$T(t) = T_m + \left(T_0 - T_m\right) e^{k(t-t_0)}. \tag{1.11}$$

STOP! Verifty this.

Exercise 1.5.2. A thermometer is taken from inside a house to the outdoors, where the air temperature is 5 degrees Fahrenheit. After 1 minute the thermometer reads 55 degrees Fahrenheit and after 5 minutes the reading is 30 degrees Fahrenheit. What was the initial temperature of the room?

MATLAB-Exercise 1.5.2. We will use **MATLAB** to investigate the temperature model given by (1.11), the situation described in Exercise 1.5.2. Open **MATLAB** and in the command line, type:

MATLAB_Heating_Cooling

Use the GUI to answer the following questions. Make certain to click on the 'Description of System' button for a description of the GUI and instructions on how to use the GUI. A typical screenshot can be found in Figure 1.3.

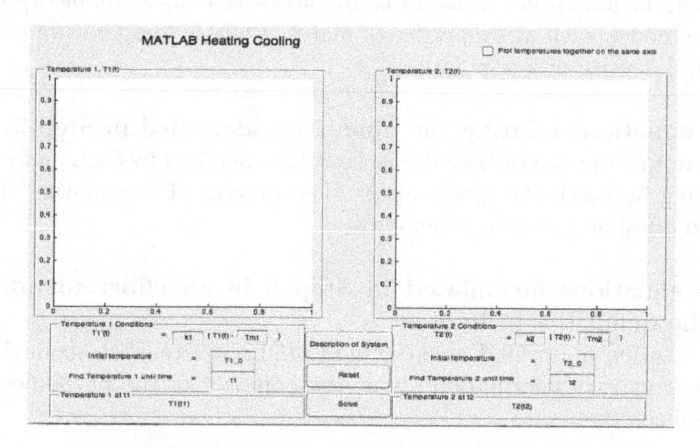

FIGURE 1.3: MATLAB Heating Cooling GUI

i.) Use your answers in Exercise 1.5.2 as parameters for the GUI to find the temperature reading on the thermometer after 8 minutes.

ii.) All measurement devices, including thermometers, are imperfect, meaning they do not necessarily output the actual measurement, but rather a value within some tolerance of the true value. Repeat Exercise 1.5.2 assuming the thermometer was reporting temperatures 1% warmer than the true values and that the outdoor temperature is truly 5 degrees Fahrenheit.

iii.) Let us refer to the situation described in Exercise 1.5.2 as *Temperature 1* and the one in ii) as *Temperature 2*. Use the GUI to plot both temperatures on the same set of axes over an eight minute time span. Comment on the shapes of the graphs. For example, do they look the same? Are they getting further apart? Closer together?

iv.) Make a conjecture for solutions to (1.11) when the parameters are a "little off."

■

A Basic Philosophy of Mathematical Modeling

Before studying more complicated models, we provide a crude outline of the underlying philosophy used when forming so-called mathematical models. Below is a list of key steps to keep in mind when embarking on a journey whose goals are to form, analyze, solve, and refine mathematical models :

1. Clearly state all assumptions.
This step constitutes the foundation of the modeling process. You must describe all

relationships that exist among the quantities being studied, as accurately as possible. Try to avoid hidden assumptions, and realize that the quality of the assumptions you impose determines the validity of the model.

2. Clearly identify variables and parameters.

The most frequently used independent variable is time, though other commonly occurring ones include position and angle. The dependent variables (that is, the quantities of interest that change with respect to change in the independent variable) are much wider in scope. Parameters include other measurable quantites that must be incorporated into the equations of the model, such as properties of material whose temperature is being studied, rate of decay, and density of a population.

3. Formulate equations relating the quantities identified in Step 2.

This step requires the use of underlying laws (geometric, physical, and economic) that govern the setting in which the model arises. The process of formulating these equations becomes more natural as you gain experience.

4. Study the equations formulated in Step 3 in an effort to understand the behavior of the quantities of interest.

The actual solution of an ODE or system of ODEs is often not attainable. But, there are other ways to extract information about the behavior of the phenomena that do not require an actual solution.

5. Test the model using actual data.

When working in the sciences, for instance, one formulates a model and then performs a few "test cases" in which the outcome is known a priori in order to test the accuracy of the model. We will not actually do this in the present text, as our main goal is the analysis of existing models.

6. Modify assumptions and formulate an enhanced, improved model.

No model is perfect. There is no single model that accounts for every possible variable that arises. And, even if such a model did exist, the complexity of the equations would render extracting useful information about the phenomena impossible. However, a typical approach is to start simple and move forward taking small steps towards a sufficiently realistic model.

We will examine a large collection of models in this text, but will rarely focus on going back to first principles when formulating them. Doing so thoroughly would require extensive knowledge of each area. Rather, our approach can be better described as a *macro-analysis* in that we establish a single framework in which to study more global features of broad collections of models.

Chapter 2

A Basic Analysis Toolbox

If you have taken an elementary differential equations course, you are familiar with various techniques used to determine solutions of differential equations, as well as possibly some methods used to describe characteristics of the solutions corresponding to different initial conditions. Pushing beyond such a rudimentary analysis with an eye toward performing a more rigorous investigation inevitably requires some mathematical sophistication. Contrary to the opinion that mathematics is all about performing calculations to solve problems, discovery of patterns, forming cogent arguments, and performing careful analysis is at the heart of the discipline. But, how does one *do mathematics* carefully and properly?

The purpose of this chapter is to provide you with a succinct, hands-on working knowledge of some concepts in elementary analysis focusing on notation, main definitions and results, and the techniques used throughout the text. Occasionally, other topics will be introduced when needed.

2.1 Some Basic Mathematical Shorthand

Symbolism is used heftily in mathematical exposition. Careful usage of some basic notation can streamline the verbiage. Some of the common symbols used are as follows.

Let P and Q be statements. (If the statement P changes depending on the value of some parameter x, we denote this dependence by writing $P(x)$.)

1.) The statement "not P," is called the *negation of* P.
2.) The statement "If P, then Q" is called an *implication*, and is denoted by "$P \Longrightarrow Q$" (read "P implies Q"). Here, P is called the *hypothesis* and Q is the *conclusion.*
3.) The statement "P if, and only if, Q" is denoted by "P iff Q" or "$P \Longleftrightarrow Q$." Precisely, this means "$(P \Longrightarrow Q)$ and $(Q \Longrightarrow P)$."
4.) The statement "$Q \Longrightarrow P$" is the *converse* of "$P \Longrightarrow Q$."
5.) The statement "not $Q \Longrightarrow$not P" is the *contrapositive* of "$P \Longrightarrow Q$." These two statements are logically equivalent.
6.) The symbol "$\exists$" is an existential quantifier and is read as "there exists" or "there is at least one."
7.) The symbol "$\forall$" is a universal quantifier and is read as "for every" or "for any."

Exercise 2.1.1. Let P, Q, R, and S be statements.

i.) Form the negation of "$\exists$ x such that $P(x)$ holds."

ii.) Form the negation of "$\forall x$, $P(x)$ holds."

Example 2.1.1. Form the negation of "P and (Q and R)."

This sort of formal exercise in logic arises when determining if a mathematical result is applicable to a given situation. For instance, if there are 3 distinct hypothesis, P, Q and R, that all must be satisfied in order to apply a theorem, all it would take to render the theorem inapplicable is for at least one of the hypotheses to not hold. That is, the negation of "P and (Q and R)" is "not P or not Q or not R."

Remark 2.1.1. Implication is a transitive relation in the sense that

$$((P \Longrightarrow Q) \text{ and } (Q \Longrightarrow R)) \Longrightarrow (P \Longrightarrow R).$$

For instance, a sequence of algebraic manipulations used to solve an equation is technically such a string of implications from which we conclude that the values of the variable obtained in the last step are the solutions of the original equation. Mathematical arguments are comprised of strings of implications, albeit of a somewhat more sophisticated nature.

A *theorem* is a statement about mathematical objects. It has a special form in that it consists of a *hypothesis* and a *conclusion.* A *mathematical proof* of a theorem is a logical argument which guarantees the conclusion holds when the hypotheses are satisfied. You have encountered many theorems in your study of the calculus, some of which are easily stated and others whose structure is more complicated because the hypotheses involve multiple conditions.

STOP! State the following calculus theorems as precisely as you can. Then, check your responses using a calculus text.

1. Product Rule for Derivatives
2. Mean Value Theorem for Derivatives
3. Fundamental Theorem of Calculus

Undoubtedly, while "how to use" each of the above theorems likely came to mind, formally and precisely stating them often poses a challenge to a beginner. It is not uncommon to overlook stating the conditions that must be satisfied in order to employ the rule suggested by each theorem, yet these conditions are crucial in the sense that the rules do not necessarily apply unless they are satisfied. For example, you cannot apply the Fundamental Theorem of Calculus to compute

$$\int_{-2}^{1} \frac{1}{x} dx$$

and you cannot apply the Mean Value Theorem to $f(x) = |x|$ on $[-2, 3]$. **(Why?)** As you work through the text, be mindful about the conditions being imposed, what they mean, and the ramifications of imposing them.

2.2 Set Algebra

Informally, a *set* can be thought of as a collection of objects (e.g., real numbers, vectors, matrices, functions, other sets, etc.); the contents of a set are referred to as its *elements.* We usually label sets using upper case letters and their elements by lower case letters. Two sets that arise often and for whom specific notation will be reserved are:

$$\mathbb{N} = \{1, 2, 3, ...\}$$
$$\mathbb{R} = \text{the set of all real numbers}$$

If P is a certain property and A is the set of all objects having property P, we write $A = \{x : x \text{ has } P\}$ or $A = \{x | x \text{ has } P\}$. A set with no elements is *empty* , denoted by $\emptyset$.

If A is not empty and a is an element of A, we denote this fact by "$a \in A$." If a is not an element of A, a fact denoted by "$a \notin A$," then where is it located? This prompts us to prescribe a universal set $\mathcal{U}$ that contains all possible objects of interest in our discussion. The following definition provides an algebra of sets.

Definition 2.2.1. Let A and B be sets.

i.) A is a *subset* of B, written $A \subset B$, whenever $x \in A \implies x \in B$.

ii.) A *equals* B, written $A = B$, whenever $(A \subset B)$ and $(B \subset A)$.

iii.) The *complement of* A *relative to* B, written $B \setminus A$, is the set $\{x | x \in B \text{ and } x \notin A\}$. The complement relative to $\mathcal{U}$ is denoted by $\widetilde{A}$.

iv.) The *union of* A *and* B is the set $A \cup B = \{x | x \in A \text{ or } x \in B\}$.

v.) The *intersection of* A *and* B is the set $A \cap B = \{x | x \in A \text{ and } x \in B\}$.

vi.) The *Cartesian product of* A *and* B is the set $A \times B = \{(a, b) | a \in A \text{ and } b \in B\}$. More generally, if $A_1, \ldots, A_n$ are sets, the Cartesian product of $A_1, \ldots, A_n$ is the set $A_1 \times \ldots \times A_n = \{(a_1, \ldots, a_n) : a_1 \in A_1, \ldots, a_n \in A_n\}$. If $A_1 = \ldots = A_n$, then denoting this common set by A, we denote the Cartesian product by simply A^n.

Verifying that two sets are equal requires that we prove two implications. Use this fact when appropriate to complete the following exercises:

Exercise 2.2.1. Let A, B, and C be sets. Illustrate each of the following pictorally and prove each of them:

i.) $A \subset B$ iff $\widetilde{B} \subset \widetilde{A}$.

ii.) $A = (A \cap B) \cup (A \setminus B)$

iii.) $A \cap (B \cup C) = (A \cap B) \cup (A \cap C)$and $A \cup (B \cap C) = (A \cup B) \cap (A \cup C)$

iv.) $\widetilde{(A \cap B)} = \widetilde{A} \cup \widetilde{B}$ and $\widetilde{(A \cup B)} = \widetilde{A} \cap \widetilde{B}$

Exercise 2.2.2. Explain how you would show two sets A and B are NOT equal.

2.3 Functions

The functions typically encountered in a calculus course are defined by explicit formulas that assign elements in one set to outputs in another. We will need to think more broadly and carefully about functions in this text and so, we introduce a more formal definition of a function below.

Definition 2.3.1. Let A and B be sets.

i.) A function from A to B is a subset $f \subset A \times B$ satisfying

a) $\forall x \in A, \exists y \in B$ such that $(x, y) \in f$,

b) $(x, y_1) \in f$ and $(x, y_2) \in f \implies y_1 = y_2$. We say f is B*–valued*, denoted by $f : A \to B$.

ii) The set A is called the *domain* of f, denoted $\text{dom}(f)$.

iii) The *range* of f, denoted by $\text{rng}(f)$, is given by $\text{rng}(f) = \{f(x) | x \in A\}$.

Remark 2.3.1. Notation: When defining a function using an explicit formula, say $y = f(x)$, the notation $x \mapsto f(x)$ is often used to denote the function. Also, we indicate the general dependence on a variable using a dot, say $f(\cdot)$. If the function depends on two independent variables, we distinguish between them by using a different number of dots for each, say $f(\cdot, \cdot\cdot)$. Also, the term *mapping* is used synonymously with the term *function.* An immediate consequence of Definition 2.3.1 is that $\text{rng}(f) \subset B$.

Exercise 2.3.1. Precisely define what it means for two functions f and g to be equal.

We sometimes wish to apply functions in succession in the following sense.

Definition 2.3.2. Suppose that $f : \text{dom}(f) \to A$ and $g : \text{dom}(g) \to B$ with $\text{rng}(g) \subset \text{dom}(f)$. The *composition of f with g*, denoted $f \circ g$, is the function $f \circ g : \text{dom}(g) \to A$ defined by $(f \circ g)(x) = f(g(x))$.

Exercise 2.3.2. Show that, in general, $f \circ g \neq g \circ f$.

Definition 2.3.3. $f : A \to B$ is called

i.) *one-to-one* if $f(x_1) = f(x_2) \Longrightarrow x_1 = x_2, \forall x_1, x_2 \in A$;
ii.) *onto* whenever $\text{rng}(f) = B$.

Exercise 2.3.3. Let $f : \text{dom}(f) \to A$ and $g : \text{dom}(g) \to B$ be such that $f \circ g$ is defined. Prove the following:

i.) If f and g are onto, then $f \circ g$ is onto.

ii.) If f and g are one-to-one, then $f \circ g$ is one-to-one.

Functions that are one-to-one and onto are invertible in the following sense.

Definition 2.3.4. Let $f : \text{dom}(f) \to \text{rng}(f)$. The function $g : \text{rng}(f) \to \text{dom}(f)$ such that

$$\begin{aligned} g(f(x)) &= x, \text{ for every } x \in \text{dom}(f), \\ f(g(y)) &= y, \text{ for every } y \in \text{rng}(f), \end{aligned}$$

is called the *inverse of f* and is denoted by f^{-1}. A function f for which such an inverse function exists is said to be *invertible.*

At times, we need to compute the functional values for all members of a subset of the domain, or perhaps determine the subset of the domain whose collection of functional values is a prescribed subset of the range. These notions are made precise below.

Definition 2.3.5. Let $f : A \to B$.

i.) For $X \subset A$, the *image of X under f* is the set $f(X) = \{f(x) | x \in X\}$.
ii.) For $Y \subset B$, the *pre-image of Y under f* is the set

$$f^{-1}(Y) = \{x \in \mathcal{A} | \; \exists y \in Y \text{ such that } y = f(x)\}.$$

We often consider functions whose domains and ranges are subsets of $\mathbb{R}$. For such functions, the notion of monotonicity is often a useful characterization.

Definition 2.3.6. Let $f : \text{dom}(f) \subset \mathbb{R} \to \mathbb{R}$ and suppose that $\varnothing \neq S \subset \text{dom}(f)$. We say that f is

i.) *nondecreasing on S* whenever $x_1, x_2 \in S$ with $x_1 < x_2 \Longrightarrow f(x_1) \leq f(x_2)$;
ii.) *nonincreasing on S* whenever $x_1, x_2 \in S$ with $x_1 < x_2 \Longrightarrow f(x_1) \geq f(x_2)$.

Remark 2.3.2. When the inequality is strict, we replace "nondecreasing" by "increasing" and "nonincreasing" by "decreasing."

The arithmetic operations of real-valued functions are defined in the natural way.

Definition 2.3.7. (Arithmetic of Functions)
Suppose that $f : \text{dom}(f) \subset \mathbb{R} \to \mathbb{R}$ and $g : \text{dom}(g) \subset \mathbb{R} \to \mathbb{R}$. The functions $f + g$, $f - g$, and $f \cdot g$ are defined as follows:

i.) $f + g : \text{dom}(f) \cap \text{dom}(g) \subset \mathbb{R} \to \mathbb{R}$ is defined by $(f + g)(x) = f(x) + g(x)$, for all $x \in \text{dom}(f) \cap \text{dom}(g)$,
ii.) $f - g : \text{dom}(f) \cap \text{dom}(g) \subset \mathbb{R} \to \mathbb{R}$ is defined by $(f - g)(x) = f(x) - g(x)$, for all $x \in \text{dom}(f) \cap \text{dom}(g)$,
iii.) $f \cdot g : \text{dom}(f) \cap \text{dom}(g) \subset \mathbb{R} \to \mathbb{R}$ is defined by $(f \cdot g)(x) = f(x) \cdot g(x)$, for all $x \in \text{dom}(f) \cap \text{dom}(g)$.

Exercise 2.3.4. Suppose that $f : \text{dom}(f) \subset \mathbb{R} \to \mathbb{R}$ and $g : \text{dom}(g) \subset \mathbb{R} \to \mathbb{R}$ are nondecreasing (resp. nonincreasing) functions on their domains.

i.) Which of the functions $f+g$, $f-g$, $f \cdot g$, and $\frac{f}{g}$, if any, are nondecreasing (resp. nonincreasing) on their domains? Explain.

ii.) Assuming that $f \circ g$ is defined, must it be nondecreasing (resp. nonincreasing) on its domain? Explain.

2.4 The Space $(\mathbb{R}, |\cdot|)$

2.4.1 Order Properties

The basic arithmetic and order features of the real number system are likely familiar. For our purposes, we shall begin with a set $\mathbb{R}$ equipped with two operations, addition and multiplication, satisfying these algebraic properties:

1. addition and multiplication are both commutative and associative;
2. multiplication distributes over addition;
3. adding zero to any real number yields the same real number;
4. multiplying a real number by one yields the same real number;
5. every real number has a unique additive inverse; and
6. every nonzero real number has a unique multiplicative inverse.

Moreover, $\mathbb{R}$ equipped with the natural "$<$" ordering is an ordered field and obeys the following properties:

Proposition 2.4.1. (Order Features of $\mathbb{R}$)
For all $x, y, z \in \mathbb{R}$, the following are true:

i.) Exactly one of the relationships $x = y,\ x < y,\ or\ y < x$ holds;
ii.) $x < y \implies x + z < y + z$;
iii.) $(x < y)$ and $(y < z) \implies x < z$;
iv.) $(x < y)$ and $(c > 0) \implies cx < cy$;
v.) $(x < y)$ and $(c < 0) \implies cx > cy$;
vi.) $(0 < x < y)$ and $(0 < w < z) \implies 0 < xw < yz$.

The following is an immediate consequence of these properties and is often the underlying principle used when verifying an inequality.

Proposition 2.4.2. If $x, y \in \mathbb{R}$ such that $x < y + \epsilon$, $\forall \epsilon > 0$, then $x \leq y$.

Proof. Suppose not; that is, $y < x$. Observe that for $\epsilon = \frac{x-y}{2} > 0$, $y + \epsilon = \frac{x+y}{2} < x$. **(Why?)** This is a contradiction. Hence, it must be the case that $x \leq y$. □

Remark 2.4.1. The above argument is a very simple example of a proof by contradiction. The strategy is to assume that the conclusion is false and then use this additional hypothesis to obtain a false statement or a contradiction of another hypothesis in the claim.

2.4.2 Absolute Value

The above is a heuristic description of the familiar algebraic structure of $\mathbb{R}$. When equipped with a distance-measuring artifice, deeper properties of $\mathbb{R}$ can be defined and studied. This is done with the help of the absolute value function.

Definition 2.4.1. For any $x \in \mathbb{R}$, the *absolute value* of x, denoted $|x|$, is defined by

$$|x| = \begin{cases} x, \, x \geq 0, \\ -x, \, x < 0. \end{cases}$$

This can be viewed as a measurement of distance between real numbers within the context of a number line. For instance, the solution set of the equation "$|x-2| = 3$" is the set of real numbers x that are "3 units away from 2," namely $\{-1, 5\}$.

Exercise 2.4.1. Determine the solution set for the following equations.

i.) $|x-3| = 0$

ii.) $|x+6| = 2$

Proposition 2.4.3. The following properties hold for all $x, y, z \in \mathbb{R}$ and $a \geq 0$:

i.) $-|x| = \min\{-x, x\} \leq x \leq \max\{-x, x\} = |x|$;
ii.) $|x| \geq 0, \forall x \in \mathbb{R}$;
iii.) $|x| = 0$ iff $x = 0$;
iv.) $\sqrt{x^2} = |x|$;
v.) $|xy| = |x|\,|y|$;
vi.) $|x| \leq a$ iff $-a \leq x \leq a$;
vii.) $|x+y| \leq |x| + |y|$;
viii.) $|x-y| \leq |x-z| + |z-y|$;
ix.) $|\,|x| - |y|\,| \leq |x-y|$;
x.) $|x-y| < \epsilon, \forall \epsilon > 0 \Longrightarrow x = y$.

Remark 2.4.2. The properties listed in Prop. 2.4.3 will be used heftily throughout the text.

1.) (vii) is called the *triangle inequality* . Its main utility lies in providing an estimate on a sum in terms of its parts.
2.) (viii) is an immediate consequence of (vii) obtained by "putting in and taking out" z and then applying (vii). Indeed observe

$$\begin{aligned} |x-y| &= |(x-z) + (z-y)| \\ &\leq |x-z| + |z-y| \end{aligned}$$

This is useful when trying to form an upper bound on $|x-y|$ in terms of quantities different from just $|x|$ and $|y|$.
3.) (x) is a useful trick to use when showing two quantities x and y are equal. In words, if the absolute value of the difference of two real numbers is smaller than every positive number, then the two numbers must be equal.

The following inequalities are sometimes useful when establishing estimates.

Proposition 2.4.4. Let $n \in \mathbb{N}$ and $x_1, x_2, \ldots, x_n, y_1, y_2, \ldots, y_n \in \mathbb{R}$.

i.) (Cauchy-Schwarz) $\sum_{i=1}^{n} x_i y_i \leq \left(\sum_{i=1}^{n} x_i^2\right)\left(\sum_{i=1}^{n} y_i^2\right)$,

ii.) (Minkowski) $\left(\sum_{i=1}^{n} (x_i + y_i)^2\right)^{\frac{1}{2}} \leq \left(\sum_{i=1}^{n} x_i^2\right)^{\frac{1}{2}} + \left(\sum_{i=1}^{n} y_i^2\right)^{\frac{1}{2}}$,

iii.) $\left|\sum_{i=1}^{n} x_i\right|^M \leq \left(\sum_{i=1}^{n} |x_i|\right)^M \leq n^{M-1} \sum_{i=1}^{n} |x_i|^M$, $\forall M \in \mathbb{N}$.

2.4.3 Completeness Property of $(\mathbb{R}, |\cdot|)$

It turns out that $\mathbb{R}$ has a fundamental and essential property referred to as *completeness,* without which the study of analysis could not proceed. We introduce some terminology needed to state certain fundamental properties of $\mathbb{R}$.

Definition 2.4.2. Let $\varnothing \neq S \subset \mathbb{R}$.

i.) S is *bounded above* if $\exists u \in \mathbb{R}$ such that $x \leq u$, $\forall x \in S$;
ii.) $u \in \mathbb{R}$ is an *upper bound* of S (ub(S)) if $x \leq u$, $\forall x \in S$;
iii.) $u_0 \in \mathbb{R}$ is the *maximum* of S (max(S)) if u_0 is an ub(S) and $u_0 \in S$;
iv.) $u_0 \in \mathbb{R}$ is the *supremum* of S (sup(S)) if u_0 is an ub(S) and $u_0 \leq u$, for any other u =ub(S);

STOP! Formulate precise definitions of the following analogous terms in Def. 2.4.2: *bounded below, lower bound* of S (lb(S)), *minimum* of S (min(S)), and *infimum* of S (inf(S)).

Exercise 2.4.2. Complete the following table. If a quantity does not exist, indicate so.

Set	max(S)	sup(S)	min(S)	inf(S)
$S = \{\pm\frac{n-1}{n} : n \in \mathbb{N}\}$				
$S = \{2^n : n \in \mathbb{N}\}$				
$S = (1, 2]$				
$S = (0, \infty)$				
$S = \emptyset$				
$S = \{\pm\frac{n-1}{n} : n \in \mathbb{N}\}$				
$S = \{x \in \mathbb{R} : x^2 < 2\}$				
$S = \{(1 - \frac{1}{n})^{-n} : n \in \mathbb{N}\}$				

Exercise 2.4.3. Let $\varnothing \neq S \subset \mathbb{R}$.

i.) How would you prove that sup(S)= ∞?

ii.) Repeat (i) for inf(S)= $-\infty$.

Definition 2.4.3. A set $\varnothing \neq S \subset \mathbb{R}$ is *bounded* if $\exists M > 0$ such that $|x| \leq M$, $\forall x \in S$.

It can be formally shown that $\mathbb{R}$ possesses the so-called *completeness property.* The importance of this concept in the present and more abstract settings cannot be overemphasized. We state it in the form of a theorem to highlight its importance.

Theorem 2.4.1. If $\varnothing \neq S \subset \mathbb{R}$ is bounded above, then $\exists u \in \mathbb{R}$ such that u =sup(S). We say $\mathbb{R}$ is complete.

Proposition 2.4.5. (Properties of inf and sup) Let $\varnothing \neq S, T \subset \mathbb{R}$.

i.) Assume S has a sup(S). Then, $\forall \epsilon > 0$, $\exists x \in S$ such that sup(S) $- \epsilon < x \leq$ sup(S).
ii.) If $S \subset T$ and $\exists$sup(T), then $\exists$sup(S) and sup(S) $\leq$ sup(T).

Remark 2.4.3. Proposition 2.4.5(i) indicates that we can get "arbitrarily close" to sup(S) with elements of S. This is especially useful in convergence arguments.

2.5 A Closer Look at Sequences in $(\mathbb{R}, |\cdot|)$

You have already encountered sequences in calculus courses, even if only in a subtle way. For instance, the motivation of the derivative as the slope of a tangent line involves constructing a sequence of secant lines that become arbitrarily close to a tangent line, and observing that the slopes must do the same. The study of integral calculus is chock full of such examples. Indeed, every application that involves setting up a Riemann sum, in turn, involves creating a sequence of objects (e.g., staircase-like regions approximating a bounded region in the plane, block-like approximating solids, and polygonal paths approximating a continuous curve). Two related questions that arise are:

1. What does it mean for a sequence of objects to "become arbitrarily close" to a fixed target?
2. What type of object is the target to which the terms of a sequence becomes arbitrarily close? Precisely, is it the same type of quantity as the terms of the sequence, or can it be different?

The answers we will provide will be somewhat of a compromise between theory and practice. The complete truth *per se* is formally unveiled in a real analysis course.

2.5.1 Sequences and Subsequences

Definition 2.5.1. A *sequence* in $\mathbb{R}$ is a function $x : \mathbb{N} \to \mathbb{R}$. We often write x_n for $x(n)$, $n \in \mathbb{N}$, called the nth*-term* of the sequence, and denote the sequence itself by $(x_n)_{n\in\mathbb{N}}$ or by enumerating the range as $x_1, x_2, x_3, \ldots$

The notions of monotonicity and boundedness given in Defs. 2.3.6 and 2.4.3 apply in particular to sequences. We formulate them in this specific setting for later reference.

Definition 2.5.2. A sequence is called

i.) *nondecreasing* whenever $x_n \leq x_{n+1}$, $\forall n \in \mathbb{N}$;
ii.) *increasing* whenever $x_n < x_{n+1}$, $\forall n \in \mathbb{N}$;
iii.) *nonincreasing* whenever $x_n \geq x_{n+1}$, $\forall n \in \mathbb{N}$;
iv.) *decreasing* whenever $x_n > x_{n+1}$, $\forall n \in \mathbb{N}$;
v.) *monotone* if any of (i) - (iv) are satisfied;
vi.) *bounded above* (resp. *below*) if $\exists M \in \mathbb{R}$ such that $x_n \leq M$ (resp. $x_n \geq M$), $\forall n \in \mathbb{N}$;
vii.) *bounded* whenever $\exists M > 0$ such that $|x_n| \leq M$, $\forall n \in \mathbb{N}$.

Exercise 2.5.1. Explain why a nondecreasing (resp. nonincreasing) sequence must be bounded below (resp. above).

Exercise 2.5.2. Determine if the following sequences exhibit each of the following characteristics: increasing, decreasing, bounded above, and bounded below.

i.) $x_n = \left(\frac{1}{3}\right)^n$, $n \geq 1$

ii.) $x_n = (-1)^n$, $n \geq 1$

iii.) $x_n = (-1)^n \left(\frac{n-1}{n}\right)$, $n \geq 1$

iv.) $x_n = \left(1 + \frac{1}{3n}\right)^{2n}$, $n \geq 1$

v.) $1, -4, \frac{1}{2}, -3, \frac{1}{3}, -4, \frac{1}{4}, -3, \ldots$

We would like to further characterize these sequences according to whether or not the terms become arbitrarily close to a single target as the index n increases.

STOP! Which, if any, of the sequences in Exercise 2.5.2 seem(s) as though their terms become arbitrarily close to a single number as the index n increases? For the others, what seems to prevent this from occurring?

2.5.2 Convergence

We now formalize the notion of terms becoming arbitrarily close to a single target as the index n increases.

Definition 2.5.3. A sequence $(x_n)_{n\in\mathbb{N}}$ has *limit* L whenever $\forall\epsilon > 0, \exists N \in \mathbb{N}$ (N depending in general on ϵ) such that

$$n \geq N \implies |x_n - L| < \epsilon.$$

In such case, we write $\lim_{n\to\infty} x_n = L$ or $x_n \longrightarrow L$ and say that $(x_n)_{n\in\mathbb{N}}$ *converges* (or *is convergent*) *to* L. Otherwise, we say $(x_n)_{n\in\mathbb{N}}$ *diverges*.

If we paraphrase Def. 2.5.3, it would read: $\lim_{n\to\infty} x_n = L$ whenever given any open interval $(L-\epsilon, L+\epsilon)$ around L (that is, no matter how small the positive number ϵ is), it is the case that $x_n \in (L-\epsilon, L+\epsilon)$ for all but possibly finitely many indices n. That is, the "tail" of the sequence ultimately gets into every open interval around L. Also note that, in general, the smaller the ϵ, the larger the index N must be used (to get deeper into the tail) since ϵ is an error gauge, namely how far the terms are from the target. We say N must be chosen "sufficiently large" as to ensure the tail behaves in this manner for the given ϵ.

In order to gain familiarity with the definition, work through the following **MATLAB** exercise.

MATLAB-Exercise 2.5.1. We will use **MATLAB** to investigate the dependence of N on ϵ in Definition 2.5.3, and then compare the theoretical N values to those obtained using **MATLAB**.

Open **MATLAB** and in the command line, type: **MATLAB_Convergence**
Use the GUI to answer the following questions. Make certain to click on the HELP button for instructions on how to use the GUI. A typical screenshot of the GUI can be found in Figure 2.1.

i.) Consider the sequence whose n^{th}-term is given by $x_n = \frac{\sin n}{n}$, $n \geq 1$.

   a.) Plot the terms of the sequence corresponding to $n = 1$ through $n = 50$.

   b.) Does the information obtained in (a) convince you that $\lim_{n\to\infty} \frac{\sin n}{n} = 0$? Explain.

   c.) Plot the terms of the sequence corresponding to $n = 500$ through $n = 600$. Does *this* information convince you that $\lim_{n\to\infty} \frac{\sin n}{n} = 0$? Explain.

   d.) In the GUI, input a limit of 0 and $\epsilon = 0.1$. Find a value of N that satisfies Definition 2.5.3.

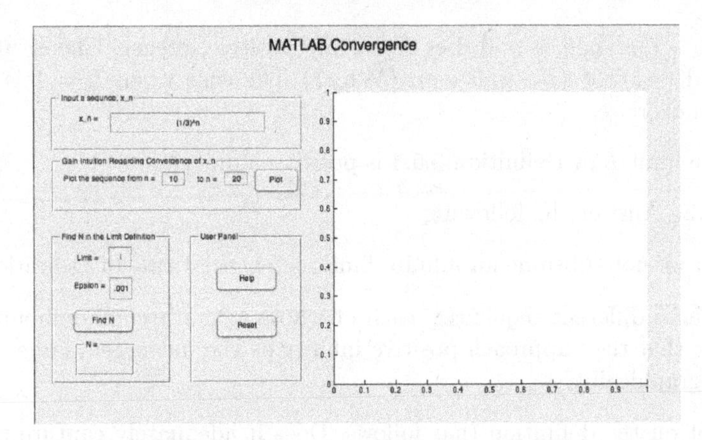

FIGURE 2.1: MATLAB Convergence GUI

e.) Repeat part (d) for $\epsilon = 0.01$, $\epsilon = 0.001$, and $\epsilon = 0.0001$.

f.) Make a conjecture for how N depends on ϵ.

g.) Using the formula for N in part (f), compute the corresponding N values for $\epsilon = 0.01$, $\epsilon = 0.001$, and $\epsilon = 0.0001$.

ii.) Consider the sequence whose n^{th}-term is given by $x_n = \frac{3n+5}{2n+7}$. Repeat parts (a) - (g) above for this sequence.

iii.) Consider the sequence whose n^{th}-term is given by $x_n = \frac{2n^2+1}{n^3}$. Repeat parts (a) - (g) above for this sequence.

■

STOP! Before moving on, answer these questions related to Definition 2.5.3.

1. Fill in the blank: As the value of ϵ decreases, the value of N should ________.

2. Prove the following assertion: $x_n \longrightarrow L$ iff $|x_n - L| \longrightarrow 0$

3. Fill in the blank: $\lim\limits_{n\to\infty} x_n \neq L$ (that is, $(x_n)_{n\in\mathbb{N}}$ does not converge to L) iff there exists an $\epsilon > 0$ such that no matter what ________ is chosen, we can find an $n \geq N$ for which ________.

Example 2.5.1. Consider the sequence $x_n = \frac{\sin n}{n}, n \in \mathbb{N}$. Claim: $\lim\limits_{n\to\infty} \frac{\sin n}{n} = 0$
Proof. We formalize the observations made in MATLAB-Exercise 2.5.1 as follows. Let $\varepsilon > 0$. Choose $N \in \mathbb{N}$ such that $N\epsilon > 1$. Observe that

$$n \geq N \Longrightarrow \left| \frac{\sin n}{n} - 0 \right| \leq \frac{|\sin n|}{n} \leq \frac{1}{n} \leq \frac{1}{N} < \epsilon.$$

Example 2.5.2. Consider the sequence $x_n = (-1)^n, n \in \mathbb{N}$. Claim: The sequence $\left((-1)^n\right)_{n\in\mathbb{N}}$ is divergent.
Proof. Let $L \in \mathbb{R}$ be an arbitrary real number. We must show there exists an $\epsilon > 0$ (which may depend on L) such that given *any* $N \in \mathbb{N}$, there exists $n \geq N$ for which $(-1)^n \notin (L - \epsilon, L + \epsilon)$. For convenience, we proceed in two cases. First, suppose $L \neq 1$.

Choose $\epsilon = |L - 1|$. Then, $\epsilon > 0$. Let N be any positive integer. Choose $n = 2N$. Then, $(-1)^n = 1$ and $(-1)^n \notin (L - \epsilon, L + \epsilon)$. **(Why?)** The case when $L = 1$ is left to you to complete as an exercise.

What if the limit L in Definition 2.5.3 is positive infinity?

Exercise 2.5.3. Answer the following:

i.) Why can we not subsume an infinite limit as a special case of Definition 2.5.3?

ii.) Create three different sequences, each of whose terms become unboundedly large in the sense that they approach positive infinity as the index gets large. Illustrate each of these graphically.

iii.) Comment on the definition that follows. Does it adequately capture the meaning of $\lim_{n \to \infty} x_n = \infty$?

> For every $M > 0$, there exists some $N \in \mathbb{N}$ such that whenever $n \geq N$, it is the case that $x_n \geq M$.

iv.) How would you modify this definition to define $\lim_{n \to \infty} x_n = -\infty$?

MATLAB-Exercise 2.5.2. We will use **MATLAB** to investigate the dependence of N on M in the definition proposed in Exercise 2.5.3.

Open **MATLAB** and in the command line, type: **MATLAB_Infinite_Limits**
Use the GUI to answer the following questions. Make certain to click on the HELP button for instructions on how to use the GUI. A typical screenshot of the GUI can by found in Figure 2.2.

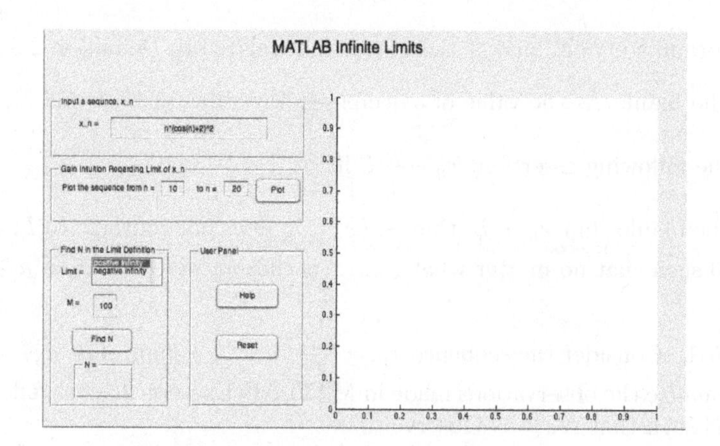

FIGURE 2.2: MATLAB Infinite Limits GUI

i.) Consider the sequence whose n^{th}-term is given by $x_n = \frac{n^2-1}{n+2}$, $n \geq 1$.

a.) Plot the terms of the sequence for index values $n = 50$ through $n = 100$.

b.) Are you convinced that $\lim_{n \to \infty} x_n = \infty$?

c.) Select "positive infinity" for limit in the GUI and choose $M = 100$. Find N.

d.) Repeat (c) for $M = 200$ and then for $M = 500$.

e.) Make a conjecture for how N depends on M.

ii.) Consider the sequence whose n^{th}-term is given by $x_n = -n(\cos n + 2)^2 \; n \geq 1$.

a.) Use the plot feature in the GUI to make a conjecture regarding the limit of x_n as $n \to \infty$.

b.) Select the appropriate choice for limit, and choose $M = 100$. Find N.

c.) Repeat (b) for $M = 200$ and then for $M = 500$.

d.) Make a conjecture for how N depends on M.

e.) Is the pattern observed in (i)(c) present in (ii)(b) and (ii)(c)? Explain.

■

2.5.3 Properties of Convergent Sequences

We now discuss the main properties of convergence. Make certain to interpret each of the properties in words and understand what each of them says.

Theorem 2.5.1. (Properties of Sequences)

i.) If $(x_n)_{n \in \mathbb{N}}$ is a convergent sequence, then its limit is unique.
ii.) If $(x_n)_{n \in \mathbb{N}}$ is a convergent sequence, then it is bounded.
iii.) (Squeeze Theorem) Let $(x_n)_{n \in \mathbb{N}}, (y_n)_{n \in \mathbb{N}},$ and $(z_n)_{n \in \mathbb{N}}$ be sequences such that

$$x_n \leq y_n \leq z_n, \forall n \in \mathbb{N}, \tag{2.1}$$

and

$$\lim_{n \to \infty} x_n = L = \lim_{n \to \infty} z_n. \tag{2.2}$$

Then, $\lim_{n \to \infty} y_n = L$.

iv.) If $\lim_{n \to \infty} x_n = L$, where $L \neq 0$, then $\exists m > 0$ and $N \in \mathbb{N}$ such that

$$|x_n| > m, \forall n \geq N.$$

v.) Suppose that $\lim_{n \to \infty} x_n = L$ and $\lim_{n \to \infty} y_n = M$. Then,

a.) $\lim_{n \to \infty} (x_n + y_n) = L + M$;
b.) $\lim_{n \to \infty} x_n y_n = LM$.
c.) $\lim_{n \to \infty} \frac{x_n}{y_n} = \frac{L}{M}$, provided $M \neq 0$

vi.) If $(x_n)_{n \in \mathbb{N}}$ is nondecreasing sequence which is bounded above, then $(x_n)_{n \in \mathbb{N}}$ converges and $\lim_{n \to \infty} x_n = \sup \{x_n | n \in \mathbb{N}\}$.

Exercise 2.5.4. Interpret each property in Theorem 2.5.1 verbally.

Exercise 2.5.5. Let $c \in \mathbb{R}$ and assume that $\lim_{n \to \infty} x_n = L$ and $\lim_{n \to \infty} y_n = M$. Prove that:

i.) $\lim_{n \to \infty} c x_n = cL$,

ii.) $\lim_{n \to \infty} (x_n - y_n) = L - M$.

Exercise 2.5.6. Argue that if $\lim_{n\to\infty} x_n = 0$ and $(y_n)_{n\in\mathbb{N}}$ is bounded, then $\lim_{n\to\infty} x_n y_n = 0$.

Exercise 2.5.7. Answer the following questions:

i.) Show that if $\lim_{n\to\infty} x_n = L$, then $\lim_{n\to\infty} |x_n| = |L|$.

ii.) Provide an example of a sequence $(x_n)_{n\in\mathbb{N}}$ for which the $\lim_{n\to\infty} |x_n|$ exists, but $\lim_{n\to\infty} x_n$ does not.

Exercise 2.5.8. Consider the following:

i.) Let $(x_k)_{n\in\mathbb{N}}$ be a sequence of nonnegative real numbers. For every $n \in \mathbb{N}$, define $s_n = \sum_{k=1}^{n} x_k$. Show the sequence $(s_n)_{n\in\mathbb{N}}$ converges iff it is bounded above.

ii.) Argue $\left(\frac{a^n}{n!}\right)_{n\in\mathbb{N}}$ converges, $\forall a \in \mathbb{R}$. In fact, $\lim_{n\to\infty} \frac{a^n}{n!} = 0$.

2.5.4 Cauchy Sequences

Definition 2.5.4. A sequence $(x_n)_{n\in\mathbb{N}}$ is a *Cauchy sequence* if $\forall \epsilon > 0, \exists N \in \mathbb{N}$ such that

$$n, m \geq N \implies |x_n - x_m| < \epsilon.$$

Intuitively, the terms of a Cauchy sequence squeeze together as the index increases. Given any positive error tolerance ϵ, there is an index past which any two terms of the sequence, no matter how greatly their indices differ, have values within the tolerance of ϵ of one another. For brevity, we often write "$(x_n)_{n\in\mathbb{N}}$ is Cauchy" instead of "$(x_n)_{n\in\mathbb{N}}$ is a Cauchy sequence."

MATLAB-Exercise 2.5.3. We will use **MATLAB** to investigate the definition of a Cauchy sequence and its relationship to the notion of convergence.

Open **MATLAB** and in the command line, type: **MATLAB_Cauchy_Sequences**
Use the GUI to answer the following questions. Make certain to click on the HELP button for instructions on how to use the GUI. A typical screenshot of the GUI can by found in Figure 2.3.

i.) Consider the sequence whose n^{th}-term is given by $x_n = \frac{\sin n}{n}$, $n \geq 1$. We have already proven that this sequence converges. But, is it also Cauchy?

a.) Use the GUI to find the appropriate N in the definition of a Cauchy sequence for $\epsilon = 0.01$ and $\epsilon = 0.001$.

b.) Use the plot to convince yourself that if terms in the tail are close to *each other*, then they should be close to *the limit of the sequence*.

ii.) Consider the sequence whose n^{th}-term is given by $x_n = (2n)^{\left(1+\frac{1}{2n}\right)} - n^{\left(1+\frac{1}{n}\right)} - n, n \geq 1$.

a.) What is the limit of this sequence? (NOTE! Admittedly, this is rather tricky. Finding the actual value of the limit and formally proving it works is quite difficult.)

b.) Use the GUI to check the definition of Cauchy for $\epsilon = 0.01$.

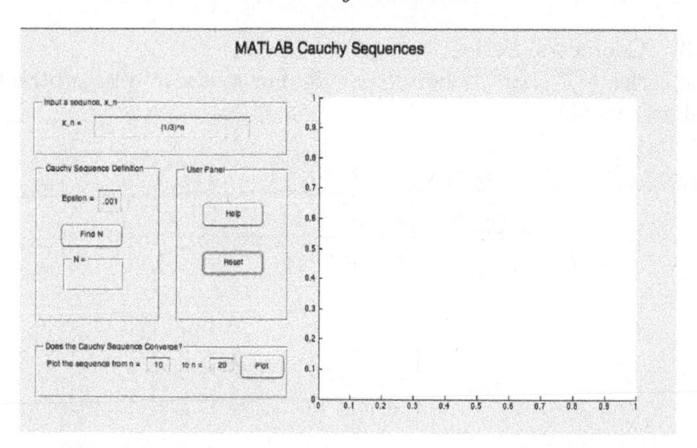

FIGURE 2.3: MATLAB Cauchy Sequences GUI

c.) Repeat (b) for $\epsilon = 0.001$ and $\epsilon = 0.0001$.

d.) Make a conjecture regarding the convergence of $(x_n)_{n\in\mathbb{N}}$.

e.) Use the plot feature to try to pinpoint the limit value of $(x_n)_{n\in\mathbb{N}}$.

■

Which of the two statements "$(x_n)_{n\in\mathbb{N}}$ is a convergent sequence" and "$(x_n)_{n\in\mathbb{N}}$ is a Cauchy sequence" seems stronger to you? Which implies which, if either? As unintuitive as it might seem, the two notions are, in fact, equivalent in $\mathbb{R}$. This fact is so important that we give it "theorem status."

Theorem 2.5.2. Every Cauchy sequence in $\mathbb{R}$ also converges in $\mathbb{R}$. We say that $\mathbb{R}$, equipped with the absolute value for measuring distance, is complete.

2.5.5 A Brief Look at Infinite Series

Sequences defined by forming partial sums of terms of a second sequence (e.g., see Exercise 2.5.9) often arise in applied analysis. You might recognize them by the name *infinite series*. We provide the bare essentials of this topic below.

Definition 2.5.5. Let $(a_n)_{n\in\mathbb{N}}$ be a sequence in $\mathbb{R}$.

i.) The sequence $(s_n)_{n\in\mathbb{N}}$ defined by $s_n = \sum_{k=1}^{n} a_k$, $n \in \mathbb{N}$ is the *sequence of partial sums of* $(a_n)_{n\in\mathbb{N}}$.

ii.) The pair $\left((a_n)_{n\in\mathbb{N}}, (s_n)_{n\in\mathbb{N}}\right)$ is called an *infinite series*, denoted by $\sum_{n=1}^{\infty} a_n$ or $\sum a_n$.

iii.) If $\lim_{n\to\infty} s_n = s$, then we say $\sum a_n$ *converges* and has *sum* s; we write $\sum a_n = s$. Otherwise, we say $\sum a_n$ *diverges.*

Remark 2.5.1. The sequence of partial sums can begin with an index n strictly larger than one. Furthermore, suppose $\sum_{k=1}^{\infty} a_k = s$ and $s_n = \sum_{k=1}^{n} a_k$. Observe that

$$s_n + \underbrace{\sum_{k=n+1}^{\infty} a_k}_{\text{Tail}} = s. \tag{2.3}$$

Since $\lim_{n\to\infty} s_n = s$, it follows that $\lim_{n\to\infty} \sum_{k=n+1}^{\infty} a_k = 0$. **(Why?)**

Example 2.5.3. (Geometric Series)

Consider the series $\sum_{k=0}^{\infty} cx^k$, where $c, x \in \mathbb{R}$. For every $n \geq 0$, subtracting the expressions for s_n and xs_n yields

$$s_n = c\left[1 + x + x^2 + \ldots + x^n\right] \tag{2.4}$$

$$\underline{-xs_n = c\left[x + x^2 + \ldots + x^n + x^{n+1}\right]} \tag{2.5}$$

$$(1-x)s_n = c\left[1 - x^{n+1}\right]$$

Hence,

$$s_n = \begin{cases} \frac{c\left[1-x^{n+1}\right]}{1-x}, & x \neq 1, \\ c(n+1), & x = 1. \end{cases}$$

If $|x| < 1$, then $\lim_{n\to\infty} x^{n+1} = 0$, so that $\lim_{n\to\infty} s_n = \frac{c}{1-x}$. Otherwise, $\lim_{n\to\infty} s_n$ does not exist. **(Tell why.)**

Exercise 2.5.9. Let $p \in \mathbb{N}$. Determine the values of x for which $\sum_{n=p}^{\infty} c(5x+1)^{3n}$ converges, and for such x, determine its sum.

Proposition 2.5.1. (n^{th} **-term Test**) If $\sum a_n$ converges, then $\lim_{n\to\infty} a_n = 0$.

A related result concerns the behavior of the "tail" of a convergent series. Intuitively, if a series converges, the contribution of terms to the overall sum become diminishingly small as the index increases. Formally, we have:

Proposition 2.5.2. Suppose $a_n \geq 0$, $\forall n \in \mathbb{N}$. If $\sum_{n=1}^{\infty} a_n$ converges, then $\sum_{n=L}^{\infty} a_n \longrightarrow 0$ as $L \to \infty$. Further, $\sum_{n=L}^{M} a_n \longrightarrow 0$ as $L, M \to \infty$.

Using the definition of a convergent sequence, we interpret this result as:

$$\forall \epsilon > 0,\ \exists L \in \mathbb{N} \text{ such that } \sum_{n=L}^{\infty} a_n < \epsilon. \tag{2.6}$$

In words, for any $\epsilon > 0$, no matter how small, a large enough index L can be found so that the sum of the terms from that index onward does not exceed ϵ.

Proposition 2.5.3. (Comparison Test)
If $a_n, b_n \geq 0$, $\forall n \in \mathbb{N}$, and $\exists c > 0$ and $N \in \mathbb{N}$ such that $a_n \leq cb_n$, $\forall n \geq N$, then

i.) $\sum b_n$ converges $\Longrightarrow \sum a_n$ converges;
ii.) $\sum a_n$ diverges $\Longrightarrow \sum b_n$ diverges.

Example. Consider the series $\sum_{n=1}^{\infty} \frac{5n}{3^n}$. Since $\lim_{n\to\infty} \frac{5n}{3^{\frac{n}{2}}} = 0$, $\exists N \in \mathbb{N}$ such that

$$n \geq N \implies \frac{5n}{3^{\frac{n}{2}}} < 1 \implies \frac{5n}{3^n} < \frac{1}{3^{\frac{n}{2}}} = \left(\frac{1}{\sqrt{3}}\right)^n. \tag{2.7}$$

(Why?) But, $\sum_{n=1}^{\infty} \left(\frac{1}{\sqrt{3}}\right)^n$ is a convergent geometric series. Thus, Prop. 2.5.3 implies that $\sum_{n=1}^{\infty} \frac{5n}{3^n}$ converges.

2.5.6 Power Series

A *(real) power series* is of the form $\sum_{n=0}^{\infty} a_n x^n$, where $(a_n)_{n\in\mathbb{N}}$ is a sequence of real numbers and the variable x takes on real values. The partial sums of such a series are really polynomial functions (in the variable x). The largest set of values of x for which this series will converge is called the *convergence set* and is found using the Ratio Test.

Why is the convergence set so important? To answer this question, consider the geometric series $\sum_{k=0}^{\infty} x^k$. This series converges if and only if $|x| < 1$, and in such case, the sum is $\frac{1}{1-x}$. Let $f(x) = \frac{1}{1-x}$ and for each $n \in \mathbb{N}$, $S_n(x) = \sum_{k=0}^{n-1} x^k$. The following plots illustrate the graphs of f, S_1, S_2, S_5, and S_{10} :

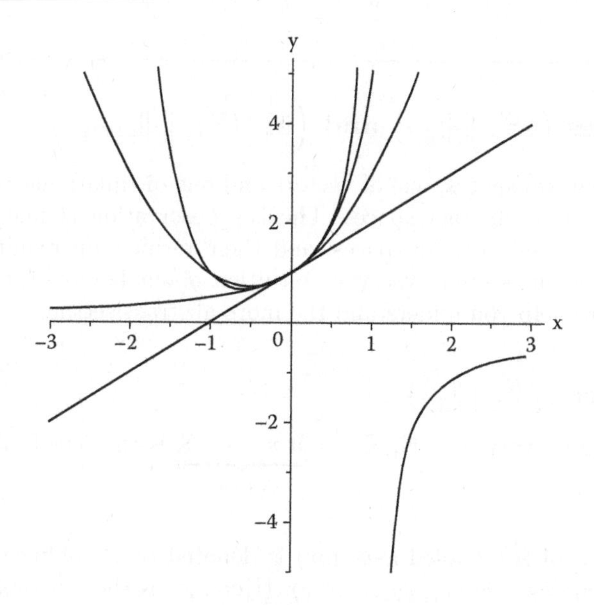

Observe that the graph of S_n gets "closer to" the graph of f for x values in the convergence set $(-1, 1)$ as $n \to \infty$, while for x outside the convergence set, the graph of S_n does not better approximate the graph of f as $n \to \infty$. We say that $(S_n)_{n\in\mathbb{N}}$ *converges uniformly to* f on the convergence set. This means

$$\lim_{n\to\infty} \underbrace{\left(\max_{x\in(-1,1)} |S_n(x) - f(x)| \right)}_{\text{As far apart as } S_n \text{ and } f \text{ can be on } (-1,1)} = 0.$$

The *interval of convergence* is therefore important because it is precisely the subset of the domain of f on which it can be approximated by a sequence of polynomials (which are much nicer to deal with). It can be shown that on the interval of convergence, power series act like finite sums in the sense that limits, derivatives, and integrals can be performed termwise. This fact is very useful when performing computations involving power series. Taylor series representations of infinitely-differentiable functions are presented in elementary calculus.

Some common examples are:

$$e^x = \lim_{N\to\infty} \sum_{n=0}^{N} \frac{x^n}{n!},\ x \in \mathbb{R}, \tag{2.8}$$

$$\sin(x) = \lim_{N\to\infty} \sum_{n=0}^{N} \frac{(-1)^n x^{2n+1}}{(2n+1)!},\ x \in \mathbb{R}, \tag{2.9}$$

$$\cos(x) = \lim_{N\to\infty} \sum_{n=0}^{N} \frac{(-1)^n x^{2n}}{(2n)!},\ x \in \mathbb{R}. \tag{2.10}$$

2.6 The Spaces $\left(\mathbb{R}^N, \|\cdot\|_{\mathbb{R}^N}\right)$ and $\left(\mathbb{M}^N(\mathbb{R}), \|\cdot\|_{\mathbb{M}^N(\mathbb{R})}\right)$

We now introduce two spaces, one of vectors and one of square matrices, as a first step toward formulating more abstract spaces. The key observation is that the characteristic properties of $\mathbb{R}$ carry over to these spaces, and their verification requires minimal effort. As you work through this section, use your intuition about how vectors in two and three dimensions behave to help you understand the more abstract setting.

2.6.1 The Space $\left(\mathbb{R}^N, \|\cdot\|_{\mathbb{R}^N}\right)$

Definition 2.6.1. For every $N \in \mathbb{N}$, $\mathbb{R}^N = \underbrace{\mathbb{R} \times \cdots \times \mathbb{R}}_{N \text{ times}}$ is the set of all ordered N-tuples of real numbers.

A typical element of $\mathbb{R}^N$ (called a *vector*) is denoted by a boldface letter, say $\mathbf{x}$, representing the ordered N-tuple $\langle x_1, x_2, \ldots, x_N \rangle$. (Here, x_k is the kth *component* of $\mathbf{x}$.) The *zero element* in $\mathbb{R}^N$ is the vector $\mathbf{0} = \underbrace{\langle 0, 0, \ldots, 0 \rangle}_{N \text{ times}}$.

STOP! The generality of N dimensions might be daunting at first. If this is the case, substitute $N = 3$ in the discussion that follows and try to digest the definitions and theorems for this particular case. This corresponds to vectors in 3-dimensional space which should be familiar from your multivariable calculus course. Once you feel comfortable with the material for $N = 3$, realize that the same computations and discussion hold no matter what value of N you use. The only difference is that you no longer have a picture when N is larger than 3.

The algebraic operations defined in $\mathbb{R}$ can be applied componentwise to define the corresponding operations in $\mathbb{R}^N$. Indeed, we have:

Definition 2.6.2. (Algebraic Operations in $\mathbb{R}^N$)
Let $\mathbf{x} = \langle x_1, x_2, \ldots, x_N \rangle$ and $\mathbf{y} = \langle y_1, y_2, \ldots, y_N \rangle$ be elements of $\mathbb{R}^N$ and $c \in \mathbb{R}$,

i.) $\mathbf{x} = \mathbf{y}$ if and only if $x_k = y_k,\ \forall k \in \{1, \ldots, N\}$,
ii.) $\mathbf{x} + \mathbf{y} = \langle x_1 + y_1, x_2 + y_2, \ldots, x_N + y_N \rangle$,
iii.) $c\mathbf{x} = \langle cx_1, cx_2, \ldots, cx_N \rangle$.

The usual properties of commutativity, associativity, and distributivity of scalar multiplication over addition carry over to this setting by applying the corresponding property in

$\mathbb{R}$ componentwise. For instance, since $x_i + y_i = y_i + x_i$, $\forall i \in \{1, \ldots, n\}$, it follows that

$$\begin{aligned}\mathbf{x}+\mathbf{y} &= \langle x_1+y_1, x_2+y_2, \ldots, x_N+y_N\rangle \\ &= \langle y_1+x_1, y_2+x_2, \ldots, y_N+x_N\rangle \\ &= \mathbf{y}+\mathbf{x}.\end{aligned} \tag{2.11}$$

Exercise 2.6.1. Establish associativity of addition and distributivity of scalar multiplication over addition in $\mathbb{R}^N$.

How do we measure distance in $\mathbb{R}^N$? There is more than one answer to this question, arguably the most natural of which is the Euclidean distance formula, defined below.

Definition 2.6.3. Let $\mathbf{x} \in \mathbb{R}^N$. The *(Euclidean) norm* of $\mathbf{x}$, denoted $\|\mathbf{x}\|_{\mathbb{R}^N}$, is defined by

$$\|\mathbf{x}\|_{\mathbb{R}^N} = \sqrt{\sum_{k=1}^{N} x_k^2}. \tag{2.12}$$

We say that the *distance between* $\mathbf{x}$ *and* $\mathbf{y}$ *in* $\mathbb{R}^N$ is given by $\|\mathbf{x}-\mathbf{y}\|_{\mathbb{R}^N}$.

Remark 2.6.1. When referring to the norm generically or as a function, we write $\|\cdot\|_{\mathbb{R}^N}$. There are other "equivalent" ways to define a norm on $\mathbb{R}^N$ that are more convenient to use in some situations. Indeed, a useful alternative norm is given by

$$\|\mathbf{x}\|_{\mathbb{R}^N-\text{alt}} = \max_{1\le i\le N} |x_i|. \tag{2.13}$$

Suffice it to say that you can choose whichever norm is more convenient to work with within a given series of computations, as long as you do not decide to use a different one halfway through. **By default, we use (2.12) unless otherwise specified.**

Exercise 2.6.2. Let $\epsilon > 0$. Provide a geometric description of these sets.

i.) $A = \left\{\mathbf{x} \in \mathbb{R}^2 \mid \|\mathbf{x}\|_{\mathbb{R}^2} < \epsilon\right\}$

ii.) $B = \left\{\mathbf{y} \in \mathbb{R}^3 \mid \|\mathbf{y} - \langle 1,0,0\rangle\|_{\mathbb{R}^3} \ge \epsilon\right\}$

iii.) $C = \left\{\mathbf{y} \in \mathbb{R}^3 \mid \|\mathbf{y} - \mathbf{x}_0\|_{\mathbb{R}^3} = 0\right\}$, where $\mathbf{x}_0 \in \mathbb{R}^3$ is prescribed.

The $\mathbb{R}^N$–norm satisfies similar properties as $|\cdot|$ (cf. Prop. 2.4.3), summarized below.

Proposition 2.6.1. Let $\mathbf{x}, \mathbf{y} \in \mathbb{R}^N$ and $c \in \mathbb{R}$. Then,

i.) $\|\mathbf{x}\|_{\mathbb{R}^N} \ge 0$,
ii.) $\|c\mathbf{x}\|_{\mathbb{R}^N} = |c| \, \|\mathbf{x}\|_{\mathbb{R}^N}$,
iii.) $\|\mathbf{x}+\mathbf{y}\|_{\mathbb{R}^N} \le \|\mathbf{x}\|_{\mathbb{R}^N} + \|\mathbf{y}\|_{\mathbb{R}^N}$,
iv.) $\mathbf{x} = \mathbf{0}$ iff $\|\mathbf{x}\|_{\mathbb{R}^N} = 0$.

STOP! Interpret the statements in Proposition 2.6.1 in words. For example, (i) can be interpreted as, "The norm of any vector is nonnegative."

Exercise 2.6.3. Let M, $p \in \mathbb{N}$ and consider the following string of inequalities:

$$\left\|\sum_{i=1}^{M} \mathbf{x}_i\right\|_{\mathbb{R}^N}^p \le \left(\sum_{i=1}^{M} \|\mathbf{x}_i\|_{\mathbb{R}^N}\right)^p \le M^{p-1} \sum_{i=1}^{M} \|\mathbf{x}_i\|_{\mathbb{R}^N}^p \tag{2.14}$$

i.) Let $p = 2$.

a.) Verify the inequalities in (2.14) when M also equals 2.

b.) How does your argument change in (a) when $M = 3$?

c.) Try to generalize your argument to the case of an arbitrary value of M.

ii.) Let $p = 3$, and repeat parts (a) - (c) from (i).

iii.) Use your work in (i) and (ii) to verify the inequalities in (2.14) for arbitrary values of p and M.

The space $(\mathbb{R}^N, \|\cdot\|_{\mathbb{R}^N})$ has an even richer geometric structure since it can be equipped with a so-called *inner product* which enables us to define orthonormality (or perpendicularity) and, by extension, the notion of angle in the space. Precisely, we have:

Definition 2.6.4. Let $\mathbf{x}, \mathbf{y} \in \mathbb{R}^N$. The *inner product of* $\mathbf{x}$ *and* $\mathbf{y}$, denoted $\langle \mathbf{x}, \mathbf{y} \rangle_{\mathbb{R}^N}$, is defined by

$$\langle \mathbf{x}, \mathbf{y} \rangle_{\mathbb{R}^N} = \sum_{i=1}^{N} x_i y_i. \tag{2.15}$$

Note that taking the inner product of any two elements of $\mathbb{R}^N$ produces a real number. This is equivalent to the dot product that you encountered when studying vectors in calculus. Some of its properties are as follows:

Proposition 2.6.2. (Properties of the Inner Product on $\mathbb{R}^N$)
Let $\mathbf{x}, \mathbf{y}, \mathbf{z} \in \mathbb{R}^N$ and $c \in \mathbb{R}$. Then,

i.) $\langle c\mathbf{x}, \mathbf{y} \rangle_{\mathbb{R}^N} = \langle \mathbf{x}, c\mathbf{y} \rangle_{\mathbb{R}^N} = c \langle \mathbf{x}, \mathbf{y} \rangle_{\mathbb{R}^N}$;
ii.) $\langle \mathbf{x} + \mathbf{y}, \mathbf{z} \rangle_{\mathbb{R}^N} = \langle \mathbf{x}, \mathbf{z} \rangle_{\mathbb{R}^N} + \langle \mathbf{y}, \mathbf{z} \rangle_{\mathbb{R}^N}$;
iii.) $\langle \mathbf{x}, \mathbf{x} \rangle_{\mathbb{R}^N} \geq 0$;
iv.) $\langle \mathbf{x}, \mathbf{x} \rangle_{\mathbb{R}^N} = 0$ iff $\mathbf{x} = \mathbf{0}$;
v.) $\langle \mathbf{x}, \mathbf{x} \rangle_{\mathbb{R}^N} = \|\mathbf{x}\|_{\mathbb{R}^N}^2$;
vi.) $\langle \mathbf{x}, \mathbf{z} \rangle_{\mathbb{R}^N} = \langle \mathbf{y}, \mathbf{z} \rangle_{\mathbb{R}^N}$, $\forall \mathbf{z} \in \mathbb{R}^N \implies \mathbf{x} = \mathbf{y}$.
vii.) (Cauchy-Schwarz Inequality) Let $\mathbf{x}, \mathbf{y} \in \mathbb{R}^N$. Then,

$$|\langle \mathbf{x}, \mathbf{y} \rangle_{\mathbb{R}^N}| \leq \|\mathbf{x}\|_{\mathbb{R}^N} \|\mathbf{y}\|_{\mathbb{R}^N} \tag{2.16}$$

Verifying these properties is straightforward. **(Try proving them!)** Property (v) is of particular importance because it asserts that an inner product generates a norm.

Exercise 2.6.4. Justify as many of the properties in Prop. 2.6.2 as you can. If need be, use $N = 2$ or $N = 3$ to slightly reduce the abstraction.

The inner product can be used to formulate a so-called *orthonormal basis* for $\mathbb{R}^N$. Precisely, let

$$\mathbf{e}_1 = \langle 1, 0, \ldots, 0 \rangle, \mathbf{e}_2 = \langle 0, 1, 0, \ldots, 0 \rangle, \ldots, \mathbf{e}_n = \langle 0, \ldots, 0, 1 \rangle,$$

and observe that

$$\|\mathbf{e}_i\|_{\mathbb{R}^N} = 1, \ \forall i \in \{1, \ldots, N\}, \tag{2.17}$$

$$\langle \mathbf{e}_i, \mathbf{e}_j \rangle_{\mathbb{R}^N} = 0, \ \text{whenever } i \neq j. \tag{2.18}$$

This is useful because it yields the following unique representation for the members of $\mathbb{R}^N$ involving the inner product.

Proposition 2.6.3. For every $\mathbf{x} \in \mathbb{R}^N$,

$$\mathbf{x} = \sum_{i=1}^{N} \langle \mathbf{x}, \mathbf{e}_i \rangle_{\mathbb{R}^N} \mathbf{e}_i. \tag{2.19}$$

If $\mathbf{x} = \langle x_1, x_2, \ldots, x_N \rangle$, then (2.19) is a succinct way of writing

$$\mathbf{x} = \langle x_1, 0, \ldots, 0 \rangle + \langle 0, x_2, \ldots, 0 \rangle + \langle 0, 0, \ldots, x_N \rangle. \tag{2.20}$$

(Tell why.) Heuristically, this representation indicates how much to "move" in the direction of each basis vector to arrive at $\mathbf{x}$.

For any $x_0 \in \mathbb{R}$ and $\epsilon > 0$, an open interval centered at x_0 with radius ϵ is defined by

$$(x_0 - \epsilon, x_0 + \epsilon) = \{x \in \mathbb{R} | \; |x - x_0| < \epsilon\}. \tag{2.21}$$

Since $\|\cdot\|_{\mathbb{R}^N}$ plays the role of $|\cdot|$ in $\mathbb{R}^N$, it is natural to define an open N-ball centered at $\mathbf{x}_0$ with radius ϵ by

$$\mathfrak{B}_{\mathbb{R}^N}(\mathbf{x}_0; \epsilon) = \left\{\mathbf{x} \in \mathbb{R}^N | \; \|\mathbf{x} - \mathbf{x}_0\|_{\mathbb{R}^N} < \epsilon\right\}. \tag{2.22}$$

Exercise 2.6.5. Interpret (2.22) geometrically in $\mathbb{R}^2$ and $\mathbb{R}^3$.

2.6.2 The Space $\left(\mathbb{M}^N(\mathbb{R}), \|\cdot\|_{\mathbb{M}^N(\mathbb{R})}\right)$

A mathematical description of certain scenarios involves considering vectors whose components are themselves vectors. Indeed, consider

$$\mathbf{A} = \langle \mathbf{x}_1, \mathbf{x}_2, \ldots, \mathbf{x}_N \rangle, \tag{2.23}$$

where

$$\mathbf{x}_i = \langle x_{i1}, x_{i2}, \ldots, x_{iN} \rangle, \; 1 \le i \le N. \tag{2.24}$$

Viewing $\mathbf{x}_i$ in column form enables us to express $\mathbf{A}$ more elegantly as the $N \times N$ matrix

$$\mathbf{A} = \begin{bmatrix} x_{11} & x_{12} & \cdots & x_{1N} \\ x_{21} & x_{22} & \cdots & x_{2N} \\ \vdots & \vdots & \vdots & \vdots \\ x_{N1} & x_{N2} & \cdots & x_{NN} \end{bmatrix}. \tag{2.25}$$

It is typical to write

$$\mathbf{A} = [x_{ij}], \text{ where } 1 \le i, j \le N, \tag{2.26}$$

and refer to x_{ij} as the ij^{th} *entry* of $\mathbf{A}$.

Definition 2.6.5. Let N be a natural number larger than 1. The space $\mathbb{M}^N(\mathbb{R})$ is the set of all $N \times N$ matrices with real entries.

The following terminology is standard in this setting.

Definition 2.6.6. Let $\mathbf{A} \in \mathbb{M}^N(\mathbb{R})$.

i.) $\mathbf{A}$ is *diagonal* if $x_{ij} = 0$, whenever $i \neq j$;
ii.) $\mathbf{A}$ is *symmetric* if $x_{ij} = x_{ji}$, $\forall i, j \in \{1, \ldots, N\}$;
iii.) The *trace* of $\mathbf{A}$ is the real number $\text{trace}(\mathbf{A}) = \sum_{i=1}^{N} x_{ii}$;

iv.) The *zero matrix*, denoted $\mathbf{0}$, is the unique member of $\mathbb{M}^N(\mathbb{R})$ for which $x_{ij} = 0$, $\forall 1 \le i, j \le N$.
v.) The *identity matrix*, denoted $\mathbf{I}$, is the unique diagonal matrix in $\mathbb{M}^N(\mathbb{R})$ for which $x_{ii} = 1, \forall 1 \le i \le N$.
vi.) The *transpose* of $\mathbf{A}$, denoted $\mathbf{A}^T$, is the matrix $\mathbf{A}^T = [x_{ji}]$. (That is, the ij^{th} entry of $\mathbf{A}^T$ is x_{ji}.)

STOP! Give two examples (using different values of N) of each of the above definitions before moving onward.

We assume familiarity with elementary matrix operations and gather some basic ones below. Note that some, but not all, of the operations are performed entrywise as they were for vector operations in $\mathbb{R}^N$.

Definition 2.6.7. (Algebraic Operations in $\mathbb{M}^N(\mathbb{R})$)
Let $\mathbf{A} = [a_{ij}]$, $\mathbf{B} = [b_{ij}]$, and $\mathbf{C} = [c_{ij}]$ be in $\mathbb{M}^N(\mathbb{R})$ and $\alpha \in \mathbb{R}$.

i.) $\mathbf{A}^0 = \mathbf{I}$,
ii.) $\alpha\mathbf{A} = [\alpha a_{ij}]$,
iii.) $\mathbf{A} + \mathbf{B} = [a_{ij} + b_{ij}]$,
iv.) $\mathbf{AB} = \left[\sum_{r=1}^{N} a_{ir} b_{rj}\right]$.

Exercise 2.6.6. Consider the operations defined in Def. 2.6.7.

i.) Does $\mathbf{A} + \mathbf{B} = \mathbf{B} + \mathbf{A}$, for all $\mathbf{A},\mathbf{B} \in \mathbb{M}^N(\mathbb{R})$?

ii.) Does $\mathbf{AB} = \mathbf{BA}$, for all $\mathbf{A},\mathbf{B} \in \mathbb{M}^N(\mathbb{R})$?

iii.) Must either $(\mathbf{A} + \mathbf{B})\,\mathbf{C} = \mathbf{AC} + \mathbf{BC}$ or $\mathbf{C}\,(\mathbf{A} + \mathbf{B}) = \mathbf{CA} + \mathbf{CB}$ hold, for all $\mathbf{A},\mathbf{B},\mathbf{C} \in \mathbb{M}^N(\mathbb{R})$?

iv.) Does $(\mathbf{AB})\,\mathbf{C} = \mathbf{A}\,(\mathbf{BC})$, for all $\mathbf{A},\mathbf{B},\mathbf{C} \in \mathbb{M}^N(\mathbb{R})$?

We assume familiarity with the basic properties of determinants of square matrices. They are used to define invertibility.

Proposition 2.6.4. For any $\mathbf{A} \in \mathbb{M}^N(\mathbb{R})$ for which $\det(\mathbf{A}) \neq 0$, there exists a unique $\mathbf{B} \in \mathbb{M}^N(\mathbb{R})$ such that $\mathbf{AB} = \mathbf{BA} = \mathbf{I}$. We say $\mathbf{A}$ is invertible and write $\mathbf{B} = \mathbf{A}^{-1}$.

The notion of an *eigenvalue* arises in the study of stability theory of ordinary differential equations (ODEs). Precisely, we have

Definition 2.6.8. Let $\mathbf{A} \in \mathbb{M}^N(\mathbb{R})$.

i.) A complex number λ_0 is an *eigenvalue* of $\mathbf{A}$ if $\det(\mathbf{A} - \lambda_0\mathbf{I}) = 0$.
ii.) An eigenvalue λ_0 has *multiplicity* M if $\det(\mathbf{A} - \lambda_0\mathbf{I}) = p(\lambda)(\lambda - \lambda_0)^M$; that is, $(\lambda - \lambda_0)^M$ divides evenly into $\det(\mathbf{A} - \lambda_0\mathbf{I})$.

Exercise 2.6.7. Let $\mathbf{A} = \begin{bmatrix} a & 0 \\ 0 & b \end{bmatrix}$, where $a, b \neq 0$.

i.) Compute the eigenvalues of $\mathbf{A}$.

ii.) Compute $\mathbf{A}^{-1}$ and its eigenvalues.

iii.) Generalize the computations in (i) and (ii) to the case of a diagonal $N \times N$ matrix $\mathbf{B}$ whose diagonal entries are all nonzero. Fill in the blank:

If λ is an eigenvalue of $\mathbf{B}$, then ________ is an eigenvalue of $\mathbf{B}^{-1}$.

Remember that a norm basically defines a distance. The Euclidean norm in $\mathbb{R}^N$ was a natural outgrowth of the so-called distance formula that you encounter when studying analytic geometry. This familiarity, coupled with the fact that we could draw a picture illustrating the information provided by the norm, helps to foster an intuition as to how the norm ought to behave. However, when dealing with a less familiar space, such as $\mathbb{M}^N(\mathbb{R})$, in which a graphical illustration is untenable, it is natural to experience difficulty with defining, interpreting, and using the notion of the norm. It turns out that we can equip $\mathbb{M}^N(\mathbb{R})$ with various norms in the spirit of those used in $\mathbb{R}^N$, each of which is useful in different situations. Let $\mathbf{A} \in \mathbb{M}^N(\mathbb{R})$. Three standard choices for $\|\mathbf{A}\|_{\mathbb{M}^N}$ are

$$\|\mathbf{A}\|_F = \left[\sum_{i=1}^{N}\sum_{j=1}^{N} |a_{ij}|^2\right]^{\frac{1}{2}}, \tag{2.27}$$

$$\|\mathbf{A}\|_S = \sum_{i=1}^{N}\sum_{j=1}^{N} |a_{ij}|, \tag{2.28}$$

$$\|\mathbf{A}\|_M = \max_{1\le i,j\le N} |a_{ij}|. \tag{2.29}$$

The following inequalities involving (2.27) - (2.29), as well as the $\mathbb{R}^{\mathrm{N}}$ norms (2.12) and (2.13), are useful.

$$\|\mathbf{Ax}\|_{\mathbb{R}^N} \le \|\mathbf{A}\|_F \|\mathbf{x}\|_{\mathbb{R}^N} \tag{2.30}$$

$$\|\mathbf{Ax}\|_{\mathbb{R}^N-\mathrm{alt}} \le \|\mathbf{A}\|_S \|\mathbf{x}\|_{\mathbb{R}^N-\mathrm{alt}} \tag{2.31}$$

$$\|\mathbf{AB}\|_{\mathbb{M}^N} \le \|\mathbf{A}\|_{\mathbb{M}^N} \|\mathbf{B}\|_{\mathbb{M}^N} \tag{2.32}$$

for all $\mathbf{A}, \mathbf{B} \in \mathbb{M}^N(\mathbb{R})$ and $\mathbf{x} \in \mathbb{R}^N$. (Note that (2.32) applies for all three norms (2.27) - (2.29).)

Going forward, when dealing with norm computations involving members of $\mathbb{M}^N(\mathbb{R})$, we will write $\|\mathbf{A}\|_{\mathbb{M}^N}$ to emphasize the fact that the underlying space is $\mathbb{M}^N(\mathbb{R})$. Typically, the norm we will use is (2.27), but in the exceptional case when we use (2.28), we will write $\|\mathbf{A}\|_S$.

2.7 Calculus of $\mathbb{R}^N$-valued and $\mathbb{M}^N(\mathbb{R})$-valued Functions

The study of single-variable calculus focuses on functions whose domain and range are subsets of $\mathbb{R}$. As we proceed through Part I of this text, the calculus operations (limits, derivatives, and integrals) will need to be performed on more general functions, specifically $\mathbb{R}^N$-valued and $\mathbb{M}^N(\mathbb{R})$-valued functions. Using the definitions developed in single-variable calculus, together with the componentwise definitions of the arithmetic operations used in $\mathbb{R}^N$ and $\mathbb{M}^N(\mathbb{R})$, leads to natural definitions. We discuss each below.

2.7.1 Notation and Interpretation

A function $\mathbf{f} : \text{dom}(\mathbf{f}) \subset \mathbb{R} \to \mathbb{R}^N$ maps each real number t in dom(**f**) to a vector in $\mathbb{R}^N$. Each component of the output vector can change with the value of t, and the components need not have anything to do with each other. As such, it is reasonable to think of each component of the output vector as being a real-valued function. That is,

$$\mathbf{f}(t) = \langle f_1(t), \ldots, f_N(t) \rangle,$$

where $f_i : \text{dom}(\mathbf{f}) \subset \mathbb{R} \to \mathbb{R}$, for all $1 \leq i \leq N$. Such a function is said to be $\mathbb{R}^N$-*valued.* It is often convenient to use the equivalent parametric form

$$\begin{cases} x_1 = f_1(t), \\ \vdots \\ x_N = f_N(t), \end{cases}$$

of the function $\mathbf{f} : \text{dom}(\mathbf{f}) \subset \mathbb{R} \to \mathbb{R}^N$. When $N = 2$ and $N = 3$, the collection of outputs of such a function can be interpreted geometrically as sweeping out a curve in $\mathbb{R}^2$ or $\mathbb{R}^3$ in the direction of increasing values of t. For instance, suppose an object's trajectory is governed by uniform circular motion of constant speed v. Assuming that this object makes 10 complete revolutions before abruptly coming to a halt, the position $(x(t), y(t))$ of this object at time t can be described parametrically by

$$x(t) = v\cos(t), \;\; y(t) = v\sin(t), \;\; 0 \leq t \leq 20\pi.$$

From this, we define the equivalent $\mathbb{R}^2$-valued function $\mathbf{f} : [0, 2\pi] \to \mathbb{R}^2$ by

$$\mathbf{f}(t) = \langle v\cos(t), v\sin(t) \rangle.$$

The picture of the path swept out as the input variable t moves from 0 to 2π, and which is subsequently retraced 9 times, is as follows:

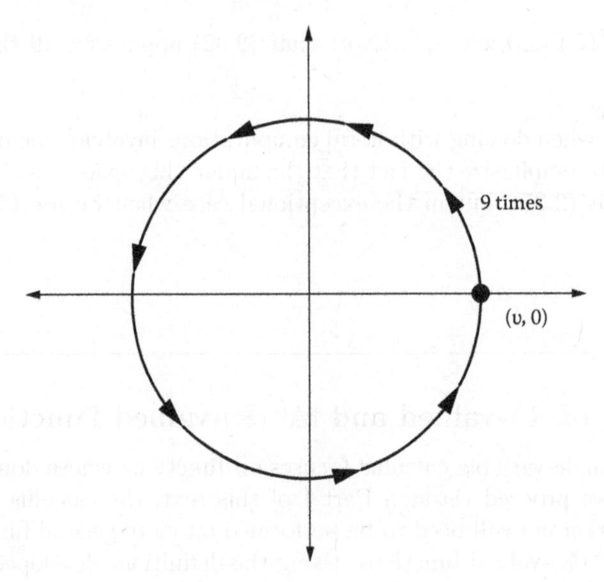

MATLAB-Exercise 2.7.1. We will use **MATLAB** to provide a graphical illustration of an $\mathbb{R}^3$-valued function.

Open **MATLAB** and in the command line, type: **MATLAB_Vector_Valued_Plot**
Use the GUI to answer the following questions. Make certain to click on the HELP button for instructions on how to use the GUI. A typical screenshot of the GUI can by found in Figure 2.4.

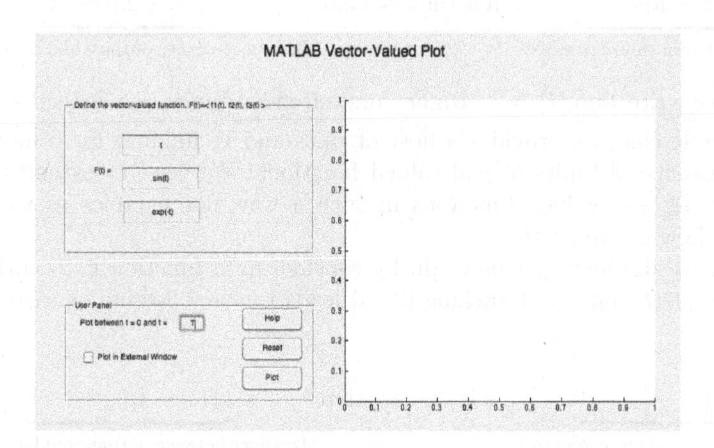

FIGURE 2.4: MATLAB Vector Valued Plot GUI

Consider the $\mathbb{R}^3$-valued function $\mathbf{f} : [0, \infty) \to \mathbb{R}^3$ defined by

$$\mathbf{f}(t) = \langle 4\cos(2t), 3\sin(2t), g(t) \rangle .$$

i.) For each of the following choices for the function $g(t)$, use the GUI to sketch the graph of $\mathbf{f}(t)$ and provide a thorough description of the curve.

  a.) $g(t) = e^{-3t}$

  b.) $g(t) = 4$

  c.) $g(t) = \pi\cos(2t)$

ii.) Explain, in general, how the corresponding graphs of the function $\mathbf{h} : [0, \infty) \to \mathbb{R}^3$ defined by $\mathbf{h}(t) = \langle 4\cos(2t), g(t), 3\sin(2t) \rangle$ compare to the graphs (a) - (c) in part (i).

■

A function $\mathbf{F} : \text{dom}(\mathbf{F}) \subset \mathbb{R} \to \mathbb{M}^N(\mathbb{R})$ maps each real number t in dom($\mathbf{F}$) to a matrix in $\mathbb{M}^N(\mathbb{R})$. As above, each component of the output vector can change with the value of t, and the components need not have anything to do with each other. As such, it is reasonable to think of each component of the output vector as being a real-valued function. That is,

$$\mathbf{F}(t) = \begin{bmatrix} F_{11}(t) & \cdots & F_{1N}(t) \\ \vdots & \ddots & \vdots \\ F_{N1}(t) & \cdots & F_{NN}(t) \end{bmatrix},$$

where $F_{ij} : \text{dom}(\mathbf{F}) \subset \mathbb{R} \to \mathbb{R}$, for all $1 \leq i, j \leq N$. Such a function is said to be $\mathbb{M}^N(\mathbb{R})$-*valued.*

2.7.2 Limits and Continuity

The notion of a limit is central to the development of calculus. Heuristically speaking, a real-valued function $g : \text{dom}(g) \subset \mathbb{R} \to \mathbb{R}$ has *limit L as t approaches a* provided that as the inputs t become "arbitrarily close to a," the corresponding functional values $g(t)$ become "arbitrarily close to L." We use the notion of distance in $\mathbb{R}$ to quantify what is meant by the phrase "arbitrarily close to" as follows:

$$\text{As} \quad \underbrace{0 < |t-a| \longrightarrow 0}_{\text{Distance between } t \text{ and } a \text{ goes to } 0}, \text{ it is the case that } \underbrace{|g(t) - L| \longrightarrow 0}_{\text{Distance between outputs and } L \text{ goes to } 0}. \tag{2.33}$$

In such case, we write $\lim_{t \to a} g(t) = L$. Right- and left-sided limits are defined similarly.

Single variable calculus provides a host of rules and techniques for computing and assessing the existence of limits of real-valued functions. We need to extend this notion to $\mathbb{R}^N$-valued and $\mathbb{M}^N(\mathbb{R})$-valued functions in such a way that enables us to tap into this reservoir. But, how do we do this?

For illustrative purposes, let us begin by considering a function $\mathbf{f} : \text{dom}(\mathbf{f}) \subset \mathbb{R} \to \mathbb{R}^2$ given by $\mathbf{f}(t) = \langle f_1(t), f_2(t) \rangle$. Mimicking (2.33) leads to the following heuristic definition of $\lim_{t \to a} \mathbf{f}(t) = \mathbf{L}$:

$$\text{As} \quad \underbrace{0 < |t-a| \longrightarrow 0}_{\text{Distance between } t \text{ and } a \text{ goes to } 0}, \text{ it is the case that } \underbrace{\|\mathbf{f}(t) - \mathbf{L}\|_{\mathbb{R}^2} \longrightarrow 0}_{\text{Distance between outputs and } \mathbf{L} \text{ goes to } 0}, \tag{2.34}$$

where $\mathbf{L} = \langle L_1, L_2 \rangle$. The condition $\|\mathbf{f}(t) - \mathbf{L}\|_{\mathbb{R}^2} \longrightarrow 0$ is equivalent to

$$\sqrt{(f_1(t) - L_1)^2 + (f_2(t) - L_2)^2} \longrightarrow 0,$$

which happens if and only if

$$(f_1(t) - L_1)^2 \longrightarrow 0 \text{ and } (f_2(t) - L_2)^2 \longrightarrow 0,$$

which, in turn, happens if and only if

$$|f_1(t) - L_1| \longrightarrow 0 \text{ and } |f_2(t) - L_2| \longrightarrow 0.$$

This suggests that $\lim_{t \to a} \mathbf{f}(t) = \mathbf{L}$ if and only if $\lim_{t \to a} f_1(t) = L_1$ and $\lim_{t \to a} f_2(t) = L_2$. In words, the limit of an $\mathbb{R}^2$-valued function as t approaches a exists if and only if each of the real-valued component functions has a limit as t approaches a; in such case, the limit is computed componentwise. This characterization enables us to use our repertoire of tools from single-variable calculus componentwise. Similar characterizations hold for general $\mathbb{R}^N$-valued and $\mathbb{M}^N(\mathbb{R})$-valued functions.

Exercise 2.7.1. Complete the following:

i.) Define what is meant by the limit of a function $\mathbf{f} : \text{dom}(\mathbf{f}) \subset \mathbb{R} \to \mathbb{R}^N$ as t approaches a.

ii.) Formulate a similar definition for the limit of a function $\mathbf{F} : \text{dom}(\mathbf{F}) \subset \mathbb{R} \to \mathbb{M}^N(\mathbb{R})$ as t approaches a.

Exercise 2.7.2. Compute the following limits, provided they exist.

i.) $\lim_{t \longrightarrow 0^+} \left\langle \frac{\sin(2t)}{t}, t\ln(t), \cos(t - \pi) \right\rangle$

ii.) $\lim_{t \longrightarrow 0} \begin{bmatrix} e^{-t} & te^{-2t} \\ 2e^{t} & e^{-4t} \end{bmatrix}$

iii.) $\lim_{t \longrightarrow \pi^-} \begin{bmatrix} \pi - t & 0 & \sin\left(\frac{1}{\pi - t}\right) \\ 1 & \pi - t & 1 \\ \sin\left(\frac{1}{\pi - t}\right) & 0 & \pi - t \end{bmatrix}$

Exercise 2.7.3. Answer the following:

i.) Define what it means for a function $\mathbf{f} : \mathrm{dom}(\mathbf{f}) \subset \mathbb{R} \to \mathbb{R}^N$ to be continuous at a.

ii.) How can you modify your response in (i) to characterize the continuity of a function $\mathbf{F} : \mathrm{dom}(\mathbf{F}) \subset \mathbb{R} \to \mathbb{M}^N(\mathbb{R})$ at a?

iii.) Suppose $\mathbf{F} : \mathrm{dom}(\mathbf{F}) \subset \mathbb{R} \to \mathbb{M}^N(\mathbb{R})$ and $h : \mathbb{R} \to \mathbb{R}$. Determine conditions under which the composition function $\mathbf{F} \circ h$ is continuous at a.

Exercise 2.7.4. Assume that

$$\alpha, \beta \in \mathbb{R}\, (\beta \neq 0)\,,\, h : \mathbb{R} \to \mathbb{R},\, \mathbf{f} : \mathbb{R} \to \mathbb{R}^N,\, \mathbf{G} : \mathbb{R} \to \mathbb{M}^N(\mathbb{R}).$$

For each of the following, determine sufficient conditions to impose on the functions in order to ensure the existence of the indicated limit at $a \in \mathbb{R}$. Then, use the limit theorems to indicate how the limit can be computed.

i.) $\lim_{t \longrightarrow a} \alpha \mathbf{G}(t)$

ii.) $\lim_{t \longrightarrow a} h(t)\mathbf{f}(t)$

iii.) $\lim_{t \longrightarrow a} h\left(\frac{t}{\beta}\right) \mathbf{G}(t)$

2.7.3 The Derivative

Measuring the rate of change of one quantity with respect to another is central to the formulation and analysis of many mathematical models. The concept is formalized in the real-valued setting via a limiting process of quantities that geometrically resemble slopes of secant lines. Precisely, a real-valued function $g : \mathrm{dom}(g) \subset \mathbb{R} \to \mathbb{R}$ is *differentiable at a* if the following limit exists and is finite :

$$\lim_{h \longrightarrow 0} \frac{g(a+h) - g(a)}{h}.$$

One-sided derivatives for real-valued functions $f : [a, b] \to \mathbb{R}$ are naturally defined using one-sided limits. We write $\frac{d^+}{dx} f(x)\big|_{x=c}$ to stand for the *right-sided derivative of f at c*, $\frac{d^-}{dx} f(x)\big|_{x=c}$ for the *left-sided derivative of f at c*, and $\frac{d}{dx} f(x)\big|_{x=c}$ the *derivative of f at c*, when they exist. The arithmetic combinations of differentiable functions are again differentiable, although some care must be taken when computing the derivative of a product and composition.

Exercise 2.7.5. Explain how you would show that $\frac{d^+}{dx} f(x)\big|_{x=c} = \infty$.

The extension of the definition to $\mathbb{R}^N$-valued and $\mathbb{M}^N(\mathbb{R})$-valued functions is an immediate consequence of the componentwise definition of limit for such functions. It is natural to expect that such a function is differentiable at a if and only if each of its component functions is differentiable at a and in such case, the derivative is computed componentwise.

Exercise 2.7.6. Verify the following two claims.

i.) A function $\mathbf{f} : \text{dom}(\mathbf{f}) \subset \mathbb{R} \to \mathbb{R}^N$ defined by $\mathbf{f}(t) = \langle f_1(t), \ldots, f_N(t) \rangle$ is differentiable at a if and only if the every component function $f_i : \mathbb{R} \longrightarrow \mathbb{R}$ is differentiable at a. In such case, $\mathbf{f}'(a) = \langle f_1'(a), \ldots, f_N'(a) \rangle$.

ii.) A function $\mathbf{F} : \text{dom}(\mathbf{F}) \subset \mathbb{R} \to \mathbb{M}^N(\mathbb{R})$ defined by

$$\mathbf{F}(t) = \begin{bmatrix} F_{11}(t) & \cdots & F_{1N}(t) \\ \vdots & \ddots & \vdots \\ F_{N1}(t) & \cdots & F_{NN}(t) \end{bmatrix}$$

is differentiable at a if and only if the every component function $F_{ij} : \text{dom}(\mathbf{F}) \subset \mathbb{R} \to \mathbb{R}$ is differentiable at a. In such case,

$$\mathbf{F}'(a) = \begin{bmatrix} F_{11}'(a) & \cdots & F_{1N}'(a) \\ \vdots & \ddots & \vdots \\ F_{N1}'(a) & \cdots & F_{NN}'(a) \end{bmatrix}.$$

As before, this characterization is useful because we can make use of all of the tools available for differentiable real-valued functions.

Exercise 2.7.7. Let $\alpha \in \mathbb{R}$ and assume $\mathbf{F} : \text{dom}(\mathbf{F}) \subset \mathbb{R} \to \mathbb{M}^N(\mathbb{R})$ is differentiable at a. Prove that $\alpha\mathbf{F}$ is differentiable at a and find a formula for $(\alpha\mathbf{F})'(a)$.

Exercise 2.7.8. Assume that

$$\alpha, \beta \in \mathbb{R}\,(\beta \neq 0)\,,\ h : \mathbb{R} \to \mathbb{R},\ \mathbf{f} : \mathbb{R} \to \mathbb{R}^N,\ \mathbf{G} : \mathbb{R} \to \mathbb{M}^N(\mathbb{R}).$$

For each of the following, determine sufficient conditions to impose on the functions in order to ensure differentiability at $a \in \mathbb{R}$. Then, use known rules to indicate how the derivative can be computed.

i.) $\alpha\mathbf{G}(t)$

ii.) $h(t)\mathbf{f}(t)$

iii.) $h\left(\frac{t}{\beta}\right)\mathbf{G}(t)$

Exercise 2.7.9. Assume that $\mathbf{F} : \text{dom}(\mathbf{F}) \subset \mathbb{R} \to \mathbb{M}^N(\mathbb{R})$ is differentiable at a. Must $\mathbf{F}$ be continuous at a? Prove your assertion.

2.7.4 The Integral

The Riemann integral is defined using the four-step process: (i) partitioning some set, (ii) approximating a quantity of interest, (iii) summing the approximations, and (iv) taking a limit in an appropriate sense. For a function $f : [a, b] \to \mathbb{R}$, the process used to define $\int_a^b f(x)dx$ is as follows:

Step 1 (Partition): Let $n \in \mathbb{N}$. Divide $[a, b]$ into n subintervals using $a = x_0 < x_1 <$

$\ldots < x_{n-1} < x_n = b$. The set $\mathcal{P} = \{x_0, x_1, \ldots, x_n\}$ is a *partition* of $[a, b]$. For convenience, let

$$\begin{aligned} \Delta x_i &= x_i - x_{i-1}, \forall 1 \le i \le n, \\ \|\mathcal{P}\| &= \max\{\Delta x_i : 1 \le i \le n\}. \end{aligned}$$

Step 2 (Approximation): For every $i \in \{1, \ldots, n\}$, choose $x_i^\star \in [x_{i-1}, x_i]$ and form the approximation

$$\underbrace{\underbrace{f(x_i^\star)}_{\text{in } \mathbb{R}} \underbrace{\Delta x_i}_{\text{in } \mathbb{R}}}_{\text{in } \mathbb{R}}. \tag{2.35}$$

Step 3 (Sum): Sum the approximations in (2.35) over $i \in \{1, \ldots, n\}$ to obtain

$$\underbrace{\sum_{i=1}^{n} \underbrace{f(x_i^\star) \Delta x_i}_{\text{in } \mathbb{R}}}_{\text{in } \mathbb{R}}. \tag{2.36}$$

Step 4 (Limit): Take the limit as $\|\mathcal{P}\| \to 0$ in (2.36) to obtain

$$\underbrace{\lim_{\|\mathcal{P}\| \to 0} \underbrace{\sum_{i=1}^{n} f(x_i^\star) \Delta x_i}_{\text{in } \mathbb{R}}}_{\text{in } \mathbb{R}}. \tag{2.37}$$

Definition 2.7.1. Let $f : [a, b] \to \mathbb{R}$. If the limit in (2.37) exists, then we say that f is *integrable* on $[a, b]$. We denote the limiting value by $\int_a^b f(x)dx$ and call it the *integral of* f *on* $[a, b]$.

The following properties hold:

Proposition 2.7.1. Assume that $f : [a, b] \to \mathbb{R}$ and $g : [a, b] \to \mathbb{R}$ are integrable on $[a, b]$. Then,

i.) $\int_a^b f(x)dx = -\int_b^a f(x)dx$;

ii.) (Linearity) $\int_a^b [\alpha f(x) + \beta g(x)]\, dx = \alpha \int_a^b f(x)dx + \beta \int_a^b g(x)dx$, $\forall \alpha, \beta \in \mathbb{R}$;

iii.) (Additivity) $\int_a^b f(x)dx = \int_a^c f(x)dx + \int_c^b f(x)dx$, $\forall c \in [a, b]$;

iv.) (Monotonicity) If $0 \le |f(x)| \le |g(x)|$, $\forall x \in [a, b]$, then $\int_a^b |f(x)|\, dx \le \int_a^b |g(x)|\, dx$.

v.) (Equality) If $f(x) = g(x)$, $\forall x \in [a, b]$, then $\int_a^b f(x)dx = \int_a^b g(x)dx$.

vi.) Assume that f is continuous on $[a, b]$. Then,

a.) $\int_a^b f(x)dx$ exists and

$$\left| \int_a^b f(x)dx \right| \le \int_a^b |f(x)|\, dx. \tag{2.38}$$

b.) The function $H : [a, b] \to \mathbb{R}$ defined by $H(x) = \int_a^x f(z)dz$ is differentiable and $H'(x) = f(x)$, $\forall x \in (a, b)$.

vii.) If f is differentiable on (a,b) and continuous on $[a, b]$, then $\int_x^y f'(z)dz = f(y) - f(x)$, $\forall [x, y] \subset (a, b)$.

It should not be surprising that the integrability of $\mathbb{R}^N$-valued and $\mathbb{M}^N(\mathbb{R})$-valued functions is defined in terms of the integrability of the component functions, and all one-variable results can be applied componentwise. **(Explain why.)**

Exercise 2.7.10. Verify the following two claims .

i.) A function $\mathbf{f} : \text{dom}(\mathbf{f}) \subset \mathbb{R} \to \mathbb{R}^N$ defined by $\mathbf{f}(t) = \langle f_1(t), \ldots, f_N(t) \rangle$ is integrable on $[a, b]$ if and only if the every component function $f_i : \mathbb{R} \longrightarrow \mathbb{R}$ is integrable on $[a, b]$. In such case,

$$\int_a^b \mathbf{f}(t) = \left\langle \int_a^b f_1(t)dt, \ldots, \int_a^b f_N(t)dt \right\rangle.$$

ii.) A function $\mathbf{F} : \text{dom}(\mathbf{F}) \subset \mathbb{R} \to \mathbb{M}^N(\mathbb{R})$ defined by

$$\mathbf{F}(t) = \begin{bmatrix} F_{11}(t) & \cdots & F_{1N}(t) \\ \vdots & \ddots & \vdots \\ F_{N1}(t) & \cdots & F_{NN}(t) \end{bmatrix},$$

is integrable on $[a, b]$ if and only if the every component function $F_{ij} : \text{dom}(\mathbf{F}) \subset \mathbb{R} \to \mathbb{R}$ is integrable on $[a, b]$. In such case,

$$\int_a^b \mathbf{F}(t)dt = \begin{bmatrix} \int_a^b F_{11}(t)dt & \cdots & \int_a^b F_{1N}(t)dt \\ \vdots & \ddots & \vdots \\ \int_a^b F_{N1}(t)dt & \cdots & \int_a^b F_{NN}(t)dt \end{bmatrix}.$$

Exercise 2.7.11. Let $t \in \mathbb{R}$. Compute the following integral:

$$\int_0^t e^{t-s} \begin{bmatrix} s & 1 & 0 \\ 1 & s & -1 \\ 0 & -1 & s \end{bmatrix} ds$$

2.7.5 Sequences in $\mathbb{R}^N$ and $\mathbb{M}^N(\mathbb{R})$

The definitions of convergent and Cauchy sequences are essentially the same as in $\mathbb{R}$ and all the results carry over without issue, requiring only that we replace $|\cdot|$ by $\|\cdot\|_{\mathbb{R}^N}$.

STOP! Convince yourself that the results from Section 2.5 extend to the $\mathbb{R}^N$ setting.

Use the fact that the algebraic operations in $\mathbb{R}^N$ are performed componentwise to help you complete the following exercise.

Exercise 2.7.12. Answer the following:

i.) Consider the two real sequences $(x_m)_{m\in\mathbb{N}}$ and $(y_m)_{m\in\mathbb{N}}$ whose m^{th} terms are given by

$$x_m = \frac{1}{m^2}, \; y_m = \frac{2m}{4m+2}, \quad m \in \mathbb{N}.$$

Show that $(\langle x_m, y_m \rangle)_{m\in\mathbb{N}}$ is convergent and $\lim_{m\to\infty} \langle x_m, y_m \rangle = \langle 0, \frac{1}{2} \rangle$.

ii.) Generally, if $\lim_{m\to\infty} x_m = p$ and $\lim_{m\to\infty} y_m = q$, what can you conclude about $(\langle x_m, y_m \rangle)_{m\in\mathbb{N}}$ in $\mathbb{R}^2$?

iii.) Consider a sequence $\left(\langle (x_1)_m, (x_2)_m, \ldots, (x_N)_m \rangle\right)_{m\in\mathbb{N}}$ in $\mathbb{R}^N$. Establish a necessary and sufficient condition for this sequence to converge in $\mathbb{R}^N$.

The following theorem is a generalization of Theorem 2.5.2.

Theorem 2.7.1. $(\mathbf{x}_n)_{n\in\mathbb{N}}$ converges in $\mathbb{R}^N$ iff $(\mathbf{x}_n)_{n\in\mathbb{N}}$is a Cauchy sequence in $\mathbb{R}^N$. We say $\left(\mathbb{R}^N, \|\cdot\|_{\mathbb{R}^N}\right)$ is complete.

MATLAB-Exercise 2.7.2. We will use **MATLAB** to provide a graphical illustration of an $\mathbb{R}^3$-valued sequences.

Open **MATLAB** and in the command line, type: **MATLAB_Vector_Sequences**
Use the GUI to answer the following questions. Make certain to click on the HELP button for instructions on how to use the GUI. A typical screenshot of the GUI can by found in Figure 2.5.

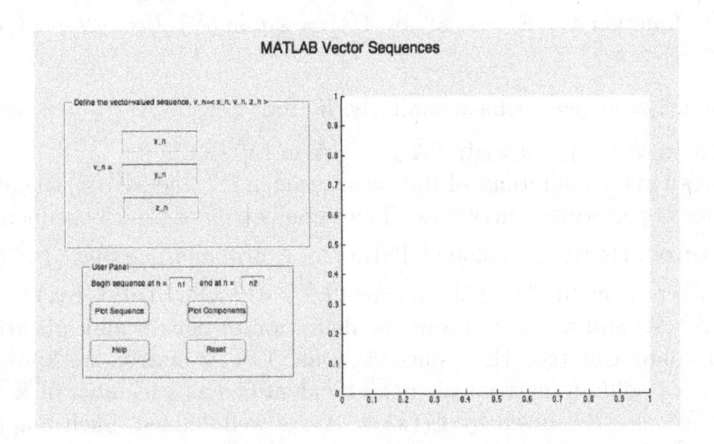

FIGURE 2.5: MATLAB Vector Sequences GUI

Since the vector-valued sequence is in $\mathbb{R}^3$ we can visualize each term in the sequence as a (x, y, z) coordinate in $\mathbb{R}^3$. Consider the sequence $(\mathbf{v}_n)_{n\in\mathbb{N}}$, where $\mathbf{v}_n = \langle x_n, y_n, z_n \rangle$, $n \in \mathbb{N}$.

i.) Suppose $x_n = \frac{\cos(n)}{n^2}$, $y_n = \frac{(-1)^n}{n}$ and $z_n = n\sin\left(\frac{1}{n}\right)$, $n \in \mathbb{N}$.

a.) Does $(\mathbf{v}_n)_{n\in\mathbb{N}}$ appear to be a Cauchy sequence? Explain.

b.) Does $(\mathbf{v}_n)_{n\in\mathbb{N}}$ appear to converge? If so, to what limit?

ii.) Suppose $x_n = (1+\frac{1}{n})^n$, $y_n = (1-\frac{1}{2n})^{-n}$ and $z_n = \sin(n)$, $n \in \mathbb{N}$.

a.) Does $(\mathbf{v}_n)_{n\in\mathbb{N}}$ appear to be a Cauchy sequence? Explain.

b.) Does $(\mathbf{v}_n)_{n\in\mathbb{N}}$ appear to converge? If so, to what limit?

iii.) Choose other $\mathbb{R}^3$-valued sequences to convince yourself of the validity of Theorem 2.7.1 and that convergence of a vector-valued sequence is completely determined by componentwise convergence.

■

Use the properties of sequences in $\mathbb{R}^N$ to complete the following exercises.

Exercise 2.7.13. Assume that $(\mathbf{x}_m)_{m\in\mathbb{N}}$ is a convergent sequence in $\mathbb{R}^N$. Prove that $\lim_{m\to\infty} \|\mathbf{x}_m\|_{\mathbb{R}^N} = \left\| \lim_{m\to\infty} \mathbf{x}_m \right\|_{\mathbb{R}^N}$. (We say that the norm $\|\cdot\|_{\mathbb{R}^N}$ is continuous.)

Exercise 2.7.14. Let $\mathbf{x} \in \mathbb{R}^N \setminus \{\mathbf{0}\}$, $a \neq 0$, and $p \in \mathbb{N}$. Must the series $\sum_{m=p}^{\infty} \left\| \frac{\mathbf{x}}{a\|\mathbf{x}\|_{\mathbb{R}^N}} \right\|_{\mathbb{R}^N}^{2m}$ converge? If so, can you determine its sum?

Exercise 2.7.15. Let $\mathbf{x} \in \mathfrak{B}_{\mathbb{R}^2}\left(\mathbf{0}; \frac{1}{3}\right)$ and $p \in \mathbb{N}$. Must the series $\sum_{m=p}^{\infty} \left| \left\langle 2\mathbf{x}, \frac{1}{4}\mathbf{x} \right\rangle_{\mathbb{R}^2} \right|^{\frac{m}{2}}$ converge? If so, can you determine its sum?

Exercise 2.7.16. Let $R > 0$ and $\langle a, b, c \rangle$ be on the boundary of the ball $\mathfrak{B}_{\mathbb{R}^3}(\mathbf{0}; R)$ and define the function $\mathbf{f} : \mathbb{R} \longrightarrow \mathbb{R}^3$ by $\mathbf{f}(t) = \left\langle a \sin\left(\frac{t}{\pi}\right), b\cos(2t + \pi), c \right\rangle$. Show that $\{\|\mathbf{f}(t)\|_{\mathbb{R}^3} : t \in \mathbb{R}\} < \infty$.

$\mathbb{M}^N(\mathbb{R})$-valued sequences behave similarly. If $\lim_{m\to\infty} \|\mathbf{A}_m - \mathbf{A}\|_{\mathbb{M}^N} = 0$, we say $(\mathbf{A}_m)_{n\in\mathbb{N}}$ *converges* to $\mathbf{A}$ in $\mathbb{M}^N(\mathbb{R})$ and write "$\mathbf{A}_m \longrightarrow \mathbf{A}$ in $\mathbb{M}^N(\mathbb{R})$."

The similarity in the definitions of the norms used in $\mathbb{R}^N$ and $\mathbb{M}^N(\mathbb{R})$ suggests that checking convergence is performed entry-wise. The same is true of Cauchy sequences in $\mathbb{M}^N(\mathbb{R})$. **(Convince yourself!)** By extension of Thrm. 2.7.1, we can argue that $\left(\mathbb{M}^N(\mathbb{R}), \|\cdot\|_{\mathbb{M}^N(\mathbb{R})}\right)$ is complete with respect to any of the norms (2.27) - (2.29). **(Tell why.)**

Let $\mathbf{A} \in \mathbb{M}^N(\mathbb{R})$ and $\mathbf{x} \in \mathbb{R}^N$. From the definition of matrix multiplication, if we view $\mathbf{x}$ as a $N \times 1$ column matrix, then since $\mathbf{A}$ is an $N \times N$ matrix, we know that $\mathbf{Ax}$ is a well-defined $N \times 1$ column matrix which can be identified as a member of $\mathbb{R}^N$. As such, the function $\mathbf{f_A} : \mathbb{R}^N \longrightarrow \mathbb{R}^N$ given by $\mathbf{f_A}(\mathbf{x}) = \mathbf{Ax}$ is well-defined. Such mappings are used frequently in Part I of this text.

Exercise 2.7.17. Prove that $\left\{ \left\| \begin{bmatrix} e^{-t} & 0 \\ 0 & e^{-2t} \end{bmatrix} \mathbf{x} \right\|_{\mathbb{R}^2} : \|\mathbf{x}\|_{\mathbb{R}^2} = 1 \text{ and } t > 0 \right\}$ is bounded.

Exercise 2.7.18. If $\left((x_{ij})_m\right)_{m\in\mathbb{N}}$ is a real Cauchy sequence for each $1 \leq i, j \leq 2$, must the sequence $(\mathbf{A}_m)_{m\in\mathbb{N}}$, where $\mathbf{A}_m = \left[(x_{ij})_m\right]$, be Cauchy in $\mathbb{M}^2(\mathbb{R})$? Prove your claim.

Exercise 2.7.19. Let $(\mathbf{A}_m)_{m\in\mathbb{N}}$ be a sequence in $\mathbb{M}^N(\mathbb{R})$.

i.) If $\lim_{m\to\infty} \mathbf{A}_m = \mathbf{0}$, what must be true about each of the N^2 sequences formed using the entries of $\mathbf{A}_m$?

ii.) If $\lim_{m\to\infty} \mathbf{A}_m = \mathbf{I}$, what must be true about each of the N^2 sequences formed using the entries of $\mathbf{A}_m$?

iii.) More generally, if $\lim_{m\to\infty} \mathbf{A}_m = \mathbf{B}$, what must be true about each of the N^2 sequences formed using the entries of $\mathbf{A}_m$?

Exercise 2.7.20. Answer the following:

i.) If $\mathbf{A}_m \longrightarrow \mathbf{A}$ in $\mathbb{M}^N(\mathbb{R})$, must $\mathbf{A}_m\mathbf{x} \longrightarrow \mathbf{A}\mathbf{x}$ in $\mathbb{R}^N$, $\forall \mathbf{x} \in \mathbb{R}^N$?

ii.) If $\mathbf{x}_m \longrightarrow \mathbf{x}$ in $\mathbb{R}^N$, must $\mathbf{A}\mathbf{x}_m \longrightarrow \mathbf{A}\mathbf{x}$ in $\mathbb{R}^N$, $\forall \mathbf{A} \in \mathbb{M}^N(\mathbb{R})$?

iii.) If $\mathbf{A}_m \longrightarrow \mathbf{A}$ in $\mathbb{M}^N(\mathbb{R})$ and $\mathbf{x}_m \longrightarrow \mathbf{x}$ in $\mathbb{R}^N$, must $(\mathbf{A}_m\mathbf{x}_m)_{m\in\mathbb{N}}$ converge in $\mathbb{R}^N$? If so, what is its limit?

2.7.6 Continuity—Revisited

Think back to your first exposure to the notion of continuity from single-variable calculus. Informally, a function f is continuous at a whenever $\lim_{x\longrightarrow a} f(x) = f(a)$. When motivating this definition, perhaps you remember moving your fingers toward each other along a graph until they met at a, and then checking to see if your fingers both landed on $f(a)$. In some sense, this process suggests that you considered a sequence (x_n) in the domain of the function that converged to a, and studied what happened to the sequence of functional values $(f(x_n))$. Checking that your fingers landed on $f(a)$ implies that you were determining if the limit of $(f(x_n))$ was indeed $f(a)$. We transform this latter process into a formal, workable definition.

Definition 2.7.2. (Sequential Definition of Continuity) A function $f : \text{dom}(f) \subset \mathbb{R} \to \mathbb{R}$ is *continuous* at $a \in \text{dom}(f)$ if for every sequence $(x_n)_{n\in\mathbb{N}} \subset \text{dom}(f)$ that converges to a, it is the case that the corresponding sequence of functional values $(f(x_n))_{n\in\mathbb{N}}$ converges to $f(a)$. In such case, we write $\lim_{x\to a} f(x_n) = f\left(\lim_{n\to\infty} x_n\right) = f(a)$. We say that f is *continuous on* $S \subset \text{dom}(f)$ if f is continuous at every element of S.

The above definition describes the notion of continuity in terms of convergent sequences. As such, all of the results of Section 2.5 apply.

STOP! Go back to Section 2.5 and review the definitions, propositions, theorems, and examples.

It is almost trivial to show $f : (0,\infty) \to \mathbb{R}, x \mapsto \frac{1}{x^2}$ is continuous on its domain because of Theorem 2.5.1. **(Why?)** Suppose $(x_n)_{n\in\mathbb{N}}$ is a sequence in the domain $(0,\infty)$, converging to some $x_0 \in (0,\infty)$. Then it follows from Theoreom 2.5.1 (v) that the sequence $(y_n)_{n\in\mathbb{N}} = \left(\frac{1}{x_n^2}\right)_{n\in\mathbb{N}}$ is a convergent sequence with limit $\frac{1}{x_0}$. Next, using the same part of theorem results in the sequence

$$f(x_n) = \frac{1}{x_n^2}, n \in \mathbb{N}$$

converging to $\frac{1}{x_0^2}$. Hence, the function is continuous on its domain.

How can we use the sequential continuity definition for a function $\mathbf{f} : dom(f) \subset \mathbb{R} \to \mathbb{R}^N$? Let us go through the given definition and see what happens.

A function $f : dom(f) \subset \mathbb{R} \to \mathbb{R}^N$ is continuous at $x_0 \in \mathbb{R}$ if every sequence $(x_n)_{n\in\mathbb{N}} \subset dom(f) \subset \mathbb{R}$ converging to $x_0 \in dom(f) \subset \mathbb{R}$ as $n \to \infty$ implies the sequence $(\mathbf{f}(x_n))_{n\in\mathbb{N}} \subset \mathbb{R}^N$ converges to $\mathbf{f}(x_0) \in \mathbb{R}^N$ as $n \to \infty$.

The only thing that has changed is that the sequences $(\mathbf{f}(x_n))_{n\in\mathbb{N}}$ and $\mathbf{f}(x_0)$ are now in $\mathbb{R}^N$ rather than $\mathbb{R}$.

STOP! What does it mean to show convergence to $\mathbf{f}(x_0) \in \mathbb{R}^N$? (Hint: Exercise 2.7.12)

Ultimately, the sequential definition of continuity will always hold provided you understand what it means to show convergence in a particular setting. The following **EXPLORE!** will give you an opportunity to see the sequential definition of continuity in action and explore new settings in which it can be used.

EXPLORE!

i.) Let $f : \mathbb{R} \to \mathbb{R}^2$ be defined by $x \mapsto \langle x, e^x \rangle$.

a.) Show f is continuous at $x_0 = 1$.

b.) Is f continuous at every $x_0 \in \mathbb{R}$?

ii.) Suppose $\mathbf{f} : \mathbb{R}^N \to \mathbb{R}^N$ and $\mathbf{x_0} \in \mathbb{R}^N$. Propose a definition using sequences to give precise meaning to continuity of $\mathbf{f}$ at $\mathbf{x_0}$ in the current situation. (Hint: Everything will look almost exactly the same except what?)

a.) Show $\mathbf{f} : \mathbb{R}^3 \to \mathbb{R}^3$ given by $\mathbf{f}(\mathbf{x}) = 2\mathbf{x} - \mathbf{e_1}$ is continuous at $\mathbf{x} = \mathbf{e_2}$, where $\mathbf{e_1} = \langle 1, 0, 0 \rangle$ and $\mathbf{e_2} = \langle 0, 1, 0 \rangle$.

b.) Is $\mathbf{f}$ continuous for every $\mathbf{x_0} \in \mathbb{R}^3$?

iii.) Suppose $f : \mathbb{R}^N \to \mathbb{R}$. Propose a definition using sequences to give precise meaning to continuity of f at $\mathbf{x_0}$ in this new situation. Then, show $f : \mathbb{R}^N \to \mathbb{R}$ defined by $\mathbf{x} \mapsto \|\mathbf{x}\|_{\mathbb{R}^N}$ is continuous at any $\mathbf{x_0} \in \mathbb{R}^N$.

iv.) Suppose $\mathbf{F} : M^N(\mathbb{R}) \to M^N(\mathbb{R})$. Propose a definition for continuity of $\mathbf{F}$ at $\mathbf{A_0}$ in this new situation. Then show the function $\mathbf{F} : M^N(\mathbb{R}) \to M^N(\mathbb{R})$ defined by $\mathbf{F}(\mathbf{A}) = \mathbf{A}^2$ is continuous at $\mathbf{I}$.

2.8 Some Elementary ODEs

Courses on elementary ordinary differential equations are chock full of techniques used to solve particular types of elementary differential equations. Within this vast toolbox are three particular scenarios that play a role in this text. We recall them informally here, along with some elementary exercises, to refresh your memory.

2.8.1 Separation of Variables

An ODE of the form $\frac{dy}{dt} = f(t)g(y)$, where $y = y(t)$, is called *separable* because symbolically the terms involving y can be gathered on one side of the equality and the terms involving t can be written on the other, thereby resulting in the equivalent equation (expressed in differential form) $\frac{1}{g(y)}dy = f(t)dt$. Integrating both sides yields an implicitly-defined function $H(y) = G(t) + C$ that satisfies the original ODE on some set and is called the *general solution* of the ODE.

Exercise 2.8.1. Determine the general solution of these ODEs:

i.) $\frac{dy}{dt} = e^{2t}\csc(\pi y)$,

ii.) $\frac{dy}{dt} = at^n$, where $n \neq -1$ and $a \in \mathbb{R} \setminus \{0\}$,

iii.) $\left(1 - y^3\right)\frac{dy}{dt} = \sum_{i=1}^{N} a_i \sin\left(b_i t\right)$, where $a_i, b_i \in \mathbb{R}$ and $n \in \mathbb{N}$.

2.8.2 First-Order Linear ODEs

Consider the *first-order linear ODE* of the form

$$\frac{dy}{dt} = ay + f(t), \text{ where } y = y(t). \tag{2.39}$$

We shall construct the solution of the initial-value problem (IVP) obtained by coupling (2.39) with the initial condition (IC)

$$y\left(t_0\right) = y_0 \tag{2.40}$$

using the so-called *variation of parameters method.*

Step 1: Solve the related homogenous equation $\frac{dy_h}{dt} - ay_h = 0$.
Separating variables and then integrating both sides over the interval (t_0, t) yields

$$\frac{dy_h}{y_h} = adt \Longrightarrow \underbrace{\ln|y_h(t)| - \ln|y_h(t_0)|}_{=\ln\left|\frac{y_h(t)}{y_h(t_0)}\right|} = \int_{t_0}^{t} ads$$

$$\Longrightarrow y_h(t) = y_h(t_0)e^{\int_{t_0}^{t} ads} = y_h(t_0)e^{a(t-t_0)}. \tag{2.41}$$

Step 2: Determine a function $C(t)$ for which $y(t) = C(t)y_h(t)$ satisfies (2.39).
Substitute this function into (2.39) to obtain

$$\begin{aligned}
\frac{d}{dt}\left[C(t)y_h(t)\right] - a\left[C(t)y_h(t)\right] &= f(t) \\
C(t)\frac{dy_h}{dt} + y_h(t)\frac{dC}{dt} - aC(t)y_h(t) &= f(t) \\
C(t)\underbrace{\left[\frac{dy_h}{dt} - ay_h(t)\right]}_{=0 \text{ by Step 1}} + y_h(t)\frac{dC}{dt} &= f(t)
\end{aligned} \tag{2.42}$$

$$\begin{aligned}
\frac{dC}{dt} &= \frac{f(t)}{y_h(t)} = b(t)\left[y_h(t_0)e^{\int_{t_0}^{t} ads}\right]^{-1} \\
C(t) &= \int_{t_0}^{t} f(s)y_h^{-1}(t_0)e^{-\int_{t_0}^{s} adt}ds + K \\
C(t) &= \int_{t_0}^{t} f(s)y_h^{-1}(t_0)e^{-a(s-t_0)}ds + K,
\end{aligned} \tag{2.43}$$

where K is an integration constant.
Step 3: Substitute (2.43) into $y(t) = C(t)y_h(t)$ and apply (2.40) to find the general solution of the IVP .

$$y(t) = y_h(t_0)e^{a(t-t_0)} \left[\int_{t_0}^{t} f(s)y_h^{-1}(t_0)e^{-a(s-t_0)}ds + K \right]$$

$$= \int_{t_0}^{t} f(s)e^{a(t-s)}ds + \underbrace{Ky_h(x_0)}_{\text{Call this } \overline{K}} e^{a(t-t_0)}. \tag{2.44}$$

Step 4: Apply the IC (2.40) to determine the solution of the IVP.
Now, apply (2.40) to see that $y(t_0) = 0 + \overline{K} = y_0$. Hence, the solution of the IVP is

$$y(t) = y_0 e^{a(t-t_0)} + \int_{t_0}^{t} f(t)e^{a(t-s)}ds \tag{2.45}$$

This extremely important and useful formula is called the *variation of parameters formula.* It tells us that the solution of any differential equation of the form (2.39) is (2.45).

Exercise 2.8.2. Justify all steps in the derivation of (2.45).

Exercise 2.8.3. Solve the IVP:

$$\begin{cases} \frac{dy}{dt} = -\frac{1}{2}y(t) + e^{-3t} \\ y(0) = \frac{1}{2}. \end{cases}$$

2.8.3 Higher-Order Linear ODEs

Higher-order linear ODEs with constant coefficients of the form

$$a_n x^{(n)} + a_{n-1}x^{(n-1)} + \ldots + a_1 x' + a_0 x = 0, \tag{2.46}$$

where $a_i \in \mathbb{R}$, $a_n \neq 0$, and $n \in \mathbb{N}$, arise in Chapters 2 - 5. We consider the special case

$$ax''(t) + bx'(t) + cx(t) = 0, \tag{2.47}$$

where $a \neq 0$ and $b, c \in \mathbb{R}$. Assuming that the solution of (2.47) is of the form $x(t) = e^{mt}$(where $m \in \mathbb{R}$) yields

$$e^{mt}\left(am^2 + bm + c\right) = 0 \Longrightarrow am^2 + bm + c = 0. \tag{2.48}$$

So, the nature of the solution of (2.47) is completely determined by the values of m. There are three distinct cases regarding the nature of the solution of (2.47) using (2.48):

Nature of the Roots of (2.48)	General Solution of (2.47)
$m_1 \neq m_2$ (real)	$x(t) = C_1e^{m_1t} + C_2e^{m_2t}$
$m_1 = m_2$ (real)	$x(t) = C_1e^{m_1t} + C_2te^{m_1t}$
$m_1, m_2 = \alpha \pm i\beta$	$x(t) = C_1e^{\alpha t}\sin(\beta t) + C_2e^{\alpha t}\cos(\beta t)$

Exercise 2.8.4. For what values of m_1 and m_2 is it guaranteed that
i.) $\lim_{t\to\infty} x(t) = 0, \forall C_1, C_2 \in \mathbb{R}$?
ii.) $x(\cdot)$ is a bounded function of t for a given $C_1, C_2 \in \mathbb{R}$?

2.9 Looking Ahead

Armed with some rudimentary tools of analysis, we are ready to embark on our journey through the world of evolution equations. We will begin by investigating some models whose mathematical description involves an extension of (2.39) to vector form. Specifically, while we know that $y(x) = Ce^{ax}$ is the general solution of the ODE $\frac{dy}{dx} = ax$, it is natural to ask what plays the role of the general solution of the equation $\frac{d\mathbf{Y}}{dx} = \mathbf{AY}$, where

$$\mathbf{Y} = \begin{bmatrix} y_1 \\ \vdots \\ y_N \end{bmatrix}$$

and $\mathbf{A} \in \mathbb{M}^N(\mathbb{R})$. It is tempting to write $\mathbf{Y}(t) = Ce^{\mathbf{A}t}$, but what does this really mean? Is this defined? What are its properties? We address these questions and much more in Chapters 3 and 4.

Chapter 3

A First Wave of Mathematical Models

Describing the behavior of natural, social, and physical phenomena quantitatively and qualitatively over time lies at the heart of research across disciplines. Phenomena that evolve over time and whose precise description involves rates of change with respect to time can be modeled mathematically using differential equations. The complexity of these equations depends on the number of quantities being introduced into the model, as well as the presence of other factors affecting the evolution of the phenomena.

3.1 Newton's Law of Heating and Cooling—Revisited

We introduced Newton's Law of Heating and Cooling in Section 1.5. There, the ambient temperature T_m of the region into which the body was being transferred to cool or warm was held constant. But, temperatures tend to rise and fall even under the most controlled circumstances. As such, it is more realistic to allow T_m to depend on time as well. Doing so leads to the slightly modified IVP

$$\begin{cases} \frac{dT(t)}{dt} = k\big(T(t) - T_m(t)\big) \\ T(t_0) = T_0 \end{cases} \tag{3.1}$$

While this is better, introducing such time dependence creates a problem unless we have the function $T_m(t)$ in hand. In such case, (3.1) will have a unique solution and there is an formula for it, but we will wait until Chapter 5 to reveal it.

It is more likely that our knowledge of the ambient temperature is somewhat less precise in that we might believe that it changes at a rate proportional to the present ambient temperature. That is,

$$\frac{dT_m(t)}{dt} = k_2 T_m(t) \tag{3.2}$$

where k_2 is a (possibly different) constant of proportionality.

STOP! Under what natural conditions would k_2 be positive? Negative? Why?

If we couple (3.2) with (3.1), we will need to impose another initial condition on $T_m(t)$ at a fixed time in order for there to be any hope of a unique solution to the resulting IVP.

STOP! Why do you think this is the case?

For convenience, we impose the initial condition at the same time as the one for the temperature $T(t)$ of the body, namely at time $t = t_0$. For simplicity, let us use the $t = 0$, the initial time at which we will begin our formal observation of the phenomenon. Doing so yields the IVP

$$\begin{cases} \frac{dT(t)}{dt} = k_1 \left(T(t) - T_m(t)\right) \\ \frac{dT_m(t)}{dt} = k_2 T_m(t) \\ T(0) = T_0, \quad T_m(0) = T_1 \end{cases} \tag{3.3}$$

Exercise 3.1.1. If a large turkey is placed in a very small room to cool, you might notice that the room in the temperature actually rises in the short term. This happens because of the heat emanating from the turkey. As such, the rate at which the ambient temperature at time t changes depends not only on the present ambient temperature, but also on the temperature of the body being transferred into the room. How would you modify (3.1) to account for this change?

There are several natural variants of the scenarios described above that can be described by similar models. For instance, suppose you take *two* items out of the oven at the same time, an 18-pound roast and a baked potato, and place them onto the table in the dining room where the surrounding temperature is 74 degrees Fahrenheit. It stands to reason that the roast and the potato will cool at different rates **(Why?)** and so, when using Newton's Law of Heating and Cooling to formulate a model this time, we must account for this difference. Identifying the temperatures of the roast and potato as $T_R(t)$ and $T_P(t)$, respectively, one such IVP is

$$\begin{cases} \frac{dT_R(t)}{dt} = k_1 \left(T_R(t) + T_P(t) - T_m(t)\right) \\ \frac{dT_P(t)}{dt} = k_2 \left(T_R(t) + T_P(t) - T_m(t)\right) \\ \frac{dT_m(t)}{dt} = k_3 T_m(t) \\ T_R(0) = T_0, T_P(t) = T_1, T_m(0) = T_2 \end{cases} \tag{3.4}$$

where k_1, k_2, and k_3 are constants of proportionality.

Exercise 3.1.2. Complete the following exercise.

i.) Comment on the relative relationship among k_1, k_2, and k_3. Which is the largest? Why? Explain.

ii.) Formulate a variant of (3.4) in the spirit of the scenario described in Exercise 3.1.1.

iii.) Compare IVP (3.4) to the following IVP. Which is more realistic? More general? Explain.

$$\begin{cases} \frac{dT_R(t)}{dt} = k_1 T_R(t) + k_2 T_P(t) - k_3 T_m(t) \\ \frac{dT_P(t)}{dt} = k_4 T_R(t) + k_5 T_P(t) - k_6 T_m(t) \\ \frac{dT_m(t)}{dt} = k_7 T_m(t) \\ T_R(0) = T_0, T_P(t) = T_1, T_m(0) = T_2 \end{cases} \tag{3.5}$$

MATLAB-Exercise 3.1.1. We will use **MATLAB** to investigate the solution to IVP (3.5).

To begin, open **MATLAB** and in the command line, type:

MATLAB_Heating_Cooling_Revisited

Use the GUI to answer the following questions. Make certain to click on the HELP button for instructions on how to use the GUI. A typical screenshot of the GUI can found in Figure 3.1.

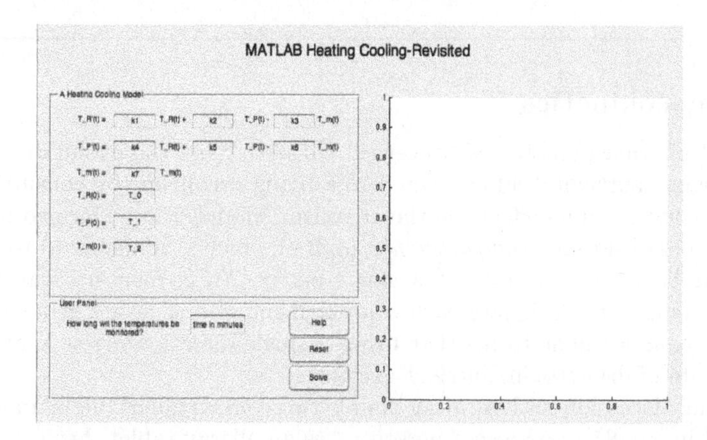

FIGURE 3.1: MATLAB Heating Cooling-Revisited GUI

i.) Suppose the initial temperatures of the roast and potato are $185°F$ and $175°F$, respectively, and that the ambient room temperature is currently $77°F$. If

$$k_1 = k_2 = k_4 = k_6 = 0.001$$
$$k_3 = 0.05$$
$$k_5 = 0.005$$
$$k_7 = 0.01$$

what will the temperatures be in 5 minutes?

ii.) Repeat (i) for 10 minutes, and then for 20 minutes. Is there a certain time at which the model breaks down?

iii.) Suppose now that the initial temperature of the roast, potato, and the room are $180°F, 177°F$, and $75°F$, respectively. Using the same proportionality constants as in (i), what are the temperatures after 5 minutes?

iv.) By how much do your answers in (iii) and (i) differ? Do you think that difference is significant?

v.) Repeat (iii), but this time, enter initial temperatures and proportionality constants different from, but close to, those prescribed in (i). Try to determine if small changes in the inputs lead to small changes in the outputs produced by the GUI.

■

Exercise 3.1.3. Consider the scenario leading to IVP (3.3), but now assume that the turkey is set to cool on the kitchen table and that the oven is turned off but its door is cracked open to help it cool down more quickly. This, in turn, affects the ambient temperature of the room.

i.) How would you incorporate this into the model? Try to formulate a modified version of IVP (3.3).

ii.) Compare and contrast IVP (3.3) and the one formulated in (i).

3.2 Pharmocokinetics

The field of pharmacokinetics is concerned with studying the evolution of a substance (e.g., drugs, toxins, nutrients) administered to a living organism (by consumption, inhalation, absorption, etc.) and its effects on the organism. Models can be formed by partitioning portions of the organism into *compartments*, each of which is understood to be a homogenous mass that behaves uniformly as a single entity. An elementary way to visualize a compartment model is to designate each compartment as a separate geometric shape and to connect one compartment to another using a single-direction arrow that indicates the direction and rate of drug/toxin/nutrient exchange.

We begin our discussion with a rather simple two compartment model motivated by the work discussed in [10, 37]. Suppose a person takes an allergy tablet. Let

$$\begin{aligned} y(t) &= \text{concentration of DRUG Y in the GI tract at time } t \\ z(t) &= \text{concentration of DRUG Y in the blood at time } t \\ a &= \text{absorption rate of DRUG Y into the bloodstream} \\ b &= \text{rate at which DRUG Y is eliminated from the blood} \end{aligned}$$

Note that a and b are positive constants. **(Why?)** The following diagram illustrates the trek of DRUG Y through the body:

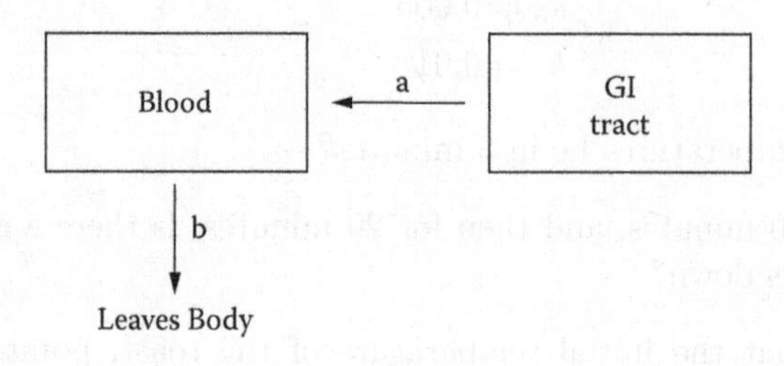

Assuming that the tablet had a concentration of y_0 milligrams once it entered the GI tract, and that there was no trace of DRUG Y in the bloodstream at the time it was administered, we can make use of the standard "Rate In - Rate Out" approach to derive the flux equations for both compartments. This yields the following IVP :

$$\begin{cases} \frac{dy(t)}{dt} = -ay(t) \\ \frac{dz(t)}{dt} = ay(t) - bz(t) \\ y(0) = y_0, z(0) = 0 \end{cases} \tag{3.6}$$

Some natural questions to ask include:

1. At what time is a prescribed level of DRUG Y reached within each compartment?
2. What happens if our measurements of the rate constants and/or the initial dosage are a little off? Are the resulting concentrations drastically different? By how much do they differ over time?
3. How do we choose the initial dosage to ensure that the concentration in the blood at time T is a certain level?

MATLAB-Exercise 3.2.1. We will use **MATLAB** to discover what happens if our measurements of the rate constants and/or the initial dosage are "a little off" in IVP (3.6).

To begin, open **MATLAB** and in the command line, type:

MATLAB_Pharmocokinetics

Use the GUI to answer the following questions. Make certain to click on the HELP button for instructions on how to use the GUI. A typical screenshot of the GUI can found in Figure 3.2.

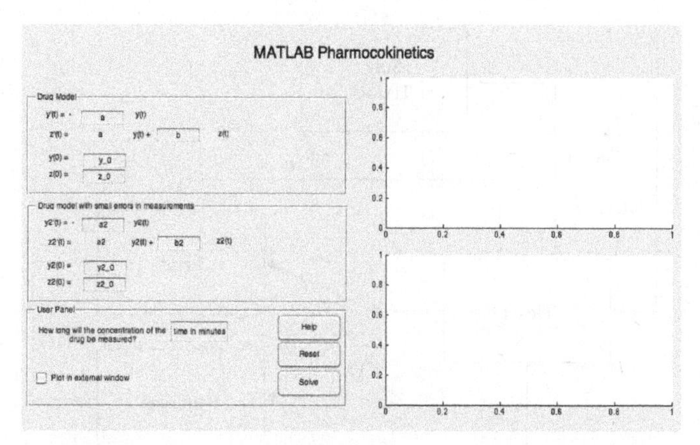

FIGURE 3.2: MATLAB Pharmocokinetics GUI

i.) We want to explore whether a small error in the parameters or initial conditions leads to a solution that is not too different than what you expect. Why would one care about this?

ii.) Suppose your measurements produce the following constants: $a = 0.17, b = 0.24, y_0 = 24, z_0 = 1$. However, in reality, the parameters are *really*: $a = 0.15, b = 0.25, y_0 = 25, z_0 = 0$. Are the solutions *different* over the initial 5-minute time interval? If so, would you describe the difference as being significant? Explain.

iii.) Repeat (ii), but now for the initial 10-minute interval.

iv.) At what time are the solutions furthest apart?

v.) Can you make a conjecture about the behavior of the solutions for large times?

vi.) Repeat (i)-(v) with new parameter choices. Pay particular attention to your answers in (iv) and (v).

■

This is just the beginning. These models can become much more sophisticated by simply accounting for more compartments. For example, let us consider a 5-compartment model related to the trek insulin takes through the body; this is a simplified version of the one mentioned in [17] and [46] . We introduce the following functions:

Compartment	Concentration at time t
Soft Tissue	$u_{ST}(t)$
Heart	$u_H(t)$
Liver	$u_L(t)$
Pancreas	$u_P(t)$
Exit Body via Periphery	$u_E(t)$

The rate of transfer from Compartment X to Compartment Y is denoted as $a_{(X,Y)}$, and the rate at which insulin leaves the body via the periphery is L. Consider the following diagram illustrating this scenario:

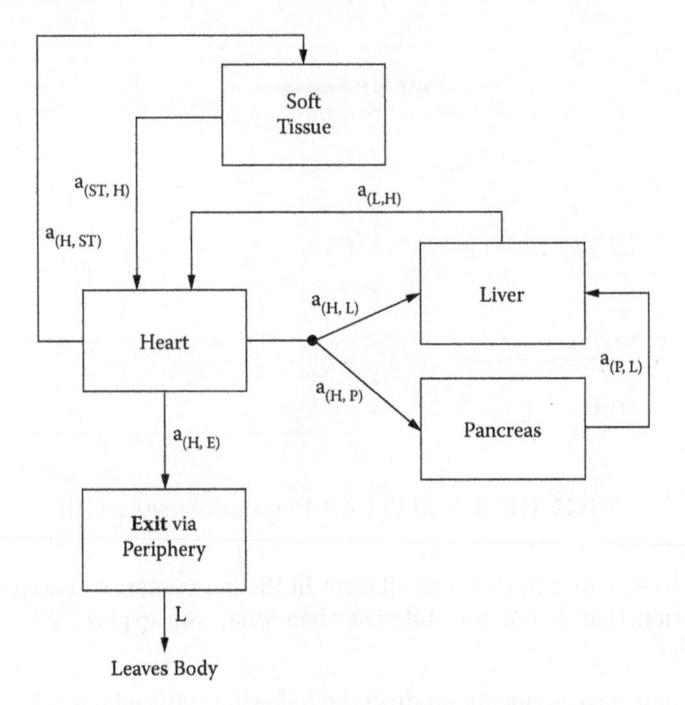

We again employ the "Rate In - Rate Out " approach to get the following flux equations, which are understandably more complex this time:

$$\begin{cases} \frac{du_{ST}(t)}{dt} &= a_{(H,ST)}u_H(t) - a_{(ST,H)}u_{ST}(t) \\ \frac{du_H(t)}{dt} &= \left[a_{(ST,H)}u_{ST}(t) + a_{(L,H)}u_L(t)\right] - \left[a_{(H,ST)} + a_{(H,E)} + a_{(H,L)} + a_{(H,P)}\right] u_H(t) \\ \frac{du_L(t)}{dt} &= \left[a_{(P,L)}u_P(t) + a_{(H,L)}u_H(t)\right] - a_{(L,H)}u_L(t) \\ \frac{du_P(t)}{dt} &= a_{(H,P)}u_H(t) - a_{(P,L)}u_P(t) \\ \frac{du_E(t)}{dt} &= a_{(H,E)}u_H(t) - Lu_E(t) \\ u_{ST}(0) &= u_1, u_H(0) = u_2, u_L(0) = u_3, u_P(0) = u_4, u_E(0) = u_5 \end{cases} \tag{3.7}$$

Insulin will leave certain compartments more readily than others, so the rate constants are compartment-dependent. The interrelationships among the constants can be specified using the underlying physiology.

Exercise 3.2.1. More realistically, we would account for a subcutaneous injection of insulin, say of w cubic centimeters, at time $t = 0$, where w is a positive real constant.

i.) How would you incorporate this into the above compartment model?

ii.) How does incorporating this into the model affect the equations in IVP (3.7)?

Exercise 3.2.2. Insulin travels to more parts of the body than those included in IVP (3.7). Suppose, in addition to those compartments, we wish to account for the GI tract as a separate compartment. Assume that insulin enters the GI tract through the bloodstream directly from the heart, and passes from the GI tract directly to the liver.

i.) Revise the initial compartment diagram to account for this.

ii.) How does incorporating this into the model affect the equations in IVP (3.7)?

We could continue refining the above models just by adding more compartments or by dissecting existing ones ad infinitum. Similar models of interest include those describing the processing of glucose and those related to processing lead and its relation to lead poisoning.

3.3 Uniform Mixing Models

Like the pharmacokinetics model introduced in Section 3.2, uniform mixing models are compartmental. In fact, there is little difference between the models mentioned in this section and those introduced in Section 3.2 other than the underlying context. See if you agree!

Suppose you are a brewer and would like to blend two beers together in order to produce a new seasonal beer. Imagine the two beers are separated and are being stored in two large tanks, T_1 and T_2, connected by a series of pipes that allow the beer to be pumped into and out of the tanks in such a way as to ensure the volume of both tanks remains constant. The following information is known.

1. T_1 initially contains 100 gallons of 6.2% alcohol by volume (ABV) beer.
2. T_2 initially contains 80 gallons of 9.0% ABV beer.

3. The mixtures in both tanks are constantly stirred so as to reasonably ensure the alcohol (and distinct flavors) are uniformly distributed throughout both tanks.
4. The rates of flow for each of the connected pipes is shown below.

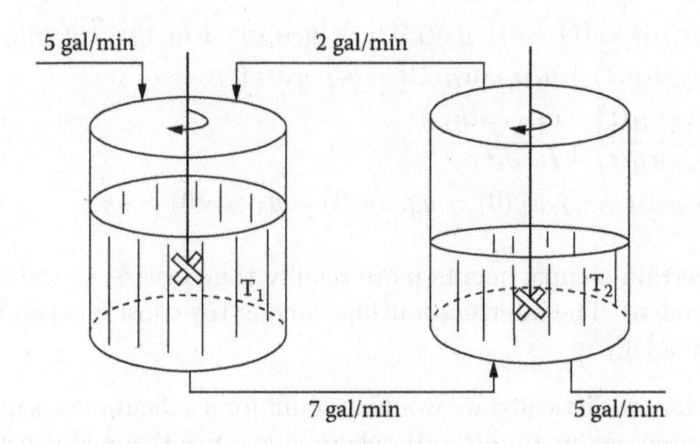

'Pure' 6.2% beer is being pumped into Tank 1. The pumps are set into motion at time $t = 0$, where t is measured in minutes. We would like to know the ABV in both tanks at time t. To this end, let us identify the following two unknown functions:

$u_1(t)$ = ABV of the beer in Tank 1;

$u_2(t)$ = ABV of the beer in Tank 2

Since ABV literally means the ratio of volume of alcohol to volume of fluid, it follows that

$$u_1(t) = \frac{alc_1(t)}{V_1(t)} \tag{3.8}$$

where $alc_1(t)$ is the volume of alcohol in Tank 1 and $V_1(t)$ is the volume of the fluid in Tank 1. To understand how the ABV changes in Tank 1 over time, we can differentiate (3.8) to obtain

$$\frac{du_1(t)}{dt} = \frac{V_1(t)\frac{d}{dt}(alc_1(t)) - alc_1(t)\frac{dV_1(t)}{dt}}{V_1^2(t)} \tag{3.9}$$

$$= \frac{\frac{d}{dt}(alc_1(t))}{V_1(t)} \tag{3.10}$$

STOP! Why did (3.9) simplify to (3.10)?

From (3.10) it is easy to see that understanding how the ABV changes over time in Tank 1 boils down to determining how the alcohol volume in Tank 1 changes, and then dividing that expression by 100 gallons (the constant volume of Tank 1). We use the "rate-in minus rate-out" scheme described eariler to determine the rate of change of alcohol volume in Tank 1. To this end the amount of alcohol flowing into Tank 1 is

$$0.062 \cdot 5 + u_2(t) \cdot 2$$

and the amount flowing out of Tank 1 is

$$u_1(t) \cdot 7.$$

Thus, applying the "rate-in minus rate-out" scheme yields

$$\frac{dalc_1(t)}{dt} = 0.062 \cdot 5 + u_2(t) \cdot 2 - u_1(t) \cdot 7$$

which, in turn, implies that the rate of change of ABV in Tank 1 is

$$\frac{du_1(t)}{dt} = \frac{0.062 \cdot 5 + u_2(t) \cdot 2 - u_1(t) \cdot 7}{100} \tag{3.11}$$

Following the same approach for Tank 2 leads to

$$\frac{du_2(t)}{dt} = \frac{u_1(t) \cdot 7 - u_2(t) \cdot 5 - u_2(t) \cdot 2}{80} \tag{3.12}$$

STOP! Verify (3.12).
Collecting (3.11), (3.12), and the initial conditions together leads to the IVP

$$\begin{cases} \frac{du_1(t)}{dt} & = \frac{-7u_1(t)+2u_2(t)+.31}{100} \\ \frac{du_2(t)}{dt} & = \frac{7u_1(t)-7u_2(t)}{80} \\ u_1(0) & = 0.062 \\ u_2(0) & = 0.09 \end{cases} \tag{3.13}$$

This is a system of first-order ODEs whose solution consists of two functions, $u_1(t)$ and $u_2(t)$. If we could somehow determine these two functions, we would understand the evolution of this system over time. We could then answer questions, such as, "How long must the beers mix until a 7.5% ABV beer is produced in Tank 2?"

MATLAB-Exercise 3.3.1. We will use **MATLAB** to solve (3.13) and determine the ABV content of the blended beers.

To begin, open **MATLAB** and in the command line, type:

MATLAB_Beer_Blending

Use the GUI to answer the following questions. Make certain to click on the 'Description of System' button for more details and instructions on how to use the GUI. A screenshot from this GUII can found in Figure 3.3.

i.) Assuming the same conditions that led to (3.13), determine the ABV content of the beer in Tank 1 and Tank 2 after 5 minutes?

ii.) Repeat (i) for 10, 30, and 60 minutes.

iii.) Describe what is happening to the ABV levels as the mixing time increases. Does this make sense physically?

iv.) Suppose in a feeble attempt to create a 'light beer' you replace the 6.2% beer that is being pumped into Tank 1 with water. What do you think will happen in the long run to the beers in Tank 1 and in Tank 2? Use the GUI to check your claim.

■

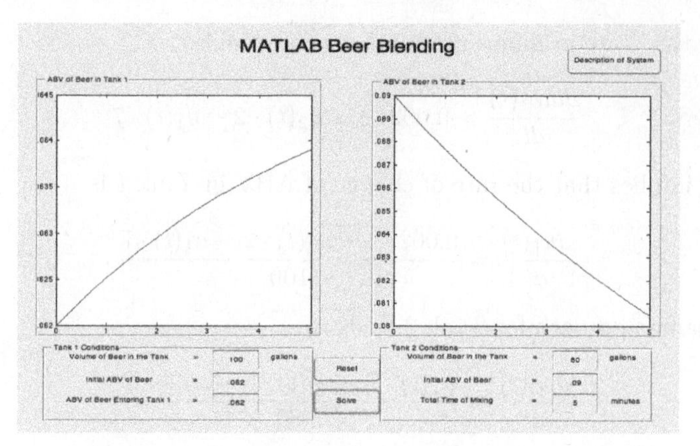

FIGURE 3.3: MATLAB Beer Blending GUI

Exercise 3.3.1. What if the volume is not held constant within the network of tanks in our beer example? Conceivably, there could be leak somewhere causing the volume to diminish slowly over time.

i.) Modify the model leading to IVP (3.13) to account for nonconstant volumes in the tanks.

ii.) For concreteness, assume Tank 1 is leaking at a rate of 1 gallon per minute and Tank 2 contains no leaks. Write an IVP, similar to (3.13), that accounts for such a leak.

iii.) Comment on the differences between IVP (3.13) and the one derived in (ii).

There are various contexts in which a uniform mixing type of model arises. For instance,

- "Concentration" of newly-printed currency that has been released into circulation;
- Concentration of nutrients in the bloodstream when someone who is sweating profusely drinks Gatorade;
- Pollution levels in a bay as pure clean water is pumped in to purify it.

3.4 Combat! Nation in Balance

This discussion is an adaptation of [7], which is based on the work of Lewis Fry Richardson.

Every nation must defend itself from its enemies. But, what feeds a nation's perception that it must increase size of its arsenal? The evening world news is chock full of stories of strife and struggle between nations and the resulting increase in armaments. During times of peace, the perceived level of threat diminishes which has an effect on the rate of growth of nations' arsenals. This ebb and flow can be modeled very naïvely using first-order ODEs of the type used throughout this chapter. A variant of one such model posed in [7] is described below.

Let $u_i(t)$ be a quantification of the size of Nation's i's weapon arsenal. We would like to formulate an expression for $\frac{du_i(t)}{dt}$. The question is what contributes to the increase and decrease of this quantity. For one, it is reasonable to think that $\frac{du_i(t)}{dt}$ will increase as tensions rise and other nations that oppose Nation i decide to increase their arsenals, or when Nation i perceives an increase in the threat of its enemies. Also, the level of disdain Nation i has for the other nations can drive its desire to increase its arsenal. On the other hand, the rate will decrease when other nations are perceived to pose less of a threat. An additional factor is the cost associated with maintaining the arsenal; as the size of the arsenal itself increases, so does the cost. This could reduce the speed at which a nation's arsenal increases despite what level of threat is perceived. We need to find a way to quantify these features.

A simple description of the "war readiness of Nation i" is $k_i u_i(t)$, where k_i is a positive constant. We can interpret $0 < k_i < 1$ as meaning that Nation i's strategic know-how and/or expertise in efficiently and effectively using the weapons in its arsenal is on the weak side. In fact, values of k_i closer to zero indicate a deficient effectiveness of Nation i to protect itself and/or to wage war. Contrarily, $k_i > 1$ indicates a degree of effectiveness in doing so.

The cost associated with maintaining an arsenal $u_i(t)$ is $c_i u_i(t)$, where c_i is a positive constant that represents an average cost per unit for Nation i to maintain such an arsenal.

For the moment, we shall describe the level of grievance Nation i feels toward another nation j using a nonnegative constant g_j . Depending on the actual Nation j, the value of g_j can be zero or positive real number.

Assume that there are N interacting nations in our system. All of the above factors can be used to formulate the following IVP, the solution of which would describe the size, over time, of each nation's weapon arsenal .

$$\begin{cases} \frac{du_1(t)}{dt} &= -c_1u_1(t) + \sum_{j\neq 1} k_j u_j(t) + \sum_{j\neq 1} g_j \\ &\vdots \\ \frac{du_i(t)}{dt} &= -c_i u_i(t) + \sum_{j\neq i} k_j u_j(t) + \sum_{j\neq i} g_j \\ &\vdots \\ \frac{du_N(t)}{dt} &= -c_N u_N(t) + \sum_{j\neq N} k_j u_j(t) + \sum_{j\neq N} g_j \\ u_1(0) &= u_1^*, \ldots, u_i(0) = u_i^*, \ldots, u_N(0) = u_N^* \end{cases} \tag{3.14}$$

where u_i^* is the initial size of the weapon arsenal of Nation i.

Exercise 3.4.1. Suppose each of the N nations was a so-called pacifist nation.

i.) Explain why in such case it is reasonable to assume the $k_j = g_j = 0$ for each $j = 1, \ldots N$.

ii.) What is the solution of (3.14) in this case?

iii.) Interpret the solution from (ii) verbally. Does this make sense?

Exercise 3.4.2. Suppose we only consider three nations, two of which are pacifist nations. Write down a simplified version of (3.14) that could reasonably describe the aresenal evolution in this scenario.

Exercise 3.4.3. Suppose we consider four nations, three of which are mutual allies and the fourth is a rogue, warring nation. How can you adapt IVP (3.14) to take into account the existence of such allies? Provide a new version of (3.14) and explain your reasoning.

3.5 Springs and Electrical Circuits—The Same, But Different

It turns out modeling springs and circuits has some unexpected applications. Not only do the models tell you something about how a spring affects the position of a mass or a current in a electrical circuit, but the ideas carry over to many other disciplines and interesting problems. For example, circuit models are used in neuroscience to study neural networks; in turn, these are used in computer science when studying artificial neural networks! A spring-mass model can be used to describe the dynamics of a building during an earthquake or even the dynamics of clothing in a computer game. In the next four sections, we will focus more on the modeling process and the derivation of the IVP corresponding to such application.

3.5.1 Spring-Mass System

Consider a spring hung vertically with one end affixed to a sturdy support, such as a ceiling or wooden board, and the other end attached to a ball. Imagine holding the ball in your hand and letting go; gravity will cause the spring to bounce up and down until the ball stops at a certain position, called the *equilibrium position*. If you were to pull the ball down below the equilibrium position and let go, what will happen? The spring will cause the ball to bob up and down until, again at some later time, the ball settles back to the equilibrium position. What is going on? What's so special about the equilibrium position? Is there a way to know the position of the ball at any given time? These are some of the questions we can answer once we have formed a mathematical model for the spring-mass system. Doing so requires a bit of elementary physics.

In order to model a spring-mass system we must first understand Newton's Second Law, which states the sum of all forces acting on a system equals mass times acceleration; that is,

$$\sum \text{Forces} = ma, \tag{3.15}$$

where m is a constant representing the mass and a represents the acceleration...***but of what?*** Before forming the model, we must explicitly state the quantity we wish to investigate. Thinking back to the spring and ball example, the ball was connected to the sturdy support by a spring. That distance can be measured by the length of the spring, sometimes called the natural length of the spring. However, once our hand lets go of the ball, the spring stretched ('gained' length) , and then compressed ('lost' length), etc. Based on this observation, we model the position of the mass relative to the natural length of the spring.

Let $x(t)$ be the position of the mass relative to the natural length of the spring in meters and t, time measured in seconds. Rewriting Newton's Second Law yields

$$\sum \text{Forces} = ma = mx''(t),$$

where $x''(t)$ is the second time derivative of $x(t)$ and m is measured in kilograms, kg. Fixing the unit of measurements results in force being measured in Newtons, denoted by N.

STOP! What does 1 N equal in terms of units of mass, position and time?

Now, we need to consider all of the forces that affect the position of the mass. For one, gravity must act on the mass. In particular, gravity will affect the mass at all times and it remains constant. So, if F_g represents the force acting on the mass due to gravity, then $F_g = mg$, where $g = 9.8\ m/s^2$.

STOP! By choosing $F_g = mg$, and not $-mg$, how do we interpret $x(t) < 0$? or $x(t) > 0$?

Another force acting on the mass through the spring is the force that naturally wants to restore the spring to its natural length. But how do we quantify this force? Physics comes to the rescue! Hooke's Law states the restoring force in the spring-mass system is proportional to its distance from the natural length. Before writing down any equations we must consider the direction in which the restoring force wants to act.

STOP! Fill in the blanks: If you pull a mass down below its equilibrium point, then the mass wants to go _______. If you compress the spring, pushing the mass above the equilibrium point, then the mass wants to go _______.

The restoring force, F_s, will always act in the opposite direction, so $F_s = -kx(t)$ where k is the proportionality constant (typically referred to as the *spring constant*). Summarizing, we have

$$\sum \text{Forces} = mg - kx(t) = mx''(t)$$

But are there any other forces acting on the mass? Well, yes! The fact that the mass moves through a medium (air, water, oil, etc.) implies there is some type of resistance force. If we assume the resistance force, F_R, is proportional to the speed of the mass, then we can express it using an equation. The tricky part occurs when we consider direction.

STOP! In which direction does F_R act? Write down an equation for F_R.

Incorporating the resistance force as one of the forces in the left side of (3.15) gives a model for the position of the mass relative to the spring's natural length. Precisely,

$$\sum Forces = ma$$

becomes

$$mg - kx(t) - rx'(t) = mx''(t), \tag{3.16}$$

where r is the proportionality constant for the resistance force. Rearranging terms in (3.16) yields

$$mx''(t) + rx'(t) + kx(t) = mg \tag{3.17}$$

Exercise 3.5.1. How many initial conditions must be prescribed in obtain a unique solution of (3.17)? In words, describe what the conditions physically represent and specify relevant units.

Exercise 3.5.2. How would the spring-mass model change if the system was on the moon?

MATLAB-Exercise 3.5.1. We will use **MATLAB** to solve (3.17) and investigate whether or not the model makes sense physically. To begin, open **MATLAB** and in the command line, type:

MATLAB_Vertical_Spring

Use the GUI to answer the following questions. Make certain to click on the 'Description of System' button for more details and instructions on how to use the GUI. A screenshot from MATLAB_Vertical_Spring can be found in Figure 3.4.

i.) Suppose the mass is at its equilibrium position and is at rest.

  a.) What do you expect to observe?
  b.) Use the GUI and its suggested parameter values to check your intuition. (Hint: $x(0) = mg/k$. Why?)

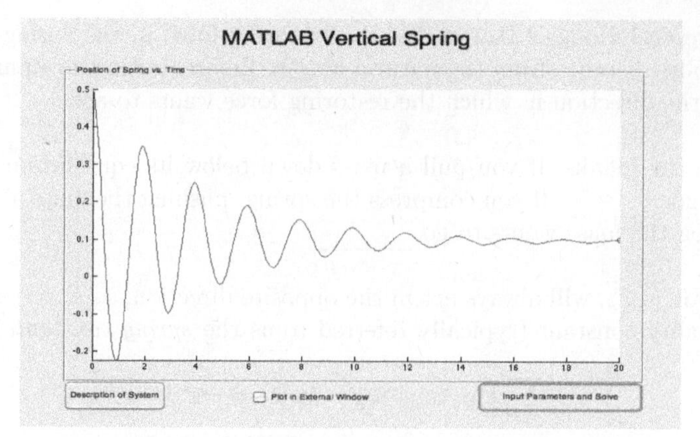

FIGURE 3.4: MATLAB Vertical Spring GUI

c.) Repeat (b), but this time use <u>any</u> nonzero parameter value of your choosing.

d.) Make a conjecture for the solution to (3.17) when both initial conditions are zero. If you are not sure, use the GUI!

ii.) Suppose r and k are both positive.

a.) What do you expect to happen in the long run? Does your conclusion depend on the initial conditions? Explain.

b.) Use the GUI to check your intuition for different parameter values.

c.) Make a conjecture for the behavior of the solution to (3.17) as time goes on.

iii.) Suppose the spring-mass system is in a vacuum (i.e. there is no air, so there is no resistance force) and the mass is displaced from its equilibrium position.

a.) What do you expect to observe?

b.) Use the GUI to check your intuition. Try different parameter values and times.

c.) Show that in the case when $r = 0$ the solution to (3.17) is given by

$$x(t) = A\cos\left(\sqrt{\frac{k}{m}}t\right) + B\sin\left(\sqrt{\frac{k}{m}}t\right) + \frac{mg}{k}$$

where A, B are constants that can be determined whenever $x(0)$ and $x'(0)$ are provided.

■

Exercise 3.5.3. Formulate a model for a horizontal spring-mass system. That is, one in which the mass rests on a surface and motion occurs only in the horizontal direction. Explain why in this case the natural length of the spring equals the equilibrium position.

In some applications, the position of several masses interconnected by springs must be described simultaneously. The series of masses could be oriented vertically, horizontally, or at some angle. The ends could be attached to a fixed support or maybe only one end is

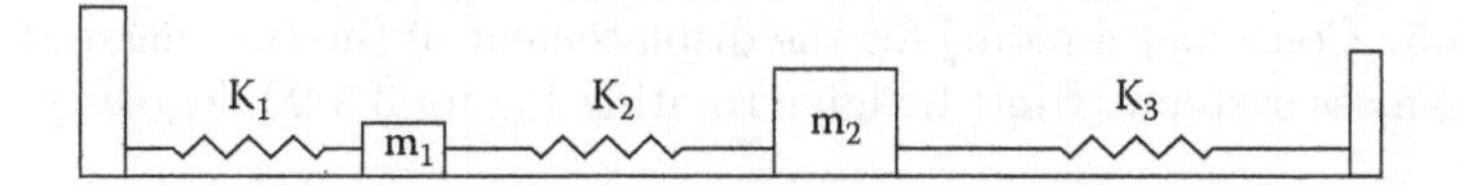

FIGURE 3.5: Horizontal Series of Springs

attached. It all depends on the situation and application. To illustrate the ideas, suppose a horizontal series of three springs connected by two masses and each end is attached to a sturdy support, as seen in Figure 3.5.

Let $x_1(t)$ and $x_2(t)$ denote the displacement from the natural positions for mass one and mass two, respectively. A positive displacement will represent displacement to the right of the natural position. As before, Hooke's Law will play an essential role in applying Newton's Second Law to each mass, but before we write down any equations, take a moment to visualize what is going on. For example, imagine moving mass one to the right of its natural position by two units and holding the second mass fixed at its equilibrium. What will happen when you let go of the masses? How many forces will be acting on each mass? What are the directions of those forces?

It is reasonable to conjecture that the first spring will stretch to account for the initial displacement. This, in turn, causes the second spring to compress because the second mass remains fixed at its natural position. The instant you let go, the stretched spring and the compressed second spring will try to restore the first mass by moving it to the left. However, the first mass will likely shoot through its natural position causing the second mass to move as well. From there, we should expect some type of back and forth movement until they both come to rest at their natural positions.

It is now time to apply Hooke's Law to each mass. Denote the masses by m_1 and m_2 and the three positive spring constants by k_1, k_2, and k_3. If we focus on applying Newton's Second Law to the first mass, then in the absence of any friction or air resistance we get

$$m_1 \frac{d^2 x_1(t)}{dt^2} = -k_1 x_1(t) + k_2\big(x_2(t) - x_1(t)\big) \tag{3.18}$$

The second term on the right-hand side of (3.18) might be unexpected, but since $x_1(t)$ and $x_2(t)$ are displacements from their natural positions we would need to subtract them in order to know the distance the spring has stretched or compressed in order to apply Hooke's Law.

STOP! Assume $x_1(0) = 2$ and $x_2(0) = 0$ as in the example described above. Does the right-hand side of (3.18) account for the proper direction of the forces when $t = 0$?

Putting it all together we arrive at the system of DEs

$$\begin{aligned} m_1 \frac{d^2 x_1(t)}{dt^2} &= -k_1 x_1(t) + k_2\big(x_2(t) - x_1(t)\big) \\ m_2 \frac{d^2 x_2(t)}{dt^2} &= -k_3 x_2(t) - k_2\big(x_2(t) - x_1(t)\big) \end{aligned} \tag{3.19}$$

as the 'frictionless' model for the coupled spring-mass system in Figure 3.5. Why is it frictionless? Did we account for any resistance forces? Could we?

Exercise 3.5.4. Propose a model that includes a resistance force, like friction, for the coupled spring-mass system in Figure 3.5.

Exercise 3.5.5. Construct a model for the displacement of the two masses in the coupled vertical spring-mass system. (Hint: Imagine rotating Figure 3.5 90 degrees.)

As mentioned before, the ideas presented here can be used in many different applications. In fact, later in this chapter you will use (3.19) to build a model describing the motion of a building during an earthquake.

3.5.2 Simple Electrical Circuits

Two commonly occurring terms in circuit analysis are: *current* and *voltage*. You probably have seen those terms on the packaging of electronic devices such as air conditioners, batteries, computers, tablets, and even cars. Both terms relate to an electric charge. The electrical charge is an essential property of subatomic particles that determine how they behave in magnetic and electric fields. The charge is measured in Coulombs (C) and is denoted by $q(t)$. The current is the flow of charges in some direction measured in amperes (A). Denoting the current as $i(t)$, we have the relationship

$$i(t) = \frac{dq(t)}{dt}$$

Voltage, or voltage drop, describes an electric potential difference between two points in a circuit and is measured in volts (V). The voltage is related to the charge by considering the work required to move a charge across a potential difference. Denoting the work by $w(t)$ we have that the voltage is equal to the time derivative of $w(t)$.

To form a circuit model we need to use appropriate laws that can steer the process. The first two laws are due to the German physicist, Gustav Kirchoff, and are as follows:

- Kirchoff's Voltage Law (KVL) : The sum of the voltage drops, at any instant, around a closed loop is zero.
- Kirchoff's Current Law (KCL) : The sum of the currents flowing into any junction equals the sum of the currents flowing out of the junction.

The term *closed loop* refers to a circuit, or portion of a circuit, for which a path can be followed that connects one point to itself without double-backing or traveling over a section of the path more than once. The circuit in Figure 3.6 has one closed loop whereas the circuit in Figure 3.7 has three.

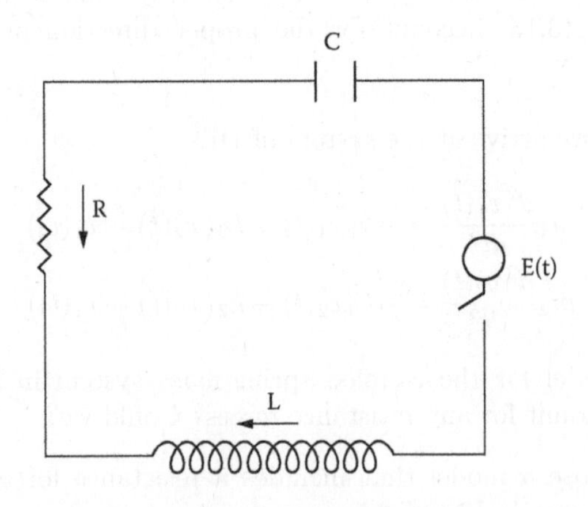

FIGURE 3.6: Single Closed Loop

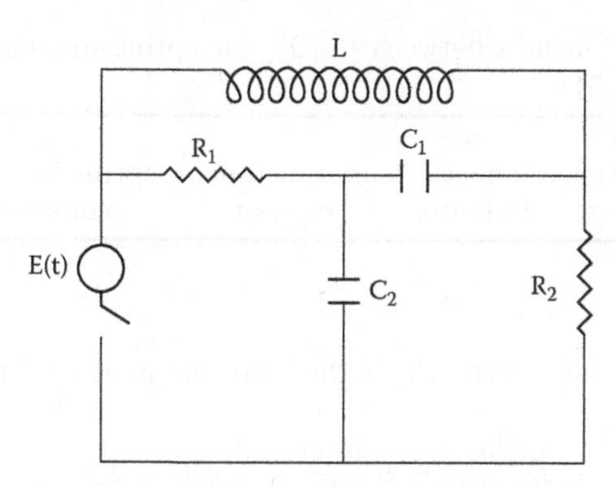

FIGURE 3.7: Three Closed Loops

As shown in both figures, a circuit is comprised of a resistor, capacitor, inductor and voltage source. A resistor is a device that dissipates electrical power. The energy loss manifests itself in a voltage drop over the resistor. A *capacitor* is an electrical circuit device that is able to store electrical charge. It can be visualized as two parallel metal plates where charges collect on one plate that will attract or repel charges on the other plate. An *inductor* is a device that stores energy stemming from the magnetic field surrounding the current. A *voltage drop* occurs across each of these circuit features, but to know what that drop equals, we need to use physical laws.

According to Ohm's Law, the voltage drop across a resistor is proportional to the current $i(t)$ so denoting the voltage drop at the resistor by $E_R(t)$ gives

$$E_R(t) = Ri(t)$$

where R is a real constant. Next, we consider the voltage drop across the inductor, $E_L(t)$. According to Lenz's Law and Faraday's Law , $E_L(t)$ is proportional to the speed of the current; that is,

$$E_L(t) = L\frac{di(t)}{dt}$$

where L a real constant. The voltage drop across the capacitor, $E_C(t)$, is proportional to the charge $q(t)$ on the capacitor. Therefore,

$$E_C(t) = \frac{1}{C}q(t)$$

where C is a non-zero real constant.

We are now in a position to use KVL and form a model for the circuit in Figure 3.6. If we let $E(t)$ represent the known voltage source *adding* potential to the system, then it follows that

$$\sum \text{voltages} = E(t) - E_R(t) - E_L(t) - E_C(t) = 0$$

or

$$\frac{1}{C}q(t) + L\frac{di(t)}{dt} + Ri(t) = E(t) \tag{3.20}$$

Perhaps at this point you have asked yourself, "What are we even modeling?!" That is a great question. Consider (3.20) and ask yourself, "Which quantities are unknown?" Every term in (3.20) is a known function or fixed constant except for $q(t)$ and $i(t)$.

STOP! What is the relationship between $i(t)$ and $q(t)$?

TABLE 3.1: Relationships between (3.22), the spring-mass model (3.17), and the simple circuit model (3.21)

Model	$y(t)$	a	b	c	$F(t)$
Spring-Mass	position	mass	air resistance	spring constant	force of gravity
Simple Circuit	charge	inductor	resistor	capacitor	voltage source

Substituting $i(t) = \frac{dq(t)}{dt}$ into (3.20) yields the following model for the charge on the capacitor in Figure 3.6:

$$L\frac{d^2q(t)}{dt} + R\frac{dq(t)}{dt} + \frac{1}{C}q(t) = E(t) \tag{3.21}$$

Exercise 3.5.6. How many initial conditions must be prescribed in order to find a unique solution to (3.21)?

Exercise 3.5.7. Explain what the initial conditions represent and indicate their units of measurement.

Exercise 3.5.8. How would you alter (3.21) to model the charge on the capacitor in the absence of an external voltage source?

Exercise 3.5.9. Explain how you can use the solution to (3.21) to find the current in the circuit.

3.5.2.1 The Same, But Different!

The spring-mass system model (3.17) describes the evolution of physical quantities different from than those in circuit model (3.21). However, the two differential equations share some striking similarities. For example, both are second-order differential equations with constant coefficients affixed to each of the unknowns. Also, each model has a known function on the right-hand side representing an external force applied to the system. In other words, both equations can be written in the form

$$ay''(t) + by'(t) + cy'(t) = F(t) \tag{3.22}$$

where $a, b, c \in \mathbb{R}$, $a \neq 0$ and $F(t)$ is a given real-valued function. This suggests that understanding the behavior of (3.22) will lead to understanding the behavior of the spring-mass system, the simple circuit system or any other system that can be written in the form (3.22).

Similarly, understanding the spring-mass system model (3.17) can help us understand the circuit model (3.21). Sometimes, visualizing and understanding the spring-mass system is easier than the circuit model, but since they are essentially described by the same DE the behavior of the solutions should be the same. Keep this mind while answering questions in this section. More details regarding springs, circuits, and their commonalities can be found in [16, 7, 50].

MATLAB-Exercise 3.5.2. We will use **MATLAB** to solve (3.21) and investigate whether or not the model makes sense physically. To begin, open **MATLAB** and in the command line, type: **MATLAB_Simple_Circuit**
Use the GUI to answer the following questions. Make certain to click on the 'Description of System' button for more details and instructions on how to use the GUI. A screenshot from this GUI can found in Figure 3.8.

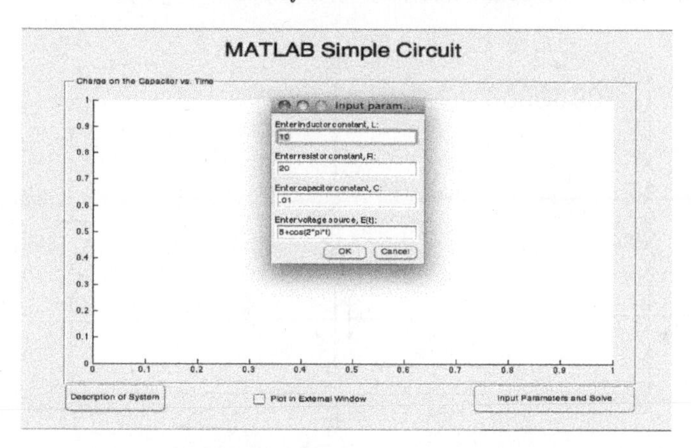

FIGURE 3.8: MATLAB Simple Circuit GUI

i.) Suppose there is no external battery source, no initial charge, and no current.

- a.) What do you expect to happen?
- b.) Use the GUI to check your intuition.
- c.) Make a conjecture for the solution to (3.21) for any choice of C, R and L when $E(t) = 0, t > 0$, $q(0) = 0 = q'(0)$.

ii.) Suppose the circuit does not possess a resistor and the battery is a *constant* voltage source.

- a.) Use the GUI to convince yourself solutions, in this case, are periodic with constant amplitudes.
- b.) Was that the case with the spring-mass model when $r = 0$?

iii.) Suppose $L = 10, r = 0, C = 0.025$ and $E(t) = 20\sin(2t)$ in model (3.21).

- a.) For different sets of initial conditions, plot the solution to (3.21) on $[0, 60]$.
- b.) What is going on? Should you expect the circuit to fail?
- c.) Use the GUI to convince yourself resistors are very beneficial in circuit design. For example, try $R = 45$.

■

Let us consider another circuit with a slightly different structure.

In Figure 3.9 you will see a circuit whose current 'breaks apart' at junction (a), then comes back together at junction (b). If you look back at Kirchoff's laws (**Do It!!**), you will see both KVL and KCL will be necessary when forming a circuit model.

Define $i_1(t)$ to be the current flowing away from the voltage source towards the capacitor and $q_1(t)$ to be the charge at the capacitor. Using KCL at junction (a) yields

$$i_1(t) = i_2(t) + i_3(t),$$

where $i_2(t)$ is the current between junction (a) and (b) through the resistor $R_2(t)$, and $i_3(t)$ the current through the inductor and resistor R_3. Using KCL again at junction (b), it must be the case that the current flowing out of junction (b) towards the voltage source equals $i_1(t)$.

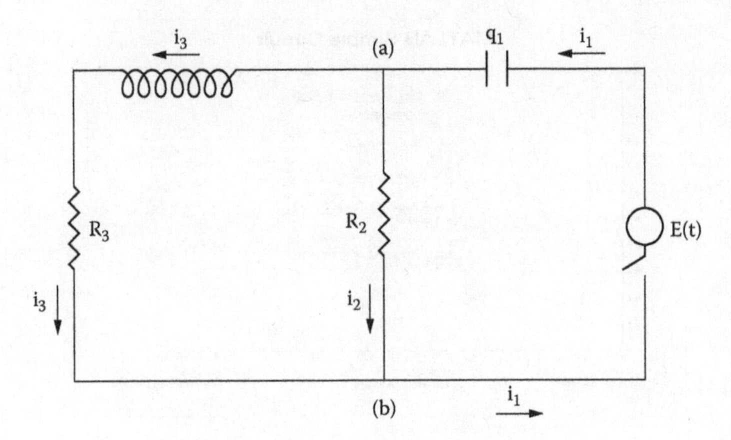

FIGURE 3.9: Circuit with junctions

STOP! Convince yourself of this!

Upon careful inspection of Figure 3.9, the circuit contains three closed loops, so we will apply KVL to each loop:

$\sum$ Voltages in Loop 1 = 0

$$E(t) - \frac{1}{C}q_1(t) - R_2 i_2(t)$$

$\sum$ Voltages in Loop 2 = 0

$$-L\frac{di_3(t)}{dt} - R_1 i_3(t) + R_2 i_2(t) = 0$$

$\sum$ Voltages in Loop 3 = 0

$$E(t) - \frac{1}{C}q_1(t) - L\frac{di_3(t)}{dt} - R_1 i_3(t) = 0$$

In Loop 2 the direction opposes the physical direction of the current $i_2(t)$, which explains why the voltage drop across the resistor R_2 term in the Loop 2 equation equals $-(-R_2 i_2(t)) = R_2 i_2(t)$. Combining the three equations from KVL and the one equation from KCL gives a great launching point for forming the circuit model.

$$\frac{1}{C}q_1(t) + R_2 i_2(t) = E(t) \tag{3.23}$$

$$L\frac{di_3(t)}{dt} + R_1 i_3(t) - R_2 i_2(t) = 0 \tag{3.24}$$

$$\frac{1}{C}q_1(t) + L\frac{di_3(t)}{dt} + R_1 i_3(t) = E(t) \tag{3.25}$$

$$i_1(t) - i_2(t) - i_3(t) = 0 \tag{3.26}$$

Again, at this point (if not earlier) you should be asking yourself, "What are we even modeling?" What are the unknowns? On the surface, it appears we have four equations and four unknowns $(q_1(t), i_1(t), i_2(t), i_3(t))$, which is normally what we want. But, subtracting

(3.25) from (3.23) results in equation (3.24).

STOP! Do the subtraction.

This means we are not gaining any extra information by including *both* (3.25) and (3.23) in our model, so we should remove one of them from the model. Ignoring the Loop 1 equation leaves us with the trimmed down system:

$$\begin{cases} L\frac{di_3(t)}{dt} + R_1 i_3(t) - R_2 i_2(t) & = 0 \\ \frac{1}{C}q_1(t) + L\frac{di_3(t)}{dt} + R_1 i_3(t) & = E(t) \\ i_1(t) - i_2(t) - i_3(t) & = 0 \end{cases}$$

This is a problem because now we have three equations and <u>four</u> unknowns. All is not lost, however, because $i_1(t)$ and $q_1(t)$ are related. In fact, we can always write one in terms of the other as follows:

$$\frac{dq_1(t)}{dt} = i_1(t) \text{ or } q_1(t) = \int_{t_0}^{t} i_1(s)ds$$

where $t_0 \in \mathbb{R}$. Also, equation (3.26) tells us that we need to only model two of the currents, since the third can always be found using (3.26).

Let us fix $i_1(t)$ and $i_3(t)$ as the quantities of interest for our model. Doing so reduces (3.24), (3.25) and (3.26) to

$$\begin{cases} \frac{1}{C}\int_{t_0}^{t} i_1(s)ds + R_2\big(i_1(t) - i_3(t)\big) & = E(t) \\ L\frac{di_3(t)}{dt} + R_1 i_3(t) - R_2\big(i_1(t) - i_3(t)\big) & = 0 \end{cases} \tag{3.27}$$

We are not quite there yet. Note that the top first equation in (3.27) is not a differential equation, but rather an *integral equation.* To formulate a differential equation, we may differentiate the first equation in (3.27) to arrive at

$$\frac{1}{C}i_1(t)ds + R_2(\frac{di_1(t)}{dt} - \frac{di_3(t)}{dt}) = E'(t) \tag{3.28}$$

$$L\frac{di_3(t)}{dt} + R_1 i_3(t) - R_2(i_1(t) - i_3(t)) = 0 \tag{3.29}$$

provided we assume $E(t)$ is differentiable.

Exercise 3.5.10. Use (3.29) to eliminate the $\frac{di_3(t)}{dt}$ term in (3.28). Then, rewrite the model (3.27) to include only one first derivative term in each equation.

Exercise 3.5.11. How many initial conditions are needed to solve (3.28)-(3.29)? Explain the physical meaning of the initial conditions and specify their units.

Exercise 3.5.12. Explain how you can use (3.28)-(3.29) to find $i_2(t)$ and $q_1(t)$.

3.6 Boom!—Chemical Kinetics

Chemical kinetics is a branch of physical chemistry that studies the rates of chemical reactions and the factors that influence the reaction rates. A *chemical reaction* is a process that transforms one or more substances into another, typically through the motion of electrons, or the breaking of chemical bonds. The initial substance, called a *reagent* or *reactant*, may go through several intermediate steps before completing the transformation to the final substance called the *product*. The rates at which these reactions occur can depend on several parameters, such as, temperature, pressure, energy, reactant concentrations, of the presence of a catalyst. For a more complete description of chemical kinetics and the mechanisms to follow refer to [36, 30, 40].

3.6.1 Reaction Models

To begin, let us consider the simple reaction

$$\mathrm{A} \xrightarrow{k} \mathrm{B} \tag{3.30}$$

Diagram (3.30) represents the situation where a reactant, A, is being transformed into some product B. We need a scientific law or rule that can guide the modeling process. For chemical reactions, the law we seek is commonly referred to as the *rate law* (historically called the *law of mass action*). The rate law describes the rate at which chemical reactants form different chemical products. It is worth noting that the rate law is more like a guiding principle that can be violated in certain situations and only applies to certain types of reactions. Physical chemists constantly go back and forth between experimental results and theoretical models in order to verify or alter their use of the law and the diagrams used to depict the reaction.

For the reaction given in (3.30), the rate law states the rate of change of the concentration of the reactant is proportional to the present concentration. The arrow in the diagram indicates direction of motion and k is the proportionality constant. Let $[A] = [A](t)$, $[B] = [B](t)$ represent the concentrations in moles per liter at time t of A and B, respectively. Then, the law of mass action yields

$$\frac{d[A]}{dt} = -k[A]$$

Since B gains whatever A loses, the rate of change of $[B]$ must equal $k[A]$, and we are left with the following reaction model

$$\begin{cases} \frac{d[A]}{dt} & = -k[A] \\ \frac{d[B]}{dt} & = k[A] \end{cases} \tag{3.31}$$

STOP! How does IVP (3.31) relate to Model I.2 *Elementary Population Growth,* (1.9)?

To slightly complicate matters, most chemical reactions can go in both directions, diagramatically expressed by,

$$\mathrm{A} \underset{k_2}{\overset{k_1}{\rightleftharpoons}} \mathrm{B}, \tag{3.32}$$

where k_1 is the proportionality constant for A transforming to B and k_2 the constant for B transforming into A. In this case, we get the model

$$\begin{cases} \frac{d[A]}{dt} & = -k_1[A] + k_2[B] \\ \frac{d[B]}{dt} & = k_1[A] - k_2[B] \end{cases} \tag{3.33}$$

Exercise 3.6.1. How many initial conditions are needed to obtain unique solutions to (3.31) and (3.33)? What do these represent in the context of the model?

Exercise 3.6.2. Explain why an initial condition of the form $[A](0) = 0$ in (3.31) would not make sense. Would such a condition be meaningful in (3.33)?

Exercise 3.6.3. Use the differential equations in (3.33) to determine the value of $[A](t) + [B](t)$ for all times during the reaction. More importantly, explain what your answer means in the context of an experiment.

Reactions may show signs of 'competition.' For example, a reactant could become three other substances

$$\begin{aligned} &\mathrm{A} \xrightarrow{k_B} \mathrm{B} \\ &\mathrm{A} \xrightarrow{k_C} \mathrm{C} \\ &\mathrm{A} \xrightarrow{k_D} \mathrm{D} \end{aligned} \tag{3.34}$$

In this case, the rate law gives

$$\begin{aligned} \frac{d[A]}{dt} &= -k_B[A] - k_C[A] - k_D[A] \\ &= -(k_B + k_C + k_D)[A] \end{aligned}$$

because the total rate of change to $[A]$ is the sum of the individual changes governed by the rate law. Finally, the change in the concentration of the products comes directly from the reactant, so that the model becomes

$$\begin{cases} \frac{d[A]}{dt} &= -(k_B + k_C + k_D)[A] \\ \frac{d[B]}{dt} &= k_B[A] \\ \frac{d[C]}{dt} &= k_C[A] \\ \frac{d[D]}{dt} &= k_D[A] \end{cases}$$

The reaction described in (3.34) is an example of a *parallel first-order reaction.* Another common reaction is a *first-order sequence reaction.* For example

$$\mathrm{A} \underset{k_-}{\overset{k_+}{\rightleftharpoons}} \mathrm{B} \underset{m_-}{\overset{m_+}{\rightleftharpoons}} \mathrm{C} \tag{3.35}$$

In diagram (3.35) you see how A can be transformed to B, but then B can be transformed to C. Along the way the reactions may also reverse. For example, after A transforms to B a portion may transform back to A whereas the remaining portion transforms to B. In this case the rate law gives

$$\begin{cases} \frac{d[A]}{dt} &= -k_+[A] + k_-[B] \\ \frac{d[B]}{dt} &= k_+[A] - (k_- + m_+)[B] + m_-[C] \\ \frac{d[C]}{dt} &= m_+[B] - m_-[C] \end{cases}$$

Some reactions need two or more reactants to come together before producing a single product. For example, the reaction diagram

$$\mathrm{A} + \mathrm{B} \underset{k_-}{\overset{k_+}{\rightleftharpoons}} \mathrm{C} \tag{3.36}$$

means one molecule of A and one molecule of B must come together to produce one molecule of C with a rate constant k_+. As before, the reactions may reverse, so in this case one molecule of C may separate to produce one molecule of A and B. The rate law for the product, C, is dependent now on the collisions between A and B. It states the rate of change of the concentration of C is proportional to the product of the concentrations of A and B. Combining this with the reverse reaction gives the following model for (3.36) :

$$\begin{cases} \frac{d[A]}{dt} &= k_-[C] - k_+[A][B] \\ \frac{d[B]}{dt} &= k_-[C] - k_+[A][B] \\ \frac{d[C]}{dt} &= k_+[A][B] - k_-[C] \end{cases} \tag{3.37}$$

MATLAB-Exercise 3.6.1. We will use **MATLAB** to investigate (3.37), specifically what happens in the long run. To begin, open **MATLAB** and in the command line, type:

MATLAB_Reactions

Use the GUI to answer the following questions. Make certain to click on the 'Description of System' button for more details and instructions on how to use the GUI. A screenshot from this GUI can found in Figure 3.10

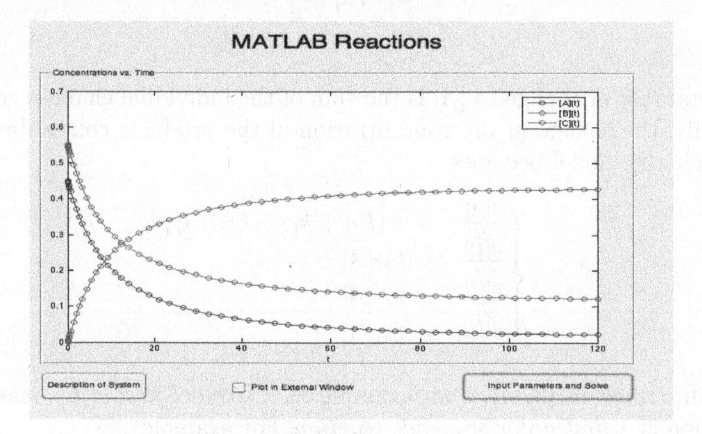

FIGURE 3.10: MATLAB Reactions GUI

i.) Explain what each of the following conditions means in the context of the reaction: $[A](0) = 0$, $[B](0) = 1$, and $[C](0) = 0$.

ii.) Use the GUI to verify your answer to (i).

iii.) Explain the applications of knowing the solution of the model as t goes to infinity.

iv.) Suppose the forward reaction rate, k_+, equals 0.2 and the backward reaction rate, k_-, is 0.001. Plot a solution for 200 seconds using $[A](0) = 0.55$, $[B](0) = 0.45$, and $[C](0) = 0$.

  a.) Do the concentrations appear to approach specific values?

  b.) Plot a solution for 400 seconds. Has your answer to (b) changed?

v.) Using the same initial conditions as in (ii) to investigate the time it takes for the reaction to stabilize when you alter k_+. For example, does increasing k_+ speed up the time it takes to stabilize? Slow down?

■

The proportionality constants will play important roles in modeling because they will essentially dictate the speed at which the reactions occur. Up to now we have assumed they are simply constants, but in reality they can depend on temperature, pressure, etc. Sometimes, the constants are neglected, approximated or even changed after noticing the model does not match experimental results. We will revisit some of these features later in Part I.

Exercise 3.6.4. Formulate a mathematical model for the reaction

$$A + B \xrightarrow{k_+} C \underset{m_-}{\overset{m_+}{\rightleftharpoons}} D$$

3.7 Going, Going, Gone! A Look at Projectile Motion

How does the space shuttle achieve orbit? Can country X's missile defense system reach country Y? Why are some fly balls in baseball home runs, but others are outs? Are the graphics in the action or sports video game realistic? All these questions have something to do with *projectile motion.* We define projectile motion as the motion of any object "near" the surface of the earth. The term near is used loosely and suggests gravity should influence the trajectory of the object.

3.7.1 Projectile Motion Models

To derive a projectile motion model we begin as we did when modeling spring-mass systems and use Newton's Second Law. Before we write any equations, we must take into consideration the fact that the projectile will be moving through a three-dimensional physical space. (Throw your pencil across the room; it's moving through a three dimensional physical space.) Such objects can move left and right, up and down, and forward and backward. So, the expression

$$\sum Forces = ma$$

really means that the sum of the forces in each direction must equal the mass of the projectile times the acceleration of the projectile in that particular direction. More precisely, let $x(t), y(t)$, and $z(t)$ represent the position of the projectile at time t in the three potential direction of motion. For definiteness, $x(t)$ represents the left/right direction of motion, $y(t)$ the forward/backward motion, and $z(t)$ the up/down direction of motion. Applying Newton's Second Law then yields

$$\sum \text{Forces in the } x\text{-direction} = m\frac{d^2x(t)}{dt^2} \tag{3.38}$$

$$\sum \text{Forces in the } y\text{-direction} = m\frac{d^2y(t)}{dt^2} \tag{3.39}$$

$$\sum \text{Forces in the } z\text{-direction} = m\frac{d^2z(t)}{dt^2} \tag{3.40}$$

where m is the mass of the projectile. As was the case with the spring-model, different types of forces can act upon the projectile. In fact, many of the same forces will act on the projectile, so the derivation of the projectile motion model will be extremely similar to the spring-mass model.

STOP! If you do not remember the ideas from the spring-mass model, go back and review Section 3.5.1.

The most obvious force acting on the projectile is gravity (in which direction does gravity act $x(t), y(t)$ or $z(t)$?). Another is the force due to air resistance, which we momentarily assume is proportional to the speed of the projectile.

STOP! In which direction does the air resistance force act? If you do not remember go back to the spring-mass model studied in Section 3.5.1.

Since the projectile can move in any of the three directions, air resistance will impede the motion of the projectile in all directions. What does this mean for our model? We must include an air resistance term in (3.38), (3.39), and (3.40). Are there any other forces? Possibly. Missiles have a propulsion system, satellites are influenced by variable gravitational forces, baseballs are affected by the wind, etc. To include all these potential external forces, we can include a generic 'forcing function' to act as the place holder for whatever additional force we wish to add to the model. Putting everything together we get

$$\begin{cases} m\frac{d^2x(t)}{dt^2} + r_x\frac{dx(t)}{dt} & = F_x(x(t), y(t), z(t), t) \\ m\frac{d^2y(t)}{dt^2} + r_y\frac{dy(t)}{dt} & = F_y(x(t), y(t), z(t), t) \\ m\frac{d^2z(t)}{dt^2} + r_z\frac{dz(t)}{dt} + mg & = F_z(x(t), y(t), z(t), t) \end{cases} \tag{3.41}$$

where r_x, r_y, and r_z are the air resistance proportionality constants, and forcing functions F_x, F_y, and F_z are written in such a way to imply they can depend on position and time. In reality, though, they can even depend on velocity, but for now we ignore that potential dependence.

To simplify the discussion, assume the forcing functions depend only on time, not position, and the gravitational force term mg is included in $F_z(t)$. Doing this transforms (3.41) into the simpler system

$$\begin{cases} m\frac{d^2x(t)}{dt^2} + r_x\frac{dx(t)}{dt} & = F_x(t) \\ m\frac{d^2y(t)}{dt^2} + r_y\frac{dy(t)}{dt} & = F_y(t) \\ m\frac{d^2z(t)}{dt^2} + r_z\frac{dz(t)}{dt} & = F_z(t) \end{cases} \tag{3.42}$$

Upon closer inspection of (3.42), you will notice that the position of the projectile, $(x(t), y(t), z(t))$, does not explicitly appear in the model. Rather, only the position's first and second time derivatives appear. This allows us to use a change of variable to reduce the system of second-order differential equations into a system of first-order differential equations. To learn how, consider the following exercise:

Exercise 3.7.1. Let $v_x(t) = \frac{dx(t)}{dt}, v_y(t) = \frac{dy(t)}{dt}$, and $v_z(t) = \frac{dz(t)}{dt}$. Using this change of variable transform (3.42) into a system of first-order differential equations for the unknowns $v_x(t), v_y(t)$, and $v_z(t)$.

In some sense, the first-order system for the directional velocities of the projectile is simpler than the second-order system for the position of the projectile. For one, there are fewer derivatives to consider and you need fewer initial conditions to solve the model. **(Why?)** However, this simplification would be useless if we did not have a way to compute the position of the projectile knowing only the directional velocities.

STOP! How would you compute the position of the projectile, $(x(t), y(t), z(t))$, knowing $(v_x(t), v_y(t), v_z(t))$?

The big assumption in our projectile motion models is that the force due to air resistance is proportional to the speed of the projectile. However, another common model assumes the air resistance is proportional to the *square* of the speed of the projectile. In order to include this assumption in a projectile motion model, we must carefully consider direction. **(Why?)** Well, remember that air resistance opposes the direction of motion, so we cannot simply say "air resistance equals a constant times velocity squared" because then air resistance would always be positive. However, velocity times the *absolute value of velocity* would allow for negative direction, and in magnitude would always equal speed squared. Even though the last sentence expresses the right idea, it is misleading because we are dealing with motion in three directions! We must proceed carefully.

Let $\mathbf{u(t)} = \langle x(t), y(t), z(t)\rangle$, so that the velocity of the projectile is

$$\frac{d\mathbf{u}(t)}{dt} = \langle\frac{dx(t)}{dt}, \frac{dy(t)}{dt}, \frac{dy(t)}{dt}\rangle.$$

We must be careful what we mean by "times" and "absolute value." What we are truly after is the size of the velocity vector, which in three dimensions is given by

$$\|\frac{d\mathbf{u}(t)}{dt}\| = \sqrt{(\frac{dx(t)}{dt})^2 + (\frac{dy(t)}{dt})^2 + (\frac{dz(t)}{dt})^2}$$

Now, multiplication makes sense and we have the following modified version of (3.42) :

$$\begin{cases} m\frac{d^2x(t)}{dt^2} + r_x\|\frac{du(t)}{dt}\|\frac{dx(t)}{dt} & = F_x(t) \\ m\frac{d^2y(t)}{dt^2} + r_y\|\frac{du(t)}{dt}\|\frac{dy(t)}{dt} & = F_y(t) \\ m\frac{d^2z(t)}{dt^2} + r_z\|\frac{du(t)}{dt}\|\frac{dz(t)}{dt} & = F_z(t) \end{cases} \tag{3.43}$$

For simplicity, we assumed the forcing functions only have time dependence. Very interesting questions regarding projectile motion can be posed using (3.41) and (3.43) or a combination of the two. For example, under what conditions will a baseball leave Fenway Park over the Green Monster in left field?

Next, we visualize solutions to the linear resistance model given by (3.42) and the quadratic resistance model given by (3.43) using **MATLAB** . We should gain some intuition regarding the differences between the two, and hopefully, determine which one is more realistic. A more complete description of the linear resistance model and the quadratic model, along with appropriate physical parameters, can be found in [15, 38, 43].

MATLAB-Exercise 3.7.1. In this exercise, you will use **MATLAB** to explore the consequence of changing the air resistance assumption from linear to quadratic. To begin, open **MATLAB** and in the command line, type:

MATLAB_Projectile_Motion

Use the GUI to answer the following questions. Make certain to click on the 'Description of System' button for more details and instructions on how to use the GUI. A screenshot from this GUI can found in Figure 3.11.

i.) In words, explain the difference between the solutions of the two models when using the default settings of the GUI.

ii.) Repeat (i) assuming that each quadratic air resistance constant is equal to one-half of the corresponding linear air resistance constant.

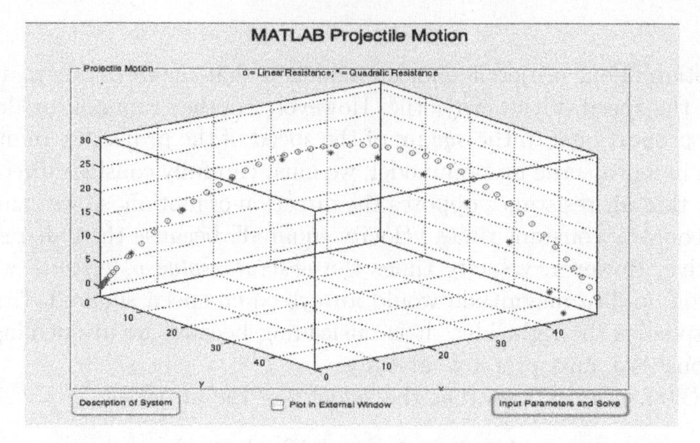

FIGURE 3.11: MATLAB Projectile Motion GUI

iii.) Explain why the linear resistance coefficients need not equal their quadratic counterpart. To illustrate, consider the following question:

a.) Suppose when a projectile has velocity vector $\langle 2,2,1\rangle$ the resistance force equals $\langle 2,1,3\rangle$. What are the resistance constants for each model?

iv.) Using any input of your choosing, explore solutions to the models and comment on the differences between the two.

■

3.8 Shake, Rattle, Roll!

Have you ever experienced an earthquake? Have you been standing or sitting in a tall building when one strikes? The bizarre sensation of horizontal motion can be quite scary, but surreal as well. You might expect a building to stay perfectly still, but think again, it moves! But, like a large tree swaying during a storm, too much motion can be devastating. If we can predict the behavior of a building during an earthquake, then we can design buildings that are safer, possibly saving numerous lives.

3.8.1 Floor Displacement Model

At the most basic level, a building is a collection of floors connected by vertical walls. If we restrict ourselves to horizontal motion only, then during an earthquake each floor can move, but that floor's motion would affect adjacent floors. **(Why?)** The vertical walls connecting the floors would themselves move. In fact, the vertical walls would be trying to *restore* the floors to an *equilibrium* position. Hmmm, restoring force, equilibrium position, etc.—this sounds familiar.

STOP! Review the Spring-Mass Section 3.5.1 before moving onward.

The idea is to model the motion of the floors as if we viewed the building as a spring-mass system. The masses are the floors and the springs the vertical walls trying to restore adjacent floors to their equilibrium positions. As in the spring-mass model, we will model a floor's displacement from its equilibrium position.

To simplify the discussion, let us assume the building has five stories plus a ground floor and that each floor's displacement and mass is denoted by $x_i(t)$ and m_i, respectively, for $i = 1, 2, \ldots 5$. A positive displacement will represent displacement to the right and negative to the left. Since an earthquake will affect the ground floor, we also need a function to represent its position, say $g(t)$. Applying Hooke's law and Newton's Second Law to the first and second floors yields

$$\begin{cases} m_1 x_1''(t) & = -k_0(x_1(t) - g(t)) + k_1(x_2(t) - x_1(t)) \\ m_2 x_2''(t) & = -k_1(x_2(t) - x_1(t)) + k_2(x_3(t) - x_2(t)) \end{cases}$$

STOP! If these equations seem 'out of the blue,' go back and read the spring-mass section.

In fact, all the floors follow the same pattern except the top floor **(Why?)** where we get

$$m_5 x_5''(t) = -k_4(x_5(t) - x_4(t))$$

Exercise 3.8.1. Write out a complete floor model for the five-story building. What should be the initial conditions? Interpret them in context.

Exercise 3.8.2. What would $g(t) = 0$ for all times t represent physically? What should the solution to the system of DEs be in this case?

Exercise 3.8.3. Adapt the model to account for 10 floors plus a ground floor.

MATLAB-Exercise 3.8.1. In this exercise, you will use **MATLAB** to explore the floor displacement model produced in Exercise 3.8.1. To begin, open **MATLAB** and in the command line, type:

MATLAB_Floor_Displacement

Use the GUI to answer the following questions. Make certain to click on the 'Description of System' button for more details and instructions on how to use the GUI. A screenshot from this GUI can found in Figure 3.12.

i.) Use the default settings of the GUI and describe the motion of the building. Specifically, address how the floors move over time. For example, do all floors move immediately, does the motion speed up, slow down, etc.?

ii.) Do the floors ever stop shaking? Repeat (i) using 300 as the input when asked how long you will observe the building.

iii.) Use the GUI to determine how the parameters affect the motion of the building. For example, what happens as the "spring" constant increase, mass decreases, the earthquakes last longer, etc.?

■

Most modern structures include shock absorbers that connect to adjacent floors to artificially add resistance and improve a building's ability to withstand an earthquake. If we assume the resistance force due to the shock absorbers is proportional to the speed of the

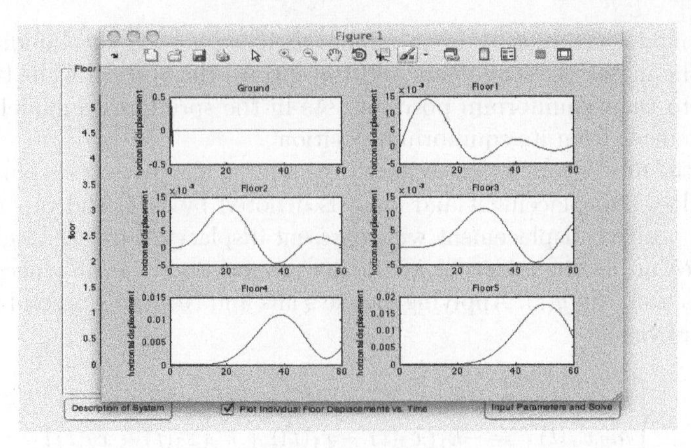

FIGURE 3.12: MATLAB Floor Displacement GUI

floors, then we can apply the exact same argument as we did in the spring-mass system. Indeed, assuming the proportionality constant for the resistance is c, the model becomes

$$\begin{cases} m_1 x_1''(t) &= -k_0(x_1(t) - g(t)) + k_1(x_2(t) - x_1(t)) - c(x_1'(t) - g'(t)) + c(x_2'(t) - x_1'(t)) \\ m_i x_i''(t) &= -k_{i-1}(x_i(t) - x_{i-1}(t)) + k_i(x_{i+1}(t) \cdots \\ & \quad -x_i(t)) - c(x_i'(t) - x_{i-1}'(t)) + c(x_{i+1}'(t) - x_i'(t)) \\ m_5 x_5''(t) &= -k_4(x_5(t) - x_4(t)) - c(x_5'(t) - x_4'(t)) \end{cases}$$

for $i = 2, 3, 4$, provided that $g(t)$ is differentiable. (See [49] and [6] for more details.) These models are used to investigate an earthquake's effect on a building, and more importantly, why shock absorbers should be included in any building design.

3.9 My Brain Hurts! A Look at Neural Networks

Theoretical neuroscience is a thriving field that combines mathematics, biology and physics to analyze neurons and other structures of the nervous system. The neuron is an amazing cell in the brain that can transmit electrical impulses over large distances in the body. The traveling impulses can be thought of as information that travels from one neuron to another. The collection of all such conduits is referred to as a *neural network*. This viewpoint has numerous applications in computer science where one uses artificial neural networking to produce learning algorithms for speech and pattern recognition, robotic control, medical imaging, forecasting, and more!

The neuron itself has a very specialized structure. First, *dendrites* receive information from other neurons' *axons* which carry information from the cell body of the neuron to other cells. Dendrites and axons of other neurons are joined by *synapses* which allow signals to pass from one neuron to another. They transmit the information by generating electrical pulses or spikes called *action potentials*. Action potentials are formed when the difference between the electrical potential on the interior of the neuron's cell membrane and the outside of the neuron's cell potential exceeds some threshold value.

The information sent and received across the neural network occur in the form of electrical signals within neurons, but most often need to be carried by special chemicals known as *neurotransmitters* . The release of neurotransmitters is controlled by ions, mainly sodium, potassium, calcium, and chloride, that move between the cells. *Ion channels* control the flow of ions across the cellular membrane, which themselves are regulated by the charges in the membrane potential.

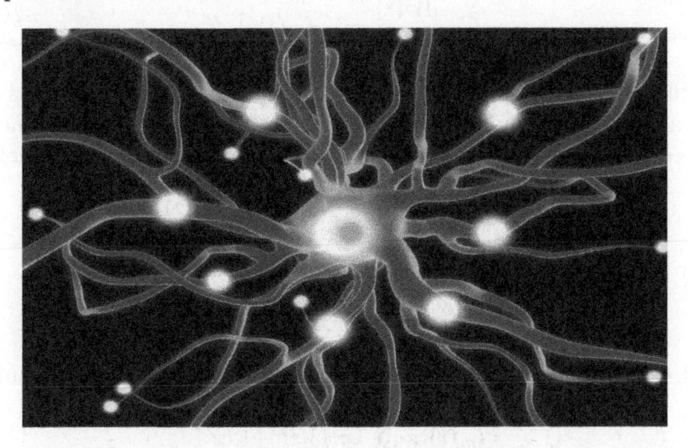

The first model we discuss is a multi-compartmental membrane potential (MCMP) model. An essential component of the MCMP model is the membrane current, I_m, of the neuron; that is, the total current passing through the cellular membrane via the ion channels. When a neuron is not sending information (sometimes called *at rest*) the cell is polarized with a negative charge on the interior of the membrane and a positive charge on the exterior. For example, in a particular species of squid the interior membrane potential is $-0.7V$ or about 70 millivolts less than that on the outside of the membrane. In addition, the polarizing effect of the membrane induces a capacitance, C_m, which will also play a significant role in the MCMP model.

Given that we have encountered the terms charge, current, and capacitance, you might expect resistance to play its part in forming a model. The cell membrane resistance, R_m, can be found experimentally by inserting an electrode into a neuron and injecting a current, I^e, so the voltage drop can then be recorded and set equal to $R_m I^e$ to find the resistance.

Since the cell membrane of a neuron functions as a capacitor, the neuron's membrane potential can modeled as an electrical circuit. Although the membrane potential varies over time and position, we simplify the situation by dividing the neuron into a set of many interconnected cylindrical compartments, called *cables* , each of which represents an electrical circuit. The neurons may contain elaborate branches of dendrites, but here we assume that the dendritic trees have only one branch. The ideas contained in the nonbranching MCMP model can be extended to include the branches.

The following neuroscience models and those contained in Section 3.10 are quite involved and should be considered advanced. Due to the significant amount of scientific knowledge necessary to perform the modeling processes in these sections, we have included only terse explanations of the modeling. We suggest reading these sections with the following references: [24, 22, 1].

To begin, let us consider a single compartment approximation of a neuron. Let $V(t)$ denote the membrane potential and $q(t)$ the charge on the capacitor (the membrane). Then,

$$V(t) = \frac{1}{C_m} q(t). \tag{3.44}$$

Differentiating each side of (3.44) with respect to time yields

$$C_m \frac{dV(t)}{dt} = \frac{dq(t)}{dt} = I(t), \tag{3.45}$$

where $I(t)$ is the total current across the membrane. Recall there are two currents of importance in the model: the membrane current, $I_m(t)$, and the current induced by the electrode, I^e. Using Kirchoff's Current Law (KCL) $I(t) = -I_m(t) + I^e$, (3.45) can be rewritten as

$$C_m \frac{dV(t)}{dt} = -I_m(t) + I^e \tag{3.46}$$

where by convention the electrode current entering the neuron is defined as positive-inward and the membrane current as positive-outward. Membrane currents are generally quantified as current per surface area. Let k be the surface area of the neuron and define

$$i_m(t) = I_m(t)/k \text{ and } c_m = C_m/k.$$

Dividing (3.46) by k yields

$$c_m \frac{dV(t)}{dt} = -i_m(t) + \frac{I^e}{k}. \tag{3.47}$$

Using KCL, Ohm's Law, and remembering how the ion channels work results in

$$i_m(t) = \sum_j g^j (V(t) - E^j)$$

where j is a particular ion channel contributing to the current, g^j is the conductance (inverse of resistance) per unit area of the channel j, and E^j is the equilibrium potential for channel j. In reality, the conductance per unit area can vary over time which significantly increases the complexity to the model. Here, we assume they are all constant so that the membrane current reduces to

$$i_m(t) = \sum_j g^j (V(t) - E^j) = g^L (V(t) - E^L)$$

where g^L and E^L are real constants. In practice, the parameters g^L and E^L are usually chosen to match the resting conductance and resting potential, respectively, of the particular cell being modeled. Substituting the membrane current back into (3.47) yields the single-compartment membrane potential model

$$c_m \frac{dv(t)}{dt} = -g^L (V(t) - E^L) + \frac{I^e}{k}. \tag{3.48}$$

Next, we extend the single-compartment idea to a multi-compartment model for the membrane potentials using our single-branch assumption on the neuron. To begin, let us assume the neuron can be approximated by three compartments as shown in Figure 3.13.

Each compartment has its own membrane potential, V_1, V_2, V_3; capacitance, C_1, C_2, C_3; membrane current, I_1^m, I_2^m, I_3^m; and surface area, k_1, k_2, k_3. Furthermore, each compartment is equipped with its own set of ion channels with equilibrium potentials, E_i^j, and resistances, r_j^i where i denotes the compartment number and j the specific ion channel. As before, an inserted electrode contributes a current to the neuron, but now we must allow that current to flow to any of the compartments. The portion of the electrode current that flows into compartment i is denoted by I_i^e. Lastly, we model the inter-compartmental current and resistances by $I_{i,i+1}$ and $r_{i,i+1}$, respectively, where the subscripts indicate the direction of the current. For example, the resistance $r_{3,2}$ represents the resistance encountered by the current $I_{3,2}$ flowing from compartment three into compartment two. The single-branch assumption plays a big role in this piece of the model.

Now it is time to derive a MCMP model for the neuron in Figure 3.13. To do so, we use KCL and the ideas presented in the single-compartmental model. Let us focus on

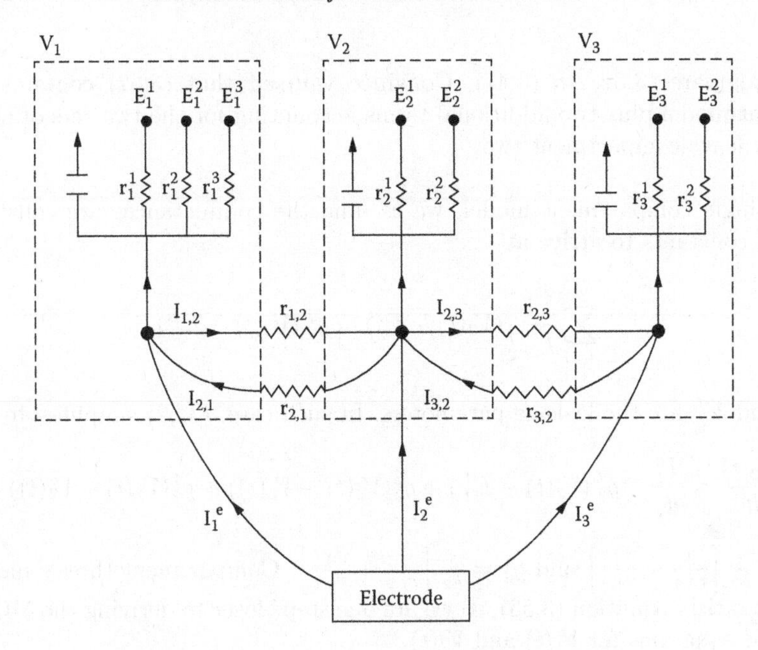

FIGURE 3.13: Multicompartmental Model

compartment two and derive a set of equations for $V_2(t)$. By KCL, it must be the case that

$$I_2^e + I_{3,2} + I_{1,2} = I_2^m + I_{2,3} + I_{2,1} \tag{3.49}$$

From our single-compartment discussion, we know that we can use (3.47) to represent I_2^m and rewrite (3.49) as

$$I_2^e + I_{3,2} + I_{1,2} = C_2 \frac{dV(t)}{dt} + \sum_{j=1}^{4} \frac{1}{r_2^j}(V_2(t) - E_2^j) + I_{2,3} + I_{2,1} \tag{3.50}$$

The remainder of the inter-compartmental currents can be determined using Ohm's Law. Doing so transforms (3.50) to

$$\begin{aligned} I_2^e + \frac{1}{r_{3,2}}(V_3(t) - V_2(t)) + \frac{1}{r_{1,2}}(V_1(t) - V_2(t)) = \\ C_2 \frac{dV_2(t)}{dt} + \sum_{j=1}^{4} \frac{1}{r_2^j}(V_2(t) - E_2^j) + \frac{1}{r_{2,3}}(V_2(t) - V_3(t)) + \frac{1}{r_{2,1}}(V_2(t) - V_1(t)) \end{aligned} \tag{3.51}$$

Rearranging terms in (3.51) and dividing every term by the surface area of compartment two yields

$$\begin{aligned} c_2 \frac{dV_2(t)}{dt} &= \frac{I_2^e}{d_2} - \sum_{j=1}^{4} \frac{1}{k_2 r_2^j}(V_2(t) - E_2^j) \\ &+ \frac{1}{k_2}\left[\frac{1}{r_{3,2}} + \frac{1}{r_{2,3}}\right](V_3(t) - V_2(t)) + \frac{1}{k_2}\left[\frac{1}{r_{1,2}} + \frac{1}{r_{2,1}}\right](V_1(t) - V_2(t)) \end{aligned} \tag{3.52}$$

where $c_2 = C_2/k_2$.

STOP! Compare (3.52) to (3.47). Convince yourself that (3.52) contains the single-compartment model plus two additional terms accounting for the two sets of currents that can enter or leave compartment two.

As in the single compartment model, we assume the conductances are constant and we fix 'leakage' constants to arrive at

$$\sum_{j=1}^{4} \frac{1}{k_2 r_2^j}(V_2(t) - E_2^j) = g_2^L(V_2(t) - E_2^L)$$

where g_2^L and E_2^L are the leakage parameters. In such case, (3.52) simplifies to

$$c_2 \frac{dV_2(t)}{dt} = \frac{I_2^e}{d_2} - g_2^L(V_2(t) - E_2^L) + g_2^3(V_3(t) - V_2(t)) + g_2^1(V_1(t) - V_2(t)) \tag{3.53}$$

where $g_2^3 = \frac{1}{k_2}\left[\frac{1}{r_{3,2}} + \frac{1}{r_{2,3}}\right]$ and $g_2^1 = \frac{1}{k_2}\left[\frac{1}{r_{1,2}} + \frac{1}{r_{2,1}}\right]$. Compartment three's membrane potential must satisfy equation (3.53), so we are one step closer to deriving the MCMP model. We still need equations for $V_1(t)$ and $V_2(t)$.

Using KCL in compartment one results in the equation

$$I_1^e + I_{2,1} = I_1^m + I_{1,2}, \tag{3.54}$$

which can be rewritten equivalently as

$$I_1^e + I_{2,1} = C_1 \frac{dV_1(t)}{dt} + \sum_{j=1}^{2} \frac{1}{r_j^1}(V_1(t) - E_1^j) + I_{1,2} \tag{3.55}$$

by (3.47).

STOP! Why do (3.54) and (3.55) have fewer terms than (3.49) and (3.50), respectively?

Using Ohm's Law, dividing by the surface area of compartment one, and rearranging terms yields

$$c_1 \frac{dV_1(t)}{dt} = \frac{I_1^e}{k_1} - \sum_{j=1}^{2} \frac{1}{k_1 r_1^j}(V_1(t) - E_1^j) + \frac{1}{k_1}\left[\frac{1}{r_{2,1}} + \frac{1}{r_{1,2}}\right](V_2(t) - V_1(t)) \tag{3.56}$$

where $c_1 = C_1/k_1$. Again, notice how (3.56) is the single compartment model plus one additional term accounting for the one set of inter-compartmental currents. Knowing the conductances are constant, we fix 'leakage' parameters g_1^L and E_1^L so that (3.56) becomes

$$c_1 \frac{dV_1(t)}{dt} = \frac{I_1^e}{k_1} - g_1^L(V_1(t) - E_1^L) + g_1^2(V_2(t) - V_1(t)) \tag{3.57}$$

where $g_1^2 = \frac{1}{k_1}\left[\frac{1}{r_{2,1}} + \frac{1}{r_{1,2}}\right]$.

We are almost there! We have one more equation to derive. Now, it's your turn.

Exercise 3.9.1. Using the ideas presented above to verify compartment three's membrane potential satisfies

$$
\begin{aligned}
c_3 \frac{dV_3(t)}{dt} &= \frac{I_3^e}{k_3} - \sum_{j=1}^{2} \frac{1}{k_3 r_3^j}(V_3(t) - E_3^j) + g_3^2(V_2(t) - V_3(t)) \\
&= \frac{I_3^e}{k_3} - g_3^L(V_3(t) - E_3^L) + g_3^2(V_2(t) - V_3(t))
\end{aligned} \tag{3.58}
$$

where $g_3^2 = \frac{1}{k_3}\left[\frac{1}{r_{2,3}} + \frac{1}{r_{3,2}}\right]$ and g_3^L, E_3^L are the leakage parameters.

Grouping (3.57), (3.53), and (3.58) yields the following MCMP model for the membrane potentials of a neuron approximated by three compartments as given in Figure 3.13:

$$
\begin{cases}
c_1 \frac{dV_1(t)}{dt} &= \frac{I_1^e}{k_1} - g_1^L(V_1(t) - E_1^L) + g_1^2(V_2(t) - V_1(t)) \\
c_2 \frac{dV_2(t)}{dt} &= \frac{I_2^e}{d_2} - g_2^L(V_2(t) - E_2^L) + g_2^3(V_3(t) - V_2(t)) + g_2^1(V_1(t) - V_2(t)) \\
c_3 \frac{dV_3(t)}{dt} &= \frac{I_3^e}{k_3} - g_3^L(V_3(t) - E_3^L) + g_3^2(V_2(t) - V_3(t))
\end{cases} \tag{3.59}
$$

Exercise 3.9.2. How many initial conditions need to be prescribed in order to obtain a unique solution of (3.59)?

Exercise 3.9.3. Suppose the neuron in Figure 3.13 has j_1 ion channels in compartment one, j_2 ion channels in compartment two, and j_3 ion channels in compartment three. Formulate a new MCMP model to account for the additional ion channels.

Exercise 3.9.4. Explain how if you use leakage parameters in Exercise 3.9.3 the model has the same 'appearance' or structure as (3.59).

Exercise 3.9.5. Propose a MCMP model for a neuron approximated by n compartments where n is a positive integer.

3.10 Breathe In, Breathe Out—A Respiratory Regulation Model

The rate at which you breathe in and breathe out depends on several factors. For example, your breathing rate while reading this book will be different than while exercising. The question then becomes, "Why do the two rates differ?" The first piece of the puzzle is the brain. Based mostly on carbon dioxide concentrations, the brain can control the ventilation rates in the lungs by activating or inhibiting the appropriate neurons. In this section, we present a very simplified version of a respiratory regulation model due to Grodins et al (see [25]). At its core, it is a compartmental model using the modeling technique 'Rate-in minus Rate out'.

STOP! Review the "Rate-in minus Rate-out" technique in Section 3.2.

The compartments in the simplified respiratory regulation model are the brain, the lungs, and the soft tissues, and the quantity of interest will be the partial pressures of CO_2 in each compartment. For those unfamiliar with partial pressures, the partial pressure of CO_2, denoted by PP_{CO_2} is defined by

$$PP_{CO_2} = x_{CO_2} P$$

where x_{CO_2} is the mole fraction of CO_2 and P is the total pressure. The reason one studies partial pressures is that when a gas (e.g. CO_2) comes into contact with a liquid (e.g. blood) some of the gas dissolves into the liquid. Furthermore, the amount of gas that dissolves in the liquid is a function of the partial pressure of the gas.

Figure 3.14 includes the first wave of simplifications and unknown quantities.

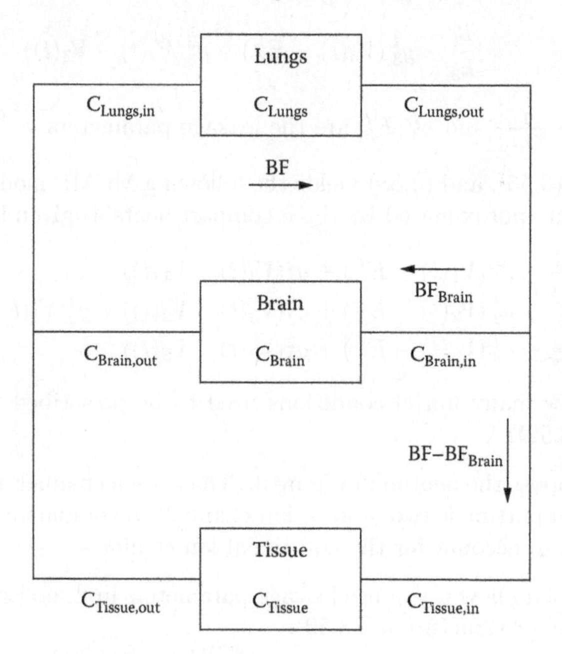

FIGURE 3.14: Simplified Respiratory Model

An uppercase C represents carbon dioxide concentration and the subscript explains where it is being measured. For example, $C_{Brain,in}$ is the concentration of CO_2 entering the brain. Blood Flow, BF, and blood flow into the brain, BF_{Brain}, are assumed to be constant. The ventilation rate, $\dot{V}$, along with the volume fraction of CO_2 leaving the lungs, F_A, when multiplied together determine the total amount of CO_2 being expelled from the lungs. These variables are all functions of time, but that dependence has been suppressed for simplicity.

The basic physiology of the Lung-Brain-Soft Tissue system is that the blood flows throughout the system providing transportation for CO_2. Some of the blood will travel to the brain where, if the CO_2 concentration is above a certain threshold, the brain tells the lungs (via neurons) to increase the ventilation rate. In turn, this increases the total amount of CO_2 being expelled from the lungs and this lowers the concentration in the blood. For a more detailed explanation of the physiology consult [25, 19, 18].

Now, it is time for a second wave of simplifications. We simplify the model by assuming the CO_2 concentrations in the brain and soft tissues are the same as those that leave their respective compartments. Using the variables of the model yields

$$C_{Brain} = C_{Brain,out} \quad C_{Tissue} = C_{Tissue,out}$$

Also, we assume the CO_2 levels that enter the brain and soft tissues are delayed versions of what left the lungs, and that what enters the lungs is combination of delayed versions of what left the brain and soft tissues. That is, there are real positive constants,

$\tau_{Brain}, \tau_{Tissues}, \tau_{Brain,out}$, and $\tau_{Tissue,out}$ such that

$$C_{Brain,in}(t) = C_{Lungs,out}(t - \tau_{Brain})$$

$$C_{Tissue,in}(t) = C_{Lungs,out}(t - \tau_{Tissue})$$

$$BF \cdot C_{Lungs,in}(t) = BF \cdot C_{Brain}(t - \tau_{Brain,out}) + (BF - BF_{Brain})C_{Tissue}(t - \tau_{Tissue,out})$$

The final assumption we impose is that the partial pressure of CO_2 expelled from the body is the same as the one that goes back into the system via the blood stream.

We can now apply the "Rate-in minus Rate-out" approach to compartmental modeling, but first note that the variable of interest P will represent the partial pressure of CO_2 and its subscript the location where it is being measured. However, for notational purposes, we use B, L, and T to denote brain, lungs, and soft tissues, respectively. Also the upper case V and M will represent volumes and metabolism rates, respectively, with subscripts used to denote locations. Converting all CO_2 concentrations to partial pressures and applying 'Rate-in minus Rate-out' yields

$$\begin{cases} V_B K_{CO_2} \frac{dP_B(t)}{dt} = M_B + BF_B K_{CO_2}(P_{L,out}(t - \tau_B) - P_B) \\ V_T K_{CO_2} \frac{dP_T(t)}{dt} = M_T + (BF - BF_B)K_{CO_2}(P_{L,out}(t - \tau_T) - P_T) \\ V_L \frac{dP_{L,out}(t)}{dt} = -\dot{V}(t)P_{L,out}(t) + \\ K_{CO_2}\left[(BF_B(P_B(t - \tau_{B,out}) - P_T(t - \tau_{T,out})) + BF(P_T(t - \tau_{T,out}) - P_T(t))\right] \end{cases} \tag{3.60}$$

where the constant K_{CO_2} arises from converting the concentrations to partial pressures. Lastly, we need to specify exactly how we want to model the ventilation rate. If we follow the idea that each of us has a normal breathing rate until the brain senses too much carbon dioxide, it makes sense to model the ventilation rate as Khoo et al. did in [25] by

$$\dot{V}(t) = \dot{V}_{base} + k \max(P_B(t) - \gamma)$$

The constant $\dot{V}_{Base}$ represents the base ventilation rate, γ the activation threshold value in the brain, and k the constant rate of increase once the brain senses too much carbon dioxide in the blood.

Phew! This 'simplified' model does not seem simple! For a moment, try not to worry about the physiology and just focus in on the equations in (3.60) and answer the following exercises.

Exercise 3.10.1. What are the unknowns in the respiratory regulation model? What are the known parameters for the model?

Exercise 3.10.2. How many initial conditions need to be specified to obtain a unique solution of the model?

Exercise 3.10.3. There is one feature to (3.60) that, up to this point, we have not encountered in a system of differential equations. Can you find it? (Hint: It has something to do with where functions are evaluated). Think about how this affects the definition of a solution to the model and the process involved in contructing one.

3.11 Looking Ahead!

Take the time to look back at the various models introduced in this chapter, and pay particular attention to identifying commonality among all of the models.

EXPLORE!
Finding a Common Link

Ideally, based on the sense that these models, while describing phenomena from disparate fields, all bear some degree of similarity to each other, we would be able to formulate or extract a common principle that binds them together, abstractly, and then formulate ONE cohesive über-theory that subsumes them all as special cases. The elegance of doing so would only be matched by its efficiency. Toward this end, attempt the following steps as a precursor to developing such a theory.

1. At this point, if we could somehow use the tools developed in Chapter 2 to show that all of the IVPs could be written in the same basic form, that would be a huge step in the right direction.
   i.) Identify the unknown functions for each of the models. Can you identify these functions as the components of a vector?
   ii.) Examine the IVPs for each model. The left-side of the IVPs is always what quantity? Specifically, what action is performed on the vector identified in (i)?
   iii.) The right-sides of the DEs in the IVPs are a little trickier to get a handle on. Is there a way to identify some entity that, when multiplied by the vector identified in (i), results in the right-side? Depending on what was accounted for in the model, there might be a couple of exceptions to this rule, but by and large, a single approach is valid.
   iv.) How would you handle the initial conditions? Do you see why we imposed initial conditions on all unknowns in a given model *at the same time t*?
2. Even if your success in completing (1) above was only marginal, still attempt this part!
   i.) Review the questions we asked in each of the models. Can you think of some questions that could be more broadly stated and applied to ALL models that could be written in the form suggested in (1)?
   ii.) Can you think of some theoretical questions that would be of interest to study for the form in (1) that would have ramifications on our analysis of each of the models?

As shocking as this might be, our guiding light in establishing a common link among the models presented in this chapter is the following simple IVP mentioned in Chapter 1.

$$\begin{cases} u'(t) &= au(t) \\ u(t_0) &= u_0, \end{cases} \tag{3.61}$$

where a is a nonzero real number, $u : [t_0, \infty) \to \mathbb{R}$ is a real-valued unknown function, and the initial condition $u_0 \in \mathbb{R}$. Let us be clear—NONE of the models explored in this chapter are equivalent to (3.61) for the simple reason that there is always more than one unknown function whereas the solution of (3.61) is a single real-valued function. But, what we are interested in is the *form* of (3.61).

With some exception, the IVPs discussed in text and exercises in this chapter can be written in the following *matrix*-form of (3.61), assuming that the pieces are all identified correctly:

$$\begin{cases} \mathbf{U}'(t) &= \mathbf{A}\mathbf{U}(t) \\ \mathbf{U}(t_0) &= \mathbf{U}_0, \end{cases} \tag{3.62}$$

where $\mathbf{U} : [t_0, \infty) \to \mathbb{R}^N$ is the unknown vector function

$$\mathbf{U}(t) = \begin{bmatrix} u_1(t) \\ \vdots \\ u_N(t) \end{bmatrix},$$

$\mathbf{A}$ is the $N \times N$ coefficient matrix

$$\begin{bmatrix} a_{11} & \cdots & a_{1N} \\ \vdots & \ddots & \vdots \\ a_{N1} & \cdots & a_{NN} \end{bmatrix},$$

and $\mathbf{U}_0 \in \mathbb{R}^N$ is the vector containing the initial conditions

$$\mathbf{U}_0 = \begin{bmatrix} u_1(t_0) \\ \vdots \\ u_N(t_0) \end{bmatrix}.$$

Is that what you observed in the previous **EXPLORE!** activity? Of course, there are a few exceptions to this form that require the addition of another term to the differential equation in (3.62) simply because one or more of its components was not simply a *linear* combination of the components of the unknown vector $\mathbf{U}(t)$. Rather, it was either a function of t or was some more complicated function of the components of the unknown vector $\mathbf{U}(t)$.

We will illustrate the process of reformulating an IVP in the form of (3.62) for one of the more complicated models for reference. Let us consider the insulin model (3.7):

$$\begin{cases} \frac{du_{ST}(t)}{dt} &= a_{(H,ST)}u_H(t) - a_{(ST,H)}u_{ST}(t) \\ \frac{du_H(t)}{dt} &= \left[a_{(ST,H)}u_{ST}(t) + a_{(L,H)}u_L(t)\right] - \left[a_{(H,ST)} + a_{(H,E)} + a_{(H,L)} + a_{(H,P)}\right]u_H(t) \\ \frac{du_L(t)}{dt} &= \left[a_{(P,L)}u_P(t) + a_{(H,L)}u_H(t)\right] - a_{(L,H)}u_L(t) \\ \frac{du_P(t)}{dt} &= a_{(H,P)}u_H(t) - a_{(P,L)}u_P(t) \\ \frac{du_E(t)}{dt} &= a_{(H,E)}u_H(t) - Lu_E(t) \\ u_{ST}(0) &= u_1, u_H(0) = u_2, u_L(0) = u_3, u_P(0) = u_4, u_E(0) = u_5 \end{cases}$$

Step 1: Identify the unknown vector $\mathbf{U}(t)$.
This vector will contain all functions for which we compute a derivative in the IVP. Here, there are five such fuctions and so, we have the following:

$$\mathbf{U}(t) = \begin{bmatrix} u_{ST}(t) \\ u_H(t) \\ u_L(t) \\ u_P(t) \\ u_E(t) \end{bmatrix} \tag{3.63}$$

Step 2: Identify the initial condition vector $\mathbf{U}(t_0)$. You must identify the *time* t_0 at which the initial data is prescribed, and then simply evaluate the vector from Step 1 at this time to get this vector. Here $t_0 = 0$ and so using the last line in this IVP yields the following:

$$\mathbf{U}(0) = \mathbf{U}_0 = \begin{bmatrix} u_1 \\ u_2 \\ u_3 \\ u_4 \\ u_5 \end{bmatrix} \tag{3.64}$$

Step 3: Identify the coefficient matrix A.
This is arguably the trickiest part of the reformulation process, but it really is not that bad as long as you keep the bookkeeping straight.

STOP! Think about how you would compute $\mathbf{AU}(t)$. What would be the components of the $N \times 1$ product vector?

Guided by the form of $\mathbf{AU}(t)$, it makes sense to simply group together all like terms in each of the five differential equations in (3.7), writing the resulting sum in the same order as the terms are arranged in the unknown vector $\mathbf{U}(\mathrm{t})$. In this example, this amounts to writing the coefficient of $u_{ST}(t)$ first, the coefficient of $u_H(t)$ second, and so on. For each of the components of $\mathbf{U}(t)$ that does not appear on the right-side of a given differential equation, insert a zero place holder; this will help when it comes time to identify the entries of the matrix **A**. We rewrite IVP (3.7) as follows:

$$\begin{cases} \frac{du_{ST}(t)}{dt} &= -a_{(ST,H)}u_{ST}(t) + a_{(H,ST)}u_H(t) + 0u_L(t) + 0u_P(t) + 0u_E(t) \\ \frac{du_H(t)}{dt} &= a_{(ST,H)}u_{ST}(t) - \left[a_{(H,ST)} + a_{(H,E)} + a_{(H,L)} + a_{(H,P)}\right]u_H(t) \\ & \quad + a_{(L,H)}u_L(t) + 0u_P(t) + 0u_E(t) \\ \frac{du_L(t)}{dt} &= 0u_{ST}(t) + a_{(H,L)}u_H(t) - a_{(L,H)}u_L(t) + a_{(P,L)}u_P(t) + 0u_E(t) \\ \frac{du_P(t)}{dt} &= 0u_{ST}(t) + a_{(H,P)}u_H(t) + 0u_L(t) - a_{(P,L)}u_P(t) + 0u_E(t) \\ \frac{du_E(t)}{dt} &= 0u_{ST}(t) + a_{(H,E)}u_H(t) + 0u_L(t) + 0u_P(t) - Lu_E(t) \\ u_{ST}(0) &= u_1, u_H(0) = u_2, u_L(0) = u_3, u_P(0) = u_4, u_E(0) = u_5 \end{cases} \tag{3.65}$$

Working backward now from what the product of $\mathbf{AU}(t)$ would be, we go one step further and rewrite the right-sides of the five differential equations appearing in (3.65) as the matrix product: $\mathbf{AU}(t)$ where

$$\mathbf{A} = \begin{bmatrix} -a_{(ST,H)} & a_{(H,ST)} & 0 & 0 & 0 \\ a_{(ST,H)} & -\left[a_{(H,ST)} + a_{(H,E)} + a_{(H,L)} + a_{(H,P)}\right] & a_{(L,H)} & 0 & 0 \\ 0 & a_{(H,L)} & -a_{(L,H)} & a_{(P,L)} & 0 \\ 0 & a_{(H,P)} & 0 & -a_{(P,L)} & 0 \\ 0 & a_{(H,E)} & 0 & 0 & -L \end{bmatrix}. \tag{3.66}$$

Step 4: Pull it all together!
Using the identification in (3.63)-(3.66), we conclude that IVP (3.7) can be formulated in the abstract form (3.62).

Now, it's your turn!

Exercise 3.11.1. Reformulate the following IVPs in the form (3.62). Make certain to show all identifications.

i.) IVP (3.3)

ii.) IVP (3.6)

There are two categories of questions that naturally arise normally—the applied context-specific questions and the *theoretical* broad-level questions. The next chapter is devoted to unveiling both types of questions, as well as providing the tools necessary to answer them.

Chapter 4

Finite-Dimensional Theory—Ground Zero: The Homogenous Case

Overview

The question of how phenomena evolve over time is central to a broad range of fields within the social, natural, and physical sciences. The behavior of such phenomena is governed by established laws in the underlying field that typically describe the rates at which it and related quantities evolve over time. A precise mathematical description involves the formulation of so-called *evolution equations* whose complexity depends largely on the realism of the model. We focus in this chapter on models in which the evolution equation is generated by a system of linear homogeneous ordinary differential equations. We are in search of an abstract paradigm into which all of these models are subsumed as special cases. Once established, we can study the rudimentary properties of the abstract paradigm and subsequently apply the results to each model.

4.1 Introducing the Homogenous Cauchy Problem (HCP)

As observed at the end of Chapter 3, the systems of linear ordinary differential equations could be written in a more compact vector form by identifying the unknown as a vector, then suitably defining a coefficient matrix by which to multiply it in order to capture the linear interactions amongst the terms, and finally identifying the initial conditions. The nature of the IVPs used to describe the scenarios and the questions posed are strikingly similar. Formally, we have

Definition 4.1.1. A system of ordinary differential equations of the form

$$\textbf{(HCP)} \begin{cases} \mathbf{U}'(t) = \mathbf{A}\mathbf{U}(t),\, t > t_0, \\ \mathbf{U}(t_0) = \mathbf{U}_0, \end{cases} \tag{4.1}$$

where $\mathbf{U} : [t_0, \infty) \to \mathbb{R}^N$ is the unknown vector function

$$\mathbf{U}(t) = \begin{bmatrix} u_1(t) \\ \vdots \\ u_N(t) \end{bmatrix},$$

$\mathbf{A}$ is the $N \times N$ coefficient matrix

$$\begin{bmatrix} a_{11} & \cdots & a_{1N} \\ \vdots & \ddots & \vdots \\ a_{N1} & \cdots & a_{NN} \end{bmatrix},$$

and $\mathbf{U}_0 \in \mathbb{R}^N$ is the vector containing the initial conditions

$$\mathbf{U}_0 = \begin{bmatrix} u_1(t_0) \\ \vdots \\ u_N(t_0) \end{bmatrix},$$

is a *homogenous linear system of ODEs with initial condition* $\mathbf{U}(t_0) = \mathbf{U}_0$. We refer to (4.1) as a *homogenous Cauchy problem* (or **(HCP)**) . A vector $\mathbf{U}(t)$ that satisfies (4.1) on an interval containing t_0 is a *solution* of **(HCP)** .

The following is an example of the type of system we intend to study:

Example 4.1.1. Consider the system of ODEs given by

$$\begin{cases} u_1'(t) & = -2u_1(t) + 3u_2(t) \\ u_2'(t) & = -u_1(t) + 2u_2(t) \\ u_1(1) & = 1 \\ u_2(1) & = -1 \end{cases} \tag{4.2}$$

Initial-value problem (4.2) can be rewritten using matrices as follows:

$$\begin{cases} \underbrace{\begin{bmatrix} u_1(t) \\ u_2(t) \end{bmatrix}'}_{\mathbf{U}'(t)} = \underbrace{\begin{bmatrix} -2 & 3 \\ -1 & 2 \end{bmatrix}}_{\mathbf{A}} \underbrace{\begin{bmatrix} u_1(t) \\ u_2(t) \end{bmatrix}}_{\mathbf{U}(t)} \\ \underbrace{\begin{bmatrix} u_1(1) \\ u_2(1) \end{bmatrix}}_{\text{Here } t_0=1} = \underbrace{\begin{bmatrix} 1 \\ -1 \end{bmatrix}}_{\mathbf{U}_0} \end{cases} \tag{4.3}$$

It is reasonable at this stage to think of a solution of (4.3) as a vector $\mathbf{U}(t) = \begin{bmatrix} u_1(t) \\ u_2(t) \end{bmatrix}$ satisfying both the equation and the initial condition in (4.3).

Exercise 4.1.1. Write the following system in matrix form:

$$\begin{cases} x'(t) &= 2x(t) + y(t) \\ y'(t) &= -x(t) + 4y(t) \\ x(2) &= 1 \\ y(2) &= -2 \end{cases}$$

4.2 Lessons Learned from a Special Case

It would be efficient to initially focus our attention on the abstract IVP (4.1) and answer as many rudimentary questions as possible regarding the existence and uniqueness of a solution, stability of the solutions with respect to initial data, etc. We could, in turn, apply those results to any model that could be viewed as a special case of (4.1). But, how do we proceed rigorously?

Initial attempts at describing population dynamics, determining the half-life of a radioactive substance, and studying other phenomena governed by exponential growth or decay all involve the first-order IVP

$$\begin{cases} u'(t) = au(t), \ t > 0, \\ u(t_0) = u_0. \end{cases} \tag{4.4}$$

It is easy to verify that $u(t) = e^{a(t-t_0)}u_0$ satisfies (4.4). **(Do so!)**

The similarity between (4.1) and (4.4) suggests that $\mathbf{U}(t) = e^{\mathbf{A}(t-t_0)}\mathbf{U}_0$ is a likely candidate for a solution of (4.1). While this *feels* right, we have at this stage done nothing more than symbol matching. Indeed, while the two IVPs "look" the same, the solutions u and $\mathbf{U}$ take values in different spaces ($\mathbb{R}$ and $\mathbb{R}^N$, respectively). This might seem like a minor difference, but when we compare the two IVPs more closely, we are immediately faced with basic structural questions regarding $e^{\mathbf{A}t}$ since $\mathbf{A}$ is a matrix: Is $e^{\mathbf{A}t}$ itself a matrix, and if so, how do we compute it? We must define what is meant by the exponential of a constant square matrix $\mathbf{A}$. And, what better way is there to proceed than to exploit our familiarity with the real-valued exponential function $y = e^x$?

We can readily verify that $y = e^x$ is a solution of the following homogenous linear IVP:

$$\begin{cases} y'(x) &= y(x), \\ y(0) &= 1. \end{cases} \tag{4.5}$$

(Do so!) Moreover, this is the *only* solution of (4.5). This fact is essentially argued in a calculus course when studying exponential functions and their derivatives. We shall use this fact, without proof, in what follows.

The uniqueness of the solution of (4.5) ensures that any other function satisfying (4.5) would necessarily equal e^x. Bearing this in mind, let us define a function $Y(x)$ by means of a power series as follows:

$$Y(x) = \sum_{k=0}^{\infty} \frac{x^k}{k!}$$

This series converges for all $x \in \mathbb{R}$. **(Why?)** So, the domain of Y is $\mathbb{R}$. As such, we can differentiate the series term-for-term to verify directly that Y is also a solution of (4.5). **(Do so!)** Consequently, we can conclude that

$$e^x = \sum_{k=0}^{\infty} \frac{x^k}{k!}. \tag{4.6}$$

We shall use (4.6) as a springboard for defining the exponential when the input is a square matrix $\mathbf{A}$ instead of a real number x.

4.3 Defining the Matrix Exponential $e^{\mathbf{A}t}$

STOP! Review the material in Sections 2.5 and 2.6 before moving onward.

4.3.1 One Approach—Taylor Series

Using (4.6), a natural definition of e^{at} is the Taylor representation $e^{at} = \sum_{k=0}^{\infty} \frac{(at)^k}{k!}$, which converges $\forall t \in \mathbb{R}$. We would like to determine if we can define $e^{\mathbf{A}t}$ in a similar manner.

Exercise 4.3.1. Explain why for any $N \times N$ matrix $\mathbf{A}$ and any $m \in \mathbb{N}$, the finite sum $\sum_{k=0}^{m} \frac{(\mathbf{A}t)^k}{k!}$ is a well-defined $N \times N$ matrix.

This is reason for optimism, but we must ensure that $\lim_{m\to\infty} \sum_{k=0}^{m} \frac{(\mathbf{A}t)^k}{k!}$ exists in order for a Taylor series-type definition to be meaningful. When is this limit guaranteed to exist? And, when it does exist, to what type of quantity does the sequence of finite sums converge? The latter question is the easier of the two to answer. Indeed, since the space $\mathbb{M}^N(\mathbb{R})$ is complete with respect to its usual norms (See Section 2.6.1), a convergent sequence of elements of $\mathbb{M}^N(\mathbb{R})$ must tend towards a member of $\mathbb{M}^N(\mathbb{R})$. This tells us that if we can prove convergence, the matrix exponential $e^{\mathbf{A}t}$ must also belong to $\mathbb{M}^N(\mathbb{R})$.

Regarding the convergence question, we cannot simply use the Ratio Test as we did for real-valued power series. However, to overcome this difficulty, for a given $\mathbf{A} \in \mathbb{M}^N(\mathbb{R})$ and $t \in \mathbb{R}$, it is sensible to prove that any sequence of finite sums $\left(\sum_{k=0}^{m} \frac{(\mathbf{A}t)^k}{k!}\right)_{m\in\mathbb{N}}$ is Cauchy, and then conclude from the completeness of $\mathbb{M}^N(\mathbb{R})$ that the series must be convergent. Doing this would prove that $\lim_{m\to\infty} \sum_{k=0}^{m} \frac{(\mathbf{A}t)^k}{k!}$ exists, meaning that $e^{\mathbf{A}t}$ is well-defined.

This is precisely the approach used to prove the following proposition.

Proposition 4.3.1. For any $\mathbf{A} \in \mathbb{M}^N(\mathbb{R})$ and $t \in \mathbb{R}$, $\lim_{m\to\infty} \sum_{k=0}^{m} \frac{(\mathbf{A}t)^k}{k!}$ exists.

Proof. Let $\mathbf{A} \in \mathbb{M}^N(\mathbb{R})$ and $t > 0$ be given. Due to the completeness of $\mathbb{M}^N(\mathbb{R})$, it suffices to show that the sequence $\left(\sum_{k=0}^{m} \frac{(\mathbf{A}t)^k}{k!}\right)_{m\in\mathbb{N}}$ is Cauchy in $\mathbb{M}^N(\mathbb{R})$. Let $\epsilon > 0$. For any $k \in \mathbb{N}$, $\frac{\left(\|\mathbf{A}\|_{\mathbb{M}^N} t\right)^k}{k!}$ is a real number and

$$\lim_{m\to\infty} \sum_{k=0}^{m} \frac{\left(\|\mathbf{A}\|_{\mathbb{M}^N} t\right)^k}{k!} = e^{\|\mathbf{A}\|_{\mathbb{M}^N} t}. \tag{4.7}$$

(Why?) So, the series $\sum_{k=0}^{\infty} \frac{\left(\|\mathbf{A}\|_{\mathbb{M}^N} t\right)^k}{k!}$ converges. From the remark directly following Definition 2.5.5, we know that

$$\sum_{k=l}^{m} \frac{\left(\|\mathbf{A}\|_{\mathbb{M}^N} t\right)^k}{k!} \longrightarrow 0 \text{ as } l, m \longrightarrow \infty.$$

So, there exists a large enough $M \in \mathbb{N}$ such that for any $k, l \geq M$ (where $k < l$), it is the case that

$$\sum_{j=k+1}^{l} \frac{\left(\|\mathbf{A}\|_{\mathbb{M}^N} t\right)^j}{j!} < \epsilon.$$

Now, going back to the original series, we see that for such k and l,

$$\left\| \sum_{j=0}^{l} \frac{(\mathbf{A}t)^j}{j!} - \sum_{j=0}^{k} \frac{(\mathbf{A}t)^j}{j!} \right\|_{\mathbb{M}^N} \leq \sum_{j=k+1}^{l} \left\| \frac{(\mathbf{A}t)^j}{j!} \right\|_{\mathbb{M}^N} \leq \sum_{j=k+1}^{l} \frac{\left(\|\mathbf{A}\|_{\mathbb{M}^N} t\right)^j}{j!} < \epsilon,$$

as desired. □

STOP! Make certain you understand where and how (4.7) was used in the proof. At key steps in the arguments involving matrices, it is often helpful to drop back to a more familiar setting (like the setting of single-variable calculus) and think about the statement in that context. This often helps you to overcome hurdles posed by abstraction.

Consequently, the following definition is meaningful.

Definition 4.3.1. For any $\mathbf{A} \in \mathbb{M}^N(\mathbb{R})$ and $t \in \mathbb{R}$, the *matrix exponential* $e^{\mathbf{A}t}$ is the unique member of $\mathbb{M}^N(\mathbb{R})$ defined by

$$e^{\mathbf{A}t} = \sum_{k=0}^{\infty} \frac{(\mathbf{A}t)^k}{k!}.$$

Before studying the properties of $e^{\mathbf{A}t}$, we consider some elementary examples illustrating how to compute it. The close relationship between $\mathbf{A}$ and $e^{\mathbf{A}t}$ suggests that it is easier to determine the components of $e^{\mathbf{A}t}$ for matrices $\mathbf{A}$ possessing a simpler structure.

Example. Let $\mathbf{A} = \begin{bmatrix} a & 0 \\ 0 & b \end{bmatrix}$, where $a, b \in \mathbb{R}$. Observe that

$$\begin{aligned} e^{\begin{bmatrix} a & 0 \\ 0 & b \end{bmatrix} t} &= \sum_{k=0}^{\infty} \frac{\left(\begin{bmatrix} a & 0 \\ 0 & b \end{bmatrix} t \right)^k}{k!} = \sum_{k=0}^{\infty} \frac{\begin{bmatrix} at & 0 \\ 0 & bt \end{bmatrix}^k}{k!} = \sum_{k=0}^{\infty} \frac{\begin{bmatrix} (at)^k & 0 \\ 0 & (bt)^k \end{bmatrix}}{k!} \\ &= \sum_{k=0}^{\infty} \begin{bmatrix} \frac{(at)^k}{k!} & 0 \\ 0 & \frac{(bt)^k}{k!} \end{bmatrix} = \begin{bmatrix} \sum_{k=0}^{\infty} \frac{(at)^k}{k!} & 0 \\ 0 & \sum_{k=0}^{\infty} \frac{(bt)^k}{k!} \end{bmatrix} = \begin{bmatrix} e^{at} & 0 \\ 0 & e^{bt} \end{bmatrix}. \end{aligned}$$

Exercise 4.3.2. Compute $e^{\mathbf{A}t}$ where $\mathbf{A}$ is a diagonal $N \times N$ matrix with real diagonal entries $a_{11}, a_{22}, \ldots, a_{NN}$. Justify all steps.

Of course, not every member of $\mathbb{M}^N(\mathbb{R})$ is diagonal. But, many are "close" in the sense introduced in the following exercise.

Exercise 4.3.3. Let $\mathbf{A} \in \mathbb{M}^N(\mathbb{R})$ and assume that there exist $\mathbf{D}, \mathbf{P} \in \mathbb{M}^N(\mathbb{R})$ such that $\mathbf{D}$ is diagonal, $\mathbf{P}$ is invertible, and $\mathbf{A} = \mathbf{P}^{-1}\mathbf{D}\mathbf{P}$. (In such case, $\mathbf{A}$ is said to be *diagonalizable.*) Show that $\forall t \in \mathbb{R}$, $e^{\mathbf{A}t} = \mathbf{P}^{-1}\mathbf{e}^{\mathbf{D}t}\mathbf{P}$.

Such matrices often arise in practice. For instance, if either $\mathbf{A}$ has n distinct real eigenvalues or $\mathbf{A}$ is symmetric, then $\mathbf{A}$ is diagonalizable. For nondiagonalizable matrices $\mathbf{A}$, even though $e^{\mathbf{A}t}$ exists, computing it can be tedious. We outline a procedure for how to compute $e^{\mathbf{A}t}$ for any $\mathbf{A} \in \mathbb{M}^N(\mathbb{R})$ in the next section.

MATLAB-Exercise 4.3.1. In this exercise, you will use **MATLAB** to aid in computing $e^{\mathbf{A}t}$ via the formula produced in Exercise 4.3.3. To begin, open **MATLAB** and in the command line, type:

MATLAB_Diagonalizable

Use the GUI to answer the following questions. Make certain to click on the 'Help' button for more details and instructions on how to use the GUI. A screenshot from this GUI can found in Figure 4.1.

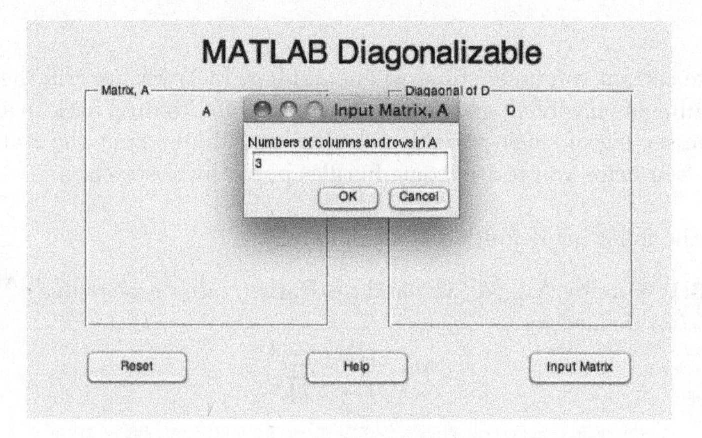

FIGURE 4.1: MATLAB Diagonalizable GUI

i.) Compute $e^{\mathbf{A}t}$

a.) $\mathbf{A} = \begin{bmatrix} 2 & 1 \\ 0 & -1 \end{bmatrix}$

b.) $\mathbf{A} = \begin{bmatrix} 1 & -1 & 4 \\ 3 & 2 & -1 \\ 2 & 1 & -1 \end{bmatrix}$

c.) $\mathbf{A} = \begin{bmatrix} 1 & 0 \\ 1 & 1 \end{bmatrix}$

ii.) Input five different symmetric matrices and determine whether or not Exercise 4.3.3 applies.

iii.) Input five different nonsymmetric matrices and determine whether or not Exercise 4.3.3 applies.

■

4.3.2 Another Approach

There is another way to define e^{At} that does not rely on power series. Instead, the limit definition of e^t presented in differential calculus is also useful. Recall that

$$e^t = \lim_{n\to\infty}\left(1+\frac{t}{n}\right)^n$$

or equivalently

$$e^t = \lim_{n\to\infty}\left(1-\frac{t}{n}\right)^{-n} \tag{4.8}$$

STOP! Show (4.8) using L'Hôpital's rule.

To motivate the limit definition of the exponential function, (4.8), consider the IVP

$$\begin{cases} y'(t) &= y(t), \ t>0 \\ y(0) &= 1 \end{cases} \tag{4.9}$$

The procedure for solving for the function, $y(t)$, satisfying (4.9) is as follows:

1. Fix $t>0$ and $n\in\mathbb{N}$;
2. Approximate the IVP by replacing the derivative with the difference quotient;
3. 'March in time to t' with step-size t/n. That is, we march from 0 to t/n by constructing the approximate solution at t/n using the fact *we know* the approximate solution at 0.
4. Solve the approximate IVP at the next step in the march using the fact we know the approximate solution at the previous step;
5. Keep marching until we reach t;
6. Determine how to use the approximate IVP solution to define the solution to (4.9).

Let $t>0$ and $n\in\mathbb{N}$, and define $\bar{y}$ to be the solution of the approximate IVP

$$\begin{cases} \frac{\bar{y}(s+\frac{t}{n})-\bar{y}(s)}{\frac{t}{n}} &= \bar{y}(s+\frac{t}{n}), \ s>0 \\ \bar{y}(0) &= 1 \end{cases} \tag{4.10}$$

Now, move to the first step in the march. Substitute $s=0$ into (4.10) to produce

$$\begin{cases} \frac{\bar{y}(\frac{t}{n})-\bar{y}(0)}{\frac{t}{n}} &= \bar{y}(\frac{t}{n}) \\ \bar{y}(0) &= 1 \end{cases}$$

Rearranging terms yields

$$(1-\frac{t}{n})\bar{y}\left(\frac{t}{n}\right) = \bar{y}(0),$$

or equivalently

$$\bar{y}\left(\frac{t}{n}\right) = \left(1-\frac{t}{n}\right)^{-1}\bar{y}(0) \tag{4.11}$$

Proceeding to the second step in the march to $\frac{2t}{n}$ leads to the approximate IVP

$$\begin{cases} \frac{\bar{y}(\frac{2t}{n})-\bar{y}(\frac{t}{n})}{\frac{t}{n}} &= \bar{y}(\frac{2t}{n}) \\ \bar{y}(0) &= 1 \end{cases} \tag{4.12}$$

Solving for $\bar{y}(\frac{2t}{n})$ yields

$$\bar{y}\left(\frac{2t}{n}\right) = \left(1 - \frac{t}{n}\right)^{-1} \bar{y}\left(\frac{t}{n}\right).$$

STOP! Do we have a formula for $\bar{y}(\frac{t}{n})$?

Replacing $\bar{y}(\frac{t}{n})$ with (4.11) gives

$$\bar{y}\left(\frac{2t}{n}\right) = \left(1 - \frac{t}{n}\right)^{-1} \left(1 - \frac{t}{n}\right)^{-1} \bar{y}(0) = \left(1 - \frac{t}{n}\right)^{-2} \bar{y}(0) \tag{4.13}$$

You might begin to see a pattern emerge at this point. If we take n steps in our march we get

$$\bar{y}(t) = \bar{y}\left(\frac{nt}{n}\right) = \left(1 - \frac{t}{n}\right)^{-n} \bar{y}(0)$$

STOP! If you did not see the pattern take another step in the march. Compute $\bar{y}(\frac{3t}{n})$. Now do you see it?

We are almost there. All that is left is to apply the initial condition and decide how to use $\bar{y}(t)$ to construct $y(t)$.

STOP! What is the relationship between (4.9) and (4.10)? What must happen to n in order for the two IVPs to equal each other?

Since the limit of the difference quotient equals the derivative, it seems reasonable that the limit of $\bar{y}(t)$ as $n \to \infty$ equals $y(t)$. With that in mind, we define

$$e^t = \lim_{n \to \infty} \left(1 - \frac{t}{n}\right)^{-n}$$

and more generally

$$e^{at} = \lim_{n \to \infty} \left(1 - \frac{at}{n}\right)^{-n} = \lim_{n \to \infty} \left(1 - \frac{t}{n}a\right)^{-n}, a \in \mathbb{R} \tag{4.14}$$

Again, this was only a motivation for the formula defining e^t and e^{at}. We must actually verify the limit definition satisfies (4.9).

Exercise 4.3.4. Show that (4.8) satisfies (4.9), assuming you can interchange limits and derivatives.

What if we replace $a \in \mathbb{R}$ by $\mathbf{A} \in \mathbb{M}^N(\mathbb{R})$? If we match symbols then (4.14) becomes

$$e^{\mathbf{A}t} = \lim_{n \to \infty} \left(\mathbf{I} - \frac{t}{n}\mathbf{A}\right)^{-n} \tag{4.15}$$

Notice the number one in (4.14) has been replaced by $\mathbf{I}$, the identity matrix in $\mathbb{M}^N(\mathbb{R})$.

STOP! Does (4.15) even make sense? Is $\mathbf{I} - \frac{t}{n}\mathbf{A}$ a square matrix? What does

$$\left(\mathbf{I} - \frac{t}{n}\mathbf{A}\right)^{-n}$$

mean?

Definition 4.3.2. Let $n \in \mathbb{N}$ and $\mathbf{B}$ an invertible matrix in $\mathbb{M}^N(\mathbb{R})$. Then $\mathbf{B}^{-n}$ is a matrix in $\mathbb{M}^N(\mathbb{R})$ given by

$$\mathbf{B}^{-n} = \left(\mathbf{B}^{-1}\right)^n.$$

In light of Definition 4.3.2, there are two facets of (4.15) that must be carefully checked. First, the matrix $\left(\mathbf{1} - \frac{t}{n}\mathbf{A}\right)$ must be invertible and the limit in (4.15) must exist. Both conditions can be shown. However, the proofs are quite technical and are omitted from the text.

Definition 4.3.3. For any $\mathbf{A} \in \mathbb{M}^N(\mathbb{R})$ and $t \in \mathbb{R}$, the *matrix exponential* $e^{\mathbf{A}t}$ is the unique member of $\mathbb{M}^N(\mathbb{R})$ defined by

$$e^{\mathbf{A}t} = \lim_{n\to\infty}\left(\mathbf{I} - \frac{t}{n}\mathbf{A}\right)^{-n}.$$

Remark 4.3.1. By uniqueness, the limit definition of the matrix exponential produces the same matrix as the power series definition. Therefore, all properties of the matrix exponential carry over when using Definition 4.3.3, and more importantly, for every $\mathbf{U_0} \in \mathbb{R}^N$, the IVP (4.1) has a unique classical solution given by $\mathbf{U}(t) = \lim_{n\to\infty}\left(\mathbf{I} - \frac{t}{n}\mathbf{A}\right)^{-n}\mathbf{U_0}$.

We illustrate Defintion 4.3.3 with some examples.

Example 4.3.1. Let $\mathbf{B} = \begin{bmatrix} a & 0 \\ 0 & b \end{bmatrix}$, where $a, b \in \mathbb{R}$. Observe

$$\begin{aligned}
e^{\begin{bmatrix} a & 0 \\ 0 & b \end{bmatrix}t} &= \lim_{n\to\infty}\left(\begin{bmatrix} 1 & 0 \\ 0 & 1 \end{bmatrix} - \frac{t}{n}\begin{bmatrix} a & 0 \\ 0 & b \end{bmatrix}\right)^{-n} \\
&= \lim_{n\to\infty}\left(\begin{bmatrix} 1-\frac{t}{n}a & 0 \\ 0 & 1-\frac{t}{n}b \end{bmatrix}^{-1}\right)^n, \ \textbf{(Why?)} \\
&= \lim_{n\to\infty}\begin{bmatrix} \left(1-\frac{t}{n}a\right)^{-1} & 0 \\ 0 & \left(1-\frac{t}{n}b\right)^{-1} \end{bmatrix}^n, \ \textbf{(Why?)} \\
&= \lim_{n\to\infty}\begin{bmatrix} \left(1-\frac{t}{n}a\right)^{-n} & 0 \\ 0 & \left(1-\frac{t}{n}b\right)^{-n} \end{bmatrix}, \ \textbf{(Why?)} \\
&= \begin{bmatrix} \lim_{n\to\infty}\left(1-\frac{t}{n}a\right)^{-n} & 0 \\ 0 & \lim_{n\to\infty}\left(1-\frac{t}{n}b\right)^{-n} \end{bmatrix}, \ \textbf{(Why?)} \\
&= \begin{bmatrix} e^{at} & 0 \\ 0 & e^{bt} \end{bmatrix}
\end{aligned}$$

As expected, the limit definition of the matrix exponential yields the same result as the power series definition for a 2×2 diagonal matrix.

Exercise 4.3.5. Compute $e^{\mathbf{A}t}$, where $\mathbf{A}$ is a diagonal matrix in $\mathbb{M}^N(\mathbb{R})$ with real diagonal entries $a_{11}, a_{22}, \ldots, a_{NN}$.

Before tackling a more complicated example, we need a few preliminary results found in the following exercise.

Exercise 4.3.6. Let a and b be nonzero real numbers.

i.) Verify

$$\begin{bmatrix} a & -b \\ 0 & a \end{bmatrix}^{-1} = \begin{bmatrix} a^{-1} & ba^{-2} \\ 0 & a^{-1} \end{bmatrix}$$

ii.) Let $n \in \mathbb{N}$. Verify

$$\begin{bmatrix} a^{-1} & ba^{-2} \\ 0 & a^{-1} \end{bmatrix}^{n} = \begin{bmatrix} a^{-n} & nba^{-(n+1)} \\ 0 & a^{-n} \end{bmatrix}$$

Example 4.3.2. Let $\mathbf{B} = \begin{bmatrix} \lambda & 1 \\ 0 & \lambda \end{bmatrix}$, where $\lambda \neq 0$. Inputting $\mathbf{B}$ in the definition of the matrix exponential yields

$$\begin{aligned} e^{\begin{bmatrix} \lambda & 1 \\ 0 & \lambda \end{bmatrix} t} &= \lim_{n\to\infty} \left(\begin{bmatrix} 1 & 0 \\ 0 & 1 \end{bmatrix} - \frac{t}{n} \begin{bmatrix} \lambda & 1 \\ 0 & \lambda \end{bmatrix} \right)^{-n} \\ &= \lim_{n\to\infty} \left(\begin{bmatrix} 1 - \frac{t}{n}\lambda & -\frac{t}{n} \\ 0 & 1 - \frac{t}{n}\lambda \end{bmatrix}^{-1} \right)^{n} \end{aligned}$$

To make the next two simplifications we use Exercise 4.3.6 (i) and (ii) with $a = 1 - \frac{t}{n}\lambda$ and $b = \frac{t}{n}$. **(Why does this exercise apply?)**

$$\begin{aligned} e^{\begin{bmatrix} \lambda & 1 \\ 0 & \lambda \end{bmatrix} t} &= \lim_{n\to\infty} \begin{bmatrix} \left(1 - \frac{t}{n}\lambda\right)^{-1} & \frac{t}{n}\left(1 - \frac{t}{n}\lambda\right)^{-2} \\ 0 & \left(1 - \frac{t}{n}\lambda\right)^{-1} \end{bmatrix}^{n} \\ &= \lim_{n\to\infty} \begin{bmatrix} \left(1 - \frac{t}{n}\lambda\right)^{-n} & t\left(1 - \frac{t}{n}\lambda\right)^{-(n+1)} \\ 0 & \left(1 - \frac{t}{n}\lambda\right)^{-n} \end{bmatrix} \end{aligned}$$

Finally, moving the limit into the matrix and dealing with each component **(Why can we do this?)**, we see

$$e^{\begin{bmatrix} \lambda & 1 \\ 0 & \lambda \end{bmatrix} t} = \begin{bmatrix} e^{\lambda t} & te^{\lambda t} \\ 0 & e^{\lambda t} \end{bmatrix}$$

We have seen two examples in which Definition 4.3.3 produces the same matrix as in Definition 4.3.1. However, just as in the case of the power series definition, computing the matrix exponential via its definition can prove difficult. When applicable, it is preferable to use Exercise 4.3.3 to compute $e^{\mathbf{A}t}$. We discuss another tool for computing the matrix exponential in the next section.

4.4 Putzer's Algorithm

The so-called *Putzer algorithm* is a useful tool when computing $e^{\mathbf{A}t}$.

STOP! How do you compute the eigenvalues of a square constant matrix?

Proposition 4.4.1. (Putzer Algorithm)
Let $\mathbf{A} \in \mathbb{M}^N(\mathbb{R})$ with eigenvalues $\lambda_1, \lambda_2, \ldots, \lambda_N$ (including multiplicity). Then,

$$e^{\mathbf{A}t} = \sum_{k=0}^{N-1} r_{k+1}(t)\mathbf{P}_k, \tag{4.16}$$

where

$$\begin{cases} \mathbf{P}_0 = \mathbf{I} \\ \mathbf{P}_j = (\mathbf{A} - \lambda_1 \mathbf{I})(\mathbf{A} - \lambda_2 \mathbf{I}) \cdots (\mathbf{A} - \lambda_j \mathbf{I}), \quad j = 1, \ldots, N, \end{cases}$$

and $r_i(t)$ is the unique solution of the IVP

$$\begin{cases} r_1'(t) = \lambda_1 r_1(t), \; r_1(0) = 1, \\ r_i'(t) = \lambda_i r_i(t) + r_{i-1}(t), \; r_i(0) = 0, \quad i = 2, \ldots, N. \end{cases} \tag{4.17}$$

The usual trick used to solve first-order linear ODEs can be used to determine the solutions of the IVPs in (4.17). We illustrate how to use the algorithm with some examples.

Example 4.4.1. Let $\mathbf{B} = \begin{bmatrix} \lambda & 1 \\ 0 & \lambda \end{bmatrix}$, where $\lambda \neq 0$. Compute $e^{\mathbf{B}t}$.
Discussion: The eigenvalues of $\mathbf{B}$ are $\lambda = \lambda_1 = \lambda_2$. From Prop. 4.4.1, we know that

$$\mathbf{P}_1 = \mathbf{B} - \lambda \mathbf{I} = \begin{bmatrix} 0 & 1 \\ 0 & 0 \end{bmatrix}$$

and

$$e^{\mathbf{B}t} = r_1(t) \begin{bmatrix} 1 & 0 \\ 0 & 1 \end{bmatrix} + r_2(t) \begin{bmatrix} 0 & 1 \\ 0 & 0 \end{bmatrix}. \tag{4.18}$$

We must solve the following two IVPs, in the order presented:

$$\begin{cases} r_1'(t) = \lambda r_1(t), \\ r_1(0) = 1, \end{cases} \tag{4.19}$$

and

$$\begin{cases} r_2'(t) = \lambda r_2(t) + r_1(t), \\ r_2(0) = 0. \end{cases} \tag{4.20}$$

The solution of (4.19) is $r_1(t) = e^{\lambda t}$. Now, substitute this expression into the right-side of (4.20) and solve the resulting equation using the variation of parameters formula (2.45) to obtain $r_2(t) = te^{\lambda t}$. Finally, substitute both of these into (4.18) to conclude that

$$e^{\mathbf{B}t} = e^{\lambda t} \begin{bmatrix} 1 & t \\ 0 & 1 \end{bmatrix} = \begin{bmatrix} e^{\lambda t} & te^{\lambda t} \\ 0 & e^{\lambda t} \end{bmatrix}.$$

(Compare with Example 4.3.2). The following example shows the Putzer algorithm in action for a 3×3 matrix.

Example 4.4.2. Let $\mathbf{B} = \begin{bmatrix} 1 & -1 & 4 \\ 3 & 2 & -1 \\ 2 & 1 & -1 \end{bmatrix}$. Compute $e^{\mathbf{B}t}$.
Discussion: The characteristic polynomial for $\mathbf{B}$ is

$$p(\lambda) = (\lambda + 2)(\lambda - 3)(\lambda - 1),$$

so the eigenvalues of $\mathbf{B}$ are $-2, 1$, and 3, which we label as λ_1, λ_2, and λ_3, respectively. From Prop. 4.4.1, we know that

$$\mathbf{P}_1 = \mathbf{B} - \lambda_1\mathbf{I} = \begin{bmatrix} 3 & -1 & 4 \\ 3 & 4 & -1 \\ 2 & 1 & 1 \end{bmatrix}$$

$$\mathbf{P}_2 = (\mathbf{B} - \lambda_1\mathbf{I})(\mathbf{B} - \lambda_2\mathbf{I}) = \begin{bmatrix} 5 & 0 & 5 \\ 10 & 0 & 10 \\ 5 & 0 & 5 \end{bmatrix}$$

and

$$e^{\mathbf{B}t} = r_1(t)\mathbf{I} + r_2(t)\mathbf{P}_1 + r_3(t)\mathbf{P}_2. \tag{4.21}$$

We must solve the following three IVPs, in the order presented:

$$\begin{cases} r_1'(t) = -2r_1(t), \\ r_1(0) = 1, \end{cases} \tag{4.22}$$

$$\begin{cases} r_2'(t) = r_2(t) + r_1(t), \\ r_2(0) = 0, \end{cases} \tag{4.23}$$

$$\begin{cases} r_3'(t) = 3r_3(t) + r_2(t), \\ r_3(0) = 0. \end{cases} \tag{4.24}$$

The solutions of IVPs (4.22) - (4.24) are as follows:

$$r_1(t) = e^{-2t} \tag{4.25}$$

$$r_2(t) = -\frac{1}{3}\left(e^{-3t} - 1\right) \tag{4.26}$$

$$r_3(t) = \frac{1}{18}\left(e^{3t} + e^{-3t}\right) - \frac{1}{9} \tag{4.27}$$

Finally, substitute (4.25) - (4.27) into (4.21) to conclude that

$$e^{\mathbf{B}t} = e^{-2t}\mathbf{I} - \frac{1}{3}\left(e^{-3t} - 1\right)\mathbf{P}_1 + \left(\frac{1}{18}\left(e^{3t} + e^{-3t}\right) - \frac{1}{9}\right)\mathbf{P}_2. \tag{4.28}$$

STOP!

i.) Verify that (4.25) - (4.27) are solutions of the IVPs (4.22) - (4.24), respectively.
ii.) Simplify (4.28) to a single 3×3 matrix.

A matrix can have complex eigenvalues as well. The computations involved in using the Putzer algorithm will inherently involve imaginary numbers, yet the final result must be a matrix containing only real entries. The following example illustrates this for a standard 2×2 matrix.

Example 4.4.3. Let $\mathbf{B} = \begin{bmatrix} 0 & \alpha \\ -\alpha & 0 \end{bmatrix}$, where $\alpha \neq 0$. Compute $e^{\mathbf{B}t}$.

Discussion: The characteristic polynomial for $\mathbf{B}$ is $p(\lambda) = \lambda^2 + \alpha^2$,so the eigenvalues of $\mathbf{B}$ are $\lambda = \pm i\alpha$. We label them as $\lambda_1 = i\alpha$ and $\lambda_2 = -i\alpha$. From Prop. 4.4.1, we know that

$$\mathbf{P}_1 = \mathbf{B} - \lambda\mathbf{I} = \begin{bmatrix} -i\alpha & \alpha \\ -\alpha & -i\alpha \end{bmatrix}$$

and

$$e^{\mathbf{B}t} = r_1(t)\begin{bmatrix} 1 & 0 \\ 0 & 1 \end{bmatrix} + r_2(t)\begin{bmatrix} -i\alpha & \alpha \\ -\alpha & -i\alpha \end{bmatrix}. \tag{4.29}$$

We must solve the following two IVPs, in the order presented:

$$\begin{cases} r_1'(t) = i\alpha r_1(t), \\ r_1(0) = 1, \end{cases} \tag{4.30}$$

and

$$\begin{cases} r_2'(t) = -i\alpha r_2(t) + r_1(t), \\ r_2(0) = 0. \end{cases} \tag{4.31}$$

The solutions of IVPs (4.30) - (4.31) are as follows:

$$r_1(t) = e^{i\alpha t} \tag{4.32}$$

$$r_2(t) = \frac{1}{2i\alpha}\left(e^{i\alpha t} - e^{-i\alpha t}\right) \tag{4.33}$$

Finally, substitute both of these into (4.29) to conclude that

$$e^{\mathbf{B}t} = e^{i\alpha t}\begin{bmatrix} 1 & 0 \\ 0 & 1 \end{bmatrix} + \frac{1}{2i\alpha}\left(e^{i\alpha t} - e^{-i\alpha t}\right)\begin{bmatrix} -i\alpha & \alpha \\ -\alpha & -i\alpha \end{bmatrix}. \tag{4.34}$$

After a bit of simplification involving Euler's formula ($e^{i\theta} = \cos\theta + i\sin\theta$, where $\theta \in \mathbb{R}$), we obtain

$$e^{\mathbf{B}t} = \frac{1}{2\alpha}\left(e^{i\alpha t} - e^{-i\alpha t}\right)\begin{bmatrix} 1 & -i \\ i & 1 \end{bmatrix} = \begin{bmatrix} \cos(\alpha t) & \sin(\alpha t) \\ -\sin(\alpha t) & \cos(\alpha t) \end{bmatrix}. \tag{4.35}$$

STOP! Verify (4.35).

Exercise 4.4.1. Use Prop. 4.4.1 to calculate $e^{\mathbf{B}t}$ for these choices of $\mathbf{B}$:

i.) $\begin{bmatrix} \alpha & \beta \\ -\beta & \alpha \end{bmatrix}$, where $\alpha \in \mathbb{R}$ and $\beta \neq 0$,

ii.) $\begin{bmatrix} \alpha & 0 \\ \beta & \alpha \end{bmatrix}$, where $\alpha \in \mathbb{R}$ and $\beta \neq 0$.

MATLAB-Exercise 4.4.1. In this exercise, you will use **MATLAB** to aid in the calculation for $e^{\mathbf{A}t}$. To begin, open **MATLAB** and in the command line, type:

MATLAB_Putzer_Help

Make certain to click on the 'Help' button for more details and instructions on how to use the GUI. A screenshot from this GUI can found in Figure 4.2.

i.) Use Prop. 4.4.1 in conjunction with the GUI to compute $e^{\mathbf{A}t}$ for these choices of $\mathbf{A}$:

a.) $\begin{bmatrix} 2 & 1 \\ -1 & 4 \end{bmatrix}$

b.) $\begin{bmatrix} 1 & 1 & 3 \\ 0 & 1 & 0 \\ 0 & 3 & 4 \end{bmatrix}$

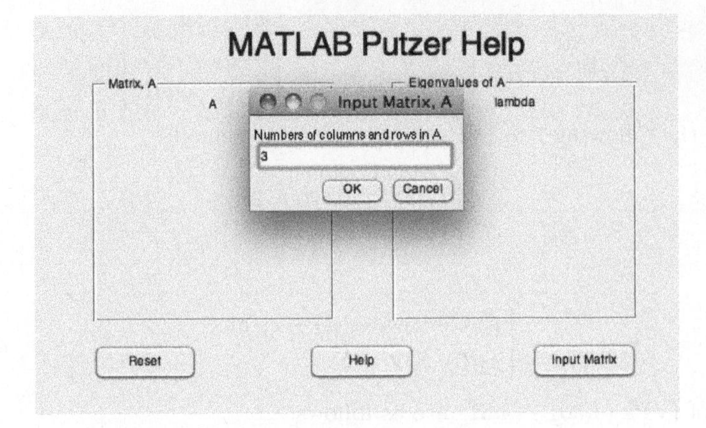

FIGURE 4.2: MATLAB Putzer Help GUI

c.) $\begin{bmatrix} 1 & 0 \\ 1 & 1 \end{bmatrix}$

ii.) Does Exercise 4.3.3 apply to any of the matrices in part (i)? If so, which ones? If not, why?

iii.) Compare and contrast the results of **MATLAB** -Exercise 4.3.1 and **MATLAB** -Exercise 4.4.1.

■

4.5 Properties of $e^{\mathbf{A}t}$

The matrix exponential satisfies many of the same arithmetic and calculus properties as the real-valued exponential function. To begin, we define a mapping involving the matrix exponential and present some basic properties, including a very useful estimate. Motivated by our discussion in Section 4.2, we expect the solution of **(HCP)** to be obtained by applying the matrix exponential to the initial condition. A natural question will be, "What happens to the solution of **(HCP)** if we tweak the initial condition a little bit?" This is where familiarity with the mapping defined below will be useful.

Proposition 4.5.1. Let $\mathbf{A} \in \mathbb{M}^N(\mathbb{R})$ and for every $t \in \mathbb{R}$, define the mapping $\boldsymbol{S}_t$: $\text{dom}(\boldsymbol{S}_t) \subset \mathbb{R}^N \to \mathbb{R}^N$ by $\boldsymbol{S}_t\mathbf{x} = e^{\mathbf{A}t}\mathbf{x}$.

i.) $\text{dom}(\boldsymbol{S}_t) = \mathbb{R}^N$;
ii.) For every $\alpha, \beta \in \mathbb{R}$ and $\mathbf{x}, \mathbf{y} \in \mathbb{R}^N$, $\boldsymbol{S}_t(\alpha\mathbf{x} + \beta\mathbf{y}) = \alpha\boldsymbol{S}_t\mathbf{x} + \beta\boldsymbol{S}_t\mathbf{y}$;
iii.) For every $\mathbf{x} \in \mathbb{R}^N$, $\|\boldsymbol{S}_t\mathbf{x}\|_{\mathbb{R}^N} = \left\|e^{\mathbf{A}t}\mathbf{x}\right\|_{\mathbb{R}^N} \le e^{|t|\|\mathbf{A}\|_{\mathbb{M}^N}} \|\mathbf{x}\|_{\mathbb{R}^N}$.

Proof. (i) follows because $e^{\mathbf{A}t} \in \mathbb{M}^N(\mathbb{R})$. **(Explain why.)**
As for (iii), property (2.32) implies that

$$\left\|e^{\mathbf{A}t}\mathbf{x}\right\|_{\mathbb{R}^N} \le \left\|e^{\mathbf{A}t}\right\|_{\mathbb{M}^N} \|\mathbf{x}\|_{\mathbb{R}^N}. \tag{4.36}$$

We now need to estimate $\|e^{\mathbf{A}t}\|_{\mathbb{M}^N}$. To do so, let us use the power series definition of $e^{\mathbf{A}t}$.

$$
\begin{aligned}
\|e^{\mathbf{A}t}\|_{\mathbb{M}^N} &= \left\| \sum_{k=0}^{\infty} \frac{\mathbf{A}^k t^k}{k!} \right\|_{\mathbb{M}^N} \\
&= \left\| \lim_{m\to\infty} \sum_{k=0}^{m} \frac{\mathbf{A}^k t^k}{k!} \right\|_{\mathbb{M}^N} \\
&= \lim_{m\to\infty} \left\| \sum_{k=0}^{m} \frac{\mathbf{A}^k t^k}{k!} \right\|_{\mathbb{M}^N} \\
&\le \lim_{m\to\infty} \sum_{k=0}^{m} \frac{\|\mathbf{A}\|_{\mathbb{M}^N}^k \, |t|^k}{k!} \\
&= \sum_{k=0}^{\infty} \frac{\left(\|\mathbf{A}\|_{\mathbb{M}^N} \, |t|\right)^k}{k!} \\
&= e^{|t| \|\mathbf{A}\|_{\mathbb{M}^N}}.
\end{aligned}
$$

(Justify the steps.) Using this estimate in (4.36) yields the result. □

MATLAB-Exercise 4.5.1. In this exercise, you will use **MATLAB** to illustrate Prop 4.5.1 and investigate new properties of the matrix exponential. To begin, open **MATLAB** and in the command line, type:

MATLAB_Exponential_Properties

A screenshot from this GUI can be found in Figure 4.3. Be sure to use the Help button for more information and instructions for the GUI.
Check the *Plot Norm Estimate* box to answer the following.

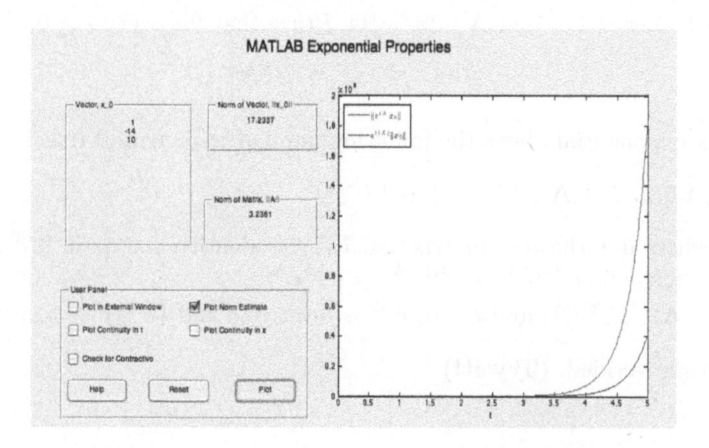

FIGURE 4.3: MATLAB Exponential Properties GUI

i.) Let $A = \begin{bmatrix} 1 & 1 & 3 \\ 0 & 1 & 0 \\ 0 & 3 & 4 \end{bmatrix}$ and any vector of your choosing. In words, explain how the plot provided by the GUI illustrates Prop 4.5.1 (iii). Choose two more matrices and verify Prop 4.5.1 (iii).

ii.) Determine a matrix **A** and non-zero vector **x** such that

$$\|e^{t\mathbf{A}}\mathbf{x}\|_{\mathbb{R}^N} = e^{t\|\mathbf{A}\|_{\mathbb{M}^N}} \|x\|_{\mathbb{R}^N}$$

for all $t \geq 0$.

Let $\mathbf{A} \in \mathbb{M}^N(\mathbb{R})$ and $t \geq 0$. The family of matrices $\{e^{\mathbf{A}t} : t \geq 0\}$ is called *contractive* if $\|e^{\mathbf{A}t}\mathbf{x}\|_{\mathbb{R}^N} \leq \|\mathbf{x}\|_{\mathbb{R}^N}$, for every $\mathbf{x} \in \mathbb{R}^N$ and $t \geq 0$.

Use a piece of paper and/or the *Check Contractive* feature of the GUI to answer the following questions.

iii.) Give an example of $\mathbf{A} \in \mathbb{M}^N(\mathbb{R})$, $N = 2, 3, 4$ for which $\{e^{\mathbf{A}t} : t \geq 0\}$ is not contractive. In each case, record the eigenvalues of the matrix.

iv.) Determine a sufficient condition that could be imposed on a diagonal matrix $\mathbf{A} \in \mathbb{M}^N(\mathbb{R})$ to ensure that $\{e^{\mathbf{A}t} : t \geq 0\}$ is contractive. (Hint: Consider the eigenvalues of **A**.)

v.) Suppose we extend the condition in (ii) to any matrix. Do you think the matrix will be contractive? Plot the contractive condition for three separate non-diagonal matrices satifying the condition derived in (ii). Explain your findings and make a conjecture regarding contractivity.

■

The following exercise provides a scenario in which certain matrix exponentials commute.

Exercise 4.5.1. Let $\alpha \in \mathbb{R}$ and $\mathbf{A} \in \mathbb{M}^N(\mathbb{R})$. Prove that for every $t \geq 0$,

$$e^{(\alpha\mathbf{A})t}e^{\mathbf{A}t} = e^{\mathbf{A}t}e^{(\alpha\mathbf{A})t}.$$

The matrix exponential obeys the following familiar exponential rules:

Proposition 4.5.2. Let $\mathbf{A} \in \mathbb{M}^N(\mathbb{R})$ and $t \geq 0$.

i.) $e^{\mathbf{0}} = \mathbf{I}$, where $\mathbf{0}$ is the zero matrix and $\mathbf{I}$ is the identity matrix in $\mathbb{M}^N(\mathbb{R})$;
ii.) For every $t, s \geq 0$, $e^{\mathbf{A}(t+s)} = e^{\mathbf{A}t}e^{\mathbf{A}s} = e^{\mathbf{A}s}e^{\mathbf{A}t}$;
iii.) For every $\mathbf{A} \in \mathbb{M}^N(\mathbb{R})$ and $t \geq 0$, $e^{\mathbf{A}t}$ is invertible and $\left(e^{\mathbf{A}t}\right)^{-1} = e^{-\mathbf{A}t}$.

Proof. (i) is easily verified. **(Try it!)**

(ii) holds because

$$\begin{aligned}
e^{\mathbf{A}(t+s)} &= \sum_{n=0}^{\infty} \frac{\mathbf{A}^n (t+s)^n}{n!} \\
&= \sum_{n=0}^{\infty} \frac{\mathbf{A}^n}{n!} \sum_{k=0}^{n} \binom{n}{k} t^k s^{n-k} \\
&= \sum_{n=0}^{\infty} \sum_{k=0}^{n} \frac{\mathbf{A}^n}{n!} \frac{n!}{k!(n-k)!} t^k s^{n-k} \\
&= \sum_{n=0}^{\infty} \sum_{k=0}^{n} \frac{\mathbf{A}^k \mathbf{A}^{n-k}}{k!(n-k)!} t^k s^{n-k} \\
&= \sum_{k=0}^{\infty} \sum_{n=k}^{\infty} \frac{\mathbf{A}^k t^k}{k!} \frac{\mathbf{A}^{n-k} s^{n-k}}{(n-k)!} \\
&= \left(\sum_{k=0}^{\infty} \frac{\mathbf{A}^k t^k}{k!} \right) \left(\sum_{n=k}^{\infty} \frac{\mathbf{A}^{n-k} s^{n-k}}{(n-k)!} \right) \\
&= \left(\sum_{k=0}^{\infty} \frac{\mathbf{A}^k t^k}{k!} \right) \left(\sum_{n=0}^{\infty} \frac{\mathbf{A}^n s^n}{n!} \right) \\
&= e^{\mathbf{A}t} e^{\mathbf{A}s}.
\end{aligned}$$

where we have used the Binomial Theorem. **(Justify the steps.)** Finally, since

$$\mathbf{I} = e^{\mathbf{0}} = e^{\mathbf{A}t + (-\mathbf{A}t)} = e^{\mathbf{A}t} e^{-\mathbf{A}t} = e^{-\mathbf{A}t} e^{\mathbf{A}t},$$

we conclude that (iii) holds. □

MATLAB-Exercise 4.5.2. In this exercise, you will use **MATLAB** to investigate continuity of the matrix exponential. To begin, open **MATLAB** and in the command line, type:

MATLAB_Exponential_Properties

A screenshot from this GUI can be found in Figure 4.4. Use the *Plot Continuity in t* or *Plot Continuity in x* feature to answer the following.

Consider the following statements:

1. For every $\mathbf{x}_0 \in \mathbb{R}^N$, the mapping $\mathbf{g}: [0, \infty) \to \mathbb{R}^N$ defined by $\mathbf{g}(t) = e^{\mathbf{A}t}\mathbf{x}_0$ is continuous;
2. For every $t_0 \geq 0$, the mapping $\mathbf{h}: \mathbb{R}^N \to \mathbb{R}^N$ defined by $\mathbf{h}(\mathbf{x}) = e^{\mathbf{A}t_0}\mathbf{x}$ is continuous.

i.) Explain, in words, what Statement 1 means. Be sure to clearly describe the independent variable.

ii.) Intuitively, if $|t - 3|$ is small and $\mathbf{g}$ is continuous at 3, what should be the case for

$$\|\mathbf{g}(t) - \mathbf{g}(3)\|_{\mathbb{R}^N}?$$

iii.) Explain, in words, what Statement 2 means. Be sure to include clearly describe the independent variable.

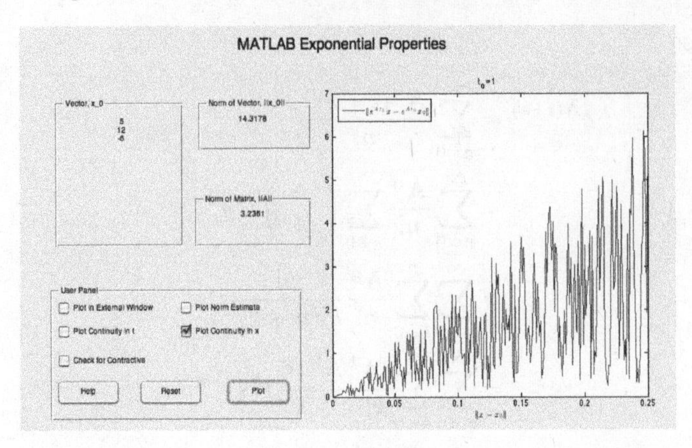

FIGURE 4.4: MATLAB Exponential Properties GUI

iv.) Intuitively, if $\|x - x_0\|_{\mathbb{R}^N}$ is small and $\mathbf{h}$ is continuous at $\mathbf{x_0}$, what should be the case for

$$\|\mathbf{h(x)} - \mathbf{h(x_0)}\|_{\mathbb{R}^N}?$$

v.) Use the GUI to gain intuition regarding the continuity of the mapping **g**. Based on your observations make a conjecture about the continuity of the mapping **g**.

vi.) Use the GUI to gain intuition regarding the continuity of the mapping **h**. Based on your observations make a conjecture about the continuity of the mapping **h**.

vii.) How can these abstract continuity properties be useful in application? (Hint: Refer to Matlab-Exercise (3.2.1).)

■

Summarizing the previous MATLAB exercise, we have

Proposition 4.5.3. (Continuity properties of $e^{\mathbf{A}t}$) Let $\mathbf{A} \in \mathbb{M}^N(\mathbb{R})$.

i.) $\lim_{t \to 0^+} \|e^{\mathbf{A}t} - \mathbf{I}\|_{\mathbb{M}^N} = 0$;

ii.) For every $\mathbf{x}_0 \in \mathbb{R}^N$, the mapping $\mathbf{g}: [0,\infty) \to \mathbb{R}^N$ defined by $\mathbf{g}(t) = e^{\mathbf{A}t}\mathbf{x}_0$ is continuous;

iii.) For every $t_0 \geq 0$, the mapping $\mathbf{h}: \mathbb{R}^N \to \mathbb{R}^N$ defined by $\mathbf{h}(\mathbf{x}) = e^{\mathbf{A}t_0}\mathbf{x}$ is continuous.

The operators $\mathbf{A}$ and $e^{\mathbf{A}t}$ are closely related, as illustrated below.

Exercise 4.5.2. Let $\mathbf{A} = \begin{bmatrix} \alpha & 0 \\ 0 & \beta \end{bmatrix}$, where $\alpha, \beta \in \mathbb{R}$. Verify that the following hold:

i.) For every $\mathbf{x}_0 \in \mathbb{R}^2$, $\lim_{t \to 0^+} \frac{(e^{\mathbf{A}t} - \mathbf{I})\mathbf{x}_0}{t} = \mathbf{A}\mathbf{x}_0$;

ii.) For every $t \geq 0$, $\mathbf{A}e^{\mathbf{A}t} = e^{\mathbf{A}t}\mathbf{A}$.

In fact, these properties hold for all $\mathbf{A} \in \mathbb{M}^N(\mathbb{R})$, as shown below.

Proposition 4.5.4. Let $\mathbf{A} \in \mathbb{M}^N(\mathbb{R})$. Then,

i.) For every $\mathbf{z} \in \mathbb{R}^N$, $\lim_{h\to 0^+} \frac{(e^{\mathbf{A}h}-\mathbf{I})\mathbf{z}}{h} = \mathbf{Az}$;

ii.) For every $t \geq 0$, $\mathbf{A}e^{\mathbf{A}t} = e^{\mathbf{A}t}\mathbf{A}$.

Proof. (ii) follows from linearity since $\mathbf{A}$ commutes with scalar multiples of $\mathbf{A}$. **(Tell why carefully.)**
For (i), let $\mathbf{z} \in \mathbb{R}^N$. Observe that

$$\begin{aligned}
0 &\leq \frac{1}{h}\left\|\left(e^{\mathbf{A}h}-\mathbf{I}\right)\mathbf{z} - \mathbf{A}h\mathbf{z}\right\|_{\mathbb{R}^N} \\
&= \frac{1}{h}\left\|\sum_{n=0}^{\infty}\frac{\mathbf{A}^n h^n}{n!}\mathbf{z} - \mathbf{Iz} - \mathbf{A}h\mathbf{z}\right\|_{\mathbb{R}^N} \\
&= \frac{1}{h}\left\|\sum_{n=2}^{\infty}\frac{\mathbf{A}^n h^n}{n!}\mathbf{z}\right\|_{\mathbb{R}^N} \\
&\leq \frac{1}{h}\sum_{n=2}^{\infty}\frac{\left(\|\mathbf{A}\|_{\mathbb{M}^N} h\right)^n}{n!}\|\mathbf{z}\|_{\mathbb{R}^N} \\
&= \frac{1}{h}\left[\sum_{n=2}^{\infty}\frac{\left(\|\mathbf{A}\|_{\mathbb{M}^N} h\right)^n}{n!} + 1 + \|\mathbf{A}\|_{\mathbb{M}^N} h - 1 - \|\mathbf{A}\|_{\mathbb{M}^N} h\right]\|\mathbf{z}\|_{\mathbb{R}^N} \\
&= \left[\frac{e^{h\|\mathbf{A}\|_{\mathbb{M}^N}}-1}{h} - \|\mathbf{A}\|_{\mathbb{M}^N}\right]\|\mathbf{z}\|_{\mathbb{R}^N}
\end{aligned}$$

(Why?) An application of l'Hopital's rule shows that the right-side goes to zero as $h \to 0^+$, so that the result follows from an application of the Squeeze Theorem. □

Making this connection precise requires the following properties.

Proposition 4.5.5. Let $\mathbf{A} \in \mathbb{M}^N(\mathbb{R})$ and $\mathbf{x} \in \mathbb{R}^N$. For every $\mathbf{x}_0 \in \mathbb{R}^N$, $\mathbf{g}: (0,\infty) \to \mathbb{R}^N$ defined by $\mathbf{g}(t) = e^{\mathbf{A}t}\mathbf{x}_0$ is differentiable and

$$\frac{d}{dt}e^{\mathbf{A}t}\mathbf{x}_0 = \mathbf{A}e^{\mathbf{A}t}\mathbf{x}_0 = e^{\mathbf{A}t}\mathbf{A}\mathbf{x}_0.$$

Remark 4.5.1. Proposition 4.5.5 not only validates the use of the chain rule for differentiating the matrix exponential, but also states that $\mathbf{A}$ commutes with the matrix exponential of $\mathbf{A}$, a result that generally does not apply with matrices.

Proof. Let $t_0 > 0$ and consider $0 < h < t_0$ such that $t_0 + h > 0$. Since

$$\frac{e^{\mathbf{A}(t_0+h)}\mathbf{x}_0 - e^{\mathbf{A}t_0}\mathbf{x}_0}{h} = \left(\frac{e^{\mathbf{A}h}-\mathbf{I}}{h}\right)e^{\mathbf{A}t_0}\mathbf{x}_0 = e^{\mathbf{A}t_0}\left(\frac{e^{\mathbf{A}h}-\mathbf{I}}{h}\right)\mathbf{x}_0,$$

we see that

$$\begin{aligned}
\lim_{h\to 0^+}\left\|\frac{e^{\mathbf{A}h}\left(e^{\mathbf{A}t_0}\mathbf{x}_0\right) - \left(e^{\mathbf{A}t_0}\mathbf{x}_0\right)}{h} - e^{\mathbf{A}t_0}\left(\mathbf{A}\mathbf{x}_0\right)\right\|_{\mathbb{R}^N} &= \lim_{h\to 0^+}\left\|e^{\mathbf{A}t_0}\left[\frac{e^{\mathbf{A}h}\mathbf{x}_0 - \mathbf{x}_0}{h} - \mathbf{A}\mathbf{x}_0\right]\right\|_{\mathbb{R}^N} \\
&= 0.
\end{aligned}$$

Hence, $\lim_{h\to 0^+}\frac{e^{\mathbf{A}h}\left(e^{\mathbf{A}t_0}\mathbf{x}_0\right)-\left(e^{\mathbf{A}t_0}\mathbf{x}_0\right)}{h}$ exists and equals both $\mathbf{A}e^{\mathbf{A}t_0}\mathbf{x}_0$ **(Why?)** and $e^{\mathbf{A}t_0}\mathbf{A}\mathbf{x}_0$ **(Why? So what?)** Argue similarly for $h < 0$. **(Tell how.)** This proves the proposition. □

Exercise 4.5.3. Verify the following identities directly for the diagonal matrices given below. Compare each to the corresponding familiar integration formulas with real-valued integrands.

$$\textbf{(I)} \int_0^t e^{\mathbf{A}s} ds = \mathbf{A}^{-1}\left(e^{\mathbf{A}t} - \mathbf{I}\right)$$

$$\textbf{(II)} \int_s^t \mathbf{A} e^{\mathbf{A}u} \mathbf{z} du = \int_s^t e^{\mathbf{A}u} \mathbf{A} \mathbf{z} du = e^{\mathbf{A}t}\mathbf{z} - e^{\mathbf{A}s}\mathbf{z}$$

i.) $\mathbf{A} = \begin{bmatrix} \lambda & 0 \\ 0 & \mu \end{bmatrix}$, where $\lambda \neq \mu$ are nonzero real numbers.

ii.) $\mathbf{A}$ is any $N \times N$ diagonal matrix.

4.6 The Homogenous Cauchy Problem: Well-posedness

The IVP (4.1) is referred to as the *homogenous Cauchy problem in* $\mathbb{R}^N$. We seek a solution in the following sense.

Definition 4.6.1. A *classical solution* of (4.1) is a function $\mathbf{U} : [t_0, \infty) \longrightarrow \mathbb{R}^N$ that is right-continuous at $t = t_0$, differentiable on (t_0, ∞), and which satisfies the ODE and IC in (4.1).

For simplicity, we begin by considering **(HCP)** with $t_0 = 0$. The properties of $e^{\mathbf{A}t}$ established in Section 4.5 enable us to establish the following result:

Theorem 4.6.1. For every $\mathbf{U}_0 \in \mathbb{R}^N$, the IVP (4.1) has a unique classical solution given by $\mathbf{U}(t) = e^{\mathbf{A}t}\mathbf{U}_0$.

Proof. Let $\mathbf{U}_0 \in \mathbb{R}^N$.
Existence: We must verify that $\mathbf{U}(t) = e^{\mathbf{A}t}\mathbf{U}_0$ satsifies Def. 4.6.1. To this end, note that Prop. 4.5.3(ii) ensures that $\mathbf{U}$ is right-continuous at $t = 0$, Prop. 4.5.5(i) ensures that $\mathbf{U}$ is differentiable on $(0, \infty)$, and $\mathbf{U}$ satisfies the first equation in (4.1). Finally, Prop. 4.5.2(i) guarantees that $\mathbf{U}$ satisfies the initial condition $\mathbf{U}(0) = \mathbf{U}_0$. **(Explain why each of these holds, carefully!)** This establishes existence.
Uniqueness: Let $\mathbf{V}$: $[0, \infty) \to \mathbb{R}^N$ be a solution to the IVP

$$\begin{cases} \mathbf{V}'(t) = \mathbf{A}\mathbf{V}(t), \ t > 0, \\ \mathbf{V}(0) = \mathbf{U}_0. \end{cases}$$

Define the function $\Psi : [0, \infty) \to \mathbb{R}^N$ by $\Psi(s) = e^{\mathbf{A}(t-s)}\mathbf{V}(s)$. Prop. 4.5.5 implies that Ψ is differentiable. **(Why?)** Let $\tau > 0$. Then, for every $s \in [0, \tau]$,

$$\begin{aligned} \frac{d\Psi}{ds}(s) &= e^{\mathbf{A}(t_0-s)}\mathbf{V}'(s) + \frac{d}{ds} e^{\mathbf{A}(t_0-s)}\mathbf{V}(s) \ \textbf{(Why?)} \\ &= e^{\mathbf{A}(t_0-s)}\left(\mathbf{A}\mathbf{V}(s)\right) - \mathbf{A}e^{\mathbf{A}(t_0-s)}\mathbf{V}(s) \ \textbf{(Why?)} \\ &= 0. \ \textbf{(Why?)} \end{aligned}$$

Hence, for every $\tau > 0$, Ψ is a constant vector in $\mathbb{R}^N$on the interval $[0, \tau]$. In particular, for every $\tau > 0$, $\Psi(0) = \Psi(\tau)$. Consequently, for every $\tau > 0$,

$$\mathbf{U}(\tau) = e^{\mathbf{A}\tau}\mathbf{U}_0 = \Psi(0) = \Psi(\tau) = \mathbf{V}(\tau),$$

thereby showing uniqueness. This completes the proof. □

What if reformulating the system of ODEs in the abstract form (4.1) requires that the initial condition be computed at some time t_0 other than zero? There was nothing special about specifying the initial condition at time $t_0 = 0$ except that perhaps the computations were slightly nicer. However, we might have information about the system's behavior at some time $t_0 \neq 0$ in the form of an initial condition $\mathbf{U}(t_0) = \mathbf{U}_0$. The existence and uniqueness of a classical solution of the translated IVP, which we originally labeled as **(HCP)**, is guaranteed. The proof above can be modified line by line to verify it.

STOP! Do so!

It is more direct, however, to establish the following corollary.

Corollary 4.6.1. For every $t_0 > 0$, the IVP

$$\begin{cases} \mathbf{U}'(t) = \mathbf{A}\mathbf{U}(t), \, t > t_0, \\ \mathbf{U}(t_0) = \mathbf{U}_0, \end{cases} \tag{4.37}$$

has a unique classical solution $\mathbf{U}\colon [t_0, \infty) \to \mathbb{R}^N$ given by $\mathbf{U}(t) = e^{\mathbf{A}(t-t_0)}\mathbf{U}_0$.

Proof. Let $t_0 > 0$ and suppose that $\mathbf{U}\colon [t_0, \infty) \to \mathbb{R}^N$ satisfies (4.37). Define the function $\mathbf{W}\colon [0, \infty) \to \mathbb{R}^N$ by $\mathbf{W}(t) = \mathbf{U}(t + t_0)$. Then

$$\mathbf{W}'(t) = \mathbf{U}'(t + t_0) = \mathbf{A}\mathbf{U}(t + t_0) \text{ and } \mathbf{W}(0) = \mathbf{U}(t_0) = \mathbf{U}_0.$$

Thus, $\mathbf{W}$ satisfies (4.1) and so, $\mathbf{W}(t) = e^{\mathbf{A}t}\mathbf{U}_0$, $\forall t > 0$. As such, the classical solution of (4.37) is given by $\mathbf{U}(t) = \mathbf{W}(t - t_0) = e^{\mathbf{A}(t-t_0)}\mathbf{U}_0$, for $t > t_0$, as desired. □

4.7 Higher-Order Linear ODEs

The theory and methods developed in this chapter can be used to solve higher-order linear homogenous ODEs with constant coefficients. As you have seen, such equations arise when modeling the behavior of spring-mass systems, electric circuits, and mechanics, to name just a few.

Definition 4.7.1. An nth-*order homogenous linear ODE with constant coefficients* is of the general form

$$a_n x^{(n)}(t) + a_{n-1}x^{(n-1)}(t) + \ldots + a_1 x'(t) + a_0 x(t) = 0, \tag{4.38}$$

where t is in some interval I and $a_n \neq 0$. If we equip (4.38) with n initial conditions of the form

$$x(t_0) = c_1, \, x'(t_0) = c_2, \ldots, x^{(n-1)}(t_0) = c_n, \tag{4.39}$$

then (4.38) - (4.39) forms an nth-*order IVP*.

We must interpret the IVP (4.38) - (4.39) as a linear system of the form **(HCP)** in order to apply the theory developed thus far. This is made possible by means of renaming the variables as described below:

First, move the terms *except* the term of degree n to the right-side of (4.38):

$$a_n x^{(n)}(t) = \underbrace{-\sum_{j=0}^{n-1} a_j x^{(j)}(t)}_{\text{Lower-order Derivatives}} \tag{4.40}$$

Now, divide both sides of (4.40) by a_n so that the leading coefficient on the left-side is 1:

$$x^{(n)}(t) = -\sum_{j=0}^{n-1} \underbrace{\left(\frac{a_j}{a_n}\right)}_{=a_j^\star} x^{(j)}(t). \tag{4.41}$$

Next, perform the following change of variable. We suppress the dependence on t throughout:

$$\begin{aligned} y_1 &= x \\ y_2 &= y_1' = x' \\ y_3 &= y_2' = y_1'' = x'' \\ &\vdots \\ y_{n-1} &= y_{n-2}' = \ldots = x^{(n-2)} \\ y_n &= y_{n-1}' = \ldots = x^{(n-1)} \end{aligned}$$

Observe that

$$y_n' = x^{(n)} = -\sum_{j=0}^{n-1} a_j^\star y_{j+1}. \tag{4.42}$$

The initial conditions in (4.39) become

$$y_1(t_0) = c_1,\ y_2(t_0) = c_2, \ldots, y_n(t_0) = c_n. \tag{4.43}$$

Combining these observations enables us to reformulate IVP (4.38) - (4.39) as the following equivalent homogenous linear system:

$$\begin{cases} \begin{bmatrix} y_1 \\ y_2 \\ \vdots \\ y_{n-1} \\ y_n \end{bmatrix}(t)' = \begin{bmatrix} 0 & 1 & 0 & \cdots & 0 & 0 \\ 0 & 0 & 1 & \cdots & 0 & 0 \\ & & & \vdots & & \\ 0 & 0 & 0 & \cdots & 0 & 1 \\ a_0^\star & a_1^\star & a_2^\star & \cdots & a_{n-1}^\star & a_n^\star \end{bmatrix} \begin{bmatrix} y_1 \\ y_2 \\ \vdots \\ y_{n-1} \\ y_n \end{bmatrix}(t) \\ \begin{bmatrix} y_1 \\ y_2 \\ \vdots \\ y_{n-1} \\ y_n \end{bmatrix}(t_0) = \begin{bmatrix} c_1 \\ c_2 \\ \vdots \\ c_{n-1} \\ c_n \end{bmatrix} \end{cases} \tag{4.44}$$

Since IVP (4.44) is of the form **(HCP)**, it follows immediately from Theorem 4.6.1 that this IVP has a unique solution $\mathbf{Y} : I \longrightarrow \mathbb{R}^n$ given by $\mathbf{Y}(t) = e^{\mathbf{A}(t-t_0)}\mathbf{Y}_0$, where $\mathbf{A}$ is the coefficient matrix and $\mathbf{Y}_0$ is the initial condition vector. The desired function that satisfies the original n^{th}-order IVP (4.38) - (4.39) is then just the first component, $y_1(t)$, of $\mathbf{Y}(t)$. **(Why?)**

We illustrate this procedure with a couple of examples below.

Example 4.7.1. Consider the following IVP which describes an elementary spring-mass system:

$$\begin{cases} x''(t) + \omega^2 x(t) = 0, \ t > 0, \\ x(0) = x_0, \ x'(0) = x_1, \end{cases} \tag{4.45}$$

where $\omega^2 = \frac{k}{m}$, x_0 is the initial position of the mass with respect to the equilibrium and x_1 is the initial speed. This is referred to as a *harmonic oscillator*. This IVP can be converted into a system of first-order ODEs by way of the change of variable

$$\begin{cases} y = x, \\ z = x' \end{cases} \text{so that} \begin{cases} y' = x' = z \\ z' = x'' = -\omega^2 x = -\omega^2 y. \end{cases} \tag{4.46}$$

Then, (4.45) can be written in the equivalent matrix form

$$\begin{cases} \begin{bmatrix} y(t) \\ z(t) \end{bmatrix}' = \begin{bmatrix} 0 & 1 \\ -\omega^2 & 0 \end{bmatrix} \begin{bmatrix} y(t) \\ z(t) \end{bmatrix}, \ t > 0, \\ \begin{bmatrix} y(0) \\ z(0) \end{bmatrix} = \begin{bmatrix} x_0 \\ x_1 \end{bmatrix}. \end{cases} \tag{4.47}$$

Certainly, $z(t)$ is redundant **(Why?)**, but this transformation is useful because it enables us to consider a second-order ODE in the same form. Theorem 4.6.1 implies that IVP (4.47) describing an elementary spring-mass system has a unique classical solution given by

$$\begin{bmatrix} x(t) \\ x'(t) \end{bmatrix} = e^{\begin{bmatrix} 0 & 1 \\ -\omega^2 & 0 \end{bmatrix} t} \begin{bmatrix} x_0 \\ x_1 \end{bmatrix}, \ t > 0.$$

Certainly, the Putzer algorithm can be used to compute the matrix exponential **(Try it!)**, but we proceed in a different manner. We begin with Def. 4.3.1 to obtain:

$$\begin{aligned} e^{\begin{bmatrix} 0 & 1 \\ -\omega^2 & 0 \end{bmatrix} t} &= \sum_{n=0}^{\infty} \begin{bmatrix} 0 & 1 \\ -\omega^2 & 0 \end{bmatrix}^n \frac{t^n}{n!} \\ &= \mathbf{I} + \begin{bmatrix} 0 & 1 \\ -\omega^2 & 0 \end{bmatrix} t - \omega^2 \mathbf{I} \frac{t^2}{2!} - \omega^2 \begin{bmatrix} 0 & 1 \\ -\omega^2 & 0 \end{bmatrix} \frac{t^3}{3!} \\ &\quad + \omega^4 \mathbf{I} \frac{t^4}{4!} + \omega^4 \begin{bmatrix} 0 & 1 \\ -\omega^2 & 0 \end{bmatrix} \frac{t^5}{5!} + \ldots \end{aligned}$$

Continuing in this manner definitely suggests a pattern. The fact that we can rearrange the order of the terms of an absolutely convergent series enables us to continue this string of equalities to obtain

$$
\begin{aligned}
&= \sum_{n=0}^{\infty}(-1)^n \omega^{2n} \frac{t^{2n}}{(2n)!}\mathbf{I} + \sum_{n=0}^{\infty}(-1)^n \omega^{2n} \frac{t^{2n+1}}{(2n+1)!}\begin{bmatrix} 0 & 1 \\ -\omega^2 & 0 \end{bmatrix} \\
&= \sum_{n=0}^{\infty}(-1)^n \frac{(\omega t)^{2n}}{(2n)!}\mathbf{I} + \sum_{n=0}^{\infty}(-1)^n \frac{(\omega t)^{2n+1}}{\omega(2n+1)!}\begin{bmatrix} 0 & 1 \\ -\omega^2 & 0 \end{bmatrix} \\
&= \cos(\omega t)\begin{bmatrix} 1 & 0 \\ 0 & 1 \end{bmatrix} + \frac{1}{\omega}\sin(\omega t)\begin{bmatrix} 0 & 1 \\ -\omega^2 & 0 \end{bmatrix} \\
&= \begin{bmatrix} \cos(\omega t) & \frac{1}{\omega}\sin(\omega t) \\ -\omega\sin(\omega t) & \cos(\omega t) \end{bmatrix},
\end{aligned}
$$

where the Taylor series for $\cos(\omega t)$ and $\sin(\omega t)$ have been used. Hence, the classical solution of (4.47) can be expressed equivalently as

$$
\begin{bmatrix} x(t) \\ x'(t) \end{bmatrix} = \begin{bmatrix} \cos(\omega t) & \frac{1}{\omega}\sin(\omega t) \\ -\omega\sin(\omega t) & \cos(\omega t) \end{bmatrix}\begin{bmatrix} x_0 \\ x_1 \end{bmatrix}.
$$

Of interest is the first row, namely $x(t) = (\cos(\omega t))\, x_0 + (\sin(\omega t))\left(\frac{x_1}{\omega}\right)$. Note that $x(t)$ is periodic with period $\frac{2\pi}{\omega}$.

A slight modification of the trick of renaming variables works for systems of coupled linear higher-order ODEs, as illustrated below:

Example 4.7.2. Solve the system

$$
\begin{cases}
x''(t) &= 4x(t) + y(t) \\
y'(t) &= y(t) \\
y(1) &= 1 \\
x(1) &= 0 \\
x'(1) &= 0
\end{cases}
\tag{4.48}
$$

The "total degree" of the system is 3 (obtained by adding the orders of both ODEs, which are 2 and 1, respectively). So, three conditions (two for x and one for y) at the same time ($t_0 = 1$) are needed to ensure (4.48) has a unique solution. Rather than introducing new names for the unknowns, we make the observation that the corresponding system must be 3×3. **(Why?)** But, what is the structure of the solution vector of the system? Precisely, what should be the components? Certainly, they should involve only the unknown functions of interest and their derivatives up to the highest degree minus one for each variable (that is, x, x', and y). As such, a logical choice for the solution vector is

$$
\mathbf{U}(t) = \begin{bmatrix} x(t) \\ x'(t) \\ y(t) \end{bmatrix}.
$$

In such case, observe that (4.48) can be written as follows:

$$
\begin{cases}
\mathbf{U}'(t) = \begin{bmatrix} 0 & x'(t) & 0 \\ 4x(t) & 0 & y(t) \\ 0 & 0 & y(t) \end{bmatrix} = \begin{bmatrix} 0 & 1 & 0 \\ 4 & 0 & 1 \\ 0 & 0 & 1 \end{bmatrix}\mathbf{U}(t) \\
\mathbf{U}(1) = \begin{bmatrix} 0 \\ 0 \\ 12 \end{bmatrix}
\end{cases}
\tag{4.49}
$$

This time, the functions that constitute the solution of the original IVP are x and y, which correspond to the first and third components of $\mathbf{U}(t)$.

STOP! Solve (4.49) to show that

$$\begin{aligned} x(t) &= -4e^{(t-1)} + 3e^{2(t-1)} + e^{-2(t-1)} \\ y(t) &= 12e^{(t-1)} \end{aligned}$$

Remark 4.7.1. The GUI MATLAB_Higher_Order_ODE may be used to *aid* in the solution process for Exercises 4.7.1-4.7.3. However, MATLAB is not required to solve any of these exercises.

Exercise 4.7.1. Consider the following IVP:

$$\begin{cases} 4y''(t) - y(t) = 0 \\ y(0) = 2 \\ y'(0) = \beta \end{cases} \tag{4.50}$$

i.) Solve (4.50).

ii.) Determine β so that $\lim_{t \longrightarrow \infty} y(t) = 0$.

Exercise 4.7.2. Consider the ODE

$$y''(t) + (3-\alpha)y'(t) - 2(\alpha - 1)y(t) = 0. \tag{4.51}$$

i.) Determine the values of α, if any, for which all solutions (corresponding to any set of initial conditions) of (4.51) tend to zero as $t \longrightarrow \infty$.

ii.) Determine the values of α, if any, for which the nonzero solutions $y(t)$ of (4.51) are unbounded. That is, there does not exist $M > 0$ such that $|y(t)| \leq M$, for all t.

Exercise 4.7.3. Consider the following IVP, where $\beta > 0$:

$$\begin{cases} y''(t) + 5y'(t) + 6y(t) = 0 \\ y(0) = 2 \\ y'(0) = \beta \end{cases} \tag{4.52}$$

i.) Solve (4.52).

ii.) For a fixed value of $\beta > 0$, determine the coordinates of the point $(t^\star, y^\star)$at which the maximum of the solution curve $y = y(t)$ occurs. (The coordinates can depend on β.)

iii.) Determine the smallest value of β for which the y-coordinate of the maximum point in (i) is greater than or equal to 4.

4.8 A Perturbed (HCP)

The act of measuring parameters (e.g., rate constants, concentrations, mass, initial data, etc.) is inevitably imprecise because it is done at the mercy of imperfect environmental conditions, faulty equipment, and human error and inconsistency. The continuity properties of mappings involving the matrix exponential are useful when dealing with small pertubations in such measurements. We consider various scenarios in this section.

We begin our discussion with the first-order real-valued IVP

$$\begin{cases} u'(t) = au(t), \; t > t_0 \\ u(t_0) = u_0 \end{cases} \tag{4.53}$$

The solution of (4.53) is $u(t) = e^{a(t-t_0)}u_0$. Now, suppose that due to measurement error, the *actual* value of the initial state is $u_0 + \epsilon$, where ϵ is a real number. The solution, denoted by u_ϵ, of the IVP obtained by replacing u_0 in (4.53) by $u_0 + \epsilon$ is $u_\epsilon(t) = e^{a(t-t_0)}(u_0 + \epsilon)$. The question of interest is, "How far apart are these solutions at a given time t?"

Graphically, we measure this by computing the vertical distance between the graphs of $y = u(t)$ and $y = u_\epsilon(t)$ at all times t in an interval on which we are studying the solutions. Since the outputs are real numbers, we measure this distance using the absolute value of the difference in the outputs:

$$|u(t) - u_\epsilon(t)| = \left| e^{a(t-t_0)}u_0 - e^{a(t-t_0)}(u_0 + \epsilon) \right| = \left| e^{a(t-t_0)}\epsilon \right| = |\epsilon| \, e^{a(t-t_0)}. \tag{4.54}$$

If $|\epsilon|$ is small, the difference in (4.54) is also small. **(Why?)** Moreover, as $|\epsilon| \longrightarrow 0$, it is the case that $|u(t) - u_\epsilon(t)| \longrightarrow 0$. **(Why?)** This suggests a continuity result with respect to the initial data and prompts us to say that the solution of (4.53) *depends continuously on the initial data* .

STOP!

i.) What is the largest distance separating these functions on the time interval $[t_0, t_0 + T]$? We denote this by $\max_{t_0 \le t \le t_0 + T} |u(t) - u_\epsilon(t)|$. (That is, find the value of t in this interval that makes the difference in (4.54) the largest possible.)

ii.) Describe what happens to this maximum value of (4.54) on $[t_0, t_0 + T]$ as $|\epsilon| \longrightarrow 0$.

We would like to use the above ideas to study the same question for a system of the form (HCP). Complete the following **EXPLORE!** project as motivation for our more general discussion. If you feel confused or are wondering why we care about this refer back to MATLAB-Exercise 3.2.1 where you have already studied the continuous dependence issue in context.

EXPLORE!

Does (HCP) Depend Continuously on the Initial Data?

Consider the IVP

$$\begin{cases} \mathbf{U}'(t) = \begin{bmatrix} \alpha & 0 \\ 0 & \beta \end{bmatrix} \mathbf{U}(t) \\ \mathbf{U}(0) = \begin{bmatrix} w_1 \\ w_2 \end{bmatrix} \end{cases} \tag{4.55}$$

where $\alpha, \beta \in \mathbb{R}$.

i.) What is the solution $\mathbf{U}$ of (4.55)? Give the actual components.

ii.) Suppose that the measurements of the first component of the initial condition vector in (4.55) are subject to error, so that the *actual* initial state vector is $\begin{bmatrix} w_1 + \epsilon \\ w_2 \end{bmatrix}$, where $|\epsilon|$ is a small positive number. Consider the corresponding IVP

$$\begin{cases} \mathbf{U}_\epsilon{}'(t) &= \begin{bmatrix} \alpha & 0 \\ 0 & \beta \end{bmatrix} \mathbf{U}_\epsilon(t) \\ \mathbf{U}_\epsilon(0) &= \begin{bmatrix} w_1 + \epsilon \\ w_2 \end{bmatrix} \end{cases} \tag{4.56}$$

a.) What is the solution $\mathbf{U}_\epsilon$ of (4.56)? Give the actual components.

b.) For definiteness, assume that $\alpha = 1, \beta = 2, w_1 = -2$ and $w_2 = 3$.

i.) Use **MATLAB** to construct the graphs of the component curve of the solution of (4.56) for the following values of ϵ: 1, 0.5, 0.1, 0.01, and 0.001. To begin, open **MATLAB** and in the command line, type:

MATLAB_Linear_ACP_Solver

Make certain to click on the 'Help' button for more details and instructions on how to use the GUI to investigate continuous dependence on the initial data. A screenshot from this GUI can found in Figure 4.5.

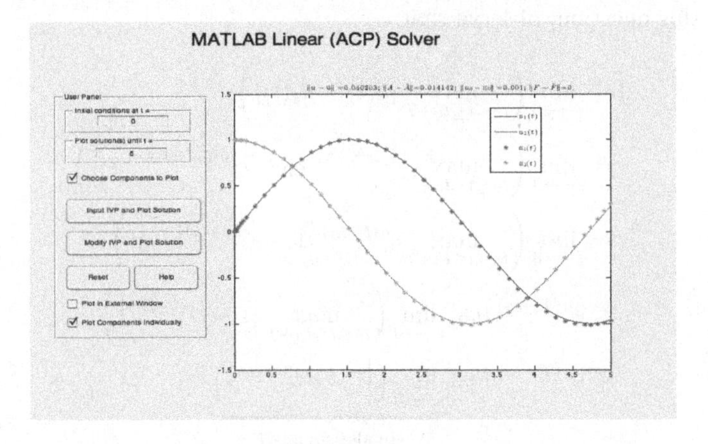

FIGURE 4.5: MATLAB Linear ACP Solver GUI

ii.) What seems to be happening to the graphs for various values of t as $|\epsilon| \longrightarrow 0$?

c.) Compute $\|\mathbf{U}(t) - \mathbf{U}_\epsilon(t)\|_{\mathbb{R}^2}$. What happens as $|\epsilon| \longrightarrow 0$ for a fixed t?

d.) Now, calculate $\max_{0 \le t \le T} \|\mathbf{U}(t) - \mathbf{U}_\epsilon(t)\|_{\mathbb{R}^2}$. (You should be able to get an exact value for any fixed T.) What happens to this value as $|\epsilon| \longrightarrow 0$?

iii.) Now, suppose that both measurements of the initial condition vector in (4.55) are subject to error, so that the actual state vector is $\begin{bmatrix} w_1 + \epsilon_1 \\ w_2 + \epsilon_2 \end{bmatrix}$, where $|\epsilon_1|$ and $|\epsilon_2|$ are small positive numbers. Replace the initial vector in (4.56) by this one and repeat parts (i) - (iv).

iv.) Generalize your findings to the case of an $N \times N$ diagonal matrix for $\mathbf{A}$ and an initial condition vector in which each of its N components is subject to error.

v.) Do you expect the same conclusion to hold when **A** is no longer a diagonal matrix? Explain.

vi.) Carefully formulate the most general result that you can muster regarding the continuous dependence of solutions of (HCP) on the initial data.

Disclaimer! The following is a more formal discussion of the ideas in the previous **EXPLORE!**.

Such imprecision is present when measuring any parameter appearing in a differential equation, not just the initial condition. For simplicity, we consider (4.53) together with the following "perturbed IVP" in which the parameter a is replaced by $(a+\epsilon)$, where ϵ is a real number:

$$\begin{cases} u_\epsilon'(t) = (a+\epsilon)u_\epsilon(t), \, t > t_0 \\ u_\epsilon\,(t_0) = u_0 \end{cases} \tag{4.57}$$

For the moment, we have kept the same initial condition in both IVPs. The solution of (4.53) is $u(t) = e^{a(t-t_0)}u_0$ and the solution of (4.57) is given by $u_\epsilon(t) = e^{(a+\epsilon)(t-t_0)}u_0$. Again, we ask the question, " What happens as $|\epsilon| \longrightarrow 0$ to $\max_{t_0 \le t \le t_0+T} |u(t) - u_\epsilon(t)|$?"

To answer this question, observe that

$$\begin{aligned}
0 &\le \lim_{\epsilon \longrightarrow 0} \left(\max_{t_0 \le t \le t_0+T} |u(t) - u_\epsilon(t)| \right) \\
&= \lim_{\epsilon \longrightarrow 0} \left(\max_{t_0 \le t \le t_0+T} \left| e^{a(t-t_0)}u_0 - e^{(a+\epsilon)(t-t_0)}u_0 \right| \right) \\
&= \lim_{\epsilon \longrightarrow 0} \left(\max_{t_0 \le t \le t_0+T} e^{a(t-t_0)} \left| 1 - e^{\epsilon(t-t_0)} \right| |u_0| \right) \\
&\le e^{a(T-t_0)} |u_0| \lim_{\epsilon \longrightarrow 0} \left(\max_{t_0 \le t \le t_0+T} \left| 1 - e^{\epsilon(t-t_0)} \right| \right) \\
&\le e^{a(T-t_0)} |u_0| \underbrace{\lim_{\epsilon \longrightarrow 0} \left(\left| 1 - e^{\epsilon(T-t_0)} \right| \right)}_{\text{equals 0 for any } T} \\
&= 0.
\end{aligned} \tag{4.58}$$

STOP! Verify each line of (4.58) before moving onward!

Exercise 4.8.1. Suppose that we now incorporate a perturbation on both a and the initial condition u_0. Precisely, consider (4.53) together with the IVP

$$\begin{cases} u_{\epsilon\delta}'(t) = (a+\epsilon)u_{\epsilon\delta}(t), \, t > t_0 \\ u_{\epsilon\delta}\,(t_0) = u_0 + \delta \end{cases} \tag{4.59}$$

where δ is a real number. Compute $\lim_{\delta \longrightarrow 0} \lim_{\epsilon \longrightarrow 0} \left(\max_{t_0 \le t \le t_0+T} |u(t) - u_{\epsilon\delta}(t)| \right)$. Interpret this result in words.

Using the above discussion as motivation, let us consider the same question for homogenous linear systems of ODEs. For simplicity, we consider a 2×2 linear system, although the discussion applies for any such $N \times N$ system. Suppose that a certain phenomenon is described by the IVP

$$\begin{cases} x'(t) = \alpha x(t) + \beta y(t), \\ y'(t) = \overline{\alpha} x(t) + \overline{\beta} y(t), \\ x(0) = x_0, \ y(0) = y_0. \end{cases} \tag{4.60}$$

In reality, due to measurement error, a more precise description would be

$$\begin{cases} x'(t) = (\alpha + \epsilon_1)\, x(t) + (\beta + \epsilon_2)\, y(t), \\ y'(t) = (\overline{\alpha} + \overline{\epsilon_1})\, x(t) + (\overline{\beta} + \overline{\epsilon_2})\, y(t), \\ x(0) = x_0, \ y(0) = y_0. \end{cases} \tag{4.61}$$

where $\epsilon_i, \overline{\epsilon_i} \in \mathbb{R}\, (i = 1, 2)$. We refer to (4.61) as a "perturbed (HCP)." Letting $\mathbf{A} = \begin{bmatrix} \alpha & \beta \\ \overline{\alpha} & \overline{\beta} \end{bmatrix}$ and $\mathbf{B}_\epsilon = \begin{bmatrix} \epsilon_1 & \epsilon_2 \\ \overline{\epsilon_1} & \overline{\epsilon_2} \end{bmatrix}$, we can rewrite (4.60) and (4.61), respectively, as

$$\begin{cases} \mathbf{U}'(t) = \mathbf{A}\mathbf{U}(t), \ t > 0, \\ \mathbf{U}(0) = \mathbf{U}_0, \end{cases} \tag{4.62}$$

and

$$\begin{cases} \mathbf{U}_\epsilon'(t) = (\mathbf{A} + \mathbf{B}_\epsilon)\, \mathbf{U}_\epsilon(t), \ t > 0, \\ \mathbf{U}_\epsilon(0) = \mathbf{U}_0. \end{cases} \tag{4.63}$$

STOP! Tell why carefully.

Since $\mathbf{A} + \mathbf{B}_\epsilon \in \mathbb{M}^2(\mathbb{R})$, the classical solution of (4.63) is $\mathbf{U}_\epsilon(t) = e^{(\mathbf{A}+\mathbf{B}_\epsilon)t}\mathbf{U}_0$. The question we asked in the one-dimensional setting translates to computing the following limit:

$$\lim_{\epsilon \longrightarrow 0} \left(\max_{0 \le t \le T} \|\mathbf{U}(t) - \mathbf{U}_\epsilon(t)\|_{\mathbb{R}^2} \right). \tag{4.64}$$

In words, what happens to the largest distance between the solutions of (4.62) and (4.63) on the interval $[0, T]$ as the size of the error shrinks to zero? It is natural to mimic the approach used above as closely as follows. To this end, observe that for every $t > 0$,

$$\|\mathbf{U}_\epsilon(t) - \mathbf{U}(t)\|_{\mathbb{R}^2} \le \left\| e^{(\mathbf{A}+\mathbf{B}_\epsilon)t} - e^{\mathbf{A}t} \right\|_{\mathbb{M}^2} \|\mathbf{U}_0\|_{\mathbb{R}^2}. \tag{4.65}$$

STOP! Verify this inequality.

We would like to further say that

$$\left\| e^{(\mathbf{A}+\mathbf{B}_\epsilon)t} - e^{\mathbf{A}t} \right\|_{\mathbb{M}^2} = \left\| e^{\mathbf{A}t} e^{\mathbf{B}_\epsilon t} - e^{\mathbf{A}t} \right\|_{\mathbb{M}^2} = \left\| e^{\mathbf{A}t} \left(e^{\mathbf{B}_\epsilon t} - \mathbf{I} \right) \right\|_{\mathbb{M}^2} \tag{4.66}$$

so that as $\mathbf{B}_\epsilon$ approaches the zero matrix $\mathbf{0}$, the norms in (4.66) also approach zero independently of t, thereby ensuring that the limit in (4.64) is zero. But, this line of reasoning hinges on the equality $e^{(\mathbf{A}+\mathbf{B}_\epsilon)t} = e^{\mathbf{A}t}e^{\mathbf{B}_\epsilon t}$ used in (4.66). **(Why?)** This certainly seems innocent enough, but does this hold, in general?

Exercise 4.8.2. Answer the following

i.) Let $T > 0$. Show that if $e^{(\mathbf{A}+\mathbf{B}_\epsilon)t} = e^{\mathbf{A}t}e^{\mathbf{B}_\epsilon t}$, for all $0 \le t \le T$, then

$$\lim_{\epsilon_i,\overline{\epsilon_i}\longrightarrow 0}\left(\max_{0\le t\le T}\|\mathbf{U}(t)-\mathbf{U}_\epsilon(t)\|_{\mathbb{R}^2}\right)=0. \tag{4.67}$$

ii.) Now, does the equality in (i) ever hold? Yes. For instance, prove that if $\mathbf{A}$ and $\mathbf{B}$ are 2×2 diagonal matrices, then $e^{\mathbf{A}+\mathbf{B}} = e^{\mathbf{A}}e^{\mathbf{B}}$. (This actually is true for any $N \times N$ diagonal matrices.)

iii.) Now, here is the bad news. Let $\mathbf{A} = \begin{bmatrix} 2 & 1 \\ 0 & 2 \end{bmatrix}$ and $\mathbf{B} = \begin{bmatrix} \frac{1}{2} & 0 \\ 0 & \frac{1}{3} \end{bmatrix}$. Show that $\mathbf{AB} \neq \mathbf{BA}$ and that $e^{\mathbf{A}}e^{\mathbf{B}} \neq e^{\mathbf{A}+\mathbf{B}}$. Hence, the desired equality (4.66) does not always hold.

The lack of commutativity of the matrices $\mathbf{A}$ and $\mathbf{B}$ presents an obstacle when trying to verify (4.66), even for a reasonably tame perturbation. However, if this hurdle is removed, then we can verify (4.66), as suggested by the following exercise.

Exercise 4.8.3.

i.) Prove that if $\mathbf{AB} = \mathbf{BA}$, then $e^{\mathbf{B}}e^{\mathbf{A}} = e^{\mathbf{A}}e^{\mathbf{B}} = e^{\mathbf{A}+\mathbf{B}}$.

ii.) Explain how (i) helps when verifying that (4.67).

As such, it is sufficient to assume that $\mathbf{A}$ commutes with all matrices $\mathbf{B}_\epsilon$ to ensure (4.67) holds. This is a very restrictive assumption that renders the theorem, while true, impractical.

Exercise 4.8.4. Suppose that the initial conditions are also perturbed, leading to the following more general version of (4.61):

$$\begin{cases} x'(t) = (\alpha+\epsilon_1)\,x(t) + (\beta+\epsilon_2)\,y(t), \\ y'(t) = (\overline{\alpha}+\overline{\epsilon_1})\,x(t) + (\overline{\beta}+\overline{\epsilon_2})\,y(t), \\ x(0) = x_0+\delta_1,\ y(0) = y_0+\delta_2. \end{cases} \tag{4.68}$$

Formulate a continuous dependence result for (4.68) in the spirit of the discussion given in this section.

APPLY IT!!

At long last, we are equipped with enough theoretical power to address many of the models introduced in Chapter 3. For each of the following IVPs, complete these:

I.) Reformulate the IVP in the abstract form (HCP).

II.) Argue that the IVP has a unique solution.

III.) Formulate and prove a continuous dependence result for the IVP in the spirit of the results established in this section.

IV.) Choose actual values for the parameters in the model. For them, illustrate the curves of each of the component functions using the **MATLAB_Linear_ACP_Solver** GUI and analyze the results in the context of the setting of each model.

i.) IVP (3.5)

$$\begin{cases} \frac{dT_R(t)}{dt} = k_1T_R(t) + k_2T_P(t) - k_3T_m(t) \\ \frac{dT_P(t)}{dt} = k_4T_R(t) + k_5T_P(t) - k_6T_m(t) \\ \frac{dT_m(t)}{dt} = k_7T_m(t) \\ T_R(0) = T_0, T_P(t) = T_1, T_m(0) = T_2 \end{cases}$$

ii.) IVP (3.6)

$$\begin{cases} \frac{dy(t)}{dt} = -ay(t) \\ \frac{dz(t)}{dt} = ay(t) - bz(t) \\ y(0) = y_0, z(0) = 0 \end{cases}$$

iii.) IVP (3.7)

$$\begin{cases} \frac{du_{ST}(t)}{dt} &= a_{(H,ST)}u_H(t) - a_{(ST,H)}u_{ST}(t) \\ \frac{du_H(t)}{dt} &= a_{(ST,H)}u_{ST}(t) + a_{(L,H)}u_L(t) - \left[a_{(H,ST)} + a_{(H,E)} + a_{(H,L)} + a_{(H,P)}\right]u_H(t) \\ \frac{du_L(t)}{dt} &= \left[a_{(P,L)}u_P(t) + a_{(H,L)}u_H(t)\right] - a_{(L,H)}u_L(t) \\ \frac{du_P(t)}{dt} &= a_{(H,P)}u_H(t) - a_{(P,L)}u_P(t) \\ \frac{du_E(t)}{dt} &= a_{(H,E)}u_H(t) - Lu_E(t) \\ u_{ST}(0) &= u_1, u_H(0) = u_2, u_L(0) = u_3, u_P(0) = u_4, u_E(0) = u_5 \end{cases}$$

iv.) IVP (3.19)

$$\begin{cases} m_1\frac{d^2x_1(t)}{dt^2} = -k_1x_1(t) + k_2(x_2(t) - x_1(t)) \\ m_2\frac{d^2x_2(t)}{dt^2} = -k_3x_2(t) - k_2(x_2(t) - x_1(t)) \\ x_1(0) = a_0; x_1'(0) = a_1; x_2(0) = b_0; x_2'(0) = b_1 \end{cases}$$

v.) IVP (3.33)

$$\begin{cases} \frac{d[A]}{dt} &= -k_1[A] + k_2[B] \\ \frac{d[B]}{dt} &= k_1[A] - k_2[B] \\ [A](0) &= A_0 \\ [B](0) &= B_0 \end{cases}$$

4.9 What Happens to Solutions of (HCP) as Time Goes On and On and On...?

We know that $\mathbf{U}(t) = e^{\mathbf{A}t}\mathbf{U}_0$ is the unique classical solution of (4.1) and is defined for all $t > 0$. But what happens to this solution as $t \to \infty$?

In the real-valued case of IVP (4.53), this question is easily answered:

1. If $a < 0$, then $\lim_{t\longrightarrow\infty} e^{at}u_0 = u_0 \cdot \lim_{t\longrightarrow\infty} e^{at} = 0$, for any value of u_0.
2. If $a = 0$, then $\lim_{t\longrightarrow\infty} e^{at}u_0 = u_0 \cdot \lim_{t\longrightarrow\infty} 1 = u_0$, for any value of u_0.

3. If $a > 0$, then $\lim_{t\longrightarrow\infty} e^{at}u_0 = u_0 \cdot \lim_{t\longrightarrow\infty} e^{at} = \begin{cases} \infty, & \text{if } u_0 > 0 \\ 0, & \text{if } u_0 = 0 \\ -\infty, & \text{if } u_0 < 0 \end{cases}$

These results constitute the *long-term behavior* of the solution of (4.53). These are the only situations that can occur for (4.53). Further, it is important to observe that the long-term behavior depends mainly on the value of a. The most "stable" case is when $a < 0$ because the solutions go to zero independent of the value of u_0. So, minor tweaks in the value of the initial condition will not have any effect on the long-term behavior for such values of a. This is not true when $a = 0$. Indeed, the limit depends heavily on the value of u_0. Worse yet is the behavior when $a > 0$ because unless u_0 is itself 0, the solutions "blow up" as time goes on, meaning that the graph of the solution curve is not bounded.

We want to study the same questions more generally for (HCP). The complication, of course, is that now $\mathbf{A}$ is a matrix, not a real number. So, the conditions on $\mathbf{A}$ that determine the nature of the long-term behavior of the solution are not simply "$\mathbf{A} > 0$" or "$\mathbf{A} < 0$." Such inequalities do not even make sense. Nevertheless, instinctually there should be a link between the long-term behavior of solutions of (HCP) and the nature of the matrix $\mathbf{A}$. But, what precisely is this link? The following **EXPLORE!** project sheds some light on this question.

EXPLORE!

How Do We Classify Long-Term Behavior for (HCP)?

We shall consider some very specific types of matrices and explore what can happen in terms of long-term behavior for the corresponding IVPs formed using them.

1. Let $\mathbf{A} = \begin{bmatrix} \alpha & 0 \\ 0 & \beta \end{bmatrix}$, where α and β are real numbers. Answer the following questions for each description of α and β that follows:

   I.) Let $t \geq 0$ be fixed. Compute $e^{\mathbf{A}t}$ and determine an upper bound for $\left\|e^{\mathbf{A}t}\right\|_{\mathbb{M}^2}$.
   II.) Let $\mathbf{x} \in \mathbb{R}^2$. Does $\lim_{t\longrightarrow\infty} \left\|e^{\mathbf{A}t}\mathbf{x}\right\|_{\mathbb{R}^2}$? Does the answer depend on the value of $\mathbf{x}$?
   III.) Does there exist $\mathbf{U}^\star \in \mathbb{R}^2$ for which $\lim_{t\to\infty} \|\mathbf{U}(t) - \mathbf{U}^\star\|_{\mathbb{R}^2} = 0$?

   i.) $\alpha, \beta < 0$
   ii.) $\alpha, \beta > 0$
   iii.) Exactly one of α and β is equal to zero and the other is negative.
   iv.) Exactly one of α and β is equal to zero and the other is positive.
   v.) $\alpha < 0 < \beta$

   For questions 2 and 3, you may choose to use MATLAB_Linear_ACP_Solver GUI to gain intuition regarding the problems. However, **MATLAB** is not required to answer the following questions. Look back at Example 4.4.3 and Exercise 4.4.1 before completing the next two questions.

2. Let $\mathbf{B} = \begin{bmatrix} 0 & \beta \\ -\beta & 0 \end{bmatrix}$, where $\beta \neq 0$. Compute $e^{\mathbf{B}t}$ and describe the long-term behavior of $e^{\mathbf{B}t}\mathbf{U}_0$, for any $\mathbf{U}_0 \in \mathbb{R}^2$.

3. More generally, let $\mathbf{C} = \begin{bmatrix} \alpha & \beta \\ -\beta & \alpha \end{bmatrix}$, where $\alpha, \beta \neq 0$.

i.) Compute $e^{\mathbf{C}t}$ and answer questions I - III from (1) under the assumption that $\alpha < 0$.

ii.) Redo (i) assuming that $\alpha > 0$.

The eigenvalues of the matrix $\mathbf{A}$ significantly impact the structure of $e^{\mathbf{A}t}$. It is known that every member of $\mathbb{M}^N(\mathbb{R})$ has N complex eigenvalues, including multiplicity. And, the matrix $\mathbf{A}$ can be written in a nice standard form using these eigenvalues. As such, it is reasonable to expect there to be a connection between the eigenvalues of $\mathbf{A}$ and the long-term behavior of $\left\{e^{\mathbf{A}t} : t \geq 0\right\}$.

Before launching this investigation, we state some definitions.

Definition 4.9.1. Let $\mathbf{A} \in \mathbb{M}^N(\mathbb{R})$ and let $\mathbf{U}$ be the solution of (HCP).

i.) $\mathbf{U}(t)$ *is globally bounded* if there exists $M > 0$ such that $\|\mathbf{U}(t)\|_{\mathbb{R}^2} \leq M$, for every $t > 0$.

ii.) $\mathbf{U}(t)$ *blows-up in finite time* if there exists $T^\star > 0$ such that

$$\lim_{t \to (T^\star)^-} \|\mathbf{U}(t)\|_{\mathbb{R}^2} = \infty.$$

iii.) $\mathbf{U}(t)$ *is time-periodic* if there exists $p > 0$ such that $\mathbf{U}(t+p) = \mathbf{U}(t)$, for every $t > 0$.

iv.) $\left\{e^{\mathbf{A}t} : t \geq 0\right\}$ is *uniformly stable* if $\lim_{t \to \infty} \left\|e^{\mathbf{A}t}\right\|_{\mathbb{M}^N} = 0$;

v.) $\left\{e^{\mathbf{A}t} : t \geq 0\right\}$ is *exponentially stable* if there exists $M \in \mathbb{N}$ and $\alpha > 0$ such that $\left\|e^{\mathbf{A}t}\right\|_{\mathbb{M}^N} \leq Me^{-\alpha t}$, for all $t \geq 0$;

vi.) $\left\{e^{\mathbf{A}t} : t \geq 0\right\}$ is *strongly stable* if $\lim_{t \to \infty} \left\|e^{\mathbf{A}t}\mathbf{z}\right\|_{\mathbb{R}^N} = 0$, for every $\mathbf{z} \in \mathbb{R}^N$.

STOP! Interpret each definition in Definition 4.9.1 in words. Give examples from the **EXPLORE!** project that illustrate each of them.

Exercise 4.9.1. Consider the matrix $\mathbf{A} = \begin{bmatrix} \alpha & 0 \\ 0 & \beta \end{bmatrix}$, whose eigenvalues are α and β. For each case listed in the **EXPLORE!** project, associate the nature of the eigenvalues with the long-term behavior of the classical solution $\mathbf{U}(t)$ of (4.1). (For instance, if α, $\beta < 0$, then the eigenvalues of $\mathbf{A}$ are both negative and in such case, $\lim_{t \to \infty} \|\mathbf{U}(t)\|_{\mathbb{R}^2} = 0$.)

The nature of the eigenvalues characterize the long-term behavior of $\left\{e^{\mathbf{A}t} : t \geq 0\right\}$, and in turn the classical solution of (4.1). Of course, we have not proven this. We have merely affirmed its truth for particular types of matrices. What if two different matrices $\mathbf{A}$ and $\mathbf{B}$ have precisely the same eigenvalues; must $\left\{e^{\mathbf{A}t} : t \geq 0\right\}$ and $\left\{e^{\mathbf{B}t} : t \geq 0\right\}$ behave in the same manner? The answer is yes, but this requires proof. In light of this discussion, the following representation theorem is very useful.

Proposition 4.9.1. Let $\mathbf{A} \in \mathbb{M}^N(\mathbb{R})$ with eigenvalues $\lambda_1, \ldots, \lambda_N$ (including multiplicity) and suppose that $\{\mathbf{e}_k : k = 1, \ldots, N\}$ is a basis for $\mathbb{R}^N$. If $\mathbf{Az} = \sum_{k=1}^N \lambda_k \langle \mathbf{z}, \mathbf{e}_k \rangle_{\mathbb{R}^N} \mathbf{e}_k$, $\forall \mathbf{z} \in \mathbb{R}^N$, then $e^{\mathbf{A}t}\mathbf{z} = \sum_{k=1}^N e^{\lambda_k t} \langle \mathbf{z}, \mathbf{e}_k \rangle_{\mathbb{R}^N} \mathbf{e}_k$.

This suggests that knowing the structure of $\mathbf{A}$ enables us to compute $e^{\mathbf{A}t}\mathbf{z}$, $\forall \mathbf{z} \in \mathbb{R}^N$ without resorting to Putzer's algorithm. The following example illustrates this result.

Example 4.9.1. Let $\mathbf{A} = \begin{bmatrix} \alpha & 0 \\ 0 & \beta \end{bmatrix}$, whose eigenvalues are α and β. Assume that $\mathbf{e}_1 = \begin{bmatrix} 1 \\ 0 \end{bmatrix}$ and $\mathbf{e}_2 = \begin{bmatrix} 0 \\ 1 \end{bmatrix}$. Let $\mathbf{z} = \begin{bmatrix} z_1 \\ z_2 \end{bmatrix} \in \mathbb{R}^2$ and observe that

$$\mathbf{Az} = \begin{bmatrix} \alpha & 0 \\ 0 & \beta \end{bmatrix} \begin{bmatrix} z_1 \\ z_2 \end{bmatrix} = \begin{bmatrix} \alpha z_1 \\ \beta z_2 \end{bmatrix} = \alpha \langle \mathbf{z}, \mathbf{e}_1 \rangle_{\mathbb{R}^2} \mathbf{e}_1 + \beta \langle \mathbf{z}, \mathbf{e}_2 \rangle_{\mathbb{R}^2} \mathbf{e}_2.$$

Hence,

$$e^{\mathbf{A}t}\mathbf{z} = \begin{bmatrix} e^{\alpha t} & 0 \\ 0 & e^{\beta t} \end{bmatrix} \begin{bmatrix} z_1 \\ z_2 \end{bmatrix} = \begin{bmatrix} e^{\alpha t} z_1 \\ e^{\beta t} z_2 \end{bmatrix} = e^{\alpha t} \langle \mathbf{z}, \mathbf{e}_1 \rangle_{\mathbb{R}^2} \mathbf{e}_1 + e^{\beta t} \langle \mathbf{z}, \mathbf{e}_2 \rangle_{\mathbb{R}^2} \mathbf{e}_2.$$

For certain cases, Prop. 4.9.1 enables us to reduce the question of the stability of $\{e^{\mathbf{A}t} : t \geq 0\}$ to examining the nature of the eigenvalues of $\mathbf{A}$. This is explored in the following exercise.

Exercise 4.9.2. Let $\mathbf{A} \in \mathbb{M}^N(\mathbb{R})$ be diagonal and assume that all eigenvalues $\lambda_1, \ldots, \lambda_N$ of $\mathbf{A}$ are negative real numbers.

i.) Prove that $\{e^{\mathbf{A}t} : t \geq 0\}$ is uniformly stable.

ii.) What happens if at least one eigenvalue is positive? Does a similar conclusion hold if some, but not all, of the eigenvalues are zero and the remaining ones are negative? Explain.

4.10 Looking Ahead

Athough many of the models introduced in Chapter 3 can be written in the abstract form (HCP), there were some that could not. For example, the beer blending IVP (3.13)

$$\begin{cases} \frac{du_1(t)}{dt} & = \frac{-7u_1(t)+2u_2(t)+0.31}{100} \\ \frac{du_2(t)}{dt} & = \frac{7u_1(t)-7u_2(t)}{80} \\ u_1(0) & = 0.062 \\ u_2(0) & = 0.09 \end{cases}$$

is one such example. Other examples from Chapter 3 include the charge on a capacitor model (3.21)

$$\begin{cases} L\frac{d^2q(t)}{dt} + R\frac{dq(t)}{dt} + \frac{1}{C}q(t) = E(t) \\ q(0) = q_0 \\ q'(0) = q_1 \end{cases}$$

and the linear air resistance projectile motion model (3.42)

$$\begin{cases} m\frac{d^2x(t)}{dt^2} + r_x\frac{dx(t)}{dt} & = F_x(t) \\ m\frac{d^2y(t)}{dt^2} + r_y\frac{dy(t)}{dt} & = F_y(t) \\ m\frac{d^2z(t)}{dt^2} + r_x\frac{dz(t)}{dt} & = F_z(t) \end{cases}$$

Exercise 4.10.1. Try to reformulate (3.13), (3.21), and (3.42) in the abstract form (HCP). What new difficulties do you encounter?

Exercise 4.10.2. Propose an additional term in abstract form (HCP) that could be a placeholder to account for the difficulties encountered in the previous exercise.

Exercise 4.10.3. Return to Chapter 3 and identify at least three more models that cannot be written in abstract form (HCP).

Do these models have solutions? Can we find them? To help answer the first question let us revisit a few **MATLAB** exercises to convince ourselves that these non-(HCP) conforming models do, in fact, have solutions.

MATLAB-Exercise 4.10.1. To begin, open **MATLAB** .

1. In the command line of **MATLAB** type:

   MATLAB_Beer_Blending

   Use the GUI to solve (3.13) at several different times.

2. In the command line of **MATLAB** type:

   MATLAB_Simple_Circuit

   Use the GUI to solve (3.21) at several different times with different initial conditions of your choosing. Suppose $L = 10, r = 0.001, C = 0.025$, and $E(t) = 20\cos(t)$.

■

How do we find solutions to non-(HCP) conforming models? Are the models mentioned above just special cases of a new abstract form? Will that new abstract form yield results similar to those in this chapter? Such questions are the focus of our investigation in the next chapter.

Chapter 5

Finite-Dimensional Theory—Next Step: The Non-Homogenous Case

Overview

Many of the models introduced in Chapter 3 illustrate what happens when external forces of various kinds are introduced into the description of phenomena involving a system of linear ODEs or a higher-order linear differential equation. For example, the DEs arising in the beer blending model (3.13), the vertical spring model (3.17), the projectile motion model (3.42), and the 5-story floor displacement model (3.8.1) all involve external forces. As in Chapter 4, our present goal is to develop a single framework that subsumes these IVPs as special cases. The theory developed is a formal extension of the theory formulated in Chapter 4.

5.1 Introducing...The Non-Homogenous Cauchy Problem (Non-CP)

All of the IVPs mentioned in the overview of this section can be rewritten in the following abstract vector form.

Definition 5.1.1. A system of ordinary differential equations of the form

$$(\textbf{Non}-\textbf{CP})\ \begin{cases}\mathbf{U}'(t)=\mathbf{A}\mathbf{U}(t)+\mathbf{f}(t),\ t>t_0,\\ \mathbf{U}(t_0)=\mathbf{U}_0,\end{cases} \tag{5.1}$$

where $\mathbf{U} : [t_0, \infty) \to \mathbb{R}^N$ is the unknown vector function

$$\mathbf{U}(t) = \begin{bmatrix} u_1(t) \\ \vdots \\ u_N(t) \end{bmatrix},$$

$\mathbf{A}$ is the $N \times N$ coefficient matrix

$$\begin{bmatrix} a_{11} & \cdots & a_{1N} \\ \vdots & \ddots & \vdots \\ a_{N1} & \cdots & a_{NN} \end{bmatrix},$$

the forcing term $\mathbf{f} : [t_0, \infty) \to \mathbb{R}^N$ has components

$$\mathbf{f}(t) = \begin{bmatrix} f_1(t) \\ \vdots \\ f_N(t) \end{bmatrix},$$

and $\mathbf{U}_0 \in \mathbb{R}^N$ is the vector containing the initial conditions

$$\mathbf{U}_0 = \begin{bmatrix} u_1(t_0) \\ \vdots \\ u_N(t_0) \end{bmatrix},$$

is a *nonhomogenous linear system of ODEs with initial condition* $\mathbf{U}(t_0) = \mathbf{U}_0$.

Converting a system into the form (Non-CP) is not difficult. Consider the following example:

Example 5.1.1. Consider the system of ODEs given by

$$\begin{cases} u_1'(t) &= -2u_1(t) + 3u_2(t) - 2t \\ u_2'(t) &= -u_1(t) + 2u_2(t) + 5 \\ u_1(1) &= 1 \\ u_2(1) &= -1 \end{cases} \tag{5.2}$$

Initial-value problem (5.2) can be rewritten using matrices as follows:

$$\begin{cases} \underbrace{\begin{bmatrix} u_1(t) \\ u_2(t) \end{bmatrix}'}_{\mathbf{U}'(t)} = \underbrace{\begin{bmatrix} -2 & 3 \\ -1 & 2 \end{bmatrix}}_{\mathbf{A}} \underbrace{\begin{bmatrix} u_1(t) \\ u_2(t) \end{bmatrix}}_{\mathbf{U}(t)} + \underbrace{\begin{bmatrix} -2t \\ 5 \end{bmatrix}}_{\mathbf{f}(t)} \\ \underbrace{\begin{bmatrix} u_1(1) \\ u_2(1) \end{bmatrix}}_{\text{Here } t_0=1} = \underbrace{\begin{bmatrix} 1 \\ -1 \end{bmatrix}}_{\mathbf{U}_0} \end{cases} \tag{5.3}$$

It is reasonable at this stage to think of a solution of (5.3) as a vector $\mathbf{U}(t) = \begin{bmatrix} u_1(t) \\ u_2(t) \end{bmatrix}$ satisfying both the equation and the initial condition in (5.3).

STOP! Revisit the IVPs in Chapter 3 and make certain you are comfortable with reformulating them in the abstract form (Non-CP).

As before, we shall formulate a theory for (5.1) and then recover the results for the particular IVPs as corollaries.

5.2 Carefully Examining the One-Dimensional Version of (Non-CP)

We begin our investigation of (5.1) in $\mathbb{R}$, so that A and U_0 are simply real numbers and $U(\cdot)$ and $f(\cdot)$ are real-valued functions. In such case, (5.1) consists of a single ODE coupled with an IC. Also we assume for simplicity, that $t_0 = 0$.

5.2.1 Solving (Non-CP)—Calculus Based

We wish to derive a representation formula for a solution of (5.1). We momentarily hold off on imposing specific hypotheses on $f(\cdot)$ and tacitly assume whatever is needed to ensure that all steps of the derivation are justified and all quantities involved are well-defined. As you work through the derivation, keep track of these conditions in preparation for formulating an existence result. Starting with (5.1), we proceed as follows:

$$
\begin{aligned}
U'(s) - AU(s) &= f(s) \\
e^{-As}\left(U'(s) - AU(s)\right) &= e^{-As}f(s) \\
\frac{d}{ds}\left(e^{-As}U(s)\right) &= e^{-As}f(s) \qquad (5.4) \\
\int_0^t \frac{d}{ds}\left(e^{-As}U(s)\right) ds &= \int_0^t e^{-As}f(s)ds \\
e^{-At}U(t) - e^{-A(0)}U(0) &= \int_0^t e^{-As}f(s)ds \\
e^{-At}U(t) &= U_0 + \int_0^t e^{-As}f(s)ds \\
U(t) &= e^{At}U_0 + e^{At}\int_0^t e^{-As}f(s)ds \\
U(t) &= e^{At}U_0 + \int_0^t e^{A(t-s)}f(s)ds \qquad (5.5)
\end{aligned}
$$

Formula (5.5) is called a *variation of parameters* formula for a solution of (5.1).

STOP! Address these three questions.

1. Carefully explain the modifications that would need to be implemented in (5.23) if a general t_0 (not simply 0) was used as the initial time.
2. What assumptions should be imposed on the forcing function $f(t)$ to warrant that all of the steps in the derivation (5.23) are meaningful?
3. Conjecture a single reasonable assumption that can be imposed on $f(t)$ to ensure that all steps of (5.23) are meaningful. While you could assume, at the extreme end of the spectrum, that f is infinitely differentiable (so that it is as smooth as possible), try to impose a condition that is just strong enough.

5.2.2 Solving (Non-CP)—Numerics Based

We derive the variation of parameters formula (5.5) using a technique similar to the one presented in Section 4.3.2. There, the alternative definition of the matrix exponential was motivated by making a discrete approximation to the homogeneous initial-value problem. We follow a similar numerical approximation approach by approximating (5.1), solving the approximate IVP, and then taking a limit of the approximate solution the solution as $n \to \infty$ to obtain the solution (5.1).

STOP! Re-read Section 4.3.2.

Let $t > 0$, $n \in \mathbb{N}$, and define $\bar{\mathrm{U}}$ to be the solution of the approximate IVP

$$\begin{cases} \frac{\bar{\mathrm{U}}(s+\frac{t}{n})-\bar{\mathrm{U}}(s)}{\frac{t}{n}} & = \mathrm{A}\bar{\mathrm{U}}(s+\frac{t}{n}) + \mathrm{f}(s), \quad s \geq 0 \\ \bar{\mathrm{U}}(0) & = \mathrm{U}_0 \end{cases} \tag{5.6}$$

Solving the first equation in (5.6) for $\bar{\mathrm{U}}(s+\frac{t}{n})$ yields

$$\bar{\mathrm{U}}\left(s+\frac{t}{n}\right) = \left(1-\frac{t}{n}\mathrm{A}\right)^{-1}\bar{\mathrm{U}}(s) + \left(1-\frac{t}{n}\mathrm{A}\right)^{-1}\mathrm{f}(s)\frac{t}{n} \tag{5.7}$$

Taking the first step in the march towards t amounts to evaluating (5.7) at $s = 0$ and realizing that

$$\bar{\mathrm{U}}\left(\frac{t}{n}\right) = \left(1-\frac{t}{n}\mathrm{A}\right)^{-1}\bar{\mathrm{U}}_0 + \left(1-\frac{t}{n}\mathrm{A}\right)^{-1}\mathrm{f}(0)\frac{t}{n}$$

As in Section 4.3.2, the second step requires us to evaluate (5.7) at $s = \frac{t}{n}$:

$$\begin{aligned} \bar{\mathrm{U}}\left(\frac{2t}{n}\right) &= \left(1-\frac{t}{n}\mathrm{A}\right)^{-1}\bar{\mathrm{U}}\left(\frac{t}{n}\right) + \left(1-\frac{t}{n}\mathrm{A}\right)^{-1}\mathrm{f}\left(\frac{t}{n}\right)\frac{t}{n} \\ &= \left(1-\frac{t}{n}\mathrm{A}\right)^{-2}\bar{\mathrm{U}}_0 + \left(\left(1-\frac{t}{n}\mathrm{A}\right)^{-2}\mathrm{f}(0) + \left(1-\frac{t}{n}\mathrm{A}\right)^{-1}\mathrm{f}\left(\frac{t}{n}\right)\right)\frac{t}{n} \end{aligned} \tag{5.8}$$

Upon closer inspection of (5.8), notice the first term is exactly the same as in the homogenous case and the second term involves a sum for which each term is multiplied by $\frac{t}{n}$.

STOP! Can you think of a mathematical object or concept that involves a sum where each term is multiplied by a number divided n?

To help identify the pattern, let us write out the next step in the march towards t. That is, evaluate (5.7) at $s = \frac{2t}{n}$ to get

$$\begin{aligned} \bar{\mathrm{U}}\left(\frac{3t}{n}\right) &= \left(1-\frac{t}{n}\mathrm{A}\right)^{-3}\mathrm{U}_0 + \\ &\left(\left(1-\frac{t}{n}\mathrm{A}\right)^{-3}\mathrm{f}(0) + \left(1-\frac{t}{n}\mathrm{A}\right)^{-2}\mathrm{f}\left(\frac{t}{n}\right) + \left(1-\frac{t}{n}\mathrm{A}\right)^{-1}\mathrm{f}\left(\frac{2t}{n}\right)\right)\frac{t}{n} \end{aligned} \tag{5.9}$$

STOP! What pattern emerges?

Taking n steps in the march we arrive at t and have

$$\bar{\mathrm{U}}(t) = \left(1-\frac{t}{n}\mathrm{A}\right)^{-n}\mathrm{U}_0 + \sum_{j=0}^{n-1}\left(1-\frac{t}{n}\mathrm{A}\right)^{-(n-j)}\mathrm{f}\left(\frac{t}{n}j\right)\frac{t}{n} \tag{5.10}$$

Remark 5.2.1. The formula found in (5.10) can used as a computer algorithm to numerically approximate (5.1).

As explained in Section 4.3.2, taking the limit of (5.10) as n goes to infinity produces the following solution for (5.1):

$$\mathrm{U}(t) = e^{\mathrm{A}t}\mathrm{U}_0 + \lim_{n\to\infty}\sum_{j=0}^{n-1}\left(1 - \frac{t}{n}\mathrm{A}\right)^{-(n-j)} \mathrm{f}\left(\frac{t}{n}j\right)\frac{t}{n}$$

STOP! Go back and look at the variation of parameters formula (5.5). What is different? the same?

We know an integral must play a role in the formula for $U(t)$, so let us apply the definition of the integral to a function very similiar to the one in the variations of parameters formula. Let T and f be real-valued functions such that their product is integrable on $[0, t]$. By definition of the integral, we have

$$\int_0^t T(t-s)f(s)ds = \lim_{n\to\infty}\sum_{k=1}^{n} T(t - s_k^*)f(s_k^*)\triangle s_k \tag{5.11}$$

where $0 = s_0 < s_1 < \cdots < s_n = t$, $\triangle s_k = s_k - s_{k-1}$, and $s_k^* \in [s_{k-1}, s_k]$. Our final task in the variation of parameters formula derivation is to show

$$\lim_{n\to\infty}\sum_{j=0}^{n-1}\left(1 - \frac{t}{n}\mathrm{A}\right)^{-(n-j)} \mathrm{f}\left(\frac{t}{n}j\right)\frac{t}{n} \tag{5.12}$$

is equivalent to

$$\int_0^t e^{\mathrm{A}(t-s)}\mathrm{f}(s)ds \tag{5.13}$$

Let $s_j = \frac{t}{n}j$ and $\triangle s_j = \frac{t}{n}$ for $j = 0, 1, \ldots, n-1$. Then

$$t - s_j = \frac{t}{n}(n-j)$$

and

$$\frac{t}{n} = \frac{t - s_j}{n - j} \tag{5.14}$$

Using (5.14) enables us to rewrite (5.12) as

$$\lim_{n\to\infty}\sum_{j=0}^{n-1}\left(1 - \frac{t - s_j}{n - j}\mathrm{A}\right)^{-(n-j)} \mathrm{f}(s_j)\triangle s_j. \tag{5.15}$$

Next, we change the index variable in the sum in order to match (5.15) with the definition of (5.13).

STOP! Compute the values of $n - j$ for $j = 0, 1, \ldots, n-1$.

Let $k = n - j$, $s_k^* = \frac{t}{n}(n-k)$, and $\Delta s_k = \frac{t}{n}$ so that

$$\lim_{n\to\infty} \sum_{j=0}^{n-1} \left(1 - \frac{t-s_j}{n-j}\mathrm{A}\right)^{-(n-j)} \mathrm{f}(s_j)\Delta s_j = \lim_{n\to\infty} \sum_{k=n}^{1} \left(1 - \frac{t-s_k^*}{k}\mathrm{A}\right)^{-k} \mathrm{f}(s_k^*)\Delta s_k \quad (5.16)$$

$$= \lim_{n\to\infty} \sum_{k=1}^{n} \left(1 - \frac{t-s_k^*}{k}\mathrm{A}\right)^{-k} \mathrm{f}(s_k^*)\Delta s_k \quad (5.17)$$

$$= \int_0^t e^{\mathrm{A}(t-s)}\mathrm{f}(s)ds$$

STOP! Justify the equalities in (5.16) and (5.17).

Thus, we have obtained the variations of parameters formula via a numerical approximation technique.

EXPLORE!

5.2.3 Building an Existence Theory for One-Dimensional (Non-CP)

What goes into developing a theory for (Non-CP)? Complete this **EXPLORE!** project to get a feel for some preliminary work that goes into building a theory in the one-dimensional case by considering the influence of the behavior of the forcing term in the existence of a solution. Consider the IVP

$$\begin{cases} y'(t) + ay(t) = f(t) \\ y(0) = y_0 \end{cases} \quad (5.18)$$

How "badly behaved" can the forcing term $f(t)$ be without hindering our derivation of the variation of parameters formula outlined in (5.23)? When does its behavior adversely affect the solvability of (5.18)? Answering these questions requires that we check that all terms involving $f(t)$ are defined, and that performing calculus operations on such terms is justified.

i.) Before tacking the general setting in IVP (5.18), consider the following specific version of IVP (5.18):

$$\begin{cases} y'(t) - 2y(t) = f(t),\ t > 0, \\ y(0) = 2, \end{cases} \quad (5.19)$$

where

$$f(t) = \begin{cases} t,\ 0 \le t < 1, \\ 0,\ t \ge 1. \end{cases}$$

Does the presence of the discontinuity of $f(\cdot)$ at $t = 1$ prevent the solution candidate for (5.19), as given by (5.5), from being continuous? Worse yet, does it prevent differentiability of the solution?

a.) Verify that the solution $y(\cdot)$ of (5.19), as expressed by (5.5), is continuous at all times $t \neq 1$.

b.) Focusing on the behavior at $t = 1$, we see that $y(\cdot)$ must satisfy the following two IVPs:

$$\begin{cases} y'(t) - 2y(t) = t, \ 0 < t < 1, \\ y(0) = 2, \end{cases} \tag{5.20}$$

$$\begin{cases} y'(t) - 2y(t) = 0, \ t \geq 1, \\ y(1) = 2e^2 + e - 2, \end{cases} \tag{5.21}$$

Verify that the solution $y(\cdot)$ of (5.19) is

$$y(t) = \begin{cases} 2e^{2t} + e^t - t - 1, \ 0 \leq t < 1, \\ \left(2 + \frac{1}{e} - \frac{2}{e^2}\right) e^{2t}, \ t \geq 1. \end{cases} \tag{5.22}$$

c.) Show that while $y(\cdot)$ given by (5.22) is continuous on $[0, \infty)$,

$$\frac{d^+}{dt} y(t)|_{t=1} \neq \frac{d^-}{dt} y(t)|_{t=1} .$$

As such, $y(\cdot)$ is not differentiable on $(0, \infty)$ and hence cannot be a solution of (5.19) in the classical sense. As such, we will need to consider different notions of what is meant by a solution to IVPs going forward. Consequently, even a single point of irregularity in the forcing term $f(\cdot)$ can create complications. What conditions can be imposed on the initial data and/or $f(\cdot)$ to ensure that the function $y(\cdot)$ given by (5.5) is continuous on $[0, \infty)$? differentiable on $(0, \infty)$?

ii.) Look back at the steps used to derive the variation of parameters formula in (5.5). In order to determine if a given function $f(t)$ is allowable, which terms must be defined?

iii.) Now, let us work backwards. Suppose you are given a function $y(t)$ that is claimed to satisfy IVP (5.18).

a.) How would you verify this claim?

b.) Does the information from (a) help you to clarify what condition(s) are sufficient to impose on $f(t)$ to facilitate the verification that $y(t)$ satisfies (5.18)?

iv.) Now, let us try to pull our discussion together. Consider the general variation of parameters formula for a solution of (5.18), namely

$$y(t) = e^{at} + \int_0^t e^{a(t-s)} f(s) ds. \tag{5.23}$$

Since the function $t \mapsto e^{at} y_0$ is differentiable at every real number, the differentiability and continuity of the solution formula for $y(t)$ is determined solely by the behavior of the mapping $t \mapsto \int_0^t e^{a(t-s)} f(s) ds$.

Observe that

$$\int_0^t e^{a(t-s)} f(s) ds = e^{at} \cdot \int_0^t e^{-as} f(s) ds. \tag{5.24}$$

a.) Using the Fundamental Theorem of Calculus, determine a sufficient condition to impose on $f(\cdot)$ to ensure the mapping $t \mapsto \int_0^t e^{-as} f(s) ds$ is differentiable.

b.) Explain why if both terms of the product on the right side of (5.24) are differentiable, then the product is differentiable.

c.) Compute the derivative of (5.24) under the assumption imposed in (a).

d.) Even if the mapping $t \mapsto \int_0^t e^{a(t-s)} f(s)ds$ is not differentiable, as long as it is defined (meaning that the integrand $s \mapsto e^{-as} f(s)$ is integrable), then the variation of parameters formula is still meaningful. Why?

5.2.4 Defining What is Meant By a Solution of (Non-CP)

The final observations from Section 5.2.3 suggest that if the mapping $t \mapsto \int_0^t e^{\mathrm{A}(t-s)}\mathrm{f}(s)ds$ is differentiable, then the variation of parameters formula for the solution is itself differentiable. In such case, we can trace backwards through the derivation in (5.4) - (5.5) to verify the formula is indeed a solution of the original IVP. However, without the differentiability of $t \mapsto \int_0^t e^{\mathrm{A}(t-s)}\mathrm{f}(s)ds$, we must be content with the fact that the variation of parameters formula defines a solution of the IVP in a weaker sense. It still describes the evolution of the unknown over time, but it cannot be differentiated. Such a notion is still very useful in practice. Actually, using the variation of parameters formula is preferred in applications and numerical approximations. This prompts us to define two notions of a solution, as follows:

Definition 5.2.1. A function $\mathbf{U}:[t_0, t_0 + T] \to \mathbb{R}$ is a

i.) *classical solution* of (5.1) if

a.) $\mathbf{U}(\cdot)$ is continuous on $[t_0, t_0 + T]$,
b.) $\mathbf{U}(\cdot)$ is differentiable on $(t_0, t_0 + T)$,
c.) $\mathbf{U}'(t) = \mathbf{A}\mathbf{U}(t) + \mathbf{f}(t)$, for all $t \in (t_0, t_0 + T)$,
d.) $\mathbf{U}(t_0) = \mathbf{U}_0$.

ii.) *mild solution* of (5.1) if

a.) $\mathbf{U}(\cdot)$ is continuous on $[t_0, t_0 + T]$,
b.) $\mathbf{U}(t) = e^{\mathbf{A}(t-t_0)}\mathbf{U}_0 + \int_{t_0}^t e^{\mathbf{A}(t-s)}\mathbf{f}(s)ds$ for all $t \in (t_0, t_0 + T)$.

Exercise 5.2.1. Is a classical solution necessarily a mild solution? How about conversely? Explain.

5.3 Existence Theory for General (Non-CP)

The discussion in the previous section provides insight into the one-dimensional case of (Non-CP). A natural question is, "To what extent does this discussion generalize to the system version of (Non-CP)?" This is the main topic of this section.

5.3.1 Constructing a Solution of (Non-CP)

Using elementary matrix algebra and the properties of the matrix exponential, it is reasonable to mimic the variation of parameters process (5.4) - (5.5) to construct the solution of the general system version of (Non-CP). We must bear in mind that all quantities are now matrices or vectors, so that the order in which terms appear in a product is nonignorable.

Exercise 5.3.1. (Variation of Parameters Formula - Does It Work?) Verify each line of (5.4) - (5.5) in the case when $\mathbf{A}$, $\mathbf{U}(t)$, $\mathbf{U_0}$, and $\mathbf{f}(t)$ are as in Definition 5.1.1. Make certain all operations are justified and carefully indicate all necessary definitions and theorems needed to justify each line of the derivation.

The next exercise is designed to get you to think ahead.

Exercise 5.3.2. Consider a function $\mathbf{U}:[t_0, t_0+T] \to \mathbb{R}^N$, for $N \in \mathbb{N}$.

i.) Interpret Def. 5.2.1 for (Non-CP) when the solution is vector-valued.

ii.) Look back at Section 5.2.3 and reexamine Question 3 for this general setting. Specifically, conjecture a condition to impose on $\mathbf{f}(t)$ that would ensure that the variation of parameters formula is differentiable and hence is a classical solution of (Non-CP).

5.3.2 Computing with the Variation of Parameters Formula

The nice thing about having the variation of parameters formula (5.5) is that we do not need to reinvent the wheel each time we want to solve a system of the form (Non-CP). We know what the solution is. However, we still need to compute the terms that comprise the formula, and this can be somewhat tedious as we experienced when using Putzer's algorithm.

For the moment, we consider several examples of systems of the form (Non-CP) for which the existence and uniqueness of at least a mild solution is guaranteed so that the variation of parameters formula is meaningful. The idea is to gain familiarity with constructing the solutions of these systems. We will develop the actual existence-uniqueness result in the next subsection.

Example 5.3.1. Solve the following uncoupled system:

$$\begin{cases} x'(t) &= 2x(t) + t \\ y'(t) &= -3y(t) - e^{2t} \\ x(2) &= 1, \; y(2) = -1 \end{cases} \tag{5.25}$$

<u>*Solution*</u>: We can rewrite (5.25) in the matrix form (Non-CP) using the following identifications:

$$\mathbf{U}(t) = \begin{bmatrix} x(t) \\ y(t) \end{bmatrix}, \; \mathbf{A} = \begin{bmatrix} 2 & 0 \\ 0 & -3 \end{bmatrix}, \; \mathbf{F}(t) = \begin{bmatrix} t \\ -e^{2t} \end{bmatrix}, \; \mathbf{U}_0 = \begin{bmatrix} 1 \\ -1 \end{bmatrix}, \; t_0 = 2.$$

The unique solution of (5.25) is given by the variation of parameters formula

$$\mathbf{U}(t) = e^{\mathbf{A}(t-2)}\mathbf{U}_0 + \int_2^t e^{\mathbf{A}(t-s)}\mathbf{F}(s)ds. \tag{5.26}$$

The first step is to compute $e^{\mathbf{A}t}$:

$$e^{\mathbf{A}t} = \begin{bmatrix} e^{2t} & 0 \\ 0 & e^{-3t} \end{bmatrix}.$$

Thus,

$$e^{\mathbf{A}(t-2)} = \begin{bmatrix} e^{2(t-2)} & 0 \\ 0 & e^{-3(t-2)} \end{bmatrix}$$

and subsequently applying it to the initial condition vector:

$$e^{\mathbf{A}(t-2)}\begin{bmatrix} 1 \\ -1 \end{bmatrix} = \begin{bmatrix} e^{2(t-2)} \\ -e^{-3(t-2)} \end{bmatrix}. \tag{5.27}$$

Also, we simplify the integrand in (5.26):

$$e^{\mathbf{A}(t-s)}\mathbf{F}(s) = \begin{bmatrix} e^{2(t-s)} & 0 \\ 0 & e^{-3(t-s)} \end{bmatrix}\begin{bmatrix} s \\ -e^{2s} \end{bmatrix} = \begin{bmatrix} se^{2(t-s)} \\ -e^{-3t+5s} \end{bmatrix}. \tag{5.28}$$

Integrating each component in (5.28) from 2 to t (by parts when necessary) yields

$$\begin{aligned} \int_2^t e^{\mathbf{A}(t-s)}\mathbf{F}(s)ds &= \int_2^t \begin{bmatrix} se^{2(t-s)} \\ -e^{-3t+5s} \end{bmatrix} ds \\ &= \begin{bmatrix} e^{2t}\int_2^t se^{-2s}ds \\ -e^{-3t}\int_2^t e^{5s}ds \end{bmatrix} \\ &= \begin{bmatrix} -\frac{1}{2}\left(t+\frac{1}{2}-\frac{5}{2e^4}e^{2t}\right) \\ -\frac{1}{5}\left(e^{2t}-10e^{-3t}\right) \end{bmatrix}. \end{aligned} \tag{5.29}$$

The solution to (5.25) is then obtained by adding the vectors (5.27) and (5.28). Doing so yields

$$\mathbf{U}(t) = \begin{bmatrix} \frac{9}{4}e^{2t-4} - \frac{1}{2}t - \frac{1}{4} \\ \left(2-e^6\right)e^{-3t} - \frac{1}{5}e^{2t} \end{bmatrix}.$$

Reading off the components, we conclude that the solution of the original system is

$$\begin{cases} x(t) &= \frac{9}{4}e^{2t-4} - \frac{1}{2}t - \frac{1}{4} \\ y(t) &= \left(2-e^6\right)e^{-3t} - \frac{1}{5}e^{2t} \end{cases}$$

The same approach works for general $N \times N$ systems as well, albeit with more tedious calculations.

Example 5.3.2. Assume that $\lambda_1, \ldots, \lambda_N$ are distinct real numbers, $c_1, \ldots, c_N$ are any real numbers, and the functions $f_1, \ldots, f_N : [0,T] \longrightarrow \mathbb{R}$ are continuous functions. Solve the following system:

$$\begin{cases} \mathbf{U}'(t) &= \begin{bmatrix} \lambda_1 & 0 & \cdots & 0 \\ 0 & \ddots & \ddots & \vdots \\ \vdots & \ddots & \ddots & 0 \\ 0 & \cdots & 0 & \lambda_N \end{bmatrix}\mathbf{U}(t) + \begin{bmatrix} f_1(t) \\ \vdots \\ \vdots \\ f_N(t) \end{bmatrix} \\ \mathbf{U}(t_0) &= \begin{bmatrix} c_1 \\ \vdots \\ \vdots \\ c_N \end{bmatrix} \end{cases} \tag{5.30}$$

<u>Solution</u>: System (5.30) is already in the form of (Non-CP). Observe that

$$e^{\mathbf{A}t} = \begin{bmatrix} e^{\lambda_1 t} & 0 & \cdots & 0 \\ 0 & \ddots & \ddots & \vdots \\ \vdots & \ddots & \ddots & 0 \\ 0 & \cdots & 0 & e^{\lambda_N t} \end{bmatrix}.$$

As such, we have

$$e^{\mathbf{A}(t-t_0)} = \begin{bmatrix} e^{\lambda_1(t-t_0)} & 0 & \cdots & 0 \\ 0 & \ddots & \ddots & \vdots \\ \vdots & \ddots & \ddots & 0 \\ 0 & \cdots & 0 & e^{\lambda_N(t-t_0)} \end{bmatrix}$$

$$e^{\mathbf{A}(t-t_0)} \begin{bmatrix} c_1 \\ \vdots \\ \vdots \\ c_N \end{bmatrix} = \begin{bmatrix} e^{\lambda_1(t-t_0)}c_1 \\ \vdots \\ \vdots \\ e^{\lambda_N(t-t_0)}c_N \end{bmatrix}$$

$$\int_{t_0}^{t} e^{\mathbf{A}(t-s)}\mathbf{F}(s)ds = \begin{bmatrix} \int_{t_0}^{t} e^{\lambda_1(t-s)} f_1(s)ds \\ \vdots \\ \vdots \\ \int_{t_0}^{t} e^{\lambda_N(t-s)} f_N(s)ds \end{bmatrix}.$$

Thus, the solution of (5.30) is

$$\mathbf{U}(t) = \begin{bmatrix} e^{\lambda_1(t-t_0)}c_1 \\ \vdots \\ \vdots \\ e^{\lambda_N(t-t_0)}c_N \end{bmatrix} + \begin{bmatrix} \int_{t_0}^{t} e^{\lambda_1(t-s)} f_1(s)ds \\ \vdots \\ \vdots \\ \int_{t_0}^{t} e^{\lambda_N(t-s)} f_N(s)ds \end{bmatrix} = \begin{bmatrix} e^{\lambda_1(t-t_0)}c_1 + \int_{t_0}^{t} e^{\lambda_1(t-s)} f_1(s)ds \\ \vdots \\ \vdots \\ e^{\lambda_N(t-t_0)}c_N + \int_{t_0}^{t} e^{\lambda_N(t-s)} f_N(s)ds \end{bmatrix}.$$

In retrospect, since all of the differential equations composing system (5.30) were completely decoupled (meaning that the right side of each ODE involved only one component of the solution vector, namely the one in the same row on the left side), we should have expected the components of the solution vector to be the one-dimensional variation of parameters formulas for the solutions of these ODEs.

Exercise 5.3.3. Solve the following systems:

i.)

$$\begin{cases} \mathbf{U}'(t) = \begin{bmatrix} -2 & 1 \\ 0 & -2 \end{bmatrix} \mathbf{U}(t) + \begin{bmatrix} 1 \\ t \end{bmatrix} \\ \mathbf{U}(0) = \begin{bmatrix} 0 \\ 1 \end{bmatrix} \end{cases} \tag{5.31}$$

ii.)

$$\begin{cases} \mathbf{U}'(t) = \begin{bmatrix} 1 & -1 & 4 \\ 3 & 2 & -1 \\ 2 & 1 & -1 \end{bmatrix} \mathbf{U}(t) + \begin{bmatrix} 1 \\ t \\ 0 \end{bmatrix} \\ \mathbf{U}(1) = \begin{bmatrix} 1 \\ 0 \\ -1 \end{bmatrix} \end{cases} \tag{5.32}$$

iii.)

$$\begin{cases} x'(t) &= 3x(t) + y(t) + \sin(t) \\ y'(t) &= -x(t) + 3y(t) + \cos(t) \\ x\left(\frac{\pi}{2}\right) &= 1,\ y\left(\frac{\pi}{2}\right) = -1 \end{cases} \tag{5.33}$$

While the calculations involved in solving these systems are not intrinsically difficult, they must be meticulously carried out. For this reason, the variation of parameters formula is arguably of more theoretical importance than it is practical. Nevertheless, familiarity with the process of constructing this solution will prove useful when studying more complicated systems.

We introduced in Chapter 4 a method by which a homogenous linear ODE of order n could be reformulated in the form (HCP). The same trick works for a nonhomogenous version of such equations, as illustrated below.

Definition 5.3.1. Suppose I is some interval in $\mathbb{R}$ and $f : I \to \mathbb{R}$.

i.) An nth-*order nonhomogenous linear ODE with constant coefficients* is of the general form

$$a_n x^{(n)}(t) + a_{n-1}x^{(n-1)}(t) + \ldots + a_1 x'(t) + a_0 x(t) = f(t), \tag{5.34}$$

where t is in I and $a_n \neq 0$.

ii.) If we equip (5.34) with n initial conditions of the form

$$x(t_0) = c_1,\ x'(t_0) = c_2, \ldots, x^{(n-1)}(t_0) = c_n, \tag{5.35}$$

then (5.34) - (5.35) forms an nth-*order IVP*.

We must interpret the IVP (5.34) - (5.35) as a linear system of the form **(Non-CP)** in order to view it as a special case of the theory we are presently developing. Fortunately, it turns out that we can make use of the same trick as in Chapter 4.

First, move the terms *except* the one of degree n to the right-side of (5.34):

$$a_n x^{(n)}(t) = f(t) \quad \underbrace{-\sum_{j=0}^{n-1} a_j x^{(j)}(t)}_{\text{Lower-order Derivatives}} \tag{5.36}$$

Now, divide both sides of (5.36) by a_n so that the leading coefficient on the left-side is 1:

$$x^{(n)}(t) = \underbrace{\frac{f(t)}{a_n}}_{=f^\star(t)} - \sum_{j=0}^{n-1} \underbrace{\left(\frac{a_j}{a_n}\right)}_{=a_j^\star} x^{(j)}(t). \tag{5.37}$$

Next, perform the following change of variable. We suppress the dependence on t throughout:

$$\begin{aligned} y_1 &= x \\ y_2 &= y_1' = x' \\ y_3 &= y_2' = y_1'' = x'' \\ &\vdots \\ y_{n-1} &= y_{n-2}' = \ldots = x^{(n-2)} \\ y_n &= y_{n-1}' = \ldots = x^{(n-1)} \end{aligned}$$

Observe that

$$y_n' = x^{(n)} = f^\star(t) - \sum_{j=0}^{n-1} a_j^\star y_{j+1}. \tag{5.38}$$

The initial conditions in (5.35) become

$$y_1(t_0) = c_1,\ y_2(t_0) = c_2, \ldots, y_n(t_0) = c_n. \tag{5.39}$$

Combining these observations enables us to reformulate IVP (5.34) - (5.35) as the following equivalent homogenous linear system:

$$\begin{cases} \begin{bmatrix} y_1 \\ y_2 \\ \vdots \\ y_{n-1} \\ y_n \end{bmatrix}'(t) = \begin{bmatrix} 0 & 1 & 0 & \cdots & 0 & 0 \\ 0 & 0 & 1 & \cdots & 0 & 0 \\ & & & \vdots & & \\ 0 & 0 & 0 & \cdots & 0 & 1 \\ a_0^\star & a_1^\star & a_2^\star & \cdots & a_{n-1}^\star & a_n^\star \end{bmatrix} \begin{bmatrix} y_1 \\ y_2 \\ \vdots \\ y_{n-1} \\ y_n \end{bmatrix}(t) + \begin{bmatrix} 0 \\ 0 \\ \vdots \\ 0 \\ f^\star(t) \end{bmatrix} \\ \begin{bmatrix} y_1 \\ y_2 \\ \vdots \\ y_{n-1} \\ y_n \end{bmatrix}(t_0) = \begin{bmatrix} c_1 \\ c_2 \\ \vdots \\ c_{n-1} \\ c_n \end{bmatrix} \end{cases} \tag{5.40}$$

The desired function that satisfies the original n^{th}-order IVP (5.34) - (5.35) is then just the first component, $y_1(t)$, of $\mathbf{U}(t)$. **(Why?)** We illustrate this procedure below.

Example 5.3.3. Solve the system

$$\begin{cases} x''(t) &= 4x(t) + y(t) + t \\ y'(t) &= y(t) \\ y(1) &= 1 \\ x(1) &= 0 \\ x'(1) &= 0 \end{cases} \tag{5.41}$$

The "total degree" of the system is 3 (obtained by adding the orders of both ODEs, which are 2 and 1, respectively). So, three conditions (two for x and one for y at the same time ($t_0 = 1$) are needed to ensure (5.41) has a unique solution. Rather than introducing new names for the unknowns, we observe that the corresponding system must be 3×3. **(Why?)** But, what is the structure of the solution vector of the system? Precisely, what should be the components? Certainly, they should involve only the unknown functions of interest and their derivatives up to the highest degree minus one for each variable (that is, x, x', and y). As such, a logical choice for the solution vector is

$$\mathbf{U}(t) = \begin{bmatrix} x(t) \\ x'(t) \\ y(t) \end{bmatrix}.$$

In such case, observe that (5.41) can be written as follows:

$$\begin{cases} \mathbf{U}'(t) = \begin{bmatrix} 0 & x'(t) & 0 \\ 4x(t) & 0 & y(t) \\ 0 & 0 & y(t) \end{bmatrix} + \begin{bmatrix} 0 \\ t \\ 0 \end{bmatrix} = \begin{bmatrix} 0 & 1 & 0 \\ 4 & 0 & 1 \\ 0 & 0 & 1 \end{bmatrix} \mathbf{U}(t) + \begin{bmatrix} 0 \\ t \\ 0 \end{bmatrix} \\ \mathbf{U}(1) = \begin{bmatrix} 0 \\ 0 \\ 12 \end{bmatrix} \end{cases} \tag{5.42}$$

This time, the functions that constitute the solution of the original IVP are x and y, which correspond to the first and third components of $\mathbf{U}(t)$.

STOP! Solve (5.42).

Exercise 5.3.4. Solve the following IVP:

$$\begin{cases} x''(t) &= 3x(t) + 2x'(t) + e^{3t} \\ x(0) &= 0 \\ x'(0) &= 0 \end{cases}$$

5.3.3 An Existence-Uniqueness Theorem for (Non-CP)

The examples in the previous subsection illustrated the construction of the variation of parameters formula of a solution on (Non-CP). We assumed that the forcing terms were sufficiently smooth as to ensure that the formula was meaningful. We now formulate and prove a result that guarantees the existence and uniqueness of solutions for (Non-CP) in the sense of Definition 5.2.1. The properties of $e^{\mathbf{A}t}$ established in Section 4.5 enable us to prove the following result:

Theorem 5.3.1. Let $\mathbf{A} \in \mathbb{M}^N(\mathbb{R})$ and $\mathbf{U}_0 \in \mathbb{R}^N$ be given, and $\mathbf{F} : \mathbb{R} \longrightarrow \mathbb{R}^N$ be a continuous function. Then, the IVP (Non-CP) has a unique classical solution given by the variation of parameters formula

$$\mathbf{U}(t) = e^{\mathbf{A}(t-t_0)}\mathbf{U}_0 + \int_{t_0}^{t} e^{\mathbf{A}(t-s)}\mathbf{F}(s)ds, \tag{5.43}$$

defined for all real numbers t.

Proof. Existence: First, the function $\mathbf{U}$ defined by (5.43) is a continuous function of t **(Why?)** and so, is a mild solution of (Non-CP). In order to show that it is a classical solution of (Non-CP), the right-side of (5.43) must be differentiable at all real numbers t. But, the differentiability of the mapping $t \mapsto e^{\mathbf{A}(t-t_0)}\mathbf{U}_0$ follows immediately from Proposition 4.5.3.

STOP! Tell why.

Next, observe that

$$\int_{t_0}^{t} e^{\mathbf{A}(t-s)}\mathbf{F}(s)ds = e^{\mathbf{A}t}\int_{t_0}^{t} e^{-\mathbf{A}s}\mathbf{F}(s)ds. \tag{5.44}$$

The first term of the product on the right side of (5.44) is differentiable by Proposition 4.5.3. And, since the integrand of the integral term is continuous, it follows from the Fundamental

Theorem of Calculus (as extended to $\mathbb{R}^N$–valued functions) that the second term of the product is also differentiable. Consequently, the entire product is differentiable. It therefore follows that the function $\mathbf{U}$, as defined by (5.43), is differentiable.

We now compute the derivatives. Using Proposition 4.5.3 yields

$$\begin{aligned} \frac{d}{dt}\left[e^{\mathbf{A}(t-t_0)}\mathbf{U}_0\right] &= \frac{d}{dt}\left[e^{\mathbf{A}(t-t_0)}\right]\mathbf{U}_0 \\ &= \left[\mathbf{A}e^{\mathbf{A}(t-t_0)}\right]\mathbf{U}_0 \\ &= \mathbf{A}\left[e^{\mathbf{A}(t-t_0)}\mathbf{U}_0\right]. \end{aligned} \tag{5.45}$$

STOP! Justify each step in (5.45). Also, why does the presence of t_0 not hinder the applicability of Proposition 4.5.3?

Differentiating the second term using the product rule yields

$$\begin{aligned} \frac{d}{dt}\int_{t_0}^{t} e^{\mathbf{A}(t-s)}\mathbf{F}(s)ds &= \frac{d}{dt}\left[e^{\mathbf{A}t}\int_{t_0}^{t} e^{-\mathbf{A}s}\mathbf{F}(s)ds\right] \\ &= e^{\mathbf{A}t}\cdot\frac{d}{dt}\left[\int_{t_0}^{t} e^{-\mathbf{A}s}\mathbf{F}(s)ds\right] + \frac{d}{dt}\left[e^{\mathbf{A}t}\right]\cdot\left[\int_{t_0}^{t} e^{-\mathbf{A}s}\mathbf{F}(s)ds\right] \\ &= \underbrace{e^{\mathbf{A}t}\cdot e^{-\mathbf{A}t}}_{\mathbf{I}}\mathbf{F}(t) + \left[\mathbf{A}e^{\mathbf{A}t}\right]\cdot\left[\int_{t_0}^{t} e^{-\mathbf{A}s}\mathbf{F}(s)ds\right] \\ &= \mathbf{F}(t) + \int_{t_0}^{t} \mathbf{A}e^{\mathbf{A}t}e^{-\mathbf{A}s}\mathbf{F}(s)ds \\ &= \mathbf{F}(t) + \int_{t_0}^{t} \mathbf{A}e^{\mathbf{A}(t-s)}\mathbf{F}(s)ds \\ &= \mathbf{F}(t) + \mathbf{A}\int_{t_0}^{t} e^{\mathbf{A}(t-s)}\mathbf{F}(s)ds. \end{aligned} \tag{5.46}$$

STOP! Justify each step of (5.46). Along the way, specifically address the following:

i.) Note the order in which the terms of the product rule in the second line are written. This was deliberate. Explain why the terms are not necessarily commutative.

ii.) Why can the terms $e^{\mathbf{A}t}$ and $\mathbf{A}$ be brought inside and outside the integral sign? Carefully explain.

Now, use (5.45) and (5.46) together to obtain

$$\begin{aligned} \mathbf{U}'(t) &= \mathbf{A}\left[e^{\mathbf{A}(t-t_0)}\mathbf{U}_0\right] + \mathbf{F}(t) + \mathbf{A}\int_{t_0}^{t} e^{\mathbf{A}(t-s)}\mathbf{F}(s)ds \\ &= \mathbf{A}\left[e^{\mathbf{A}(t-t_0)}\mathbf{U}_0\right] + \mathbf{A}\int_{t_0}^{t} e^{\mathbf{A}(t-s)}\mathbf{F}(s)ds + \mathbf{F}(t) \\ &= \mathbf{A}\left[e^{\mathbf{A}(t-t_0)}\mathbf{U}_0 + \int_{t_0}^{t} e^{\mathbf{A}(t-s)}\mathbf{F}(s)ds\right] + \mathbf{F}(t) \\ &= \mathbf{A}\mathbf{U}(t) + \mathbf{F}(t). \end{aligned} \tag{5.47}$$

STOP! Justify each line of (5.47).

Next, the variation of parameters formula satisfies the initial condition in (Non-CP) because

$$\begin{aligned}\mathbf{U}(t_0) &= e^{\mathbf{A}(t_0-t_0)}\mathbf{U}_0 + \int_{t_0}^{t_0} e^{\mathbf{A}(t_0-s)}\mathbf{F}(s)ds \\ &= \underbrace{e^{\mathbf{A}(0)}}_{=\mathbf{I}}\mathbf{U}_0 + \underbrace{\int_{t_0}^{t_0} e^{\mathbf{A}(t_0-s)}\mathbf{F}(s)ds}_{\mathbf{0}} \\ &= \mathbf{U}_0.\end{aligned}$$

This shows that (5.43) is a classical solution of (Non-CP), thereby completing the existence portion of the argument.

Uniqueness: Let $\mathbf{V}$: $[t_0, \infty) \to \mathbb{R}^N$ be any classical solution of (Non-CP). Since both $\mathbf{U}$ and $\mathbf{V}$ satisfy (Non-CP), we can subtract the two equations to arrive at the IVP

$$\begin{cases}\mathbf{U}'(t) - \mathbf{V}'(t) = \mathbf{A}\mathbf{U}(t) - \mathbf{A}\mathbf{V}(t),\ t > 0, \\ \mathbf{U}(t_0) - \mathbf{V}(t_0) = \mathbf{U}_0 - \mathbf{U}_0 = \mathbf{0}\end{cases}$$

which simplifies to

$$\begin{cases}(\mathbf{U}-\mathbf{V})'(t) = \mathbf{A}(\mathbf{U}-\mathbf{V})(t),\ t > 0, \\ (\mathbf{U}-\mathbf{V})(t_0) = \mathbf{0}.\end{cases} \tag{5.48}$$

It follows from Theorem 4.6.1 that the unique solution of (5.48) is

$$(\mathbf{U}-\mathbf{V})(t) = e^{\mathbf{A}t}\mathbf{0} = \mathbf{0},\ \text{for all } t \in \mathbb{R}.$$

Thus, $\mathbf{U}(t) = \mathbf{V}(t)$, for all $t \in \mathbb{R}$. This says that any function $\mathbf{V}$ that satisfies the definition of a classical solution on (Non-CP) must coincide with $\mathbf{U}$ given by (5.43). Hence, we conclude that the classical solution of (Non-CP) is unique.

This completes the proof. □

Exercise 5.3.5. Answer the following:

i.) Identify a specific example of a function $\mathbf{F} : \mathbb{R} \longrightarrow \mathbb{R}^N$ for which (Non-CP) has a unique mild solution which is not a classical solution.

ii.) Try to identify a specific example of a function $\mathbf{F} : \mathbb{R} \longrightarrow \mathbb{R}^N$ for which (Non-CP) does not even have a mild solution. (Hint: What cannot be true about the forcing function $\mathbf{F} : \mathbb{R} \longrightarrow \mathbb{R}^N$?)

5.4 Dealing with a Perturbed (Non-CP)

When describing a mathematical model using an IVP of the form (Non-CP), the tacit assumption is that the precise nature of the forcing term is known at every time. However, this is unrealistic because of imprecision inherent to measurement. Technically, we can only say that for any time t, the *actual value* of the forcing term, denoted $\overline{\mathbf{f}}(t)$, lies somewhere in the interval $(\mathbf{f}(t) - \epsilon_1, \mathbf{f}(t) + \epsilon_1)$, where $\epsilon_1 > 0$. Similarly, the actual value of the initial condition, denoted $\overline{\mathbf{U}}_0$, lies somewhere in the interval $(\mathbf{U}_0 - \epsilon_2, \mathbf{U}_0 + \epsilon_2)$, where $\epsilon_2 > 0$. Even worse, the error terms $\epsilon_i (i = 1, 2)$ can change over time. What are the implications of such error in the nature of solutions of (Non-CP)?

Consider (Non-CP), together with the related IVP

$$\begin{cases} \overline{\mathbf{U}}'(t) = \mathbf{A}\overline{\mathbf{U}}(t) + \overline{\mathbf{f}}(t), \ t > t_0, \\ \overline{\mathbf{U}}(t_0) = \overline{\mathbf{U}}_0. \end{cases} \tag{5.49}$$

We are hopeful that for any time interval $[t_0, T]$, the norm of the difference between the solutions $\mathbf{U}$ and $\overline{\mathbf{U}}$ of (Non-CP) and (5.49), respectively, can be controlled by ensuring the error terms $\epsilon_i (i = 1, 2)$ are sufficiently small. In such case, we would say that the solution of (Non-CP) depends continuously on the data.

This is analogous to the problems considered for (HCP) in Section 4.8, but the presence of the forcing term complicates the matter. We begin by studying this problem for the one-dimensional version of (Non-CP). Precisely, consider the following two related IVPs:

$$\begin{cases} u'(t) = au(t) + f(t), \ t > t_0 \\ u(t_0) = u_0, \end{cases} \tag{5.50}$$

$$\begin{cases} \overline{u}'(t) = a\overline{u}(t) + \overline{f}(t), \ t > t_0 \\ \overline{u}(t_0) = \overline{u}_0, \end{cases} \tag{5.51}$$

where we assume there exist $\delta_1, \delta_2 > 0$ for which

$$|u_0 - \overline{u}_0| < \delta_1 \text{ and } \sup_{t_0 \le t \le T} \left| f(t) - \overline{f}(t) \right| < \delta_2. \tag{5.52}$$

The solutions of (5.50) and (5.51) are respectively given by

$$u(t) = e^{a(t-t_0)} u_0 + \int_{t_0}^{t} e^{a(t-s)} f(s) ds, \tag{5.53}$$

$$\overline{u}(t) = e^{a(t-t_0)} \overline{u}_0 + \int_{t_0}^{t} e^{a(t-s)} \overline{f}(s) ds. \tag{5.54}$$

The question of interest is, "How far apart are these solutions at a given time t?" Graphically, we measure this by computing the vertical distance between the graphs of $y = u(t)$ and $y = u_\epsilon(t)$ at all times t in an interval on which we are studying the solutions. Since the outputs are real numbers, we measure this distance using the absolute value of the difference in the outputs. In the absence of f and $\overline{f}$, we answered this question in Section 4.8.

STOP! Re-read Section 4.8 before going further.

EXPLORE!

Does (Non-CP) Depend Continuously on the Initial Data?

1. For concreteness, open **MATLAB** and in the command line, type:

   MATLAB_Linear_ACP_Solver

   GUI instructions can be found by clicking the Help button. In the User Panel, choose

$t = 0$ for initial conditions and plot solutions until $t = 5$. Click the *Input IVP and Plot Solution* button to enter the following IVP:

$$\begin{cases} \mathbf{U}'(t) &= \begin{bmatrix} 1 & 0 \\ 0 & -1 \end{bmatrix} \mathbf{U}(t) + \begin{bmatrix} \cos(2\pi t) \\ \sin(2\pi t) \end{bmatrix} \\ \mathbf{U}(0) &= \begin{bmatrix} 1 \\ 1 \end{bmatrix} \end{cases} \tag{5.55}$$

Suppose that the measurements of the first component of the initial condition vector in (5.55) and both components of the nonhomogenous term are subject to error. The corresponding IVP actually becomes

$$\begin{cases} \mathbf{U}_\epsilon'(t) &= \begin{bmatrix} 1 & 0 \\ 0 & -1 \end{bmatrix} \mathbf{U}_\epsilon(t) + \begin{bmatrix} \cos(2\pi t) + \delta_1 \sin(2\pi t) \\ \sin(2\pi t) - \delta_2 \cos(2\pi t) \end{bmatrix} \\ \mathbf{U}_\epsilon(0) &= \begin{bmatrix} 1+\epsilon \\ 1 \end{bmatrix} \end{cases} \tag{5.56}$$

i.) Click the *Modify IVP and Plot Solution* solution button and input (5.56) for $\delta_1 = 0.1, \delta_2 = 0.08$, and $\epsilon = 0.01$. Are the solutions to (5.55) and (5.56) close to each other? How close? How far?

ii.) Repeat a) for $\delta_1 = 0.01, \delta_2 = 0.008$ and $\epsilon = 0.001$.

iii.) Repeat a) for $\delta_1 = 0.001, \delta_2 = 0.0008$ and $\epsilon = 0.0001$.

iv.) Describe the patterns found in the graphs and norms of the corresponding IVPs.

v.) Repeat question 1 for any pair of IVPs that satisfies the forms provided in (5.57) and (5.58).

2. Consider the IVP

$$\begin{cases} \mathbf{U}'(t) &= \begin{bmatrix} \alpha & 0 \\ 0 & \beta \end{bmatrix} \mathbf{U}(t) + \begin{bmatrix} f_1(t) \\ f_2(t) \end{bmatrix} \\ \mathbf{U}(0) &= \begin{bmatrix} w_1 \\ w_2 \end{bmatrix} \end{cases} \tag{5.57}$$

where $\alpha, \beta \in \mathbb{R}$.

i.) What is the solution $\mathbf{U}$ of (5.57)? Give the actual components.

ii.) Suppose that the measurements of the first component of the initial condition vector in (5.57) are subject to error, so that the *actual* initial state vector is $\begin{bmatrix} w_1+\epsilon \\ w_2 \end{bmatrix}$, where $|\epsilon|$ is a small positive number. Moreover, assume that there exist $\delta_1, \delta_2 > 0$ such that

$$\sup_{0 \le t \le T} \left| f_i(t) - \overline{f}_i(t) \right| < \delta_i \; (i = 1, 2). \tag{5.58}$$

Consider the corresponding IVP

$$\begin{cases} \mathbf{U}_\epsilon'(t) &= \begin{bmatrix} \alpha & 0 \\ 0 & \beta \end{bmatrix} \mathbf{U}_\epsilon(t) + \begin{bmatrix} \overline{f}_1(t) \\ \overline{f}_2(t) \end{bmatrix} \\ \mathbf{U}_\epsilon(0) &= \begin{bmatrix} w_1+\epsilon \\ w_2 \end{bmatrix} \end{cases} \tag{5.59}$$

a.) What is the solution $\mathbf{U}_\epsilon$ of (5.59)? Give the actual components.

b.) Compute $\|\mathbf{U}(t) - \mathbf{U}_\epsilon(t)\|_{\mathbb{R}^2}$. Make a conjecture for what happens as $|\epsilon|, \delta_1, \delta_2 \longrightarrow 0$ for a fixed t?

3. Now that we have an intuitive sense of what should happen when $|\epsilon|, \delta_1, \delta_2 \to 0$, let us return to (5.57) and (5.59) to compute

$$\sup_{0 \le t \le T} \|\mathbf{U}(t) - \mathbf{U}_\epsilon(t)\|_{\mathbb{R}^2}.$$

(You should be able to get an exact value for any fixed T.) What happens to this value as $|\epsilon|, \delta_1, \delta_2 \longrightarrow 0$?

4. Now, suppose that both measurements of the initial condition vector in (5.57) are subject to error, so that the actual state vector is $\begin{bmatrix} w_1 + \epsilon_1 \\ w_2 + \epsilon_2 \end{bmatrix}$, where $|\epsilon_1|$ and $|\epsilon_2|$ are small positive numbers. Replace the initial vector in (5.59) by this one and repeat questions 1 and 3. If you need help visualizing this new situation, repeat question 2 as well.

5. Generalize your findings to the case of an $N \times N$ diagonal matrix for $\mathbf{A}$ and an initial condition vector in which each of its N components is subject to error.

6. Do you expect the same conclusion to hold when $\mathbf{A}$ is no longer a diagonal matrix? Explain.

7. Carefully formulate the most general result that you can muster regarding the continuous dependence of solutions of (Non-CP) on the initial data.

In light of the work done in Section 4.8, the answer to the above question for (Non-CP) rests solely on handling the forcing terms.

Disclaimer! The following is a more formal discussion of the ideas in the previous **EXPLORE!**

Proceeding as before, we subtract the two solution formulas, and estimate using the properties of absolute value. This yields the following for all $t \in [t_0, T]$.

$$\begin{aligned}
|u(t) - \overline{u}(t)| &= \left| \left(e^{a(t-t_0)} u_0 - e^{a(t-t_0)} \overline{u}_0 \right) + \left(\int_{t_0}^t e^{a(t-s)} f(s) ds - \int_{t_0}^t e^{a(t-s)} \overline{f}(s) ds \right) \right| \\
&= \left| e^{a(t-t_0)} \left(u_0 - \overline{u}_0 \right) + \int_{t_0}^t e^{a(t-s)} \left(f(s) - \overline{f}(s) \right) ds \right| \\
&\le \left| e^{a(t-t_0)} \left(u_0 - \overline{u}_0 \right) \right| + \left| \int_{t_0}^t e^{a(t-s)} \left(f(s) - \overline{f}(s) \right) ds \right| \\
&\le \left| e^{a(t-t_0)} \right| \cdot \left| u_0 - \overline{u}_0 \right| + \left| \int_{t_0}^t e^{a(t-s)} \left(f(s) - \overline{f}(s) \right) ds \right| \\
&\le e^{a(T-t_0)} \delta_1 + \left| \int_{t_0}^t e^{a(t-s)} \left(f(s) - \overline{f}(s) \right) ds \right|
\end{aligned} \tag{5.60}$$

The lingering problem is the second term in the last line in (5.60). How do we handle an absolute value of an integral term?

STOP! Review the properties of the integral in Proposition 2.7.1 before moving onward.

Applying the properties of the integral yields the following string of inequalities, using (5.60):

$$
\begin{aligned}
|u(t)-\overline{u}(t)| &= \left| \left(e^{a(t-t_0)}u_0 - e^{a(t-t_0)}\overline{u_0} \right) + \left(\int_{t_0}^{t} e^{a(t-s)}f(s)ds - \int_{t_0}^{t} e^{a(t-s)}\overline{f}(s)ds \right) \right| \\
&\leq \vdots \\
&\leq e^{a(T-t_0)}\delta_1 + \left| \int_{t_0}^{t} e^{a(t-s)}\left(f(s)-\overline{f}(s)\right)ds \right| \\
&\leq e^{a(T-t_0)}\delta_1 + \int_{t_0}^{t} \left| e^{a(t-s)}\left(f(s)-\overline{f}(s)\right) \right| ds \\
&\leq e^{a(T-t_0)}\delta_1 + e^{a(t-t_0)} \int_{t_0}^{t} \left| f(s)-\overline{f}(s) \right| ds \qquad (5.61) \\
&\leq e^{a(T-t_0)}\delta_1 + e^{a(t-t_0)} \int_{t_0}^{t} \sup_{t_0\leq s\leq T} \left| f(s)-\overline{f}(s) \right| ds \\
&\leq e^{a(T-t_0)}\delta_1 + e^{a(t-t_0)} \int_{t_0}^{t} \delta_2 ds \\
&\leq e^{a(T-t_0)}\delta_1 + e^{a(t-t_0)}\delta_2(t-t_0) \\
&\leq e^{a(T-t_0)}\delta_1 + e^{a(T-t_0)}\delta_2(T-t_0) \\
&= e^{a(T-t_0)}\left(\delta_1+\delta_2(T-t_0)\right).
\end{aligned}
$$

STOP! Justify each line of (5.61).

Such imprecision is present when measuring any parameter appearing in a differential equation, not just the initial condition. For simplicity, we consider (5.50) together with the following "perturbed IVP" in which the parameter a is replaced by $(a+\epsilon)$, where ϵ is a real number:

$$
\begin{cases} \overline{u}'_\epsilon(t) = (a+\epsilon)\overline{u}_\epsilon(t) + \overline{f}(t),\ t > t_0 \\ \overline{u}_\epsilon(t_0) = \overline{u}_0 \end{cases} \qquad (5.62)
$$

For the moment, we have kept the same initial condition in both IVPs. The solution of (5.62) is given by

$$
\overline{u}_\epsilon(t) = e^{(a+\epsilon)(t-t_0)}\overline{u}_0 + \int_{t_0}^{t} e^{(a+\epsilon)(t-s)}\overline{f}(s)ds. \qquad (5.63)
$$

Exercise 5.4.1. What happens as $|\epsilon|, \delta_2 \longrightarrow 0$ to $\sup_{t_0\leq t\leq T} |u(t)-\overline{u}_\epsilon(t)|$? Show all details in forming the estimates that lead to your conclusion.

APPLY IT!!

At long last, we are equipped with enough theoretical power to be able to address several more models studied in Chapter 3. For each of the following IVPs, complete these questions:

I.) Reformulate the IVP in the abstract form (Non-CP).

II.) Argue that the IVP has a unique solution.

III.) Formulate a continuous dependence result for the IVP in the spirit of the results established in this section.

IV.) Choose actual values for the parameters in the model. For them, illustrate the curves of each of the component functions, study the IVP using the **MATLAB_Linear_ACP_Solver** GUI, and analyze the results in the context of the setting of each model.

i.) The beer mixing model (3.13)

$$\begin{cases} \frac{du_1(t)}{dt} &= \frac{-7u_1(t)+2u_2(t)+0.31}{100} \\ \frac{du_2(t)}{dt} &= \frac{7u_1(t)-7u_2(t)}{80} \\ u_1(0) &= 0.062 \\ u_2(0) &= 0.09 \end{cases}$$

ii.) The vertical spring-mass model (3.17)

$$\begin{cases} mx''(t) + rx'(t) + kx(t) = mg \\ x(0) = x_0, x'(0) = x_1 \end{cases}$$

iii.) The electrical circuit model (3.21)

$$\begin{cases} L\frac{d^2q(t)}{dt} + R\frac{dq(t)}{dt} + \frac{1}{C}q(t) = E(t) \\ q(0) = q_0 \\ q'(0) = q_1 \end{cases}$$

iv.) The projectile motion model (3.42)

$$\begin{cases} m\frac{d^2x(t)}{dt^2} + r_x\frac{dx(t)}{dt} = F_x(t) \\ m\frac{d^2y(t)}{dt^2} + r_y\frac{dy(t)}{dt} = F_y(t) \\ m\frac{d^2z(t)}{dt^2} + r_x\frac{dz(t)}{dt} = F_z(t) \\ x(0) = x_0, y(0) = y_0, z(0) = z_0 \end{cases}$$

v.) The 5-story floor displacement model

$$\begin{cases} m_1x_1''(t) &= -k_0(x_1(t) - g(t)) + k_1(x_2(t) - x_1(t)) \\ m_ix_i''(t) &= -k_{i-1}(x_i(t) - x_{i-1}(t)) + k_i(x_{i+1}(t) - x_i(t)), i = 2, 3, 4 \\ m_5x_5''(t) &= -k_4(x_5(t) - x_4(t)) \\ x_i(0) &= (x_0)_i \quad (i = 1, 2, 3, 4, 5) \end{cases}$$

5.5 What Happens to Solutions of (Non-CP) as Time Goes On and On and On...?

We know that if the forcing term is sufficiently nice, the unique mild solution of (Non-CP) is given by (5.5), for all real numbers t. We investigate the behavior of this solution as $t \longrightarrow \infty$, starting with an investigation of the one-dimensional case.

EXPLORE!

Long-Term Behavior of Solutions of One-Dimensional (Non-CP)

We investigate the impact adding a function $f(t)$ to the right side of (HCP) has on the long-term behavior results developed in Section 4.9.

1. Recall the long-term behavior of solutions of (4.4) for different values of a:

   i.) If $a < 0$, then $\lim_{t \longrightarrow \infty} u(t) =$ ____________

   ii.) If $a = 0$, then $\lim_{t \longrightarrow \infty} u(t) =$ ____________

   iii.) If $a > 0$, then $\lim_{t \longrightarrow \infty} u(t) =$ ____________

2. Now consider the nonhomogenous IVP

$$\begin{cases} y'(t) + ay(t) = f(t) \\ y(0) = y_0 \end{cases} \tag{5.64}$$

Explore the long-term behavior of (5.64) by completing the following table. Use the GUI **MATLAB_Linear_ACP_Solver** to illustrate specific examples that can guide you in the process.

$f(t)$	a	Solution $y(t)$ of (5.64)	Behavior of $y(t)$ as $t \longrightarrow \infty$
a nonzero constant	$a < 0$		
	$a = 0$		
	$a > 0$		
$t^p (p \in \mathbb{N})$	$a < 0$		
	$a = 0$		
	$a > 0$		
$e^{-pt} (p > 0)$	$a < 0$		
	$a = 0$		
	$a > 0$		
$\omega \sin(pt)\ (\omega, p \neq 0)$	$a < 0$		
	$a = 0$		
	$a > 0$		
$\begin{cases} 0, & 0 \le t \le T_1 \\ c, & T_1 < t < T_2 \\ 0, & t \ge T_2 \end{cases}$	$a < 0$		
	$a = 0$		
	$a > 0$		

3. Conjecture as many connections of the following form as possible regarding the long-term behavior of the solutions of (5.64) in terms of a and $f(t)$:
   If $f(t)$ does ________________ and a ___ 0, then the long-term behavior as $t \longrightarrow \infty$ of $y(t)$ is ______________.

We now extend the above discussion to the more general case of a 2×2 nonhomogenous linear system in the following EXPLORE! activity:

EXPLORE!

Long-term Behavior of the Two-Dimensional (Non-CP)

1. Consider the following three matrices in $\mathbb{M}^2(\mathbb{R})$:

$$\mathbf{A}_1 = \begin{bmatrix} \alpha & 0 \\ 0 & \beta \end{bmatrix}, \ \alpha \neq \beta \neq 0,$$
$$\mathbf{A}_2 = \begin{bmatrix} \alpha & 1 \\ 0 & \alpha \end{bmatrix}, \ \alpha \neq 0, \tag{5.65}$$
$$\mathbf{A}_3 = \begin{bmatrix} \alpha & \beta \\ \beta & \alpha \end{bmatrix}, \ \alpha \neq \beta \neq 0.$$

   i.) Calculate $e^{\mathbf{A}_i}t$, for $i = 1, 2, 3$.

   ii.) In each case, what must be true in order to ensure that

$$\lim_{n \to \infty} \left\| e^{\mathbf{A}_i t} \mathbf{y} \right\|_{\mathbb{R}^2} = 0, \text{ for all } \mathbf{y} \in \mathbb{R}^2? \tag{5.66}$$

   iii.) In light of your responses in (ii) and assuming that $\mathbf{f}$ is bounded on $[0, \infty)$, determine a sufficient condition guaranteeing that $\int_0^t e^{\mathbf{A}_i(t-s)}\mathbf{f}(s)ds < \infty$, $\forall t > 0$.

   iv.) Let $\mathbf{U}(\cdot)$ be a mild solution of (Non-CP) and assume that $\mathbf{f}$ is bounded on $[0, \infty)$. Formulate sufficient conditions to ensure that $\lim_{t \to \infty} \|\mathbf{U}(t)\|_{\mathbb{R}^2} < \infty$.

   MATLAB_Linear_ACP_Solver may prove useful when answering questions 2 and 3. However, the GUI cannot plot solutions for arbitrarily large t values, so it can only be used to gain insight into the problems.

2. Assume that $\mathbf{A}_1$ in (5.65) is such that (5.66) holds. Define $\mathbf{f} : [0, \infty) \to \mathbb{R}^2$ by

$$\mathbf{f}(t) = \begin{bmatrix} 1 + e^{-t} \\ te^{-2t} - 1 \end{bmatrix}.$$

   i.) Determine a constant vector $\mathbf{f}_0 \in \mathbb{R}^2$ for which $\lim_{t \to \infty} \|\mathbf{f}(t) - \mathbf{f}_0\|_{\mathbb{R}^2} = 0$.

   ii.) Simplify (5.5) for (Non-CP) corresponding to $\mathbf{A}_1$ and $\mathbf{f}(\cdot)$ as above.

   iii.) Compute $\lim_{t \to \infty} \mathbf{U}(t)$.

3. Consider $\mathbf{A}_3 = \begin{bmatrix} -2 & 3 \\ 3 & -2 \end{bmatrix}$ and $\mathbf{f}(\cdot)$ as in (2) above. (Look back at Exercise 4.4.1)

i.) Simplify (5.5) for (Non-CP) corresponding to $\mathbf{A}_3$ and $\mathbf{f}(\cdot)$.

ii.) Compute $\lim_{t\to\infty} \mathbf{U}(t)$.

iii.) Describe the nature of $\lim_{t\to\infty} \mathbf{U}(t)$ for the more general matrix $\mathbf{A}_3$ in (5.65).

Chapter 6

A Second Wave of Mathematical Models—Now, with Nonlinear Interactions

Overview

The non-homogeneous models considered in the previous two chapters constitute a first attempt at incorporating external forces affecting a system into a description of its evolution. While it is the case that some perturbations can be described accurately and meaningfully using such simple function of t, the truth is the development of the theory for (Non-CP) primarily serves as a hub in the formulation of a richer theory of so-called *semilinear* Cauchy problems in which the forcing term describes nonlinear interaction effects among the various components of the solution vector. We will use the theory developed in Chapters 4 and 5 as a springboard in our discussion in the next two chapters. For the moment, we introduce specific scenarios, identify a common connective theme, and then uncover subtleties that throw a veritable wrench into the works as we formulate a generalization of the theory in Chapter 5 to handle these new models.

6.1 Newton's Law of Heating and Cooling Subjected to Polynomial Effects

The following most simplistic model describing the evolution of the temperature of a single entity placed into an environment whose ambient temperature is T_m degrees Fahrenheit is obtained using Newton's Law of Heating and Cooling

$$\begin{cases} \frac{dT(t)}{dt} = k\left(T(t) - T_m(t)\right) \\ T(t_0) = T_0 \end{cases} \tag{6.1}$$

But, what if, hypothetically speaking, the rate of change of the temperature with respect to time was *not* proportional to $(T(t) - T_m)$, but rather its square? This modification results in the following IVP :

$$\begin{cases} \frac{dT(t)}{dt} = k\left(T(t) - T_m(t)\right)^2 \\ T(t_0) = T_0 \end{cases} \tag{6.2}$$

STOP! Spend some time trying to rewrite (6.2) in the form of (Non-CP) for some forcing function $f(t)$. What do you notice about this function that is different from those encountered in Chapters 4 and 5?

More generally, we could consider a variant of (6.2) in which the right-side is replaced by an nth degree polynomial of $(T(t) - T_m)$, as follows :

$$\begin{cases} \frac{dT(t)}{dt} = \sum_{i=1}^{n} k_i \left(T(t) - T_m(t)\right)^i \\ T(t_0) = T_0 \end{cases} \tag{6.3}$$

where $k_i (i = 1, 2, \ldots, n)$ are constants of proportionality. Clearly, (6.1) is a special case of (6.3), so if we are able to solve (6.3) and understand the evolution of the behavior of its solution $T(t)$ over time, then we would also understand the behavior of the solution to the simpler Cauchy problem (6.1).

STOP! Try rewriting (6.3) in the form of (Non-CP) for some function $f(t)$. Again, pay attention to the nature of the forcing term.

The forcing terms arising in the different scenarios considered in Section 3.11 and Chapter 5 can be duly modified in the same manner, thereby resulting in more general Cauchy problems.

Such modifications can be introduced into such models in which there is more than one entity whose temperature is of interest. For instance, suppose you take *two* items out of the oven at the same time, an 18-pound roast and a baked potato, and place them onto the table in the dining room where the surrounding temperature is 74 degrees Fahrenheit. It stands to reason that the roast and the potato will cool at different rates **(Why?)** and so, when using Newton's Law of Heating and Cooling to formulate a model this time, we must account for this difference. Identifying the temperatures of the roast and potato as $T_R(t)$ and $T_P(t)$, respectively, and assuming that the rates of changes of both temperatures are independently proportional to different polynomials of the terms

$$(T_R(t) - T_P(t) - T_m)$$

one such IVP is

$$\begin{cases} \frac{dT_R(t)}{dt} = \sum_{i=1}^{n_1} k_i \left(T_R(t) + T_P(t) - T_m(t)\right)^i \\ \frac{dT_P(t)}{dt} = \sum_{j=1}^{n_2} \bar{k_j} \left(T_R(t) + T_P(t) - T_m(t)\right)^i \\ T_R(0) = T_0, T_P(t) = T_1 \end{cases} \tag{6.4}$$

where $k_i (i = 1, \ldots, n_1)$ and $\bar{k_j} (j = 1, \ldots, n_2)$ are constants of proportionality.

STOP! Try rewriting (6.4) in the form of a two-dimensional (Non-CP).

6.2 Pharmocokinetics with Concentration-Dependent Dosing

The rate of absorption of a drug into the bloodstream is affected by the dosage $D(t)$ at any time $t > 0$. In turn, the dosage might depend on the various concentrations of the drug in the GI tract, bloodstream, and/or other compartments accounted for in the model. And, these concentrations vary with time. We describe such nonlinear effects generically using functions whose arguments depend on time t and the concentrations of the drug in the various compartments. For instance, the simplistic model described by IVP (3.6) would be modified as follows :

$$\begin{cases} \frac{dy(t)}{dt} = -ay(t) + D\left(t, y(t), z(t)\right) \\ \frac{dz(t)}{dt} = ay(t) - bz(t) + E((t, y(t), z(t)) \\ y(0) = y_0, z(0) = 0 \end{cases} \tag{6.5}$$

where the term $E\left(t, y(t), z(t)\right)$ has been added to the right-side of the second equation to account for potentially different dosing effects in the bloodstream versus the GI tract. Certainly, in order for IVP (6.5) to be of any practical use, we would need to either define the functions D and E or at least provide some sort of description of its salient characteristics.

MATLAB-Exercise 6.2.1. We will use **MATLAB** to investigate a continuous dependence result for (6.5) similar to those described in Chapters 4 and 5. To begin, open **MATLAB** and in the command line, type:

MATLAB_Pharmocokinetics_Semilinear

Use the GUI to answer the following questions. Make certain to click on the HELP button for instructions on how to use the GUI. A typical screenshot of the GUI can found in Figure 6.1.

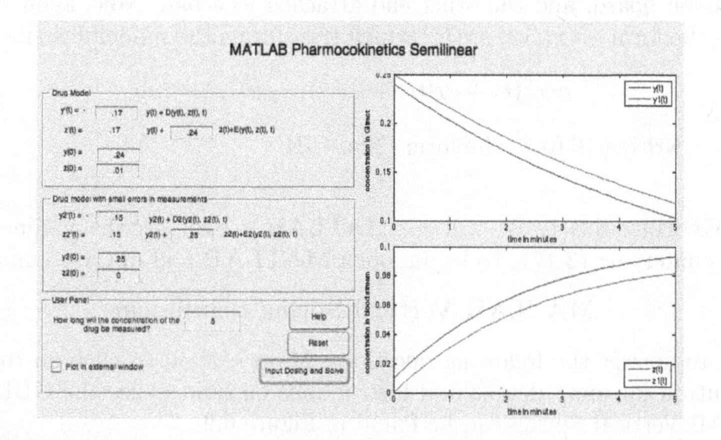

FIGURE 6.1: MATLAB Pharmocokinetics Semilinear GUI

i.) In this question use dosing functions $D\left(t, y(t), z(t)\right) = 0$ and $E\left(t, y(t), z(t)\right) = z(t)^2$.

a.) Suppose your measurements produce the following constants: $a = 0.17, b = 0.24, y_0 = 0.24, z_0 = 0.01$. However, in reality, the parameters are *actually*:

$a = 0.15, b = 0.25, y_0 = 0.25, z_0 = 0$. Are the solutions *different* over the initial 5-minute time interval? If so, would you describe the difference as being significant? Explain.

b.) Repeat (ii), but now for the initial 10-minute interval.

c.) When are the solutions furthest apart?

d.) Does the model appear to depend continuously on the parameters? initial conditions?

ii.) Suppose the dosing functions are in reality $D(t, y(t), z(t)) = c$ and $E(t, y(t), z(t)) = z(t)^2 - d\sin(y)$, for positive constants c and d.

a.) Repeat (i) parts (a-d) for $c = 0.1$ and $d = 0.01$.

b.) Repeat (i) parts (a-d) for $c = 0.01$ and $d = 0.001$.

c.) Does the IVP appear to depend continuously on the nonlinear forcing terms? If necessary, choose new values for c and d to aid in answering this question.

■

6.3 Springs with Nonlinear Restoring Forces

We thoroughly discussed several variants of the spring-mass system in Section 3.5. From horizontal to vertical to coupled systems, each model was crafted using Newton's Second Law and Hooke's Law to describe the restoring force applied to the mass due to the spring. However, not all restoring forces follow Hooke's Law. For example, when large vibrations are involved, the restoring force is actually nonlinear.

Consider a spring hung vertically with one end attached to a sturdy support, such as a ceiling or wooden board, and the other end attached to a ball. Now, assume the restoring force, F_s, has the form $-kx(t) + \mu x(t)^3$ which transforms the model described in (3.17) to

$$mx''(t) + rx'(t) + kx(t) - \mu x(t)^3 = mg \tag{6.6}$$

STOP! Try rewriting (6.6) in the form (Non-CP).

MATLAB-Exercise 6.3.1. We will use **MATLAB** to compare the nonlinear model (6.6) to its linear counterpart (3.17). To begin, open **MATLAB** and in the command line, type:

MATLAB_Vertical_Spring_Semilinear

Use the GUI to answer the following questions. Make certain to click on the 'Description of System' button for more details and instructions on how to use the GUI. A screenshot from MATLAB_Vertical_Spring can be found in Figure 6.2.

i.) Compare the two models using the parameters suggested by the GUI. Are they drastically different? the same?

ii.) Repeat (i) assuming the air resistance constant, r, is equal to zero.

iii.) Can you describe a pattern as μ tends to zero? Does your pattern depend on the other parameters?

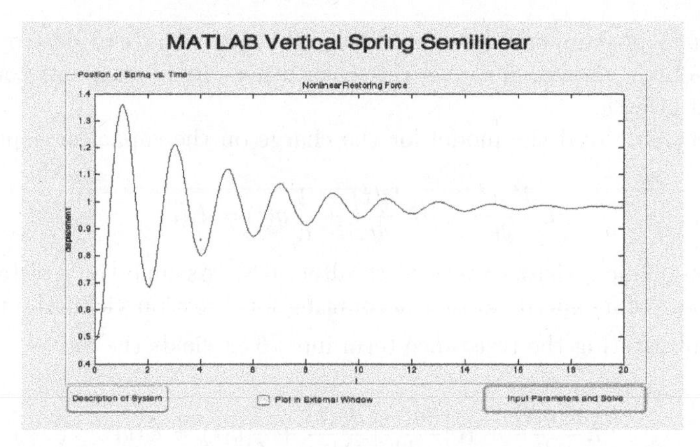

FIGURE 6.2: MATLAB Vertical Spring Semilinear GUI

iv.) What happens as μ gets larger? (Try $\mu = 20, 25, 30, \ldots$) How is this phenomena different than what was observed in earlier chapters?

■

Exercise 6.3.1. Create a model for a horizontal spring-mass system using the restoring force form $-kx + \mu x^3$.

As described in Chapter 3, many applications require a model for the position of several masses all connected by springs. The series of masses could be oriented vertically, horizontally, or at some angle. The ends could be attached to a fixed support or maybe only one end is attached; it all depends on the situation and application.

Here, let us focus on IVP (3.19), but this time, we assume the restoring force is given by $-kx + \mu x^3$. Let $x_1(t)$ and $x_2(t)$ denote the displacement from the natural positions for mass one and mass two respectively and denote the masses by m_1 and m_2; and the three positive spring constants by k_1, k_2 and k_3. In this case, IVP (3.19) becomes

$$\begin{cases} m_1 \frac{d^2 x_1(t)}{dt^2} & = -k_1 x_1(t) + \mu x_1(t)^3 + k_2(x_2(t) - x_1(t)) - k_2(x_2(t) - x_1(t))^3 \\ m_2 \frac{d^2 x_2(t)}{dt^2} & = -k_3 x_2(t) + \mu x_2(t)^3 - k_2(x_2(t) - x_1(t)) + k_2(x_2(t) - x_1(t))^3 \\ x_1(0) = x_0; \quad x_1'(0) = x_1 \\ x_2(0) = y_0; \quad x_2'(0) = y_1 \end{cases} \tag{6.7}$$

Exercise 6.3.2. Try to write IVP (6.7) in (Non-CP) form. Again, pay special attention to the forcing term.

Exercise 6.3.3. Propose a model that includes a resistance force, like friction, for the coupled spring-mass system in IVP (6.7).

6.4 Circuits with Quadratic Resistors

Some circuits are not described using Ohm's Law. In fact, non-ohmic resistance is quite important in electrical engineering. At times, linear resistance (a resistor that follows Ohm's

Law) is inadequate in dampening current flow, which in turn, can destroy a circuit. By inserting a non-ohmic resistor into a series with other circuit elements, one can control potential current growth.

In Chapter 3 we derived the model for the charge on the capacitor depicted in Figure 3.6 as

$$L\frac{d^2q(t)}{dt} + R\frac{dq(t)}{dt} + \frac{1}{C}q(t) = E(t) \tag{6.8}$$

However, for non-ohmic resistors we need to alter (6.8). Assume the resistance is proportional to the square of its speed, so that accounting for direction yields the resistance term $R\left|\frac{dq(t)}{dt}\right|\frac{dq(t)}{dt}$. Substituting the resistance term into (6.8) yields the model

$$L\frac{d^2q(t)}{dt} + R\left|\frac{dq(t)}{dt}\right|\frac{dq(t)}{dt} + \frac{1}{C}q(t) = E(t) \tag{6.9}$$

STOP! Can (6.9) be written in the form (Non-CP)? If not, which term gets in the way?

6.5 Enyzme Catalysts

In chapter 3 we considered several different types of reactions and their corresponding models. For example, the reaction

$$\mathrm{A} + \mathrm{B} \underset{k_-}{\overset{k_+}{\rightleftharpoons}} \mathrm{C} \tag{6.10}$$

was modeled by

$$\begin{cases} \frac{d[A]}{dt} = k_-[C] - k_+[A][B] \\ \frac{d[B]}{dt} = k_-[C] - k_+[A][B] \\ \frac{d[C]}{dt} = k_+[A][B] - k_-[C] \end{cases} \tag{6.11}$$

STOP! Try to write (6.11) with appropriate initial conditions in (Non-CP) form.

The reactions considered in chapter 3 typically involved two or more reactants coming together to form a product. This involved the reactant and product concentrations to change throughout the experiment. In some reactions, a substance can influence a reaction, yet the total amount of the substance remains unchanged throughout the reaction. These substances are called *catalysts* (increases the reaction rate) or *inhibitors* (decreases the reaction rate). By definition, we can view them as both a reactant and product with the extra condition that the total concentration remains constant throughout the reaction.

Among the large class of catalyzed reactions we focus on the Michaelis-Menten kinetic model for an enzyme-catalyzed reaction. Suppose E represents an enzyme catalyst and P a product. Then, the mechanism

$$\mathrm{E} + \mathrm{S} \underset{k_-}{\overset{k_+}{\rightleftharpoons}} \mathrm{ES} \xrightarrow{k_2} \mathrm{E} + \mathrm{P} \tag{6.12}$$

represents a reaction in which a catalyst interacts with a reactant S to form an intermediate product ES, called a *substrate*, before the catalytic step in which the enzyme catalyst breaks apart leaving a final product, P. If we follow the rate law, we obtain the following equations corresponding to (6.12):

$$\begin{cases} \frac{d[E](t)}{dt} &= -k_+[E](t)[S](t) + k_-[ES](t) + k_2[ES](t) \\ \frac{d[S](t)}{dt} &= -k_+[E](t)[S](t) + k_-[ES](t) \\ \frac{d[ES](t)}{dt} &= k_+[E](t)[S](t) - k_-[ES](t) - k_2[ES](t) \\ \frac{d[P](t)}{dt} &= k_2[ES](t) \end{cases} \tag{6.13}$$

Using the crucial assumption that the total concentration of the enzyme catalyst remains constant, we know

$$\frac{d}{dt}\Big(E[t] + [ES](t)\Big) = 0.$$

Thus, if $[E]_0$ represents the initial concentration of the enzyme catalyst we have

$$[E](t) = [E]_0 - [ES](t) \tag{6.14}$$

for all times during the reaction **(Why?)**. Using (6.14) in (6.13) yields:

$$\begin{cases} \frac{d[E](t)}{dt} &= -k_+\Big([E]_0 - [ES](t)\Big)[S](t) + k_-[ES](t) + k_2[ES](t) \\ \frac{d[S](t)}{dt} &= -k_+\Big([E]_0 - [ES](t)\Big)[S](t) + k_-[ES](t) \\ \frac{d[ES](t)}{dt} &= k_+\Big([E]_0 - [ES](t)\Big)[S](t) - k_-[ES](t) - k_2[ES](t) \\ \frac{d[P](t)}{dt} &= k_2[ES](t) \end{cases} \tag{6.15}$$

The Michaelis-Menten model assumes that the rate of change of the substrate is much slower than that of the enzyme catalyst and product, so the model imposes a quasi-steady state assumption; that is,

$$\frac{d[ES](t)}{dt} = 0.$$

Using the quasi-steady state assumption in the third equation of (6.15) yields

$$[ES](t) = \frac{[E]_0[S](t)}{(\frac{k_-+k_2}{k_+}) + [S](t)} \tag{6.16}$$

for all times.

STOP! Verify (6.16).

Substituting (6.16) into (6.15) yields the following Michaelis-Merten enzyme catalyst model:

$$\begin{cases} \frac{d[S](t)}{dt} &= -k_2\frac{[E]_0[S](t)}{(\frac{k_-+k_2}{k_+})+[S](t)} \\ \frac{d[P](t)}{dt} &= k_2\frac{[E]_0[S](t)}{(\frac{k_-+k_2}{k_+})+[S](t)} \end{cases} \tag{6.17}$$

where $[E]_0$ is the intitial concentration of the enzyme catalyst.

Exercise 6.5.1. Verify (6.17) and explain why the differential equations for $[E](t)$ and $[ES](t)$ do not appear in the model.

Exercise 6.5.2. Consider the Michaelis-Merten model with appropriate initial conditions.

i.) Verify that the model cannot be written in (Non-CP) form and the matrix **A** is the zero matrix.

ii.) Do any other models in this chapter have **A** as the zero matrix?

Have you developed the sense that the presence of matrix **A** in (HCP) or (Non-CP) is good thing? The matrix gives us a way to compute the matrix exponential and thus *solve* differential equations. When confronted with a model in which such a matrix **A** cannot be identified, a common approach is to use asymptotic expansions to force one to appear. For example, consider the differential equation

$$\frac{dy(t)}{dt} = \frac{1}{1 - y(t)} \tag{6.18}$$

Exercise 6.5.3. Verify that (6.18) cannot be written in (Non-CP) form and that the matrix **A** is the zero matrix. (A one-dimensional matrix is a constant)

Suppose that due to the physical constraints of the application, you know that solutions are "small", maybe percentages, ratios, or concentrations all less than one in absolute value. Using this extra assumption enables us to write

$$\frac{1}{1 - y(t)} = \sum_{n=0}^{\infty} y(t)^n \tag{6.19}$$

(Why?). Furthermore, if $|y(t)| << 1$ (ie, significantly smaller than one in absolute value) it makes sense to truncate the infinite sum since the terms in the sum are approaching zero very quickly. In this case, we could approximate (6.18) by

$$\frac{dy(t)}{dt} = 1 + y(t) \tag{6.20}$$

or, if we truncated at a later point in the sum, by

$$\frac{dy(t)}{dt} = 1 + y(t) + y(t)^2. \tag{6.21}$$

Of course, the more terms you retain the more accurate your model will be, but at the same time the more complicated the model becomes.

Exercise 6.5.4. Consider the models (6.20) and (6.21).

i.) Try to write (6.20) in (Non-CP) form.

ii.) Verify (6.21) cannot be written in (Non-CP) but a non-zero matrix **A** is present.

Let us revisit the Michaelis-Merten enzyme model and apply this asymptotic expansion approach. First, let us define $K_M = (k_- + k_2)/k_+$ and $V = k_2[E]_0$. Factoring K_M out of the denominators in (6.17) yields

$$\begin{cases} \frac{d[S](t)}{dt} & = -V\frac{[S](t)}{K_M}\frac{1}{1-\left(-\frac{[S](t)}{K_M}\right)} \\ \frac{d[P](t)}{dt} & = V\frac{[S](t)}{K_M}\frac{1}{1-\left(-\frac{[S](t)}{K_M}\right)} \end{cases} \tag{6.22}$$

Upon inspection of (6.22), there is a term with the exact structure of $(1 - y(t))^{-1}$ **(Which one?)** If we impose the assumption

$$\left|\frac{[S](t)}{K_M}\right| << 1$$

then not only can we use (6.19), but it is also reasonable to truncate the infinite sum after two terms. Therefore, we can approximate the Michaelis-Merton model by

$$\begin{cases} \frac{d[S](t)}{dt} &= -V\frac{[S](t)}{K_M}(1 - \frac{[S](t)}{K_M}) \\ \frac{d[P](t)}{dt} &= V\frac{[S](t)}{K_M}(1 - \frac{[S](t)}{K_M}) \end{cases} \tag{6.23}$$

Exercise 6.5.5. In physical terms and the parameters of the model, how could you ensure that

$$\left|\frac{[S](t)}{K_M}\right| << 1?$$

Exercise 6.5.6. Verify that (6.23) cannot be written (Non-CP) form, but that it *can* be written in the form

$$\frac{d\mathbf{U}(t)}{dt} = \mathbf{A}\mathbf{U}(t) + \mathbf{f}(t, \mathbf{U}(t)),$$

where $\mathbf{A}$ is a non-zero matrix.

MATLAB-Exercise 6.5.1. We will use **MATLAB** to investigate the differences between (6.17) and (6.23). To begin, open **MATLAB** and in the command line, type: **MATLAB_Reactions_Semilinear**

Use the GUI to answer the following questions. Make certain to click on the 'Description of System' button for more details and instructions on how to use the GUI. A screenshot from this GUI can found in Figure 6.3.

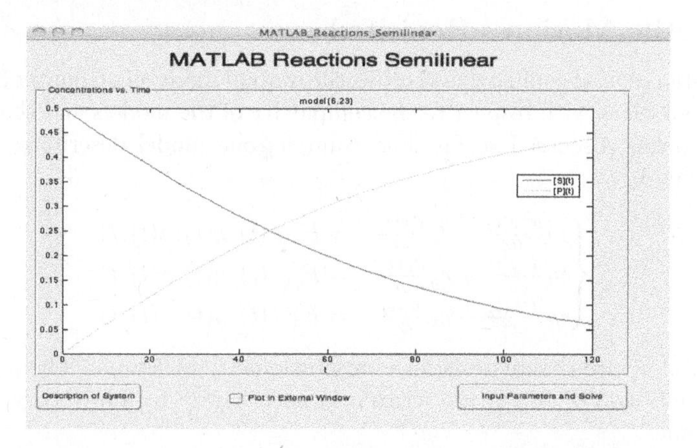

FIGURE 6.3: MATLAB Reactions Semilinear GUI

i.) Suppose $k_+ = 0.2, k_- = 0.18, k_2 = 0.1$, $[S]_0 = 0.5$ and $[E]_0 = 0.5$.

a.) Plot the solutions to (6.17) and (6.23) for a reaction lasting 120 seconds.

b.) How would you describe the differences between the two solutions? What is going on in the reaction at the time at which the models differ the most?

c.) Do the models' solutions get closer together in the long run?

d.) From the graphs of the solutions, convince yourself that $|[S](t)| \leq [S](0)$ for all times t and then compute an upper bound for

$$\left|\frac{[S](t)}{K_M}\right|.$$

e.) Did the parameters in this problem satisfy the assumptions of model (6.23)?

ii.) Suppose $k_+ = 0.4, k_- = 0.18, k_2 = 0.02$, $[S]_0 = 0.5$ and $[E]_0 = 0.5$.

a.) Plot the solutions to (6.17) and (6.23) for a reaction lasting 120 seconds.

b.) How would you describe the differences between the two solutions? What is going on in the reaction at the time at which the models differ the most?

c.) Do the models' solutions get closer together in the long run?

d.) Did the parameters given in this problem satisfy the assumptions of model (6.23)?

iii.) Suppose $k_+ = 0.1, k_- = 0.039, k_2 = 0.001$, $[S]_0 = 0.5$ and $[E]_0 = 0.5$.

a.) Plot the solutions to (6.17) and (6.23) for a reaction lasting 120 seconds.

b.) How would you describe the differences between the two solutions? Something very different is going on with the solution to (6.23)? What is it?

c.) Do the models' solutions get closer together in the long run?

d.) Did the parameters in this problem satisfy the assumptions of model (6.23)? ■

6.6 Projectile Motion—Revisited

We considered several sophisticated projectile motion models in Chapter 3. At this point, we can orient ourselves with respect to the complexity of the models and the different types of Cauchy problems discussed so far. For example, one model describing the position of projectile is given by

$$\begin{cases} m\frac{d^2x(t)}{dt^2} + r_x\frac{dx(t)}{dt} &= F_x(x(t), y(t), z(t), t) \\ m\frac{d^2y(t)}{dt^2} + r_y\frac{dy(t)}{dt} &= F_y(x(t), y(t), z(t), t) \\ m\frac{d^2z(t)}{dt^2} + r_z\frac{dz(t)}{dt} &= F_z(x(t), y(t), z(t), t) \end{cases} \tag{6.24}$$

where r_x, r_y, and r_z are the air resistance proportionality constants. The forcing functions on the right-hand side of (6.24) are written in such a way to imply they can depend on position and time.

STOP! Explain why the theories developed in Chapters 4 and 5 do not apply to (6.24).

We simplified the discussion by removing the state dependence in the forcing terms,

$$\begin{cases} m\frac{d^2x(t)}{dt^2} + r_x\frac{dx(t)}{dt} &= F_x(t) \\ m\frac{d^2y(t)}{dt^2} + r_y\frac{dy(t)}{dt} &= F_y(t) \\ m\frac{d^2z(t)}{dt^2} + r_z\frac{dz(t)}{dt} &= F_z(t) \end{cases} \tag{6.25}$$

Then we altered the air resistance assumption to obtain the model

$$\begin{cases} m\frac{d^2x(t)}{dt^2} + r_x\left\|\frac{d\mathbf{u}(t)}{dt}\right\|\frac{dx(t)}{dt} &= F_x(t) \\ m\frac{d^2y(t)}{dt^2} + r_y\left\|\frac{d\mathbf{u}(t)}{dt}\right\|\frac{dy(t)}{dt} &= F_y(t) \\ m\frac{d^2z(t)}{dt^2} + r_z\left\|\frac{d\mathbf{u}(t)}{dt}\right\|\frac{dz(t)}{dt} &= F_z(t) \end{cases} \tag{6.26}$$

where $\|\frac{d\mathbf{u}(t)}{dt}\| = \sqrt{(\frac{dx(t)}{dt})^2 + (\frac{dy(t)}{dt})^2 + (\frac{dz(t)}{dt})^2}$.

STOP! Can (6.26), when combined with the six necessary initial conditions, be written in (Non-CP) form?

In summary, our (HCP) and (Non-CP) theories cannot be used to study the evolution of the solutions of (6.24) and (6.26). However, we have seen instances in which they adequately model a projectile's motion. (See **MATLAB_Projectile_Motion** GUI.) It is reasonable to expect a theory similar to those formed in Chapters 4 and 5, but just as (Non-CP) added a layer of complexity to (HCP), semilinear models will add a layer of complexity on top of (Non-CP).

6.7 Floor Displacement Model with Nonlinear Shock Absorbers

We introduced two floor displacement models, one with shock absorbers and one without in Chapter 3. Recall in our discussion of circuits, we observed that not accounting for resistance or not having appropriate resistors could cause devastating damage to the circuit. The same applies to floor displacement. **(Why?)** Since circuits behave like springs, and the floor displacement model was guided by spring-mass systems, the physics that caused damage in the circuit would potentially cause damage in a building. Shock absorbers now play the role of resistors and having nonlinear shock absorbers can be more effective in stabilizing the floors of a building during an earthquake.

A common nonlinear shock absorber is one for which the resistance to motion is proportional to the square of the speed of the floor.

STOP! Does this type of resistance sound familiar? If so, into which models was it incorporated?

If we use this type of shock absorber, our first shock absorber model

$$\begin{aligned}
m_1 x_1''(t) &= -k_0(x_1(t) - g(t)) + k_1(x_2(t) - x_1(t)) - c_0(x_1'(t) - g'(t)) + c_1(x_2'(t) - x_1'(t)) \\
m_i x_i''(t) &= -k_{i-1}(x_i(t) - x_{i-1}(t)) + k_i(x_{i+1}(t) - x_i(t)) - c_{i-1}(x_i'(t) - x_{i-1}'(t)) \\
&\quad + c_i(x_{i+1}'(t) - x_i'(t)) \\
m_5 x_5''(t) &= -k_4(x_5(t) - x_4(t)) - c_4(x_5'(t) - x_4'(t))
\end{aligned}$$

will become

$$\begin{aligned}
m_1 x_1''(t) = -k_0(x_1(t) - g(t)) + k_1(x_2(t) - x_1(t)) - c_0\Big(|x_1'(t)|x_1'(t) - |g'(t)|g'(t)\Big) \\
+ c_1\Big(|x_2'(t)|x_2'(t) - |x_1'(t)|x_1'(t)\Big)
\end{aligned} \tag{6.27}$$

$$\begin{aligned}
m_i x_i''(t) = -k_{i-1}(x_i(t) - x_{i-1}(t)) + k_i(x_{i+1}(t) - x_i(t)) - c_{i-1}\Big(|x_i'(t)|x_i'(t) - |x_{i-1}'(t)|x_{i-1}'(t)\Big) \\
+ c_i\Big(|x_{i+1}'(t)|x_{i+1}'(t) - |x_i'(t)|x_i'(t)\Big)
\end{aligned} \tag{6.28}$$

$$m_5 x_5''(t) = -k_4\left(x_5(t) - x_4(t)\right) - c_4\left(|x_5'(t)|x_5'(t) - |x_4'(t)|x_4'(t)\right) \tag{6.29}$$

for $i = 2, 3, 4$ where $c_i (i = 0, \ldots, 4)$ are the proportionality constants for the absorbers. We will refer to (6.27)-(6.29) as the 5-story floor displacement model with nonlinear shock absorbers.

Exercise 6.7.1. Consider the 5-story floor displacement model with nonlinear shock absorbers.

i.) Verify the model's equations.

ii.) How does the model simplify if we assume that all the proportionality constants are the same?

iii.) Try to write the model in (Non-CP) form. What happens?

Chapter 7

Finite-Dimensional Theory—Last Step: The Semi-Linear Case

Overview

The models introduced in Chapter 6 are all examples of what happens when the external forces appearing in a system of differential equations or a higher-order differential equation involve nonlinear interactions among the components of the solution vector. Our present goal is to develop an extension of the theory formulated for (Non-CP) in Chapter 5 that enables us to handle large classes of such nonlinear forcing terms. The theory developed will subsume the theory of Chapters 4 and 5 as special cases.

7.1 Introducing the Even-More General Semi-Linear Cauchy Problem (Semi-CP)

As observed at the end of Chapter 6, homogenous systems of linear differential equations, when equipped with time-dependent external forcing terms involving nonlinear interactions among the components of the solution vector, can be written in the following form.

Definition 7.1.1. A system of ordinary differential equations of the form

$$\textbf{(Semi}-\textbf{CP)}\quad \begin{cases} \mathbf{U}'(t) = \mathbf{A}\mathbf{U}(t) + \mathbf{f}(t, \mathbf{U}(t)),\ t > t_0, \\ \mathbf{U}(t_0) = \mathbf{U}_0, \end{cases} \tag{7.1}$$

where $\mathbf{U} : [t_0, \infty) \to \mathbb{R}^N$ is the unknown vector function

$$\mathbf{U}(t) = \begin{bmatrix} u_1(t) \\ \vdots \\ u_N(t) \end{bmatrix},$$

$\mathbf{A}$ is the $N \times N$ coefficient matrix

$$\begin{bmatrix} a_{11} & \cdots & a_{1N} \\ \vdots & \ddots & \vdots \\ a_{N1} & \cdots & a_{NN} \end{bmatrix},$$

the nonlinear forcing term $\mathbf{f} : [t_0, \infty) \times R^N \to \mathbb{R}^N$ has components

$$\mathbf{f}(t) = \begin{bmatrix} f_1(t, \mathbf{U}(t)) \\ \vdots \\ f_N(t, \mathbf{U}(t)) \end{bmatrix} = \begin{bmatrix} f_1(t, u_1(t), ..., u_N(t)) \\ \vdots \\ f_N(t, u_1(t), ..., u_N(t)) \end{bmatrix},$$

and $\mathbf{U}_0 \in \mathbb{R}^N$ is the vector containing the initial conditions

$$\mathbf{U}_0 = \begin{bmatrix} u_1(t_0) \\ \vdots \\ u_N(t_0) \end{bmatrix},$$

is a *semi-linear system of ODEs with initial condition* $\mathbf{U}(t_0) = \mathbf{U}_0$.

Converting a system into the form (Semi-CP) is not difficult. Consider the following example:

Example 7.1.1. Consider the system of ODEs given by

$$\begin{cases} u_1'(t) &= -2u_1(t) + 3u_2(t) - 5u_1^2(t) + 2u_2^4(t) \\ u_2'(t) &= -u_1(t) + 2u_2(t) + t - \sin\left(u_1^2(t)\right) \\ u_1(1) &= 1 \\ u_2(1) &= -1 \end{cases} \tag{7.2}$$

Initial-value problem (7.2) can be rewritten using matrices as follows:

$$\begin{cases} \underbrace{\begin{bmatrix} u_1(t) \\ u_2(t) \end{bmatrix}'}_{\mathbf{U}'(t)} = \underbrace{\begin{bmatrix} -2 & 3 \\ -1 & 2 \end{bmatrix}}_{\mathbf{A}} \underbrace{\begin{bmatrix} u_1(t) \\ u_2(t) \end{bmatrix}}_{\mathbf{U}(t)} + \underbrace{\begin{bmatrix} -5u_1^2(t) + 2u_2^4(t) \\ t - \sin\left(u_1^2(t)\right) \end{bmatrix}}_{\mathbf{f}(t, \mathbf{U}(t))} \\ \underbrace{\begin{bmatrix} u_1(1) \\ u_2(1) \end{bmatrix}}_{\text{Here } t_0 = 1} = \underbrace{\begin{bmatrix} 1 \\ -1 \end{bmatrix}}_{\mathbf{U}_0} \end{cases} \tag{7.3}$$

It is reasonable at this stage to think of a solution of (7.3) as a vector $\mathbf{U}(t) = \begin{bmatrix} u_1(t) \\ u_2(t) \end{bmatrix}$ satisfying both the equation and the initial condition in (7.3).

STOP! Revisit the IVPs in Chapter 6 and make certain you are comfortable with reformulating them in the abstract form (Semi-CP).

Exercise 7.1.1. Reformulate the following IVPs in the form of (Semi-CP).

$$\text{i.)}\begin{cases} x'(t) &= 2y(t) - tx^3(t) \\ y'(t) &= y(t) - 3z(t) + 1 \\ z'(t) &= -3x(t) - e^{-y(t)} + (1 - tx(t))^3 \\ x(1) &= -1 \\ y(1) &= -4 \\ z(1) &= -2 \end{cases}$$

$$\text{ii.)}\begin{cases} x'''(t) &= x''(t) - x(t) + y^2(t) - 3 \\ y''(t) &= y'(t) - 2\cos\left(y(t) + z'(t)\right) - \sin t \\ z''(t) &= z(t) + (1 - x''(t))^3 \\ x(-1) &= x'(-1) = x''(-1) = 0 \\ y(-1) &= y'(-1) = 2 \\ z(-1) &= z'(-1) = -5 \end{cases}$$

7.2 New Challenges

You might, quite reasonably, conclude from a cursory comparison of (Non-CP) and (Semi-CP) that the two Cauchy problems really are not that different. After all, the processes used to reformulate non-homogenous systems of linear ODEs and semi-linear systems of ODEs as abstract Cauchy problems are virtually identical, with the only exception being that the forcing term for semi-linear problems depends on more arguments. While this is true, the fact that the arguments in the forcing term can be the components of the solution vector can cause trouble! A brief illustration of the problems that can arise is provided in the following exercises.

Exercise 7.2.1. For simplicity, we consider some special cases of the one-dimensional version of (Semi-CP) with $\mathbf{A} = 0$.

i.) Show that the following IVP has more than one solution:

$$\begin{cases} x'(t) &= 3\left[x(t)\right]^{\frac{5}{8}}, \ t > 0, \\ x(0) &= 0. \end{cases} \tag{7.4}$$

ii.) Consider the IVP

$$\begin{cases} x'(t) = (1 + 2x(t))^4, \ t > 0, \\ x(0) = x_0. \end{cases} \tag{7.5}$$

a.) Show that a solution of (7.5) is given by

$$x(t) = \frac{1}{2}\left[-1 + \left((2x_0 + 1)^{-3} - 6t\right)^{-\frac{1}{3}}\right].$$

b.) Let $T = \frac{1}{6}(2x_0+1)^{-3}$. Assuming that $x_0 > -\frac{1}{2}$, show that $\lim_{t\to T^-} |x(t)| = \infty$. As such, (7.5) does not have a continuous solution on the entire interval $[0,\infty)$.

Note that the right-side $f(t,x(t))$ of both IVPs in Exercise 7.2.1 is a continuous function of both variables. **(Why?)** This was sufficient to guarantee the existence and uniqueness of a mild solution of (Non-CP) on $[0,\infty)$ when $f(t,x(t)) = f(t)$. However, this is false for (7.4), and in a big way for (7.5). And, this is the case for *one*-dimensional problems. We do not expect the situation to improve for the systems-version of (Semi-CP). As such, we are faced with several questions right off the bat:

1. How is a mild solution of (Semi-CP) defined? How about a classical solution?
2. When does (Semi-CP) have a unique solution on a given interval?
3. Under what conditions is a mild solution also a classical solution?
4. What is the largest interval on which a mild solution exists?

We shall explore these questions and along the way develop various strategies of attack that will be used throughout the remainder of the text.

7.3 Behind the Scenes: Issues and Resolutions Arising in the Study of (Semi-CP)

We introduced two notions of a solution of (Non-CP) in Chapter 5—mild and classical—in response to observing that the smoothness of the forcing term entered in a nontrivial way when proving the differentiability of the solution candidate given by the variation of parameters formula. Specifically, the existence and uniqueness of a classical solution of (Non-CP) was not guaranteed unless the forcing term was continuous. The present situation for (Semi-CP) is starting off more bleak in the sense that continuity of the forcing term apparently falls woefully short of "saving the day" as it once did in Chapter 5. This begs the question as to whether or not there is a reasonable condition that, when imposed on the forcing term $f(t,\mathbf{U}(t))$, does come to the rescue and enables us to formulate an existence-uniqueness result for (Semi-CP).

For the moment, we begin with the assertion that the following definition is reasonable.

Definition 7.3.1. A function $\mathbf{U}:[t_0,t_0+T]\to\mathbb{R}$ is a

i.) *classical solution* of (Semi-CP) if

a.) $\mathbf{U}(\cdot)$ is continuous on $[t_0,t_0+T]$,
b.) $\mathbf{U}(\cdot)$ is differentiable on (t_0,t_0+T),
c.) $\mathbf{U}'(t) = \mathbf{A}\mathbf{U}(t) + \mathbf{f}(t,\mathbf{U}(t))$, for all $t\in(t_0,t_0+T)$,
d.) $\mathbf{U}(t_0) = \mathbf{U}_0$.

ii.) *mild solution* of (5.1) if

a.) $\mathbf{U}(\cdot)$ is continuous on $[t_0,t_0+T]$,
b.) $\mathbf{U}(t) = e^{\mathbf{A}(t-t_0)}\mathbf{U}_0 + \int_{t_0}^{t} e^{\mathbf{A}(t-s)}\mathbf{f}(s,\mathbf{U}(s))ds$, for all $t\in(t_0,t_0+T)$.

STOP! Why is Definition 7.3.1 reasonable?

Of course, actually establishing the existence of solutions of either type is another story. The situation is more complicated because the "solution" $\mathbf{U}(t)$ occurs on both sides of the variation of parameters formula in (ii)(b), thereby making the solution implicitly-defined.

STOP! Go back to the discussion surrounding the proof of Theorem 5.3.1. Try to identify specific parts of the argument that become problematic, or simply false, when replacing $\mathbf{f}(t)$ by $\mathbf{f}(t, \mathbf{U}(t))$.

In light of this we need to devise a new approach to establishing the existence and uniqueness of mild solutions. We investigate one such approach in the following **EXPLORE!** activity. Devising a new approach to establishing the existence of a mild solution of (Semi-CP) and identifying a suitable manner in which to restrict the behavior of the forcing term are intertwined tasks. Beginning with (HCP) and working our way up in the stages, we shall attempt to solve this mystery.

EXPLORE!

1. An Exploration of (HCP)

We begin by presenting an alternate approach to constructing the solution of (HCP) called the *method of successive approximations.* Broadly speaking, the underlying idea of a convergence argument is to define a sequence of functions whose limit (taken in an appropriate sense and space) is a solution of the IVP under investigation. We have encountered a version of this method in the proofs of various approximation theorems. **(Tell where.)** We shall illustrate such a convergence approach, which can be adapted to the present setting, by considering the following homogenous IVP in $\mathbb{R}^N$:

$$\begin{cases} \mathbf{U}'(t) = \mathbf{A}\mathbf{U}(t), \ 0 \leq t \leq T, \\ \mathbf{U}(0) = \mathbf{U}_0, \end{cases} \tag{7.6}$$

where $\mathbf{U}_0 \in \mathbb{R}^N$and $\mathbf{A} \in \mathbb{M}^N(\mathbb{R})$. The desired solution $\mathbf{U}(t) = e^{\mathbf{A}t}\mathbf{U}_0$ is constructed using the iteration scheme described below.

We begin with the integrated form of (7.6), namely

$$\mathbf{U}(t) = \mathbf{U}_0 + \int_0^t \mathbf{A}\mathbf{U}(s)ds, \ 0 \leq t \leq T. \tag{7.7}$$

We must overcome the self-referential nature of (7.7). One way to do so is to replace $\mathbf{U}(s)$ on the right-side of (7.7) by an approximation of $\mathbf{U}(s)$. Presently, the only knowledge about $\mathbf{U}$ that we have is its value at $t = 0$, namely $\mathbf{U}_0$. So, naturally we use $\mathbf{U}(s) = \mathbf{U}_0$ as an initial approximation. Making this substitution yields the following crude approximation of $\mathbf{U}(t)$:

$$\mathbf{U}(t) \approx \mathbf{U}_0 + \int_0^t \mathbf{A}\mathbf{U}_0 ds = \mathbf{U}_0 + \mathbf{A}\mathbf{U}_0 t, \ 0 \leq t \leq T. \tag{7.8}$$

Let $\mathbf{U}_1(t) = \mathbf{U}_0 + \mathbf{A}\mathbf{U}_0 t$. Now, in order to improve the approximation, we can replace

$\mathbf{U}(s)$ on the right-side of (7.7) by $\mathbf{U}_1(s)$ to obtain

$$
\begin{aligned}
\mathbf{U}(t) &\approx \mathbf{U}_0 + \int_0^t \mathbf{A}\mathbf{U}_1(s)ds \\
&= \mathbf{U}_0 + \int_0^t \mathbf{A}\left[\mathbf{U}_0 + \int_0^s \mathbf{A}\mathbf{U}_0 d\tau\right] ds \\
&= \mathbf{U}_0 + \int_0^t \mathbf{A}\mathbf{U}_0 ds + \int_0^t \mathbf{A}\left(\int_0^s \mathbf{A}\mathbf{U}_0 d\tau\right) ds \\
&= \mathbf{U}_0 + \mathbf{A}\mathbf{U}_0 t + \mathbf{A}^2\mathbf{U}_0 \frac{t^2}{2} \\
&= \left(\mathbf{A}^0 t^0 + \mathbf{A}t + \mathbf{A}^2 \frac{t^2}{2}\right)\mathbf{U}_0.
\end{aligned}
\tag{7.9}
$$

i.) Justify all steps in (7.9).

The above sequence of "successively better" approximations can be formally described by the recursive sequence

$$\mathbf{U}_m(t) = \mathbf{U}_0 + \int_0^t \mathbf{A}\mathbf{U}_{m-1}(s)ds, \; m \in \mathbb{N}. \tag{7.10}$$

Proceeding as in (7.9) leads to the following explicit formula for $\mathbf{U}_m(t)$:

$$\mathbf{U}_m(t) = \sum_{k=0}^{m} \left(\frac{\mathbf{A}^k t^k}{k!}\right)\mathbf{U}_0. \tag{7.11}$$

Moreover,

$$\lim_{m\to\infty}\left(\sup_{0\le t\le T} \left\|\mathbf{U}_m(t) - e^{\mathbf{A}t}\mathbf{U}_0\right\|_{\mathbb{R}^N}\right) = 0. \tag{7.12}$$

ii.) Justify (7.11) and (7.12).

iii.) For a fixed $m_0 \in \mathbb{N}$, what does the quantity $\sup_{0\le t\le T} \left\|\mathbf{U}_{m_0}(t) - e^{\mathbf{A}t}\mathbf{U}_0\right\|_{\mathbb{R}^N}$ tell you?

Of course, this is old news since we already proved that $\mathbf{U}(t) = e^{\mathbf{A}t}\mathbf{U}_0$ is the unique mild solution of (7.6) in Chapter 4. But, it does suggest that studying the convergence of the sequence

$$\mathbf{U}_n(t) = e^{\mathbf{A}t}\mathbf{U}_0 + \int_0^t e^{\mathbf{A}(t-s)} f(s, \mathbf{U}_{n-1}(s))ds, \; 0 \le t \le T, \tag{7.13}$$

is a viable approach to establishing the existence of a mild solution of (Semi-CP).

iv.) Why must the limit u of (7.13) be continuous?

2. Extension to (Non-CP)
Modify each line of the above argument suitably to take into account the forcing term. The result of applying this procedure should be the variation of parameters formula for the mild solution of (Non-CP).

The discussion in the above **EXPLORE!** activity suggests that the variation of parameters formula $\mathbf{U}(t) = e^{\mathbf{A}(t-t_0)}\mathbf{U}_0 + \int_{t_0}^{t} e^{\mathbf{A}(t-s)}\mathbf{f}(s,\mathbf{U}(s))ds$ in Definition 7.3.1 is reasonable. However, the self-referential nature of the formula—that is, the fact that the unknown occurs on both sides—poses a new challenge. The following numerical scheme, however, enables us to effectively use this formula.

STOP! Re-read Section 5.2.2.

Let $t > 0$, $n \in \mathbb{N}$ and define $\bar{\mathbf{U}}$ as the solution to the approximate semilinear IVP

$$\begin{cases} \frac{\bar{\mathbf{U}}(s+\frac{t}{n})-\bar{\mathbf{U}}(s)}{\frac{t}{n}} &= \mathbf{A}\bar{\mathbf{U}}(s+\frac{t}{n}) + \mathbf{f}(s,\bar{\mathbf{U}}(s)), \;\; s \geq 0 \\ \bar{\mathbf{U}}(0) &= \mathbf{U}_0 \end{cases} \tag{7.14}$$

Solving the first equation in (7.14) for $\bar{\mathbf{U}}(s+\frac{t}{n})$ yields

$$\bar{\mathbf{U}}\left(s+\frac{t}{n}\right) = \left(1-\frac{t}{n}\mathbf{A}\right)^{-1}\bar{\mathbf{U}}(s) + \left(1-\frac{t}{n}\mathbf{A}\right)^{-1}\mathbf{f}(s,\bar{\mathbf{U}}(s))\frac{t}{n} \tag{7.15}$$

Taking one step towards t amounts to evaluating (7.15) at $s = 0$ and realizing that

$$\begin{aligned} \bar{\mathbf{U}}\left(\frac{t}{n}\right) &= \left(1-\frac{t}{n}\mathbf{A}\right)^{-1}\bar{\mathbf{U}}_0 + \left(1-\frac{t}{n}\mathbf{A}\right)^{-1}\mathbf{f}\left(0,\bar{\mathbf{U}}(0)\right)\frac{t}{n} \\ &= \left(1-\frac{t}{n}\mathbf{A}\right)^{-1}\bar{\mathbf{U}}_0 + \left(1-\frac{t}{n}\mathbf{A}\right)^{-1}\mathbf{f}\left(0,\mathbf{U}_0)\right)\frac{t}{n} \end{aligned} \tag{7.16}$$

Notice the right side of (7.16) is known. That is, the formula for $\bar{\mathbf{U}}(\frac{t}{n})$ is explicit, not self-referential, meaning we can use it to compute a solution. Continuing as in Section 5.2.2, evaluate (7.15) at $s = \frac{t}{n}$ to get

$$\begin{aligned} \bar{\mathbf{U}}\left(\frac{2t}{n}\right) &= \left(1-\frac{t}{n}\mathbf{A}\right)^{-1}\bar{\mathbf{U}}\left(\frac{t}{n}\right) + \left(1-\frac{t}{n}\mathbf{A}\right)^{-1}\mathbf{f}\left(\frac{t}{n},\mathbf{U}\left(\frac{t}{n}\right)\right)\frac{t}{n} \\ &= \left(1-\frac{t}{n}\mathbf{A}\right)^{-2}\bar{\mathbf{U}}_0 \\ &+ \left(\left(1-\frac{t}{n}\mathbf{A}\right)^{-2}\mathbf{f}\left(0,\bar{\mathbf{U}}(0)\right) + \left(1-\frac{t}{n}\mathbf{A}\right)^{-1}\mathbf{f}\left(\frac{t}{n},\bar{\mathbf{U}}\left(\frac{t}{n}\right)\right)\right)\frac{t}{n} \end{aligned} \tag{7.17}$$

Like before, the right side of (7.17) is explicit because $\bar{\mathbf{U}}(\frac{t}{n})$ is given by (7.16), and thus an applicable numeric scheme for the solution. Taking n steps to arrive at t yields

$$\bar{\mathbf{U}}(t) = \left(1-\frac{t}{n}\mathbf{A}\right)^{-n}\mathbf{U}_0 + \sum_{j=0}^{n-1}\left(1-\frac{t}{n}\mathbf{A}\right)^{-(n-j)}\mathbf{f}\left(\frac{t}{n}j,\bar{\mathbf{U}}\left(\frac{t}{n}j\right)\right)\frac{t}{n} \tag{7.18}$$

As was the case in Section 5.2.2, under the appropriate assumptions taking the limit of the approximate IVP solution (7.18) yields the solution to the original IVP. More importantly, we recover the mild solution formula found in Definition 7.3.1. To convince yourself of this, return to the argument given in Section 5.2.2 and everywhere you see $\mathbf{f}(s)$ replace it with $\mathbf{f}(s,\mathbf{U}(s))$.

STOP! Go back and do it!

Remark 7.3.1. The numerical scheme found in (7.18) allows us to effectively use the formula in Definition 7.3.1.

We have the variations of parameters formula via a numerical approximation technique. Going back to applying the method of successive approximations to (Semi-CP), the main struggle with which we are faced is ensuring the convergence of the sequence. As usual, this boils down to the behavior of the mapping $t \mapsto \int_{t_0}^{t} e^{\mathbf{A}(t-s)}\mathbf{f}(s, \mathbf{U}(s))ds$. Simply put, if the mapping $\mathbf{f}$ is "sufficiently well-behaved," we will be able to show convergence. That's great, but we must be precise as to what constitutes "sufficiently well-behaved." We do not want to impose ridiculously stringent conditions on $\mathbf{f}$ because while doing so will facilitate establishing convergence, the downside will be that the class of problems to which the resulting theorem applies will be disappointingly small. On the other hand, if we do not impose a strong enough restriction on $\mathbf{f}$, then showing convergence might very well be impossible. We need to somehow strike a balance.

We will explore this in the next section with an eye toward naturally facilitating the convergence argument while paying attention to circumventing the self-referential nature of the variation of parameters formula without making the class of problems to which the resulting theory applies too small.

7.4 Lipschitz to the Rescue!

We shall introduce different ways of controlling the behavior of the nonlinear forcing terms arising in this chapter.

Exercise 7.2.1 revealed that mere continuity of $\mathbf{f}$ in both variables is insufficient to guarantee uniqueness of a mild solution of (Semi-CP) on $[0, \infty)$, even in the one-dimensional setting. As such, it is sensible to ask what conditions could be coupled with continuity to ensure the existence (and possibly uniqueness) of a mild solution of (Semi-CP) on at least some interval $[0, T_0)$. There is a plentiful supply of such conditions that can be imposed on $\mathbf{f}$ which further control its "growth." We introduce a common one in this section.

At the restrictive end of the spectrum, we can require $\mathbf{f}$ to have a smooth derivative on its domain (that is, $\mathbf{f}$ is continuously differentiable on its domain). In such case, it would follow that for a given continuous function $\mathbf{u}$, the mapping

$$t \mapsto \int_0^t e^{\mathbf{A}(t-s)}\mathbf{f}(s, \mathbf{u}(s))ds$$

is continuously differentiable. **(Why?)** Hence, the function $\mathbf{u}(\cdot)$ given by (Semi-CP) would be differentiable, and so has a chance at being a classical solution of (Semi-CP).

While imposing this level of regularity leads to a desirable conclusion, it prevents us from considering many forcing terms that arise naturally in practice (e.g., $f(u) = |u|$). A close investigation of (7.4) reveals that the curve corresponding to the forcing term was sufficiently steep in a vicinity of (0,0) as to enable us to construct a sequence of chord lines, all passing through (0,0), whose slopes became infinitely large.

Exercise 7.4.1. Show that the sequence of chord line slopes for $f(x) = 3x^{\frac{5}{8}}$ connecting (0,0) to $\left(\frac{1}{n}, f\left(\frac{1}{n}\right)\right)$ approaches infinity as $n \to \infty$.

As such, close to the initial starting point, the behavior of f changes very quickly. This, in turn, affects the behavior of x' in a short interval of time. Moreover, this worsens the closer you get to the origin. Thus, if we were to try to generate the solution path on a given time interval $[0,T]$ numerically, refining the partition of $[0,T]$ (in order to increase the number of time points used to construct the approximate solution) would subsequently result in a sequence of paths which does not approach a single recognizable continuous curve.

The presence of the cusp in the graph is the troublemaker! Certainly, continuously differentiable functions cannot exhibit such behavior, but can we somehow control the chord line slopes without demanding that f be so nice? The search for such control over chord line slopes prompts us to make the following definition.

Definition 7.4.1. A function $f:\mathbb{R}\to\mathbb{R}$ is *globally Lipschitz on* $\mathcal{D}\subset\mathbb{R}$ if there exists $M_f>0$ such that

$$|f(x)-f(y)|\le M_f\,|x-y|\,,\ \forall x,y\in\mathcal{D}. \tag{7.19}$$

(M_f is called a *Lipschitz constant* for f.)

This definition is easily adapted to functions of more than one independent variable, but we must carefully indicate to which of the independent variables we intend the condition to apply. Functions of the form $\mathbf{g}:[0,T]\times\mathbb{R}^N\to\mathbb{R}^N$ commonly arise in practice. We introduce the following modification of Def. 7.4.1 as it applies to such functions.

Definition 7.4.2. A function $\mathbf{g}:[0,T]\times\mathbb{R}^N\to\mathbb{R}^N$ is *globally Lipschitz on* $\mathcal{D}\subset\mathbb{R}^N$ *(uniformly in t)* if there exists $M_g>0$ (independent of t) such that

$$\|\mathbf{g}(t,\mathbf{x})-\mathbf{g}(t,\mathbf{y})\|_{\mathbb{R}^N}\le M_{\mathbf{g}}\,\|\mathbf{x}-\mathbf{y}\|_{\mathbb{R}^N}\,,\ \forall t\in[0,T]\ \text{and}\ \mathbf{x},\mathbf{y}\in\mathcal{D}. \tag{7.20}$$

MATLAB-Exercise 7.4.1. We will use **MATLAB** to provide a graphical illustration of the definition of Lipschitz continuity. Open **MATLAB** and in the command line, type:

MATLAB_Lipschitz

Use the GUI to answer the following questions. Make certain to click on the HELP button for instructions on how to use the GUI. A typical screenshot of the GUI can be found in Figure 7.1.

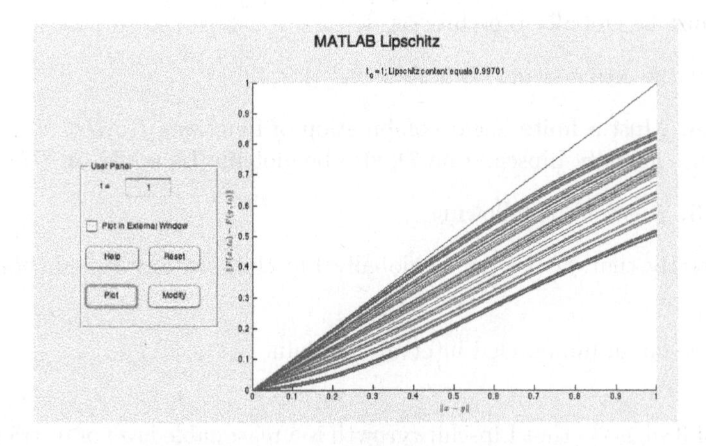

FIGURE 7.1: MATLAB Lipschitz GUI

i.) For each of the following choices for the function $\mathbf{g} : [0,T] \times \mathbb{R}^N \to \mathbb{R}^N$ use the GUI to investigate Lipshitz continuity at t_0 near the provided $\mathbf{x} = \langle x_1, x_2, \ldots, x_N \rangle \in \mathbb{R}^N$.

a.) $g(t,x) = xe^{-3t}$ at $t_0 = 0$ near $x = 0$. Repeat for $t_0 = 1,2,3$. Did the Lipschitz constant change? If so, how?

b.) $g(t,x) = tx^2$ at $t_0 = 1$ near $x = 1$. Repeat for $t_0 = 2,3,4$. Did the Lipschitz constant change? If so, how?

c.) $\mathbf{g}(t,\mathbf{x}) = \langle \cos(2\pi x_1), \sin(2\pi x_2) \rangle$ at $t_0 = 0$ near $\mathbf{x} = \langle 1,1 \rangle$. Repeat for $t_0 = 0.1, 0.25, 0.5$. Did the Lipschitz constant change? If so, how?

d.) $\mathbf{g}(t,\mathbf{x}) = \langle \frac{1}{1+x_1^2}, \sqrt{x_1^2 + x_2^2 + x_3^2}, e^{-t} \rangle$ at $t_0 = 0$ near $\mathbf{x} = \langle 1,1,1 \rangle$. Repeat for $t_0 = 0.5, 0.1, 50$. Did the Lipschitz constant change? If so, how?

ii.) Use the GUI and your observations in (i) to convince yourself of the following claims:

a.) The function $g : [0,\infty) \times [0,1] \to \mathbb{R}$ given by $g(t,x) = xe^{-3t}$ is globally Lipschitz on $[0,1]$ (uniformly in t).

b.) The function $g : [0,\infty) \times \mathbb{R} \to \mathbb{R}$ given by $g(t,x) = tx^2$ is not globally Lipschitz on $\mathbb{R}$ (uniformly in t).

c.) The function $\mathbf{g} : [0,0.5] \times \mathbb{R} \to \mathbb{R}$ given by $\mathbf{g}(t,\mathbf{x}) = \langle \cos(2\pi x_1), \sin(2\pi x_2) \rangle$ is globally Lipschitz on $\mathbb{R}$ (uniformly in t).

d.) The function $\mathbf{g} : [0,\infty] \times \mathbb{R} \to \mathbb{R}$ given by $\mathbf{g}(t,\mathbf{x}) = \langle \frac{1}{1+x_1^2}, \sqrt{x_1^2 + x_2^2 + x_3^2}, e^{-t} \rangle$ is globally Lipschitz on $\mathbb{R}$ (uniformly in t).

■

Exercise 7.4.2. Interpret (7.20) geometrically. For simplicity, assume that $N = 1$. How does this interpretation change if $M_{\mathbf{g}}$ depends on t?

Exercise 7.4.3. Let $f : \mathbb{R} \to \mathbb{R}$ be given.

i.) If f is globally Lipschitz, show that f is continuous.

ii.) Provide a counterexample to show the converse of (i) is false.

iii.) If f is continuously differentiable on $\mathbb{R}$ and its partial derivatives are bounded, show that f must be globally Lipschitz on $\mathbb{R}$.

iv.) Show that the converse of (iii) is false.

Exercise 7.4.4. Must a finite linear combination of functions $f_i : \mathcal{D} \subset \mathbb{R} \to \mathbb{R}$, $1 \le i \le n$, each of which are globally Lipschitz on $\mathcal{D}$, also be globally Lipschitz on $\mathcal{D}$?

Exercise 7.4.5. Answer the following:

i.) Show directly that $f(x) = x^2$ is globally Lipschitz on any closed, bounded interval $[a,b]$.

ii.) Is this true on an unbounded interval? Explain.

Exercise 7.4.3 suggests that Lipschitz growth is a reasonable level of restriction to impose on a function's behavior. But, will it help to facilitate our use of the method of successive approximations? In order to do this, it would seem that it should enable us to circumvent the self-referential nature of the variation of parameters formula. But, does it?

For the sake of argument, assume that the functions in the variation of parameters formula are all real-valued and that the mapping $t \mapsto f(t, U(t))$ is globally Lipschitz in the second variable with Lipschitz constant M. Let us focus on showing the uniqueness of a mild solution of (Semi-CP) under this assumption. To this end, write down the variation of parameters formula for two different functions U and V:

$$U(t) = e^{\mathrm{A}(t-t_0)}U_0 + \int_{t_0}^{t} e^{\mathrm{A}(t-s)} f(s, U(s))ds$$

$$V(t) = e^{\mathrm{A}(t-t_0)}U_0 + \int_{t_0}^{t} e^{\mathrm{A}(t-s)} f(s, V(s))ds$$

Mimicking the uniqueness proof for a mild solution of (Non-CP) we note that subtracting these two equations and then taking absolute values on both sides yields

$$U(t) - V(t) = \int_{t_0}^{t} e^{\mathrm{A}(t-s)} \left(f(s, U(s)) - f(s, V(s))\right) ds$$

$$|U(t) - V(t)| = \left| \int_{t_0}^{t} e^{\mathrm{A}(t-s)} \left(f(s, U(s)) - f(s, V(s))\right) ds \right|.$$

Using properties of the integral and the fact that f is Lipschitz then yields the following estimate

$$\begin{aligned} |U(t) - V(t)| &\le \int_{t_0}^{t} \left| e^{\mathrm{A}(t-s)} \left(f(s, U(s)) - f(s, V(s))\right) \right| ds \\ &\le \int_{t_0}^{t} \underbrace{e^{\mathrm{A}(t-s)}}_{\le M^\star} \underbrace{|f(s, U(s)) - f(s, V(s))|}_{\le M|U(s)-V(s)|} ds \\ &\le \int_{t_0}^{t} M^\star M \, |U(s) - V(s)| \, ds. \end{aligned}$$

When the dust clears, we are left with the inequality

$$|U(t) - V(t)| \le K \int_{t_0}^{t} |U(s) - V(s)| \, ds, \tag{7.21}$$

for all $t \in (t_0, t_0 + T)$, where K is a positive constant. Certainly, the Lipschitz assumption has had a simplifying effect. But, we are not out of the woods yet. If we want to prove uniqueness, we need to argue that the left-side of (7.21) equals zero, for all $t \in (t_0, t_0 + T)$. Interestingly, this actually turns out to be true and without needing to impose additional assumptions. But, this is not obvious.

7.5 Gronwall's Lemma

The following classical inequality, from which many others are derived, is the key to showing uniqueness of a mild solution to (Semi-CP).

Theorem 7.5.1. Gronwall's Lemma
Let $t_0 \in (-\infty, T)$, $x \in \mathbb{C}([t_0, T]; \mathbb{R})$, $K \in \mathbb{C}([t_0, T]; [0, \infty))$, and M be a real constant. If

$$x(t) \le M + \int_{t_0}^{t} K(s)x(s)ds, \; t_0 \le t \le T, \tag{7.22}$$

then

$$x(t) \leq Me^{\int_{t_0}^{t} K(s)ds},\ t_0 \leq t \leq T. \tag{7.23}$$

Exercise 7.5.1. Explain why Gronwall's lemma enables us to conclude from (7.21) that $U(t) = V(t)$, for all $t_0 \leq t \leq T$.

7.6 The Existence and Uniqueness of a Mild Solution for (Semi-CP)

We now possess the tools necessary to prove the following existence-uniqueness theorem for a mild solution of (Semi-CP).

Theorem 7.6.1. Let $\mathbf{A} \in \mathbb{M}^N(\mathbb{R})$ and $\mathbf{U}_0 \in \mathbb{R}^N$ be given, and $\mathbf{F} : [t_0, T] \times \mathbb{R}^N \longrightarrow \mathbb{R}^N$ be continuous in t and globally Lipschitz in the spatial variables (uniformly in t). Then, the IVP (Semi-CP) has a unique mild solution given by the variation of parameters formula

$$\mathbf{U}(t) = e^{\mathbf{A}(t-t_0)}\mathbf{U}_0 + \int_{t_0}^{t} e^{\mathbf{A}(t-s)}\mathbf{F}(s, \mathbf{U}(s))ds, \tag{7.24}$$

defined for all real numbers $t_0 \leq t \leq T$.

If interested, you can find the proof of Theorem 7.6.1 in Appendix A.

7.7 Dealing with a Perturbed (Semi-CP)

We formulate a generalization of the results and discussion presented for (Non-CP) in Chapter 5.

STOP! Carefully review the discussion in Chapter 5 and as you work through the present section, compare the developments line by line to understand the nature of the modifications implemented.

Consider (Semi-CP), together with the related IVP

$$\begin{cases} \overline{\mathbf{U}}'(t) = \mathbf{A}\overline{\mathbf{U}}(t) + \overline{\mathbf{f}}(t, \overline{\mathbf{U}}(t)),\ t > t_0, \\ \overline{\mathbf{U}}(t_0) = \overline{\mathbf{U}}_0. \end{cases} \tag{7.25}$$

We are hopeful that for any time interval $[t_0, T]$, the norm of the difference between the solutions $\mathbf{U}$ and $\overline{\mathbf{U}}$ of (Semi-CP) and (7.25), respectively, can be controlled by ensuring the error terms $\epsilon_i (i = 1, 2)$ are sufficiently small. In such case, we would say that the solution of (Semi-CP) depends continuously on the data.

This is analogous to the problems considered for (HCP) in Section 4.8, but the presence of the forcing term complicates the matter. We begin by studying this problem for the one-dimensional version of (Semi-CP). Precisely, consider the following two related IVPs:

$$\begin{cases} u'(t) = au(t) + f(t, u(t)), \ t > t_0 \\ u(t_0) = u_0, \end{cases} \tag{7.26}$$

$$\begin{cases} \overline{u}'(t) = a\overline{u}(t) + \overline{f}(t, \overline{u}(t)), \ t > t_0 \\ \overline{u}(t_0) = \overline{u}_0, \end{cases} \tag{7.27}$$

where we assume there exist $\delta_1, \delta_2 > 0$ for which

$$|u_0 - \overline{u}_0| < \delta_1 \text{ and } \sup_{t_0 \le t \le T} \left| f(t,x) - \overline{f}(t,x) \right| < \delta_2, \text{ uniformly for } x \in \mathbb{R}, \tag{7.28}$$

and there exist Lipschitz constants $M, \overline{M} > 0$ such that

$$\begin{aligned} |f(t,x) - f(t,y)| &\le M|x-y|, \ \forall x,y \in \mathbb{R} \text{ (uniformly in } t), \\ \left|\overline{f}(t,x) - \overline{f}(t,y)\right| &\le \overline{M}|x-y|, \ \forall x,y \in \mathbb{R} \text{ (uniformly in } t). \end{aligned} \tag{7.29}$$

The solutions of (7.26) and (7.27) are respectively given by

$$u(t) = e^{a(t-t_0)}u_0 + \int_{t_0}^t e^{a(t-s)} f(s, u(s))ds, \tag{7.30}$$

$$\overline{u}(t) = e^{a(t-t_0)}\overline{u}_0 + \int_{t_0}^t e^{a(t-s)} \overline{f}(s, \overline{u}(s))ds. \tag{7.31}$$

The question of interest is, "How far apart are these solutions at a given time t?" Graphically, we measure this by computing the vertical distance between the graphs of $y = u(t)$ and $y = u_\epsilon(t)$ at all times t in an interval on which we are studying the solutions. Since the outputs are real numbers, we measure this distance using the absolute value of the difference in the outputs.

Disclaimer! The following is a more formal discussion of the ideas presented in Section 5.4, yet extended to apply to a specific (Semi-CP). If you skipped the previous two **Disclaimer!** subsections, please skip to the next **EXPLORE!**. If you choose to proceed, re-read Section 5.4 before going further.

In light of the work done in Section 5.4, the answer to the above question for (Non-CP) rests solely on handling the forcing terms. Proceeding as before, we subtract the two solution formulas and estimate the difference using the properties of absolute value. This yields the following string of inequalities that hold for all $t \in [t_0, T]$.

$$\begin{aligned} |u(t) - \overline{u}(t)| &= \left| e^{a(t-t_0)}u_0 - e^{a(t-t_0)}\overline{u}_0 + \int_{t_0}^t e^{a(t-s)} f(s,u(s))ds - \int_{t_0}^t e^{a(t-s)} \overline{f}(s,\overline{u}(s))ds \right| \\ &= \left| e^{a(t-t_0)}(u_0 - \overline{u}_0) + \int_{t_0}^t e^{a(t-s)} \left(f(s,u(s)) - \overline{f}(s,\overline{u}(s)) \right) ds \right| \\ &\le \left| e^{a(t-t_0)}(u_0 - \overline{u}_0) \right| + \left| \int_{t_0}^t e^{a(t-s)} \left(f(s,u(s)) - \overline{f}(s,\overline{u}(s)) \right) ds \right| \\ &\le \left| e^{a(t-t_0)} \right| \cdot |u_0 - \overline{u}_0| + \left| \int_{t_0}^t e^{a(t-s)} \left(f(s,u(s)) - \overline{f}(s,\overline{u}(s)) \right) ds \right| \\ &\le e^{a(T-t_0)}\delta_1 + e^{a(t-t_0)} \underbrace{\int_{t_0}^t \left| f(s,u(s)) - \overline{f}(s,\overline{u}(s)) \right| ds}_{=I_{t_0}^t(f-\overline{f})} \end{aligned} \tag{7.32}$$

Again, the struggle is with the second term on the right-side of (7.32), but the nature of this struggle is different from before. Not only are the forcing functions different, but also the second variables of each of them are different. The problem is that each of the conditions (7.28) and (7.29) only allows for one of these changes to occur, not both simultaneously.

We have used the trick of adding and subtracting the same term within a given expression and subsequently grouping terms in a manner which enabled us to apply our hypotheses.

STOP! How exactly do you suspect we will handle this in (7.32)?

Concentrating on just the integral in the second-term on the right-side ($I_{t_0}^t(f-\overline{f})$) of (7.32), we have the following:

$$\begin{aligned} I_{t_0}^t(f-\overline{f}) &\leq \int_{t_0}^t \left|(f(s,u(s)) - f(s,\overline{u}(s))) + \left(f(s,\overline{u}(s)) - \overline{f}(s,\overline{u}(s))\right)\right| ds \\ &\leq \int_{t_0}^t |f(s,u(s)) - f(s,\overline{u}(s))|\, ds + \int_{t_0}^t \left|f(s,\overline{u}(s)) - \overline{f}(s,\overline{u}(s))\right| ds \\ &\leq \int_{t_0}^t M\,|u(s) - \overline{u}(s)|\, ds + \int_{t_0}^t \sup_{s\in[t_0,T]} \left|f(s,\overline{u}(s)) - \overline{f}(s,\overline{u}(s))\right| ds \\ &\leq \int_{t_0}^t M\,|u(s) - \overline{u}(s)|\, ds + \delta_2\,|T - t_0|\,. \end{aligned} \tag{7.33}$$

STOP! Justify each line in (7.33).

Now, substitute (7.33) into (7.32) to obtain the following estimate:

$$\begin{aligned} |u(t) - \overline{u}(t)| &\leq e^{a(T-t_0)}\delta_1 + e^{a(t-t_0)}\left[\int_{t_0}^t M\,|u(s) - \overline{u}(s)|\, ds + \delta_2\,|T - t_0|\right] \\ &= e^{a(T-t_0)}\left(\delta_1 + \delta_2\,|T - t_0|\right) + Me^{a(t-t_0)}\int_{t_0}^t |u(s) - \overline{u}(s)|\, ds. \end{aligned} \tag{7.34}$$

Finally, we apply Gronwall's lemma in (7.34) to arrive at

$$|u(t) - \overline{u}(t)| \leq e^{a(T-t_0)}\left(\delta_1 + \delta_2\,|T - t_0|\right) e^{Me^{a(t-t_0)^2}}, \tag{7.35}$$

for all $t_0 \leq t \leq T$. Observe that as δ_1 and $\delta_2 \to 0$, the right-side of (7.35) goes to 0 for all t, as needed. **(Why?)**

As in Section 5.4, we would like to use the above ideas to study the same question, but now for a system of such equations of the form (Semi-CP). Below, we modify the development in Section 5.4 slightly to account for the presence of a forcing term.

EXPLORE!

Does (Semi-CP) Depend Continuously on the Initial Data?

Note: A starred question assumes you read the **Disclaimer!** subsection found on the previous pages.

1. Consider the IVP

$$\begin{cases} \mathbf{U}'(t) &= \begin{bmatrix} \alpha & 0 \\ 0 & \beta \end{bmatrix} \mathbf{U}(t) + \begin{bmatrix} f_1(t, \mathbf{U}(t)) \\ f_2(t, \mathbf{U}(t)) \end{bmatrix} \\ \mathbf{U}(0) &= \begin{bmatrix} w_1 \\ w_2 \end{bmatrix} \end{cases} \tag{7.36}$$

where $\alpha, \beta \in \mathbb{R}$ and there exist Lipschitz constants $M_i > 0$ $(i = 1, 2)$ such that

$$|f_i(t, x) - f_i(t, y)| \leq M\,|x - y|\,, \forall x, y \in \mathbb{R}\, (\text{uniformly in } t)$$

   i.) What is the solution $\mathbf{U}$ of (7.36)? Give the actual components.

   ii.) Suppose that the measurements of the first component of the initial condition vector in (7.36) are subject to error, so that the *actual* initial state vector is $\begin{bmatrix} w_1 + \epsilon \\ w_2 \end{bmatrix}$, where $|\epsilon|$ is a small positive number.

   Consider the corresponding IVP

$$\begin{cases} \mathbf{U}_\epsilon{}'(t) &= \begin{bmatrix} \alpha & 0 \\ 0 & \beta \end{bmatrix} \mathbf{U}_\epsilon(t) + \begin{bmatrix} f_1(t, \mathbf{U}_\epsilon(t)) \\ f_2(t, \mathbf{U}_\epsilon(t)) \end{bmatrix} \\ \mathbf{U}_\epsilon(0) &= \begin{bmatrix} w_1 + \epsilon \\ w_2 \end{bmatrix} \end{cases} \tag{7.37}$$

   a.) What is the solution $\mathbf{U}_\epsilon$ of (7.37)?

   b.*) Compute $\|\mathbf{U}(t) - \mathbf{U}_\epsilon(t)\|_{\mathbb{R}^2}$. Make a conjecture for what happens as $|\epsilon| \longrightarrow 0$ for a fixed t. (Hint: Follow (7.32) and (7.33). Notice that in this example $\bar{f} = f$.)

2. For concreteness, open **MATLAB** and in the command line, type:

MATLAB_SemiLinear_ACP_Solver

GUI instructions can be found by clicking the Help button. In the User Panel, choose $t = 0$ for initial conditions and plot solutions until $t = .5$. Click the 'Input IVP and Plot Solution' button to enter the following IVP:

$$\begin{cases} \mathbf{U}'(t) &= \begin{bmatrix} 1 & 0 \\ 0 & -1 \end{bmatrix} \mathbf{U}(t) + \begin{bmatrix} \cos(2\pi u_1(t)) \\ \sin(2\pi u_2(t)) \end{bmatrix} \\ \mathbf{U}(0) &= \begin{bmatrix} 1 \\ 1 \end{bmatrix} \end{cases} \tag{7.38}$$

where $\mathbf{U}(t) = [u_1(t), u_2(t)]$. Suppose that the measurement of the first component of the initial condition vector in (7.38) is subject to error. The corresponding IVP actually becomes

$$\begin{cases} \mathbf{U}_\epsilon{}'(t) &= \begin{bmatrix} 1 & 0 \\ 0 & -1 \end{bmatrix} \mathbf{U}_\epsilon(t) + \begin{bmatrix} \cos(2\pi u_1(t)) \\ \sin(2\pi u_2(t)) \end{bmatrix} \\ \mathbf{U}_\epsilon(0) &= \begin{bmatrix} 1 + \epsilon \\ 1 \end{bmatrix} \end{cases} \tag{7.39}$$

i.) Click the 'Modify IVP and Plot Solution' solution button and input (7.39) for $\epsilon = .01$. Are the solutions to (7.38) and (7.39) close to each other? How close? How far?

ii.) Repeat a) for $\epsilon = 0.001$.

iii.) Repeat a) for $\epsilon = 0.0001$.

iv.) Describe the patterns found in the graphs and norms of the corresponding IVPs.

v.) Repeat question 2 for any pair of IVPs that satisfy the form provided in (7.36) and (7.37).

3.* Now that we have an intuitive sense of what should happen when $|\epsilon| \to 0$, let us return to (7.36) and (7.37) to compute

$$\sup_{t_0 \leq t \leq T} \|\mathbf{U}(t) - \mathbf{U}_\epsilon(t)\|_{\mathbb{R}^2}$$

What happens to this value as $|\epsilon| \longrightarrow 0$? If you get stuck, compute instead

$$\sup_{0 \leq t \leq 0.5} \|\mathbf{U}(t) - \mathbf{U}_\epsilon(t)\|_{\mathbb{R}^2}$$

for (7.38) and (7.39). (Hint: Use (7.34) and Gronwall's lemma.)

4. Now, suppose that both measurements of the initial condition vector in (7.36) are subject to error, so that the actual state vector is $\begin{bmatrix} w_1 + \epsilon_1 \\ w_2 + \epsilon_2 \end{bmatrix}$, where $|\epsilon_1|$ and $|\epsilon_2|$ are small positive numbers.

i.) Replace the initial vector in (7.37) by this one and repeat questions 2. Choose values for ϵ_1 and ϵ_2 near the ϵ values provided in question 2.

ii.*) Replace the initial vector in (7.37) by this one and repeat questions 1 and 3.

5. Generalize your findings to the case of an $N \times N$ diagonal matrix for $\mathbf{A}$ and an initial condition vector in which each of its N components is subject to error.

i.) Do you expect the same conclusion to hold when $\mathbf{A}$ is no longer a diagonal matrix? Explain.

ii.) Carefully formulate the most general result that you can muster regarding the continuous dependence of solutions of (Semi-CP) on the initial data.

APPLY IT!!

At long last, we are equipped with enough theoretical power to be able to address all of the models studied in Chapter 6.

I.) Reformulate the IVP in the abstract form (Non-CP).

II.) Choose actual values for the parameters in the model. For them, illustrate the curves of each of the component functions, study the IVP using the **MATLAB_SemiLinear_ACP_Solver** GUI, and analyze the results in the context of the setting of each model.

i.) The pharmocokinetics model (6.5)

$$\begin{cases} \frac{dy(t)}{dt} = -ay(t) + D\left(t, y(t), z(t)\right) \\ \frac{dz(t)}{dt} = ay(t) - bz(t) + \bar{D}((t, y(t), z(t)) \\ y(0) = y_0, z(0) = 0 \end{cases}$$

ii.) The coupled spring-mass system model (6.7)

$$\begin{cases} m_1 \frac{d^2 x_1(t)}{dt^2} & = -k_1 x_1(t) + \mu x_1(t)^3 + k_2(x_2(t) - x_1(t)) - k_2(x_2(t) - x_1(t))^3 \\ m_2 \frac{d^2 x_2(t)}{dt^2} & = -k_3 x_2(t) + \mu x_2(t)^3 - k_2(x_2(t) - x_1(t)) + k_2(x_2(t) - x_1(t))^3 \\ x_1(0) = x_0; & x_1'(0) = x_1 \\ x_2(0) = y_0; & x_2'(0) = y_1 \end{cases}$$

iii.) The electrical circuit model (6.9)

$$\begin{cases} L \frac{d^2 q(t)}{dt} + R \left| \frac{dq(t)}{dt} \right| \frac{dq(t)}{dt} + \frac{1}{C} q(t) = E(t) \\ q(0) = q_0 \\ q'(0) = q_1 \end{cases}$$

iv.) The chemical kinetics model (6.23), equipped with appropriate initial conditions

$$\begin{cases} \frac{d[S](t)}{dt} & = -V \frac{[S](t)}{K_M} (1 - \frac{[S](t)}{K_M}) \\ \frac{d[P](t)}{dt} & = V \frac{[S](t)}{K_M} (1 - \frac{[S](t)}{K_M}) \end{cases}$$

v.) The projectile motion model (6.26), equipped with appropriate initial conditions

$$\begin{cases} m \frac{d^2 x(t)}{dt^2} + r_x \| \frac{d\mathbf{u}(t)}{dt} \| \frac{dx(t)}{dt} & = F_x(t) \\ m \frac{d^2 y(t)}{dt^2} + r_y \| \frac{d\mathbf{u}(t)}{dt} \| \frac{dy(t)}{dt} & = F_y(t) \\ m \frac{d^2 z(t)}{dt^2} + r_z \| \frac{d\mathbf{u}(t)}{dt} \| \frac{dz(t)}{dt} & = F_z(t) \end{cases}$$

where $\| \frac{d\mathbf{u}(t)}{dt} \| = \sqrt{(\frac{dx(t)}{dt})^2 + (\frac{dy(t)}{dt})^2 + (\frac{dz(t)}{dt})^2}$.

Part II

Abstract Ordinary Differential Equations

Chapter 8

Getting the Lay of a New Land

Overview

You were introduced to linear, nonhomogeneous, and semilinear systems of ODEs in $\mathbb{R}^N$ in Part I. The theory developed there applied to an eclectic collection of contextually unrelated mathematical models. Alas, not all phenomena are described by a mathematical model that can be captured under this theoretical umbrella. There are various reasons for this, even within the realm of ODEs. The primary departure on which we shall now focus is the dependence of the unknown function (or "solution") on *more than one* independent variable. As such, the equations used to form the mathematical models now involve partial derivatives, not ordinary ones. The purpose of Part II is to probe this issue more deeply and try to answer the question, "Can we somehow rewrite our mathematical models in a manner that resembles the Cauchy problems (HCP), (Non-CP) and (Semi-CP) from Part I?"

8.1 A Hot Example

Let us jump in with both feet and consider a classic model that loosely describes how heat moves, or diffuses, through a medium. This example will help us to illustrate many of the key points without being overly technical. The model itself will be more carefully developed as you proceed through Part II.

Consider a one-dimensional wire of length a with uniform properties and cross-sections. Assuming that no heat is generated and the surface is insulated, the homogeneous heat equation describes the evolution of temperature throughout the wire over time. This equation, coupled with the initial profile, can be described by the IVP

$$\begin{cases} \frac{\partial}{\partial t}z(x,t) = k\frac{\partial^2}{\partial x^2}z(x,t), \ 0 < x < a, \ t > 0, \\ z(x,0) = z_0(x), \ 0 < x < a, \end{cases} \tag{8.1}$$

where $z(x,t)$ represents the temperature at position x along the wire at time t and k is a proportionality constant depending on the thermal conductivity and material density. We will derive this equation in the next chapter. A complete description of this phenomenon requires that we prescribe what happens to the temperature on the boundary of the wire. This can be done in many naturally-occurring ways, some of which are described below:

1. Temperature is held constant along the boundary of the wire:
$$z(0,t) = C_1 \text{ and } z(a,t) = C_2, \ \forall t > 0. \tag{8.2}$$
2. Temperature is controlled along the boundary of the wire, but changes with time:
$$z(0,t) = C_1(t) \text{ and } z(a,t) = C_2(t), \ \forall t > 0. \tag{8.3}$$
3. Heat flow rate is controlled along the boundary of the wire:
$$\frac{\partial z}{\partial x}(0,t) = C_1(t) \text{ and } \frac{\partial z}{\partial x}(a,t) = C_2(t), \ \forall t > 0. \tag{8.4}$$
4. Convection (governed by Newton's Law of heating and cooling):
$$\begin{aligned} C_1 z(0,t) + C_2\frac{\partial z}{\partial x}(0,t) &= C_3(t), \ \forall t > 0, \\ \overline{C_1} z(a,t) + \overline{C_2}\frac{\partial z}{\partial x}(a,t) &= \overline{C_3}(t), \ \forall t > 0. \end{aligned} \tag{8.5}$$

STOP! Read about this in [39] for some contextual background.

Boundary conditions (BCs) of the forms (8.2) and (8.3) are called *Dirichlet* BCs, while those of type (8.4) are called *Neumann* BCs. If the constants/functions C_i are zero, the BCs are called *homogeneous*; otherwise, they are *nonhomogeneous.* We can use a combination of the types of BCs in the formulation of an IBVP. For instance, a homogeneous Dirichlet BC can be imposed at one end of the wire and a nonhomogeneous Neumann BC at the other end.

We first consider the IBVP formed by coupling (8.1) with the homogeneous Dirichlet BCs
$$z(0,t) = z(a,t) = 0, \ \forall t > 0. \tag{8.6}$$
A solution of the IBVP can be constructed using the standard *separation of variables method,* as follows. Let us assume the solution $z(x,t)$ is of the form
$$z(x,t) = X(x)T(t). \tag{8.7}$$
Assume that $X(x) \neq 0$, for all x. Substituting (8.7) into the first equation in (8.1) yields

$$\underbrace{\underbrace{\frac{X''(x)}{X(x)}}_{\text{Function of only } x} = \lambda =}_{X''(x)-\lambda X(x)=0} \overbrace{\underbrace{\frac{T'(t)}{kT(t)}}_{\text{Function of only } t}}^{T'(t)-k\lambda T(t)=0}$$

where λ is a constant. The general solution of $T'(t) - k\lambda T(t) = 0$ is

$$T(t) = C_\lambda e^{\lambda kt},\ t > 0, \tag{8.8}$$

where C_λ is a constant. **(Why?)** We seek all values of λ (referred to as *eigenvalues*) for which the resulting BVP

$$\begin{cases} X''(x) - \lambda X(x) = 0 \\ X(0) = X(a) = 0, \end{cases} \tag{8.9}$$

has a nonzero solution.

STOP! Show that when $\lambda \geq 0$, the only solution of (8.9) is the zero solution. (Hint: Consider the cases when $\lambda = 0$ and $\lambda > 0$ separately.)

When $\lambda < 0$, which we denote as $\lambda = -n^2$ for convenience, the general solutions of (8.9) are

$$X(x) = C_1 \cos(nx) + C_2 \sin(nx). \tag{8.10}$$

Applying the BCs given in (8.9) to (8.10) yields the two conditions

$$C_1 = 0 \text{ and } C_2 \sin(na) = 0. \tag{8.11}$$

Since choosing $C_2 = 0$ would result in the zero solution in (8.10) and hence would not contribute meaningfully to the construction of the general solution of the IBVP, we assume instead that $\sin(na) = 0$, which is satisfied when $n = \frac{m\pi}{a}$, $\forall m \in \mathbb{N}$. As such, the desired eigenvalues are given by

$$\lambda_m = -\left(\frac{m\pi}{a}\right)^2,\ \forall m \in \mathbb{N}. \tag{8.12}$$

For each $m \in \mathbb{N}$, the function

$$z_m(x,t) = b_m X_m(x) T_m(t) = b_m e^{-\left(\frac{m\pi}{a}\right)^2 kt} \sin\left(\frac{m\pi}{a}x\right), \tag{8.13}$$

where b_m is an appropriate constant, satisfies both the heat equation and the BCs. Hence, for every $N \in \mathbb{N}$, the finite sum

$$\sum_{m=1}^{N} b_m e^{-\left(\frac{m\pi}{a}\right)^2 kt} \sin\left(\frac{m\pi}{a}x\right)$$

satisfies them. **(Why?)** Moreover, it can be shown that the function

$$z(x,t) = \sum_{m=1}^{\infty} b_m e^{-\left(\frac{m\pi}{a}\right)^2 kt} \sin\left(\frac{m\pi}{a}x\right) \tag{8.14}$$

satisfies the heat equation and the BCs. (See Figure 8.1 for examples that define (8.14) at a particular time.) Applying the initial condition (IC) in (8.1) to determine the constants (b_m) yields

$$z_0(x) = z(x,0) = \sum_{m=1}^{\infty} b_m \sin\left(\frac{m\pi}{a}x\right). \tag{8.15}$$

We must determine the constants (b_m). This requires the use of so-called *Fourier series.* It can be shown that as long as $z_0(\cdot)$ is piecewise-continuous, b_m is given by

$$b_m = \frac{2}{a} \int_0^a z_0(w) \sin\left(\frac{m\pi}{a}w\right) dw. \tag{8.16}$$

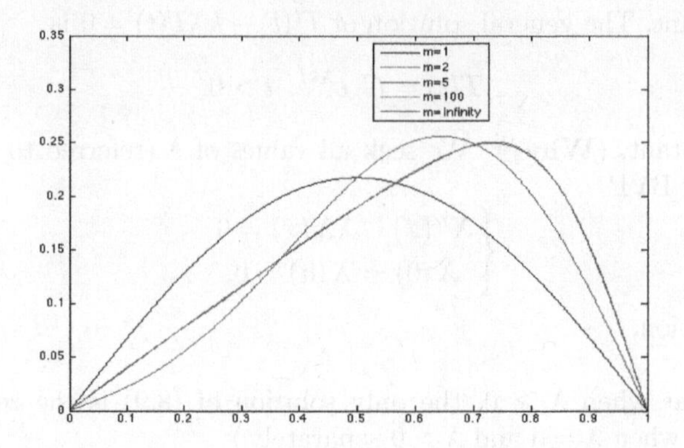

FIGURE 8.1: Examples of partial sums defining (8.14) at a fixed t.

Substituting (8.16) into (8.14) renders the solution of the IBVP as

$$z(x,t) = \sum_{m=1}^{\infty} \left(\frac{2}{a} \int_0^a z_0(w) \sin\left(\frac{m\pi}{a} w\right) dw \right) e^{-\left(\frac{m\pi}{a}\right)^2 kt} \sin\left(\frac{m\pi}{a} x\right) \tag{8.17}$$

where $0 < x < a$ and $t > 0$. Observe that the right side of (8.17) involves three different inputs, namely x, t, and $z_0(\cdot)$. To emphasize these distinct dependencies, we introduce the notation

$$z(x,t) = S(t)\,[z_0]\,(x). \tag{8.18}$$

Let us get a feel for how this function works as we change one variable at a time. First, suppose we fix the spatial variable x and consider values of $z(x,t)$ for different values of t.

$$\begin{aligned} z(x,0) &= S(0)\,[z_0]\,(x) = z_0(x) \\ z(x,1) &= S(1)\,[z_0]\,(x) = \text{Level curve at } t = 1 \\ z(x,2) &= S(2)\,[z_0]\,(x) = \text{Level curve at } t = 2 \end{aligned}$$

This is pictured below, as follows:

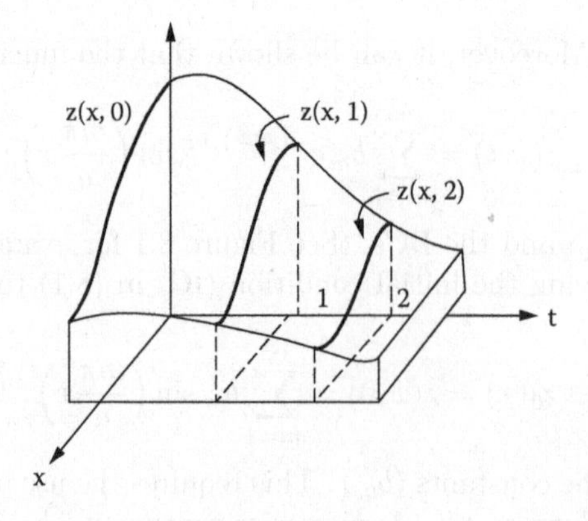

Next, suppose we fix the time variable t and consider values of $z(x,t)$ for different values of x (which correspond to different positions along the wire).

$$z(0,t) = S(t)\,[z_0]\,(0)$$
$$z(\frac{a}{2},t) = S(t)\,[z_0]\,(\frac{a}{2})$$
$$z(a,t) = S(t)\,[z_0]\,(a)$$

Each of these describes the evolution of the temperature at specific locations along the wire over time.

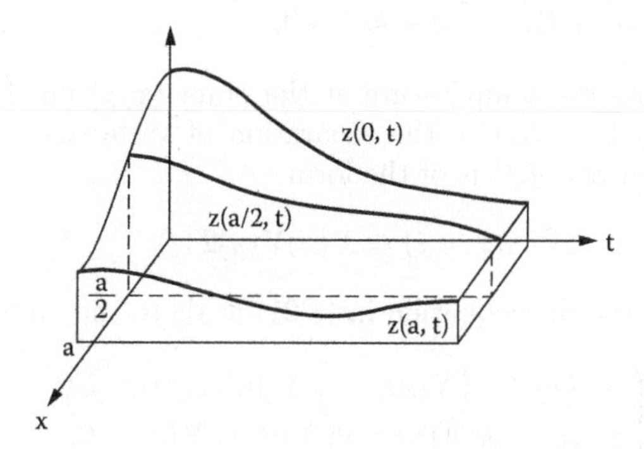

STOP! Verify that for sufficiently smooth functions v and for $x \in [0, L]$ and t, t_1, $t_2 \geq 0$, the function $z(x, t)$ in (8.18) satisfies the following properties:

1. $S(0)[v](x) = v(x)$
2. $S\left(t_1 + t_2\right)[v](x) = S\left(t_1\right)\left(S\left(t_2\right)[v]\right)(x)$
3. $S\left(t\right)\left[v_1 + v_2\right](x) = S\left(t\right)\left[v_1\right](x) + S\left(t\right)\left[v_2\right](x)$

Do these properties look familiar? They resemble the exponential properties exhibited by the matrix exponential $\left(e^{\mathbf{A}t} : t \geq 0\right)$when $\mathbf{A} \in \mathbb{M}^N(\mathbb{R})$. This begs the question, "What plays the role of $\mathbf{A}$ in the above mathematical model?" It is certainly not a matrix. We'll come back to this later.

Exercise 8.1.1.

i.) Use the separation of variables method to show that the solution of the IBVP obtained by coupling (8.1) instead with the homogeneous Neumann BCs

$$\frac{\partial z}{\partial x}(0,t) = \frac{\partial z}{\partial x}(a,t) = 0, \ \forall t > 0, \tag{8.19}$$

is given by

$$z(x,t) = \sum_{m=0}^{\infty} \left(\frac{2}{a}\int_0^a z_0(w)\cos\left(\frac{m\pi}{a}w\right)dw\right)e^{-\left(\frac{m\pi}{a}\right)^2 kt}\cos\left(\frac{m\pi}{a}x\right). \tag{8.20}$$

ii.) Simplify (8.20) when $z_0(x) = x$.

iii.) Graph typical samples of the following level curves of the solution surface: $z(x,0)$, $z(x,1)$, $z(x,2)$ and $z(0,t)$, $z(\frac{a}{2},t)$, $z(a,t)$. Describe verbally what they represent as each of the unfixed variable changes in its domain.

Exercise 8.1.2. Express the function (8.20) in the form (8.18) and verify that it satisfies the three properties that immediately follow.

Next, we consider a similar model for heat conduction in a two-dimensional rectangular plate composed of an isotropic, uniform material. Assuming that the temperature is zero along the boundary of the rectangle, the IBVP describing the transient temperature at every point on the plate over time is given by

$$\begin{cases} \frac{\partial z}{\partial t}(x,y,t) = k\left(\frac{\partial^2 z}{\partial x^2}(x,y,t) + \frac{\partial^2 z}{\partial y^2}(x,y,t)\right), \ 0 < x < a, \ 0 < y < b, \ t > 0, \\ z(x,y,0) = z_0(x,y), \ 0 < x < a, \ 0 < y < b, \\ z(x,0,t) = 0 = z(x,b,t), \ 0 < x < a, \ t > 0, \\ z(0,y,t) = 0 = z(a,y,t), \ 0 < y < b, \ t > 0, \end{cases} \tag{8.21}$$

where $z(x,y,t)$ represents the temperature at the point (x,y) on the plate at time t. We apply a suitably-modified version of the separation of variables method to solve (8.21). Assume that the solution $z(x,y,t)$ is of the form

$$z(x,y,t) = X(x)Y(y)T(t). \tag{8.22}$$

Substituting (8.22) into the first equation in (8.21) leads to the system of ODEs

$$\begin{cases} X''(x) + \lambda_x^2 X(x) = 0, \ X(0) = X(a) = 0, \\ Y''(y) + \lambda_y^2 Y(y) = 0, \ Y(0) = Y(b) = 0, \\ T'(t) + k\lambda^2 T(t) = 0, \ t > 0, \end{cases} \tag{8.23}$$

where the eigenvalues λ_x^2 and λ_y^2 are related by $\lambda^2 = \lambda_x^2 + \lambda_y^2$. **(Tell why.)** Proceeding as before, the following functions satisfy (8.23), $\forall m, \ n \in \mathbb{N}$ and constants C_{mn} :

$$\begin{cases} X_m(x) = \sin\left(\frac{m\pi}{a}x\right), \ 0 < x < a, \\ Y_n(y) = \sin\left(\frac{n\pi}{b}y\right), \ 0 < y < b, \\ T_{mn}(t) = C_{mn}e^{-\left(\left(\frac{m\pi}{a}\right)^2 + \left(\frac{n\pi}{b}\right)^2\right)kt}, \ t > 0. \end{cases} \tag{8.24}$$

STOP! Verify this.

For each $m, \ n \in \mathbb{N}$ and constants b_{mn} , the function

$$z_{mn}(x,y,t) = b_{mn}X_m(x)Y_n(y)T_{mn}(t) \tag{8.25}$$

satisfies the heat equation and the BCs in (8.21). **(Why?)** It can be shown that

$$z(x,y,t) = \sum_{m=1}^{\infty}\sum_{n=1}^{\infty} b_{mn}X_m(x)Y_n(y)T_{mn}(t) \tag{8.26}$$

also satisfies them. Applying the ICs in (8.21) to determine the constants (b_{mn}) yields

$$z_0(x,y) = z(x,y,0) = \sum_{m=1}^{\infty}\sum_{n=1}^{\infty} b_{mn}X_m(x)Y_n(y), \ 0 < x < a, \ 0 < y < b. \tag{8.27}$$

Assuming that $z_0(\cdot)$ is sufficiently smooth, we conclude that

$$b_{mn} = \frac{4}{ab}\int_0^b\int_0^a z_0(w,v)\sin\left(\frac{m\pi}{a}w\right)\sin\left(\frac{n\pi}{b}v\right)dwdv, \ m, \ n \in \mathbb{N}, \tag{8.28}$$

which are double Fourier series coefficients. Finally, substituting (8.28) into (8.26) and using (8.24) yields the solution of the IBVP.

Exercise 8.1.3. Replace the BCs in (8.21) by the homogeneous Neumann BCs

$$\begin{aligned} \frac{\partial z}{\partial x}(0,y,t) &= \frac{\partial z}{\partial x}(a,y,t) = 0,\ 0<y<b, t>0, \\ \frac{\partial z}{\partial y}(x,0,t) &= \frac{\partial z}{\partial y}(x,b,t) = 0,\ 0<x<a, t>0. \end{aligned} \tag{8.29}$$

i.) Solve the resulting IBVP using the separation of variables method.

ii.) Formulate the solution (8.27) (with the coefficients given by (8.28) substituted into the formula) as a function of the form (8.18), suitably modified. Then, verify that it satisfies the three properties immediately following (8.18).

Exercise 8.1.4. Construct an IBVP for heat conduction on an n-dimensional rectangular plate $[0,a_1] \times \ldots \times [0,a_n]$ equipped with homogeneous Neumann BCs.

i.) Without going through all of the computations, conjecture a form of the solution.

ii.) Formulate the solution as a function of the form (8.18), suitably modified. Then, verify that it satisfies the three properties immediately following (8.18).

8.2 The Hunt for a New Abstract Paradigm

Although the various IBVPs are formulated in different spatial dimensions and are equipped with different boundary conditions, there is a sense of similarity amongst them which suggests that we might be able to treat them all as special cases of a slightly more abstract problem. We explore this possibility in this section.

We would like to be able to view IBVP (9.1) (with BCs given by (8.6)) as a Cauchy problem of the form (HCP) because we have a good basis of understanding of this type of problem from Part I. A hurdle that we immediately encounter, however, is that the equation now involves partial derivatives and so, it is not clear if it can be viewed as a "constant matrix times the unknown function" as it was in Part I. In fact, the gut instinct is probably not. Moreover, our unknown function now depends on two independent variables, x and t, not just t as it did in Part I. Thus, if we are to have any hope of viewing the IBVP in the form (HCP), we must identify a way in which to subsume the dependence on x into the formulation.

Let us try to reformulate the IBVP (8.1) as an abstract initial-value problem (called an *abstract evolution equation*) of the form

$$(\mathbf{A-HCP}) \begin{cases} \frac{d}{dt}(u(t)) = A(u(t)),\ t>0, \\ u(0) = u_0, \end{cases} \tag{8.30}$$

for some quantity A. What happens along the way? To begin, let us simply try to naïvely symbol-match terms between (8.1) and (8.30). Knowing nothing else, the following is a natural set of identifications:

	IBVP (8.1)	**Abstract IVP** (8.30)	
Solution	$z : [0, a] \times [0, \infty) \to \mathbb{R}$ given by $z(x, t)$	$u : [0, \infty) \to \_$ given by $u(t)$	(8.31)
Initial Condition	$z_0(x)$	u_0	(8.32)
Left Side	$\frac{\partial}{\partial t} \underbrace{(\ \cdot\ )}_{\text{function of } x,\, t}$	$\frac{d}{dt} \underbrace{(\ \cdot\ )}_{\text{function of } t}$	(8.33)
Right Side	$k \frac{\partial^2}{\partial x^2} \underbrace{(\ \cdot\ )}_{\text{function of } x,\, t}$	$A \underbrace{(\ \cdot\ )}_{\text{function of } t}$	(8.34)
Boundary Condition	$z(0, t) = z(L, 0) = 0$	None	(8.35)

Since we are attempting to reformulate a *partial* differential equation (whose solution by its very nature depends upon time and at least one other variable) as an abstract *ordinary* differential equation (whose solution depends only on time), identification (8.31) suggests that for each time $t_0 \geq 0$, the term $u(t_0)$ must "contain" the information for the entire trajectory $\{z(x, t_0) : x \in [0, a]\}$. For instance, $u(0)$ " = " $\{z(x, 0) : x \in [0, a]\}$. Consulting our graph of the solution surface, this means that $u(0)$ is the entire trajectory, highlighted below:

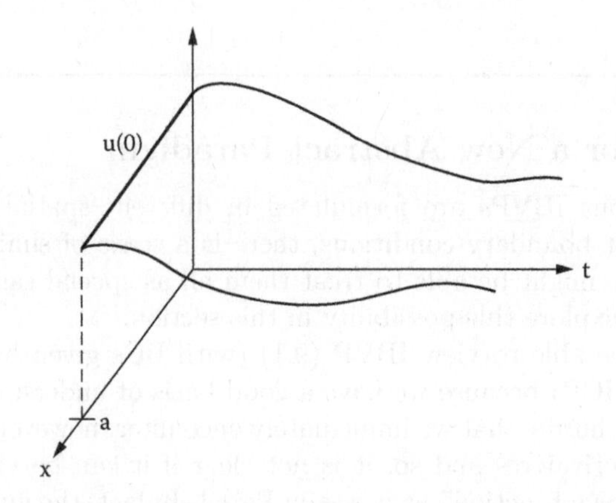

Similarly, $u(1)$ " = " $\{z(x, 1) : x \in [0, a]\}$, which is another entire trajectory, but now at a different time, $t = 1$, as shown below:

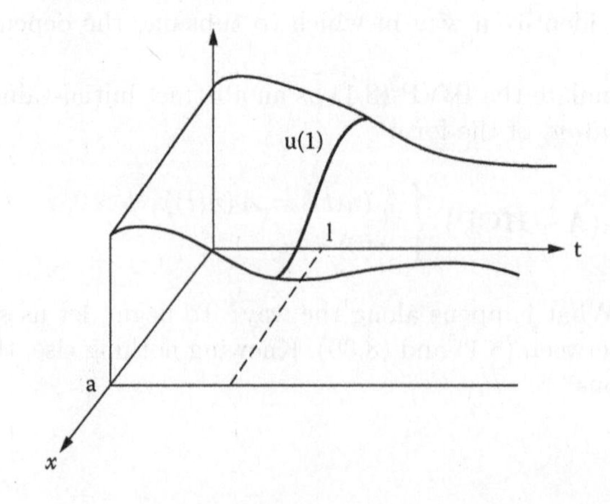

Generally speaking, $u(t_0)$ given by

$$u(t_0)[x] = z(x, t_0), \ x \in [0, a] \tag{8.36}$$

is an entire trajectory that can be visualized as a level curve on the solution surface.

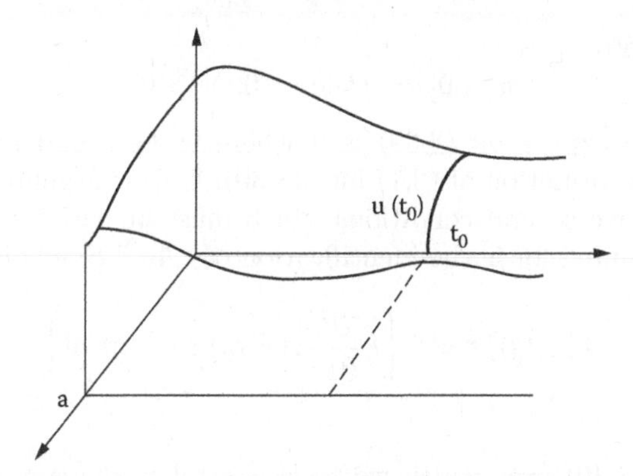

Remark 8.2.1. Brackets are often used to enclose the input z while **parentheses** are used to enclose t_0 to emphasize the distinction between the function $u(t_0)$ and *its* input x. Later in the text, once we have acquired a reasonable level of comfort with these notions, the brackets will be suppressed.

Note that (8.32) follows from (8.36) because

$$u_0[x] = u(0)[x] = z(x, 0) = z_0(x), \ x \in [0, a]. \tag{8.37}$$

Given that $u(t)$ corresponds to an entire trajectory, it makes sense that the range of $u(\cdot)$ mentioned in (8.31) must be a space of functions. But, *which* space exactly? This is a very delicate and important question to answer because the minute we cast the die, the degree of smoothness of the solution for which we are hunting is determined. Providing a completely precise answer would require machinery from real analysis, but let us discuss this issue loosely. Observe that when $t = 0$, $u(0)$ corresponds to the initial condition function $z_0(\cdot)$. Honing in on (8.15) - (8.16) of the solution process, the output $u(0)$ must be sufficiently smooth in order to guarantee that we can define the coefficients in the manner we did. It turns out that as long as the function $z_0(\cdot)$ is *square-integrable on* $[0, a]$, meaning that $\int_0^a z_0^2(x)dx < \infty$, everything will work out (at least for the models we shall consider). Notice that this imposes a restriction on the range of the function u because not every function satisfies this condition.

STOP! Why? (Hint: Consider the function $z(x) = \frac{1}{\sqrt{x}}$ on $(0, 1)$.)

We shall denote the set of all real-valued functions that are square-integrable on $[0, a]$ by $\mathbb{L}^2(0, a; \mathbb{R})$. From the above discussion, it follows that the range of the function $u(\cdot)$ must be contained within $\mathbb{L}^2(0, a; \mathbb{R})$. Therefore, we can complete the right-side entry in (8.31) with $u : [0, \infty) \to \mathbb{L}^2(0, a; \mathbb{R})$. How does this range change impact our work? Take a moment to compare this to (HCP) from Part I. There, the unknown was a vector-valued function $\mathbf{u} : [0, \infty) \to \mathbb{R}^N$, meaning that the outputs $\mathbf{u}(t)$ were vectors in $\mathbb{R}^N$. In the present setting of the heat equation, instead of the outputs being vectors in $\mathbb{R}^N$, they are now actual functions in $\mathbb{L}^2(0, a; \mathbb{R})$.

Based on the discussion leading to (8.36) and (8.37), it is quite natural that the *partial* derivative (in t) for the real-valued function z should be transformed into an *ordinary* derivative (also in t) for the $\mathbb{L}^2(0,a;\mathbb{R})$-valued function u. So, (8.33) is reasonable **(Why?)**, and the derivative is also an $\mathbb{L}^2(0,a;\mathbb{R})$-valued function. **(Why?)**

Finally, we must handle (8.34) and (8.35). Judging from (8.36), the boundary condition

$$z(0,t) = z(a,t) = 0, \ \forall t > 0, \tag{8.38}$$

is easily transformed into

$$u(t)[0] = u(t)[a] = 0, \ \forall t > 0. \tag{8.39}$$

(Why?) However, the expression (8.39) is nowhere to be found in (8.30), yet we must account for in the transformation of (8.1) into (8.30). This is arguably the trickiest part of the formulation. We have several conditions which must be satisfied and the only part of (A-HCP) not yet accounted for is the identification of "Au." Symbolically, we know that

$$A\left(u(t_0)\right) \text{ `` = '' } \left\{ k\frac{\partial^2}{\partial x^2} z(x,t_0) : x \in [0,a] \right\} \tag{8.40}$$

where $u \in \mathbb{L}^2(0,a;\mathbb{R})$.

The only part of (8.40) over which we have control is the space $\mathbb{L}^2(0,a;\mathbb{R})$. Earlier, we introduced the use of space as the range because we needed to guarantee the initial condition z_0 had a unique Fourier representation. But now, we need to further hone the collection of functions that we will allow in our formulation. But how?

For one, observe that not every function in $\mathbb{L}^2(0,a;\mathbb{R})$ satisfies the desired boundary condition (8.38). For instance, just take

$$z(x,t) = 1, \text{ where } x \in [0,a],\ t > 0.$$

Clearly, this function could not be a viable solution candidate for the IBVP (8.1) even though it does belong to $\mathbb{L}^2(0,a;\mathbb{R})$ because it does not satisfy the BC. This suggests that we whittle $\mathbb{L}^2(0,a;\mathbb{R})$ down a bit to exclude from consideration those functions that do not satisfy the BC. More precisely, instead of using the entire space $\mathbb{L}^2(0,a;\mathbb{R})$, let us allow the operator A to only operate on functions in the following subspace of $\mathbb{L}^2(0,a;\mathbb{R})$:

$$Z_1 = \left\{ f \in \mathbb{L}^2(0,a;\mathbb{R}) : f(0) = f(a) = 0 \right\} \tag{8.41}$$

Doing this takes care of the BC (8.38), but it is not a complete fix. We need (8.39) to be defined. This leads to the question, "Is Az defined for all $z \in Z_1$?" Examining that space, we see that we only imposed the condition that the $\mathbb{L}^2(0,a;\mathbb{R})$-function pass through two specific points, but in order for $A\left(u(t_0)\right)$ in (8.40) to be defined, the trajectory $\{z(x,t_0) : x \in [0,a]\}$ must also be twice differentiable in the variable x on the interval $[0,a]$. Again, not every function in the space Z_1 is twice differentiable, so we need to even further restrict this subspace to esnure that the set of functions on which we allow A to operate satisfies this condition. This prompts us to consider the following even more refined space for the domain of A:

$$Z_2 = \left\{ f \in \mathbb{L}^2(0,a;\mathbb{R}) \,\middle|\, \frac{df}{dx}, \frac{d^2 f}{dx^2} \in \mathbb{L}^2(0,L;\mathbb{R}), \text{ and } f(0) = f(a) = 0 \right\}.$$

Naturally, you may be wondering why we require the derivatives to belong to the space $\mathbb{L}^2(0,a;\mathbb{R})$. This is a very good question, but one whose answer requires a bit of technical background. We need the space Z_2 to be a **closed** subspace of $\mathbb{L}^2(0,a;\mathbb{R})$ in order for us to even define e^{At} in this setting. We encourage you to read about these details in [33] if interested. For now, we shall be content to identify Z_2 as the domain of the operator A, and formally write the following: $A : \text{dom}(A) \subset \mathbb{L}^2(0,a;\mathbb{R}) \to \mathbb{L}^2(0,a;\mathbb{R})$ by

$$A[f] = k\frac{d^2}{dx^2}[f], \tag{8.42}$$

$$\text{dom}(A) = \left\{ f \in \mathbb{L}^2(0,a;\mathbb{R}) \,\middle|\, \frac{df}{dx}, \frac{d^2f}{dx^2} \in \mathbb{L}^2(0,L;\mathbb{R}), \text{ and } f(0) = f(a) = 0 \right\}$$

Using all of the above identifications enables us to view the original IBVP in the form (**A-HCP**).

Exercise 8.2.1. Answer the following questions.

i.) Explain how to rewrite the IBVP from Exercise 8.1.1 as the abstract IVP (**A-HCP**).

ii.) How does the process of formulating an IBVP in the abstract form (**A-HCP**) change for the IBVP considered in Exercise 8.1.3?

8.3 A Small Dose of Functional Analysis

The discussion in Sections 8.1 and 8.2 was enlightening and provided some direction as to what to try in order to mimic the approach developed in Part I in our study of mathematical models involving partial differential equations. We now make our discussion a little more precise by introducing new terminology and a few concepts from analysis. To this end, we shall expand our analysis toolbox from Part I to now include some essential elements of so-called *functional analysis.*

Many spaces, like $\mathbb{L}^2(0,a;\mathbb{R})$, possess the same salient features regarding norms, inner products, and completeness exhibited by $\left(\mathbb{R}^N, \|\cdot\|_{\mathbb{R}^N}\right)$ and $\left(\mathbb{M}^N(\mathbb{R}), \|\cdot\|_{\mathbb{M}^N(\mathbb{R})}\right)$. Momentarily we would need to verify these properties separately for each such space that we encountered. As you will see, the space necessarily changes from one IBVP to another, so this approach would be inefficient and tedious. It would be beneficial to examine a more abstract structure possessing these characteristics and establish results directly for *them.* We do so in this section.

8.3.1 Moving Beyond Just $\mathbb{R}^N$ and $\mathbb{M}^N$: Introducing the Notions of Banach Space and Hilbert Space

We begin with the notion of a linear space over $\mathbb{R}$ that you encountered in a linear algebra course.

Definition 8.3.1. A *real linear space* X is a set equipped with addition and scalar multiplication by real numbers satisfying the following properties:

i.) $x + y = y + x$, $\forall x, y \in X$;
ii.) $x + (y + z) = (x + y) + z$, $\forall x, y, z \in X$;
iii.) There exists a unique element $0 \in X$ such that $x + 0 = 0 + x$, $\forall x \in X$;
iv.) For every $x \in X$, there exists a unique element $-x \in X$ such that

$$x + (-x) = (-x) + x = 0, \ \forall x \in X;$$

v.) $a(bx) = (ab)x, \forall x \in X$ and $a, b \in \mathbb{R}$;
vi.) $a(x+y) = ax + ay, \forall x, y \in X$ and $a \in \mathbb{R}$;
vii.) $(a+b)x = ax + bx, \forall x \in X$ and $a, b \in \mathbb{R}$.

Restricting attention to a subset Y of elements of a linear space X that possesses the same structure as the larger space leads to the following notion.

Definition 8.3.2. Let X be a real linear space. A subset $Y \subset X$, equipped with the same operations as X, is a *linear subspace* of X if

i.) $x, y \in Y \implies x + y \in Y$,
ii.) $x \in Y \implies ax \in Y, \forall a \in \mathbb{R}$.

Exercise 8.3.1.

i.) Verify that $(\mathbb{R}, |\cdot|)$, $(\mathbb{R}^N, \|\cdot\|_{\mathbb{R}^N})$, and $\left(\mathbb{M}^N(\mathbb{R}), \|\cdot\|_{\mathbb{M}^N(\mathbb{R})}\right)$ are linear spaces.

ii.) Is $Y = \{\mathbf{A} \in \mathbb{M}^N(\mathbb{R}) : \mathbf{A}\,\text{is diagonal}\}$ a linear subspace of $\mathbb{M}^N(\mathbb{R})$?

We can enhance the structure of a real linear space by equipping it with a norm, used to measure distance, in the following sense.

Definition 8.3.3. Let X be a real linear space. A real-valued function $\|\cdot\|_X : X \longrightarrow \mathbb{R}$ is a *norm* on X if $\forall x, y \in X$ and $a \in \mathbb{R}$,

i.) $\|x\|_X \geq 0$,
ii.) $\|ax\|_X = |a| \, \|x\|_X$,
iii.) $\|x + y\|_X \leq \|x\|_X + \|y\|_X$,
iv.) $x = 0$ iff $\|x\|_X = 0$.

We say that the *distance* between x and y is $\|x - y\|_X$. Introducing a norm enables us to perform limit operations in the same sense that the role of the absolute value played in the limit definition of real-valued functions and sequences. We use this to obtain the following richer abstract structure.

Definition 8.3.4. A real linear space X equipped with a norm $\|\cdot\|_X$ is called a *(real) normed linear space.*

The spaces $(\mathbb{R}, |\cdot|)$, $(\mathbb{R}^N, \|\cdot\|_{\mathbb{R}^N})$, and $\left(\mathbb{M}^N(\mathbb{R}), \|\cdot\|_{\mathbb{M}^N(\mathbb{R})}\right)$ are all normed linear spaces. Many of the normed linear spaces that we will encounter are collections of functions satisfying certain properties. Some standard function spaces and typical norms with which they are equipped are listed below. We shall assume that these spaces satisfy all properties listed above - see [27] for details.

Some Common Function Spaces: Let $I \subset \mathbb{R}$ and X be a normed linear space. (For our purposes, X is typically taken to be $\mathbb{R}$ or $\mathbb{R}^N$, or some subset thereof.)

1. $\mathbb{C}(I; X) = \{f : I \longrightarrow X \mid f \text{ is continuous on } I\}$ equipped with the *sup norm*

$$\|f\|_{\mathbb{C}} = \sup_{t \in I} \|f(t)\|_X \tag{8.43}$$

2. $\mathbb{L}^2(I; \mathbb{R}) = \{f : I \longrightarrow \mathbb{R} | f^2 \text{ is integrable on } I\}$ equipped with

$$\|f\|_{\mathbb{L}^2} = \left[\int_I |f(t)|^2 \, dt\right]^{\frac{1}{2}} \tag{8.44}$$

3. $\mathbb{H}^2(I;\mathbb{R}) = \{f \in \mathbb{L}^2(I;\mathbb{R}) | f', f''\text{exist and } f'' \in \mathbb{L}^2(I;\mathbb{R})\}$equipped with

$$\|f\|_{\mathbb{H}^2} = \left[\int_I |f(t)|^2 \, dt\right]^{\frac{1}{2}} \tag{8.45}$$

4. $\mathbb{H}_0^1(a,b;\mathbb{R}) = \{f : (a,b) \longrightarrow \mathbb{R} | f'\text{exists and } f(a) = f(b) = 0\}$ equipped with

$$\|f\|_{\mathbb{H}_0^1} = \left[\int_I \left(|f(t)|^2 + |f'(t)|^2\right) dt\right]^{\frac{1}{2}} \tag{8.46}$$

5. Let $m \in \mathbb{N}$. $\mathbb{W}^{2,m}(I;\mathbb{R}) = \{f \in \mathbb{L}^2(I;\mathbb{R}) | f^{(k)} \in \mathbb{L}^2(I;\mathbb{R}),\ \forall k = 1,\ldots,m\}$ equipped with

$$\|f\|_{\mathbb{W}^{2,m}} = \left[\int_I \left(|f(t)|^2 + |f'(t)|^2 + \ldots + \left|f^{(m)}(t)\right|^2\right) dt\right]^{\frac{1}{2}}. \tag{8.47}$$

Here, $f^{(k)}$ represents the kth- order derivative of f.

8.3.1.1 The Notion of Convergence Revisited

The notions of convergent and Cauchy sequences are defined in any normed linear space in the same manner as in Part I. For example, we say that an X-valued sequence $(x_n)_{n\in\mathbb{N}}$ *converges to x in X* if $\lim_{n\to\infty} \|x_n - x\|_X = 0$.

Exercise 8.3.2.

i.) Interpret the statement "$\lim_{n\to\infty} x_n = x$ in X" for these specific choices of X:

   a.) (8.43)
   b.) (8.44)
   c.) (8.47)

ii.) Interpret the statement "$(x_n)_{n\in\mathbb{N}}$ is Cauchy in X" for the same choices of X.

When working with specific function spaces, knowing when Cauchy sequences in X must converge to an element *in* X is often crucial. In other words, we need to know if a space is complete in the following sense.

Definition 8.3.5. (Completeness)

i.) A normed linear space X is *complete* if every Cauchy sequence in X converges to an element of X.

ii.) A complete normed linear space is called a *Banach space.*

It can be shown that all spaces above are complete with respect to the associated norms. We shall routinely work with sequences of continuous functions in $\mathbb{C}(I;\mathcal{X})$, where $\mathcal{X}$ is a Banach space. We now focus on the terminology and some particular results for this space. (We have already encountered these notions when $\mathcal{X}$ was $\mathbb{R}$, $\mathbb{R}^N$, or $\mathbb{M}^N$.)

Definition 8.3.6. Suppose that $\varnothing \neq S \subset D \subset \mathbb{R}$ and $f_n, f : D \longrightarrow \mathcal{X}$, $n \in \mathbb{N}$.

i.) $(f_n)_{n\in\mathbb{N}}$ *converges uniformly to f on S* whenever $\forall \epsilon > 0$, $\exists N \in \mathbb{N}$ such that

$$n \geq N \Longrightarrow \sup_{x\in S} \|f_n(x) - f(x)\|_{\mathcal{X}} < \epsilon.$$

(We write "$f_n \longrightarrow f$ uniformly on S.")

ii.) $(f_n)_{n\in\mathbb{N}}$ *converges pointwise to* f *on* S whenever $\lim_{n\to\infty} f_n(x) = f(x), \forall x \in S$.

iii.) $(f_n)_{n\in\mathbb{N}}$ is *uniformly bounded on* S whenever there exists $M > 0$ such that

$$\sup_{n\in\mathbb{N}} \sup_{x\in S} \|f_n(x)\|_{\mathcal{X}} \leq M.$$

iv.) $\sum_{k=1}^{\infty} f_k(x)$ *converges uniformly to* $f(x)$ *on* S whenever $s_n \longrightarrow f$ uniformly on S, where $s_n(x) = \sum_{k=1}^{n} f_k(x)$.

Remark 8.3.1. Taylor series representations of infinitely-differentiable functions are presented in elementary calculus. Some common examples are:

$$e^x = \lim_{N\to\infty} \sum_{n=0}^{N} \frac{x^n}{n!}, \ x \in \mathbb{R}, \tag{8.48}$$

$$\sin(x) = \lim_{N\to\infty} \sum_{n=0}^{N} \frac{(-1)^n x^{2n+1}}{(2n+1)!}, \ x \in \mathbb{R}, \tag{8.49}$$

$$\cos(x) = \lim_{N\to\infty} \sum_{n=0}^{N} \frac{(-1)^n x^{2n}}{(2n)!}, \ x \in \mathbb{R}. \tag{8.50}$$

It can be shown that the convergence in each case is uniform on all closed bounded intervals of $\mathbb{R}$. The benefit of such a representation is the uniform approximation of the function on the left-side by the sequence of nicely-behaved polynomials on the right-side. Generalizations of these formulae to more abstract settings will be a key tool throughout the rest of the text.

8.3.1.2 An Important Topological Notion—Closed Sets

The notion of convergence of sequences can be used to define a very important topological concept called *closure*. You encountered the concept when considering closed intervals $[a, b]$ along the real line. But, what makes $[0, 1]$ closed, but $[0, 1)$ not closed? Naïvely, it is because 1 is not included in the latter interval. Think of this as a tiny open door through which a sequence of points inside $[0, 1)$ can escape (by converging to 1). For instance, the sequence $\left(1 - \frac{1}{n}\right)_{n\in\mathbb{N}}$ is entirely inside $[0, 1)$, but its limit as $n \to \infty$ does not belong to this interval.

Similar behavior can occur in the more general setting of a Banach space and is not desirable from the viewpoint of studying mathematical models because it can lead to undefined situations. Just think, what if we clevely formulated a sequence of nicely-defined functions whose limit was the solution we sought, but in the end, the limit did not belong to the nice space from which the sequence was formed. We would be in a bad way, right? As such, we would like to build in the impossibility of this behavior occurring from the beginning of our analysis. To this end, we define the notion of closed set in a Banach space below.

Definition 8.3.7. Let $\mathcal{X}$ be a Banach space and $D \subset \mathcal{X}$. We say that D is *closed in* $\mathcal{X}$ provided that if a sequence $(x_n) \subset D$ converges to L, then L must belong to D.

Exercise 8.3.3. Determine if these sets are closed in the indicated spaces.

i.) $(-\infty, -2] \cup \{3\}$ in $\mathbb{R}$

ii.) $\{\mathbf{x} \in \mathbb{R}^2 : \|\mathbf{x} - \mathbf{x_0}\| = 1\}$

iii.) $\{\mathbf{x} \in \mathbb{R}^3 : \|\mathbf{x} - \mathbf{x_0}\| > 2\}$

Even if a set D is not closed in X, we can build a set using D that *is* closed in X. The idea is simple. Start with D and tack on all limits of convergent sequences of elements of D. This effectively "closes" all doors through which sequences can escape the set via convergence. We call this set the *closure of* D in X and denote it by $cl_X(D)$.

Exercise 8.3.4. Compute the closures of the sets in Exercise 8.3.3.

We use the following notation:

$$\begin{aligned} \mathfrak{B}_{\mathcal{X}}(x_0;\epsilon) &= \{x \in \mathcal{X} \mid \|x - x_0\|_{\mathcal{X}} < \epsilon\} \\ \mathrm{cl}_{\mathcal{X}}(\mathcal{Z}) &= \text{closure of } \mathcal{Z} \text{ (in the sense of } \|\cdot\|_{\mathcal{X}}). \end{aligned} \tag{8.51}$$

Exercise 8.3.5. Describe the elements of the ball $\mathfrak{B}_{\mathbb{C}([0,2];\mathbb{R})}\left(x^2;1\right)$.

The need to restrict our attention to a particular subspace of a function space whose elements satisfy some special characteristic arises often. But, can we be certain that we remain in the subspace upon performing limiting operations involving its elements? Put differently, must a subspace of a Banach space be complete? The answer is provided in the following exercise.

Exercise 8.3.6. Let Y be the subspace $((0,2]\,;|\cdot|)$ of $\mathbb{R}$. Prove that $\left(\frac{2}{n}\right)_{n\in\mathbb{N}}$ is Cauchy in Y, but that there does not exist $y \in Y$ to which $\left(\frac{2}{n}\right)_{n\in\mathbb{N}}$ converges.

If the subspace had been closed in the topological sense, would it have made a difference? It turns out that it would have indeed, as suggested by the following result:

Proposition 8.3.1. A closed subspace $\mathcal{Y}$ of a Banach space $\mathcal{X}$ is complete.

Exercise 8.3.7. Prove Prop. 8.3.1.

Exercise 8.3.8. Let $(\mathcal{X}, \|\cdot\|_{\mathcal{X}})$ and $(\mathcal{Y}, \|\cdot\|_{\mathcal{Y}})$ be real Banach spaces. Prove that $(\mathcal{X}\times\mathcal{Y}; \|\cdot\|_1)$ and $(\mathcal{X}\times\mathcal{Y}; \|\cdot\|_2)$ are also Banach spaces, where

$$\|(x,y)\|_1 = \|x\|_{\mathcal{X}} + \|y\|_{\mathcal{Y}}, \tag{8.52}$$

$$\|(x,y)\|_2 = \left(\|x\|_{\mathcal{X}}^2 + \|y\|_{\mathcal{Y}}^2\right)^{\frac{1}{2}}. \tag{8.53}$$

8.3.2 Hilbert Spaces

Equipping $\mathbb{R}^N$ with a dot product enhanced its structure by introducing the notion of orthogonality. This prompts us to define the general notion of an inner product on a linear space.

Definition 8.3.8. Let X be a real linear space. A real-valued function $\langle\cdot,\cdot\cdot\rangle_X : X\times X \longrightarrow \mathbb{R}$ is an *inner product on* X if $\forall x,y,z \in X$ and $a \in \mathbb{R}$,

i.) $\langle x,y\rangle_X = \langle y,x\rangle_X$,
ii.) $\langle ax,y\rangle_X = a\langle x,y\rangle_X$,
iii.) $\langle x+y,z\rangle_X = \langle x,z\rangle_X + \langle y,z\rangle_X$,
iv.) $\langle x,x\rangle_X > 0$ iff $x \neq 0$.

The pair $(X,\langle\cdot,\cdot\cdot\rangle_X)$ is called a *(real) inner product space.*

Some Common Inner Product Spaces:

1. $\mathbb{R}^N$ equipped with the usual "dot product."
2. $\mathbb{C}([a,b];\mathbb{R})$ equipped with

$$\langle f,g\rangle_{\mathbb{C}} = \int_a^b f(t)g(t)dt. \tag{8.54}$$

3. $\mathbb{L}^2(a,b;\mathbb{R})$ equipped with (8.54).

4. $\mathbb{W}^{2,m}(a,b;\mathbb{R})$ equipped with

$$\langle f,g\rangle_{\mathbb{W}^{2,k}} = \int_a^b \left[f(t)g(t)+f'(t)g'(t)+\ldots+f^{(m)}(t)g^{(m)}(t)\right]dt. \tag{8.55}$$

An inner product on X induces a norm on X via the relationship

$$\langle x,x\rangle_X^{\frac{1}{2}} = \|x\|_X. \tag{8.56}$$

Exercise 8.3.9. Prove that the usual norms in $\mathbb{C}([a,b];\mathbb{R})$ and $\mathbb{W}^{2,m}(a,b;\mathbb{R})$ can be obtained from their respective inner products (8.54) and (8.55).

STOP! The remainder of this section is included primarily to be able to formally define the notion of a Fourier series representation for a function in $\mathbb{L}^2(a,b;\mathbb{R})$. If you are comfortable with Fourier series and do not feel the need for the theoretical underpinnings, feel free to skip this part of the section.

Proposition 8.3.2. Let $(X,\langle\cdot,\cdot\cdot\rangle_X)$ be an inner product space and suppose $\|\cdot\|_X$ is given by (8.56). Then, $\forall x,y\in X$ and $a\in\mathbb{R}$,

i.) $\langle x,ay\rangle_X = a\langle x,y\rangle_X$;
ii.) $\langle x,z\rangle_X = \langle y,z\rangle_X,\ \forall z\in X \Longrightarrow x=y$;
iii.) $\|ax\|_X = |a|\,\|x\|_X$;
iv.) (Cauchy-Schwarz)$|\langle x,y\rangle_X| \le \|x\|_X\,\|y\|_X$;
v.) (Minkowski) $\|x+y\|_X \le \|x\|_X + \|y\|_X$;
vi.) If $x_n \longrightarrow x$ and $y_n \longrightarrow y$ in X, then $\langle x_n,y_n\rangle_X \longrightarrow \langle x,y\rangle_X$.

Exercise 8.3.10. Interpret Prop. 8.3.2(iv) specifically for the space $\mathbb{L}^2(a,b;\mathbb{R})$. (This is a special case of the so-called *Hölder inequality.*)

Since inner product spaces come equipped with a norm, it makes sense to further characterize them using completeness.

Definition 8.3.9. A *Hilbert space* is a complete inner product space.

Both $\mathbb{R}^N$ and $\mathbb{L}^2(a,b;\mathbb{R})$ equipped with their usual norms are Hilbert spaces, while $\mathbb{C}([a,b];\mathbb{R})$ equipped with (8.54) is not. Again, the underlying norm plays a crucial role.

The notion of a *basis* encountered in linear algebra can be made precise in the Hilbert space setting and plays a central role in formulating representation formulae for elements of the space. We begin with the following definition.

Definition 8.3.10. Let $\mathcal{H}$ be a Hilbert space and $\mathfrak{B} = \{e_n | n\in K\subset\mathbb{N}\}$.

i.) The *span of* $\mathfrak{B}$ is given by span$(\mathfrak{B}) = \left\{\sum_{n\in K}\alpha_n e_n |\ \alpha_n\in\mathbb{R}, \forall n\in K\right\}$;
ii.) If $\langle e_n,e_m\rangle_{\mathcal{H}} = 0$, then e_n and e_m are *orthogonal*;
iii.) The members of $\mathfrak{B}$ are *linearly independent* if

$$\sum_{n\in K}\alpha_n e_n = 0 \Longrightarrow \alpha_n = 0,\ \forall n\in K;$$

iv.) $\mathfrak{B}$ is an *orthonormal set* if

a.) $\|e_n\|_{\mathcal{H}} = 1, \forall n \in K$,

b.) $\langle e_n, e_m \rangle_{\mathcal{H}} = \begin{cases} 0, & \text{if } n \neq m, \\ 1, & \text{if } n = m. \end{cases}$

v.) $\mathfrak{B}$ is a *complete set* if $(\langle x, e_n \rangle_{\mathcal{H}} = 0, \forall n \in K) \Longrightarrow (x = 0, \forall x \in \mathcal{H})$;
vi.) A complete orthonormal subset of $\mathcal{H}$ is a *basis* for $\mathcal{H}$.

The utility of a basis $\mathfrak{B}$ of a Hilbert space $\mathcal{H}$ is that every element of $\mathcal{H}$ can be decomposed into a linear combination of the members of $\mathfrak{B}$. For general Hilbert spaces, specifically those that are not finite dimensional like $\mathbb{R}^N$, the existence of a basis is not guaranteed. There are, however, sufficiency results that indicate when a basis must exist. For instance, consider

Definition 8.3.11. An inner product space is *separable* if it contains a countable dense subset $\mathfrak{D}$.

Remark. A set $\mathfrak{D}$ is *countable* if a one-to-one function $f : \mathfrak{D} \to \mathbb{N}$ exists. In such case, the elements of $\mathfrak{D}$ can be matched in a one-to-one manner with those of $\mathbb{N}$. Intuitively, $\mathfrak{D}$ has no more elements than $\mathbb{N}$.

Theorem 8.3.1. Any separable inner product space has a basis.

Example. The set

$$\mathfrak{B} = \left\{ \frac{1}{\sqrt{2\pi}} \right\} \cup \left\{ \frac{\cos(nt)}{\sqrt{\pi}} | n \in \mathbb{N} \right\} \cup \left\{ \frac{\sin(nt)}{\sqrt{\pi}} | n \in \mathbb{N} \right\} \tag{8.57}$$

is an orthonormal, dense subset of $\mathbb{L}^2(-\pi, \pi; \mathbb{R})$ equipped with inner product (8.54). Here, $\cos(n\cdot)$ means "$\cos(nt)$, $-\pi \leq t \leq \pi$."

Exercise 8.3.11. Answer the following:

i.) Prove that $\|f\|_{\mathbb{L}^2} = 1, \forall f \in \mathfrak{B}$.

ii.) Prove that $\langle f, g \rangle_{\mathbb{L}^2} = 0, \forall f \neq g \in \mathfrak{B}$.

iii.) How would you adapt the set defined in (8.57) for $\mathbb{L}^2(a, b; \mathbb{R})$, where $a < b$, such that properties (i) and (ii) remain true?

Proposition 8.3.3. (Properties of Orthonormal Sets)
Let $\mathcal{H}$ be an inner product space and $\mathcal{Y} = \{y_1, \ldots, y_n\}$ an orthonormal set in $\mathcal{H}$. Then,

i.) $\left\|\sum_{i=1}^n y_i\right\|_{\mathcal{H}}^2 = \sum_{i=1}^n \|y_i\|_{\mathcal{H}}^2$;
ii.) The elements of $\mathcal{Y}$ are linearly independent;
iii.) If $x \in \text{span}(\mathcal{Y})$, then $x = \sum_{i=1}^n \langle x, y_i \rangle_{\mathcal{H}} y_i$;
iv.) If $x \in H$, then $\langle x - \sum_{i=1}^n \langle x, y_i \rangle_{\mathcal{H}} y_i, y_k \rangle_{\mathcal{H}} = 0, \forall k \in \{1, \ldots, n\}$.

The following result is the "big deal!"

Theorem 8.3.2. (Representation Theorem for a Hilbert Space)
Let $\mathcal{H}$ be a Hilbert space and $\mathfrak{B} = \{e_n | n \in \mathbb{N}\}$ a basis for $\mathcal{H}$. Then,

i.) For every $x \in \mathcal{H}$, $\sum_{k=1}^\infty |\langle x, e_k \rangle_{\mathcal{H}}|^2 \leq \|x\|_{\mathcal{H}}^2$;

ii.) $\lim_{N\to\infty} \left\| \sum_{k=1}^{N} \langle x, e_k \rangle_{\mathcal{H}} e_k - x \right\|_{\mathcal{H}} = 0$, and we write $x = \sum_{k=1}^{\infty} \langle x, e_k \rangle_{\mathcal{H}} e_k$.

Outline of Proof:

Proof of (i): Observe that $\forall N \in \mathbb{N}$,

$$\begin{aligned} 0 &\le \left\| \sum_{k=1}^{n} \langle x, e_k \rangle_{\mathcal{H}} e_k - x \right\|_{\mathcal{H}}^2 \\ &= \left\langle \sum_{k=1}^{n} \langle x, e_k \rangle_{\mathcal{H}} e_k - x, \sum_{k=1}^{n} \langle x, e_k \rangle_{\mathcal{H}} e_k - x \right\rangle_{\mathcal{H}} \\ &= \|x\|_{\mathcal{H}}^2 - \sum_{k=1}^{n} |\langle x, e_k \rangle_{\mathcal{H}}|^2. \end{aligned}$$

(Why?) Thus, $\sum_{k=1}^{n} |\langle x, e_k \rangle_{\mathcal{H}}|^2 \le \|x\|_{\mathcal{H}}^2$, $\forall n \in \mathbb{N}$. The result then follows because $\left(\sum_{k=1}^{n} |\langle x, e_k \rangle_{\mathcal{H}}|^2 \right)_{n\in\mathbb{N}}$ is an increasing sequence bounded above. **(Why and so what?)**

Proof of (ii): For each $N \in \mathbb{N}$, let $S_N = \sum_{k=1}^{N} \langle x, e_k \rangle_{\mathcal{H}} e_k$. The fact that $(S_N)_{N\in\mathbb{N}}$ is a Cauchy sequence in $\mathcal{H}$ follows from Prop. 8.3.3 and part (i) of this theorem. **(How?)** Moreover, $(S_N)_{N\in\mathbb{N}}$ must converge since $\mathcal{H}$ is complete. The fact that the limit is x follows from the completeness of $\mathfrak{B}$. **(Tell how.)** □

Remark 8.3.2. (Fourier Series) An important application of Thrm. 8.3.2 occurs in the study of Fourier series. This involves identifying a sufficiently smooth function with a unique series representation defined using a family of sines and cosines (cf. Section 3.2). To this end, we infer from the example directly following Thrm. 8.3.1 that every $f \in \mathbb{L}^2(-\pi, \pi; \mathbb{R})$ can be expressed uniquely as

$$\begin{aligned} f(t) &= \sum_{n=1}^{\infty} \langle f(\cdot), e_n \rangle_{\mathbb{L}^2} e_n \\ &= \left\langle f(\cdot), \frac{1}{\sqrt{2\pi}} \right\rangle_{\mathbb{L}^2} \frac{1}{\sqrt{2\pi}} + \sum_{n=1}^{\infty} \left\langle f(\cdot), \frac{\cos(n\cdot)}{\sqrt{\pi}} \right\rangle_{\mathbb{L}^2} \frac{\cos(nt)}{\sqrt{\pi}} \\ &+ \sum_{n=1}^{\infty} \left\langle f(\cdot), \frac{\sin(n\cdot)}{\sqrt{\pi}} \right\rangle_{\mathbb{L}^2} \frac{\sin(nt)}{\sqrt{\pi}}, \ -\pi \le t \le \pi, \end{aligned} \tag{8.58}$$

where the convergence of the series is in the $\mathbb{L}^2$–sense. For brevity, let

$$a_0 = \left\langle f(\cdot), \frac{1}{\sqrt{2\pi}} \right\rangle_{\mathbb{L}^2} = \frac{1}{\sqrt{2\pi}} \int_{-\pi}^{\pi} f(t)dt \tag{8.59}$$

$$a_n = \left\langle f(\cdot), \frac{\cos(n\cdot)}{\sqrt{\pi}} \right\rangle_{\mathbb{L}^2} = \frac{1}{\sqrt{\pi}} \int_{-\pi}^{\pi} f(t)\cos(nt)dt, \ n \in \mathbb{N}, \tag{8.60}$$

$$b_n = \left\langle f(\cdot), \frac{\sin(n\cdot)}{\sqrt{\pi}} \right\rangle_{\mathbb{L}^2} = \frac{1}{\sqrt{\pi}} \int_{-\pi}^{\pi} f(t)\sin(nt)dt, \ n \in \mathbb{N}. \tag{8.61}$$

Then, (8.58) can be written as

$$f(t) = \frac{a_0}{\sqrt{2\pi}} + \sum_{n=1}^{\infty} \left[a_n \frac{\cos(nt)}{\sqrt{\pi}} + b_n \frac{\sin(nt)}{\sqrt{\pi}} \right], -\pi \le t \le \pi.$$

8.3.3 Linear Operators

We conjectured, primarily by symbol matching, that the quantity A arising in (A-HCP) must be a function that maps one space of functions into another. We must consider this much more carefully. We begin by considering the prototypical example arising in Part I.

Proposition 8.3.4. Let $\mathbf{A} \in \mathbb{M}^N(\mathbb{R})$ and define $\mathcal{B} : \text{dom}(\mathcal{B}) \subset \mathbb{R}^N \to \mathbb{R}^N$ by $\mathcal{B}\mathbf{x} = \mathbf{A}\mathbf{x}$.

i.) $\text{dom}(\mathcal{B}) = \mathbb{R}^N$;
ii.) For every $\alpha, \beta \in \mathbb{R}$ and $\mathbf{x}, \mathbf{y} \in \mathbb{R}^N$, $\mathcal{B}(\alpha\mathbf{x} + \beta\mathbf{y}) = \alpha\mathcal{B}\mathbf{x} + \beta\mathcal{B}\mathbf{y}$;
iii.) For every $\mathbf{x} \in \mathbb{R}^N$, $\|\mathcal{B}\mathbf{x}\|_{\mathbb{R}^N} \leq \|\mathbf{A}\|_{\mathbb{M}^N} \|\mathbf{x}\|_{\mathbb{R}^N}$.

Proof. (i) holds since the product $\mathbf{A}\mathbf{x}$ is defined, $\forall \mathbf{x} \in \mathbb{R}^N$, and (ii) follows from the properties of matrix multiplication. **(Tell how.)** As for (iii), for simplicity we prove the result for $N = 2$. Let $\mathbf{x} = \begin{bmatrix} x_1 \\ x_2 \end{bmatrix}$. Observe that

$$\begin{aligned}
\|\mathbf{A}\mathbf{x}\|_{\mathbb{R}^2} &= \left\| \begin{bmatrix} a_{11} & a_{12} \\ a_{21} & a_{22} \end{bmatrix} \cdot \begin{bmatrix} x_1 \\ x_2 \end{bmatrix} \right\|_{\mathbb{R}^2} = \left\| \begin{bmatrix} a_{11}x_1 + a_{12}x_2 \\ a_{21}x_1 + a_{22}x_2 \end{bmatrix} \right\|_{\mathbb{R}^2} \\
&= |a_{11}x_1 + a_{12}x_2| + |a_{21}x_1 + a_{22}x_2| \\
&\leq (|a_{11}| + |a_{21}|)\,|x_1| + (|a_{12}| + |a_{22}|)\,|x_2| \\
&\leq \left(\sum_{i=1}^{2} \sum_{j=1}^{2} |a_{ij}| \right) |x_1| + \left(\sum_{i=1}^{2} \sum_{j=1}^{2} |a_{ij}| \right) |x_2| \\
&= \left(\sum_{i=1}^{2} \sum_{j=1}^{2} |a_{ij}| \right) (|x_1| + |x_2|) \\
&= \|\mathbf{A}\|_{\mathbb{M}^2} \|\mathbf{x}\|_{\mathbb{R}^2}.
\end{aligned}$$

STOP! Justify the steps and try proving the general case. □

Proposition 8.3.5. Let $\mathbf{A} \in \mathbb{M}^N(\mathbb{R})$ and $\forall t \in \mathbb{R}$, define $\boldsymbol{S}_t : \text{dom}(\boldsymbol{S}_t) \subset \mathbb{R}^N \to \mathbb{R}^N$ by $\boldsymbol{S}_t\mathbf{x} = e^{\mathbf{A}t}\mathbf{x}$.

i.) $\text{dom}(\boldsymbol{S}_t) = \mathbb{R}^N$;
ii.) For every $\alpha, \beta \in \mathbb{R}$ and $\mathbf{x}, \mathbf{y} \in \mathbb{R}^N$, $\boldsymbol{S}_t(\alpha\mathbf{x} + \beta\mathbf{y}) = \alpha\boldsymbol{S}_t\mathbf{x} + \beta\boldsymbol{S}_t\mathbf{y}$;
iii.) For every $\mathbf{x} \in \mathbb{R}^N$, $\|\boldsymbol{S}_t\mathbf{x}\|_{\mathbb{R}^N} = \|e^{\mathbf{A}t}\mathbf{x}\|_{\mathbb{R}^N} \leq e^{t\|\mathbf{A}\|_{\mathbb{M}^N}} \|\mathbf{x}\|_{\mathbb{R}^N}$.

Proof. (i) and (ii) follow since $e^{\mathbf{A}t} \in \mathbb{M}^N(\mathbb{R})$. As for (iii), Prop. 8.3.4 implies that

$$\|e^{\mathbf{A}t}\mathbf{x}\|_{\mathbb{R}^N} \leq \|e^{\mathbf{A}t}\|_{\mathbb{M}^N} \|\mathbf{x}\|_{\mathbb{R}^N}. \tag{8.62}$$

Continuing, we see

$$\begin{aligned}
\|e^{\mathbf{A}t}\|_{\mathbb{M}^N} &= \left\| \sum_{k=0}^{\infty} \frac{\mathbf{A}^k t^k}{k!} \right\|_{\mathbb{M}^N} = \left\| \lim_{m\to\infty} \sum_{k=0}^{m} \frac{\mathbf{A}^k t^k}{k!} \right\|_{\mathbb{M}^N} = \lim_{m\to\infty} \left\| \sum_{k=0}^{m} \frac{\mathbf{A}^k t^k}{k!} \right\|_{\mathbb{M}^N} \\
&\leq \lim_{m\to\infty} \sum_{k=0}^{m} \frac{\|\mathbf{A}\|_{\mathbb{M}^N}^k |t|^k}{k!} = \sum_{k=0}^{\infty} \frac{(\|\mathbf{A}\|_{\mathbb{M}^N} |t|)^k}{k!} = e^{|t|\|\mathbf{A}\|_{\mathbb{M}^N}}.
\end{aligned}$$

(Justify the steps.) Using this estimate in (8.62) yields the result. □

Both mappings satisfied the same basic properties. Mappings that satisfy this collection of properties arise often enough in different contexts that we single them out as follows.

Definition 8.3.12. Let $(\mathcal{X}, \|\cdot\|_{\mathcal{X}})$ and $(\mathcal{Y}, \|\cdot\|_{\mathcal{Y}})$ be real Banach spaces.

i.) A *bounded linear operator* from $(\mathcal{X}, \|\cdot\|_{\mathcal{X}})$ into $(\mathcal{Y}, \|\cdot\|_{\mathcal{Y}})$ is a mapping $\mathcal{F} : \mathcal{X} \to \mathcal{Y}$ such that

a.) (linear) $\mathcal{F}(\alpha x + \beta y) = \alpha \mathcal{F}(x) + \beta \mathcal{F}(y)$, $\forall \alpha, \beta \in \mathbb{R}$ and $x, y \in \mathcal{X}$,

b.) (bounded) There exists $m \geq 0$ such that $\|\mathcal{F}(x)\|_{\mathcal{Y}} \leq m \|x\|_{\mathcal{X}}$, $\forall x \in \mathcal{X}$.

We denote the set of all such operators by $\mathbb{B}(\mathcal{X}, \mathcal{Y})$. If $\mathcal{X} = \mathcal{Y}$, we write $\mathbb{B}(\mathcal{X})$ and refer to its members as "bounded linear operators on $\mathcal{X}$."

ii.) If there does not exist $m \geq 0$ such that $\|\mathcal{F}(x)\|_{\mathcal{Y}} \leq m \|x\|_{\mathcal{X}}$, $\forall x \in \mathcal{X}$, we say that $\mathcal{F}$ is *unbounded.*

The terminology "bounded operator" may seem to be somewhat of a misnomer in comparison to the notion of a bounded real-valued function, but the name arose because such operators map norm-bounded subsets of $\mathcal{X}$ into norm-bounded subsets of $\mathcal{Y}$. We must simply contend with this nomenclature issue on a contextual basis. Also, regarding the notation, the quantity $\mathcal{F}(x)$ is often written more succinctly as $\mathcal{F}x$ (with parentheses suppressed) as in the context of matrix multiplication.

Now, consider the various solution formulas arising in Section 8.1 when solving various heat-related IBVPs. For the IBVPs formulated for a one-dimensional region through which heat propogated, the mapping $S(t) : \mathbb{L}^2(0, a; \mathbb{R}) \to \mathbb{L}^2(0, a; \mathbb{R})$ defined by

$$S(t)[u_0](x) = u(x, t) \tag{8.63}$$

arose. We claim that $S(t)$ is a bounded linear operator on $\mathbb{L}^2(0, a; \mathbb{R})$, for every $t > 0$.

Proposition 8.3.6. The mapping $S(t) : \mathbb{L}^2(0, a; \mathbb{R}) \to \mathbb{L}^2(0, a; \mathbb{R})$ defined by (8.63) associated with boundary conditions given by either (8.6) or (8.19) is a bounded linear operator.

Proof. For simplicity, assume that the BCs are given by (8.6). You have already shown the linearity in Section 8.1. To verify the boundedness, let $f \in \mathbb{L}^2(0, a; \mathbb{R})$ and observe that

$$\begin{aligned}
\|S(t)[f]\|_{\mathbb{L}^2(0,a;\mathbb{R})} &= \int_0^a \left[\sum_{m=1}^{\infty} \left(\frac{2}{a} \int_0^a |f(w)| \sin\left(\frac{m\pi}{a} w\right) dw \right) e^{-\left(\frac{m\pi}{a}\right)^2 kt} \sin\left(\frac{m\pi}{a} x\right) \right]^2 dx \\
&\leq \frac{4}{a^2} \int_0^a \underbrace{\left[\sum_{m=1}^{\infty} \left(\int_0^a |f(w)|\, dw \right) e^{-\left(\frac{m\pi}{a}\right)^2 kt} \right]^2}_{\text{Independent of } x} dx \qquad (8.64) \\
&= \frac{4}{a} \left[\sum_{m=1}^{\infty} \left(\int_0^a |f(w)|\, dw \right) e^{-\left(\frac{m\pi}{a}\right)^2 kt} \right]^2 \\
&\leq \frac{4}{a} \sum_{m=1}^{\infty} \left(\int_0^a |f(w)|\, dw \right)^2 e^{-2\left(\frac{m\pi}{a}\right)^2 kt}
\end{aligned}$$

where we used the fact that $\left(\sum c_n\right)^2 \leq \sum c_n^2$ when, $c_n \geq 0$, for all values of n. Now, note that Hölder's inequality yields

$$\begin{aligned}
\left[\int_0^a f(z) dz \right]^2 &\leq \left[\left(\int_0^a f^2(z) dz \right)^{\frac{1}{2}} \left(\int_0^a 1 dz \right)^{\frac{1}{2}} \right]^2 \\
&= \left[\|f\|_{\mathbb{L}^2(0,a;\mathbb{R})} \sqrt{a} \right]^2 \qquad (8.65) \\
&= \|f\|^2_{\mathbb{L}^2(0,a;\mathbb{R})}\, a
\end{aligned}$$

Using (8.65) in (8.64) yields

$$\begin{aligned}\|S(t)[f]\|_{\mathbb{L}^2(0,a;\mathbb{R})} &\leq \frac{4}{a}\cdot\|f\|^2_{\mathbb{L}^2(0,a;\mathbb{R})}\, a\sum_{m=1}^{\infty} e^{-2\left(\frac{m\pi}{a}\right)^2 kt} \\ &= 4\,\|f\|^2_{\mathbb{L}^2(0,a;\mathbb{R})} \underbrace{\sum_{m=1}^{\infty} e^{-\left(\frac{2kt\pi^2}{a^2}\right)m^2}}_{\leq M} \\ &= M\,\|f\|^2_{\mathbb{L}^2(0,a;\mathbb{R})} \end{aligned} \tag{8.66}$$

The last step in (8.66) holds because the series $\sum_{m=1}^{\infty} e^{-\left(\frac{2kt\pi^2}{a^2}\right)m^2}$ is dominated by a convergent geometric series. □

Exercise 8.3.12. Verify that the properties in Proposition 8.3.6 hold for the version of (8.63) arising from the two-dimensional IBVP.

The notions of domain and range are the same for any mapping. One nice feature of an operator $\mathcal{F}\in\mathbb{B}(\mathcal{X},\mathcal{Y})$ is that both $\text{dom}(\mathcal{F})$ and $\text{rng}(\mathcal{F})$ are vector subspaces of $\mathcal{X}$ and $\mathcal{Y}$, respectively. The need to compare two operators and to consider the restriction of a given operator to a subset of its domain arise often. These notions are made precise below.

Definition 8.3.13. Let $\mathcal{F},\mathcal{G}\in\mathbb{B}(\mathcal{X},\mathcal{Y})$.

i.) We say $\mathcal{F}$ *equals* $\mathcal{G}$, written $\mathcal{F}=\mathcal{G}$, if
   a.) $\text{dom}(\mathcal{F}) = \text{dom}(\mathcal{G})$,
   b.) $\mathcal{F}x = \mathcal{G}x$, for all x in the common domain.

ii.) Let $\mathcal{Z}\subset\text{dom}(\mathcal{F})$. The operator $\mathcal{F}|_{\mathcal{Z}} : \mathcal{Z}\subset\mathcal{X}\to\mathcal{Y}$ defined by $\mathcal{F}|_{\mathcal{Z}}(z) = \mathcal{F}(z)$, $\forall z\in\mathcal{Z}$, is called the *restriction* of $\mathcal{F}$ to $\mathcal{Z}$.

At this point, let us revisit the actual operator A introduced in Section 8.2 in (8.42), namely $A : \text{dom}(A)\subset\mathbb{L}^2(0,a;\mathbb{R})\to\mathbb{L}^2(0,a;\mathbb{R})$ by

$$\begin{aligned} A[f] &= k\frac{d^2}{dx^2}[f], \\ \text{dom}(A) &= \left\{ f\in\mathbb{L}^2(0,a;\mathbb{R}) \,\middle|\, \frac{df}{dx}, \frac{d^2 f}{dx^2}\in\mathbb{L}^2(0,L;\mathbb{R}), \text{ and } f(0)=f(a)=0 \right\}. \end{aligned}$$

STOP! Show that A is a linear operator.

Remark 8.3.3. While A is linear, it can be shown that it is NOT bounded on the space $\mathbb{L}^2(0,a;\mathbb{R})$ equipped with its usual norm. This argument is technical and requires a firm understanding of functional analysis. We simply need the result here, but provide you with a MATLAB-Exercise to affirm its truth. There are also other more technical properties satisfied by A that are very important, namley its so-called graph $\{(x,Ax) : x\in\text{dom}(A)\}$is closed in $\mathbb{L}^2(0,a;\mathbb{R})$ and that its domain is dense in $\mathbb{L}^2(0,a;\mathbb{R})$, meaning that $\text{cl}_X(\text{dom}(A)) = X$.

MATLAB-Exercise 8.3.1. In this exercise, we use MATLAB to visualize the definition of an unbounded operator, which is defined as *not* bounded (see Definition 8.3.12). We focus on the second derivative operator using different domains. To begin, open **MATLAB** and in the command line, type:

MATLAB_Unbounded_Operators

Use the GUI to answer the following questions. A typical screenshot of the GUI can be found in Figure 8.2, and be sure to use the Help button to find more information about the MATLAB Unbounded operator GUI.

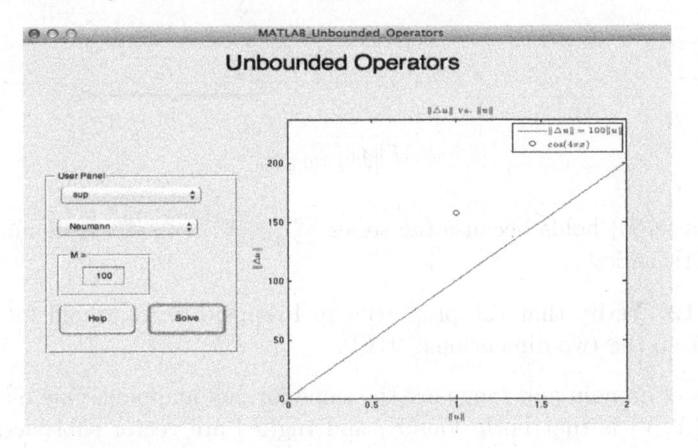

FIGURE 8.2: MATLAB Unbounded Operator GUI

i.) Consider the operator

$$A[f] = \frac{d^2}{dx^2}[f],$$

$$\mathrm{dom}(A) = \left\{ f \in \mathbb{L}^2(0,1;\mathbb{R}) \,\middle|\, \frac{df}{dx}, \frac{d^2 f}{dx^2} \in \mathbb{L}^2(0,L;\mathbb{R}), \text{ and } f(0) = f(1) = 0 \right\}.$$

We will show this operator is unbounded on $\mathbb{L}^2(0,1;\mathbb{R})$ by showing that for any $M > 0$ there is a function f in *dom*(A) such that

$$\|A[f]\|_{\mathbb{L}^2(0,1;\mathbb{R})} > M\|f\|_{\mathbb{L}^2(0,1;\mathbb{R})}. \tag{8.67}$$

a.) Find a function f that satisfies (8.67) for $M = 5$.
b.) Repeat (a) for different increasing values of M.
c.) Repeat (a) and (b) using Neumann BCs in the definition of *dom*(A).
d.) Were you always able to find an f satisfying (8.67)?

ii.) Consider the operator

$$A[f] = \frac{d^2}{dx^2}[f],$$

$$\mathrm{dom}(A) = \left\{ f \in C^2([0,1],\mathbb{R}) \,|\, f(0) = f(1) = 0 \right\}.$$

We will show this operator is unbounded when the sup-norm is used by showing that for any $M > 0$ there is a function f in *dom*(A) such that

$$\|A[f]\|_{\mathbb{C}(0,1;\mathbb{R})} > M\|f\|_{\mathbb{C}(0,1;\mathbb{R})}. \tag{8.68}$$

a.) Find a function f that satisfies (8.68) for $M = 5$.
b.) Repeat (a) for different increasing values of M.
c.) Repeat (a) and (b) using Neumann BCs in the definition of *dom*(A).
d.) Were you always able to find an f satisfying (8.68)?

■

8.3.4 Calculus in Abstract Spaces

The plan is to provide the formal definitions of these notions, together with their properties and important main results. The discussion is a terse outline, and you are encouraged to review these topics to fill in the gaps. Pay particular attention to the comparison between the real-valued setting and the more abstract setting of a normed space.

8.3.4.1 Limits

We begin with the extension of the notion of convergence (as defined for sequences) to the function setting.

Definition 8.3.14. A function $f : \text{dom}(f) \subset \mathcal{X} \to \mathcal{Y}$ has *limit L (in $\mathcal{Y}$) at $x = a \in (\text{dom}(f))'$* if for every sequence $(x_n)_{n\in\mathbb{N}} \subset \text{dom}(f)$ for which $\lim_{n\to\infty} \|x_n - a\|_{\mathcal{X}} = 0$, it is the case that $\lim_{n\to\infty} \|f(x_n) - L\|_{\mathcal{Y}} = 0$. We write $\lim_{x\to a} f(x) = L$ or equivalently, "$\|f(x_n) - L\|_{\mathcal{Y}} \to 0$ as $x \to a$."

Loosely speaking, the interpretation of Def. 8.3.14 for $\mathcal{X} = \mathcal{Y} = \mathbb{R}$ is that as the inputs approach a in any manner possible (i.e., via any sequence in $\text{dom}(f)$ convergent to a), the corresponding functional values approach L. The benefit of this particular definition is that the limit rules follow easily from the corresponding sequence properties.

An alternate definition equivalent to Def. 8.3.14, which is often more convenient to work with when involving certain norm estimates in an argument, is as follows:

Definition 8.3.15. A function $f : \text{dom}(f) \subset \mathcal{X} \to \mathcal{Y}$ has *limit L (in $\mathcal{Y}$) at $x = a \in (\text{dom}(f))'$* if every $\epsilon > 0$, there exists a $\delta > 0$ for which

$$x \in \text{dom}(f) \text{ and } 0 < \|x - a\|_{\mathcal{X}} < \delta \implies \|f(x) - L\|_{\mathcal{Y}} < \epsilon. \tag{8.69}$$

The notion of *one-sided limits* for real-valued functions arises occasionally, especially when limits are taken as the inputs approach the endpoints of an interval. Definition 8.3.14 can be naturally modified in such case, with the only changes occurring regarding which inputs near a are considered.

STOP! Formulate such definitions.

We denote the *right-limit at a* by $\lim_{x\to a^+} f(x)$, meaning that all inputs chosen when forming sequences that approach a are comprised of values that are greater than or equal to a. Likewise, we denote the *left-limit at a* by $\lim_{x\to a^-} f(x)$.

8.3.4.2 Continuity

Understanding the nature of continuous functions is crucial since much of our work in this text is performed in the space $\mathbb{C}(I; X)$. To begin, we need only to slightly modify Defs. 8.3.14 and 8.3.15 to arrive at the following stronger notion of *(norm) continuity.*

Definition 8.3.16. A function $f : \text{dom}(f) \subset \mathcal{X} \to \mathcal{Y}$ is *continuous* at $a \in \text{dom}(f)$ if either of these two equivalent statements hold:

i.) For every sequence $(x_n)_{n\in\mathbb{N}} \subset \text{dom}(f)$ for which $\lim_{n\to\infty} \|x_n - a\|_{\mathcal{X}} = 0$, it is the case that $\lim_{n\to\infty} \|f(x_n) - f(a)\|_{\mathcal{Y}} = 0$. We write $\lim_{x\to a} f(x_n) = f\left(\lim_{n\to\infty} x_n\right) = f(a)$.

ii.) For every $\epsilon > 0$, there exists a $\delta > 0$ for which

$$x \in \text{dom}(f) \text{ and } \|x - a\|_{\mathcal{X}} < \delta \implies \|f(x) - f(a)\|_{\mathcal{Y}} < \epsilon. \tag{8.70}$$

We say f is *continuous on* $S \subset \text{dom}(f)$ if f is continuous at every element of S.

"Continuity at a" is a strengthening of merely "having a limit at a" because the limit candidate being $f(a)$ requires that a be in the domain of f. It follows that the arithmetic combinations of continuous functions preserve continuity. **(Tell how.)**

Exercise 8.3.13. Prove that $f : \text{dom}(f) \subset \mathcal{X} \to \mathbb{R}$ defined by $f(x) = \|x\|_{\mathcal{X}}$ is continuous.

We will frequently consider functions defined on a product space, such as $f : \mathcal{X}_1 \times \mathcal{X}_2 \to \mathcal{Y}$. Interpreting Def. 8.3.16 for such a function requires that we use $\mathcal{X}_1 \times \mathcal{X}_2$ as the space $\mathcal{X}$. This raises the question as to what is meant by the phrases "$\|(x_1, x_2) - (a, b)\|_{\mathcal{X}_1 \times \mathcal{X}_2} < \delta$" or "$((x_1^n, x_2^n))_{n\in\mathbb{N}} \to (a, b)$ in $\mathcal{X}_1 \times \mathcal{X}_2$." One typical product space norm is

$$\|(x_1, x_2)\|_{\mathcal{X}_1 \times \mathcal{X}_2} = \|x_1\|_{\mathcal{X}_1} + \|x_2\|_{\mathcal{X}_2}. \tag{8.71}$$

Both conditions can be loosely interpreted by focusing on controlling each of the components of the members of the product space.

STOP! Make this precise.

Exercise 8.3.14.

i.) Explain what it means for a function $f : [a, b] \times \mathcal{X} \times \mathcal{X} \to \mathcal{X}$ to be continuous on $\mathcal{X} \times \mathcal{X}$ uniformly on $[a, b]$.

ii.) Explain what it means for a mapping $\Phi : \mathbb{C}([a, b]; \mathcal{X}) \to \mathbb{C}([a, b]; \mathcal{X})$ to be continuous.

iii.) Interpret Def. 8.3.16 for functions of the form $f : [a, b] \to \mathbb{M}^N(\mathbb{R})$.

The notion of continuity for real-valued functions can be modified to give meaning to *left-* and *right-sided continuity* in a manner similar to one-sided limits. All continuity results also hold for one-sided continuity. A function that possesses both left- and right-limits at $x = a$, but for which these limits are different, is said to have a *jump discontinuity* at $x = a$.

We now define a concept which is stronger than continuity in the sense that for a given $\epsilon > 0$, there exists a *single* $\delta > 0$ which "works" for every point in the set. Precisely,

Definition 8.3.17. A function $f : \mathrm{S} \subset \text{dom}(f) \subset \mathcal{X} \to \mathcal{Y}$ is *uniformly continuous (UC) on* S provided that for every $\epsilon > 0$, there exists a $\delta > 0$ for which

$$x, y \in S \text{ and } \|x - y\|_{\mathcal{X}} < \delta \Longrightarrow \|f(x) - f(y)\|_{\mathcal{Y}} < \epsilon.$$

Remark. The critical feature of uniform continuity on S is that the δ depends on the ϵ only, and not on the actual points $x, y \in S$ at which we are located. That is, given any $\epsilon > 0$, there exists $\delta > 0$ such that $\|f(x) - f(y)\|_{\mathcal{Y}} < \epsilon$, for any pair of points $x, y \in S$ with $\|x - y\|_{\mathcal{X}} < \delta$, no matter where they are located in S. In this sense, uniform continuity of f on S is a "global" property, whereas mere continuity at $a \in S$ is a "local" property.

8.3.4.3 The Derivative

Measuring the rate of change of one quantity with respect to another is central to the formulation and analysis of many mathematical models. The concept is formalized in the real-valued setting via a limiting process of quantities that geometrically resemble slopes of secant lines. We can extend this definition to $\mathcal{X}-$valued functions by making use of the norm on $\mathcal{X}$. This leads to

Definition 8.3.18. A function $f:(a,b)\to\mathcal{X}$ is *differentiable* at $x_0\in(a,b)$ if there exists a member of $\mathcal{X}$, denoted by $f'(x_0)$, such that

$$\lim_{h\to 0}\left\|\frac{f(x_0+h)-f(x_0)}{h}-f'(x_0)\right\|_{\mathcal{X}}=0. \tag{8.72}$$

The number $f'(x_0)$ is called the *derivative of f at x_0*. We say f is *differentiable on S* if f is differentiable at every element of S.

Exercise 8.3.15. Interpret Def. 8.3.18 using the formal notion of a limit.

The notion of differentiability is more restrictive than continuity, a fact illustrated for real-valued functions by examining the behavior of $f(x)=|x|$ at $x=0$. Indeed, differentiable functions are necessarily continuous, but not vice versa, in the abstract setting of $\mathcal{X}$−valued functions. Further, if the derivative of a function f is itself differentiable, we say f has a *second* derivative. Such a function has a "higher degree of regularity" than one that is merely differentiable. The pattern continues with each order of derivative, from which the following string of inclusions is derived, $\forall n\in\mathbb{N}$:

$$\mathbb{C}^n(I;\mathcal{X})\subset\mathbb{C}^{n-1}(I;\mathcal{X})\subset\ldots\subset\mathbb{C}^1(I;\mathcal{X})\subset\mathbb{C}(I;\mathcal{X}). \tag{8.73}$$

Here, inclusion means that a space further to the left in the string is a closed linear subspace of all those occurring to its right.

We shall often work with real-valued differentiable functions. The arithmetic combinations of differentiable functions are again differentiable, although some care must be taken when computing the derivative of a product and composition. The following result provides two especially nice features of real-valued differentiable functions.

Proposition 8.3.7. (Properties of Real-Valued Differentiable Functions)

i.) (Mean Value Theorem) If $f:[a,b]\to\mathbb{R}$ is differentiable on (a,b) and continuous on $[a,b]$, then there exists $c\in(a,b)$ for which

$$f(b)-f(a)=f'(c)(b-a). \tag{8.74}$$

ii.) (Intermediate Value Theorem) If $f:I\to\mathbb{R}$ is differentiable on $[a,b]\subset I$ and $f'(a)<f'(b)$, then for all $z\in(f'(a),f'(b))$ there exists $c\in(a,b)$ such that $f'(c)=z$.

8.3.4.4 The Integral

Actually, saying "the" integral is misleading since there are many different notions of the integral, all loosely based on the four-step process: (i) partitioning some set, (ii) approximating a quantity of interest, (iii) summing the approximations, and (iv) taking a limit in an appropriate sense. We assume familiarity with the development and properties of the Riemann integral, and focus on the formulation of two abstract extensions of this integral.

The first integral that we define is a natural generalization of the Riemann integral applicable to $\mathcal{X}$-valued functions. A thorough treatment reveals that this integral satisfies the same basic properties as the Riemann integral. Indeed, for a function $f:[a,b]\to\mathcal{X}$, the process used to define $\int_a^b f(x)dx$ is the same as the approach outlined in Section 2.7.4, and the properties listed in Proposition 2.7.1 also hold in this more abstract setting. We omit the details.

Exercise 8.3.16. Assume that $f,u\in\mathbb{C}([a,b];\mathcal{X})$ and that $g:[a,b]\times\mathcal{X}\to\mathcal{X}$ is a continuous mapping for which there exist positive real numbers $M_1$and M_2 such that

$$\|g(s,z)\|_{\mathcal{X}}\le M_1\|z\|_{\mathcal{X}}+M_2,\ \forall s\in[a,b],\ z\in\mathcal{X}. \tag{8.75}$$

i.) Prove that the set $\left\{f(x)+\int_a^x g(z,u(z))dz : x\in[a,b]\right\}$ is uniformly bounded above (with respect to $\|\cdot\|_{\mathcal{X}}$) and provide an upper bound.

ii.) Let $N\in\mathbb{N}$. Determine upper bounds for these:

a.) $\left\|f(x)+\int_a^x g(z,u(z))dz\right\|_{\mathcal{X}}^N$
b.) $\int_a^b \left\|f(x)+\int_a^x g(z,u(z))dz\right\|_{\mathcal{X}}^N dx$

8.4 Looking Ahead

Armed with some rudimentary tools of functional analysis, we are ready to embark on our journey through the world of abstract evolution equations. We will begin by investigating some models whose mathematical description lends itself to reformulation as (A-HCP). Then, we will proceed to formulate a theory for (A-HCP) that mimics as closely as possible the analogous theory for (HCP) established in Chapter 4.

Chapter 9

Three New Mathematical Models

Overview

We study three classical models in this section. Several key elements of the theory are motivated by extensive **EXPLORE!** activities.

9.1 Turning Up the Heat—Variants of the Heat Equation

Classical diffusion theory originated in 1855 with the work of the physiologist Adolf Fick. The premise is simply that a diffusion substance (e.g., heat, gas, virus) will move from areas of high level of concentration toward areas of lower concentration. As such, in the absence of other factors (like advection or external forcing terms), we expect that if the substance diffuses only over a bounded region, its concentration would, over time, become uniformly distributed throughout the region.

This phenomenon arises in many disparate settings. Some common areas include inter-symbol distortion of a pulse transmitted along a cable [47, 32], pheromone transport (emitted by certain species to identify mates) [4], migratory patterns of moving herds [14, 42], the spread of infectious disease through populated areas [21, 48], and the dispersion of salt through water [31]. We will consider different interpretations of diffusion (with added complexity) as the opportunity arises. We begin with a well-known classical model of heat conduction in one and two dimensions. We provide a very heuristic standard derivation of this equation from basic physical principles.

9.1.1 One-Dimensional Diffusion Equation

Consider a one-dimensional wire of length L that is comprised of a uniform material and assume that the temperature at any point of a given cross-section at a fixed position $0 < x < L$ is the same. At any position x along the wire and at time $t > 0$, the temperature is given by the function $z = z(x,t)$. Assume that no heat is generated by the wire itself and that the surface is insulated, meaning no heat enters or escapes through its lateral boundary. Further, assume that heat flow is one-dimensional in the sense that it can flow to the left or to the right, and we shall assume that the Fourier Law governs heat conduction in this wire. This means that the heat flux $\phi(x,t)$ from left to right is given by

$$\phi(x,t) = -K\frac{\partial z}{\partial x}(x,t), \tag{9.1}$$

where the thermal conductivity constant K is between 0 and 1. This means that heat propogates toward the left boundary (at $x = 0$) when $\frac{\partial z}{\partial x}(x,t) > 0$ and propogates toward the right boundary (at $x = L$) when $\frac{\partial z}{\partial x}(x,t) < 0$. It is known that

$$\phi(x,t) = \lambda m z(x,t), \tag{9.2}$$

where λ is the specific heat and m is the mass.

To derive the equation, we consider an infinitesimally small section of the wire, say between x and $x + \Delta x$, where Δx is small, as pictured below:

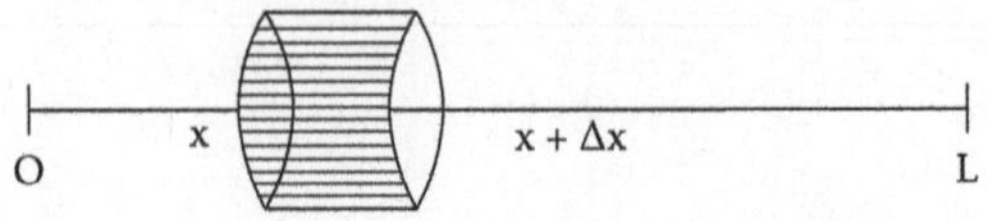

The infinitesimal volume ΔV of this portion of the wire is $S\Delta x$. Assuming a constant density c throughout the wire, the infinitesimal mass is given by

$$\Delta m = cS\Delta x. \tag{9.3}$$

Using (9.2) and (9.3) yields

$$\phi(x,t) = \lambda\,(cS\Delta x)\,z(x,t). \tag{9.4}$$

Since heat is conserved in this wire, it follows that $\frac{\partial \phi}{\partial t}(x,t)$ is computed by subtracting the rate at which heat flows out of the point at x from the rate at which heat flows into that position of the wire; that is,

$$\frac{\partial \phi}{\partial t}(x,t) = \phi(x,t)S - \phi(x+\Delta x,t)S. \tag{9.5}$$

We also know from (9.4) that

$$\frac{\partial \phi}{\partial t}(x,t) = \lambda\,(cS\Delta x)\,\frac{\partial z}{\partial t}(x,t). \tag{9.6}$$

Equating (9.5) and (9.6), we obtain

$$\lambda\,(cS\Delta x)\,\frac{\partial z}{\partial t}(x,t) = S\left(\phi(x,t) - \phi(x+\Delta x,t)\right). \tag{9.7}$$

Dividing both sides by $S\triangle x$ yields

$$\lambda c \frac{\partial z}{\partial t}(x,t) = -\frac{(\phi(x+\triangle x,t) - \phi(x,t))}{\triangle x}, \tag{9.8}$$

so that letting $\triangle x \longrightarrow 0$ yields

$$\lambda c \frac{\partial z}{\partial t}(x,t) = -\frac{\partial \phi}{\partial x}(x,t). \tag{9.9}$$

Now, invoking the Fourier Law (9.1) leads to the so-called *diffusion equation*

$$\frac{\partial z}{\partial t}(x,t) = \frac{K}{\lambda c}\frac{\partial^2}{\partial x^2} z(x,t). \tag{9.10}$$

The quantity $\frac{K}{\lambda c}$ is nonnegative and is often relabeled as k^2, which is called the *diffusivity parameter* of the wire.

Remarks.

1. The relationship $k^2 = \frac{K}{\lambda c}$ implies that the diffusivity of the wire is proportional to the conductivity. This means that the speed at which heat spreads through the material behaves proportionally to how well the material conducts heats. The same equation implies that the diffusivity is inversely proportional to the density, meaning that if the heat spreads quickly, the material is not very dense.
2. The *boundary conditions* (BCs) prescribe what happens to the temperature at the endpoints of the wire. Four standard boundary conditions are described in (8.2)–(8.5).
3. The *initial condition* (IC) is the initial temperature distribution along all points of the wire just before we start the "experiment."

A standard initial-boundary value problem (IBVP) governing diffusion in this context, assuming homogeneous Dirichlet BCs, is given by

$$\begin{cases} \frac{\partial}{\partial t} z(x,t) = k^2 \frac{\partial^2}{\partial x^2} z(x,t), \ 0 < x < L, \ t > 0, \\ z(x,0) = z_0(x), \ 0 < x < L, \\ z(0,t) = z(L,t) = 0, \ t > 0. \end{cases} \tag{9.11}$$

We constructed a solution of this IBVP in Chapter 8 using the separation of variables method. Assuming that $z_0(\cdot)$ is sufficiently smooth to ensure the existence of a unique Fourier representation, it was shown that

$$z(x,t) = \sum_{m=1}^{\infty} b_m e^{-\left(\frac{m\pi}{L}\right)^2 kt} \sin\left(\frac{m\pi}{L}x\right) \tag{9.12}$$

where $b_m, m \in \mathbb{N}$, is given by

$$b_m = \left\langle z_0(\cdot), \sin\left(\frac{m\pi}{L}\cdot\right)\right\rangle_{\mathbb{L}^2(0,L;\mathbb{R})} = \frac{2}{L}\int_0^L z_0(w)\sin\left(\frac{m\pi}{L}w\right)dw. \tag{9.13}$$

Substituting (9.13) into (9.12) renders the solution of the IBVP as

$$\begin{aligned} z(x,t) &= \sum_{m=1}^{\infty}\left(\frac{2}{L}\int_0^L z_0(w)\sin\left(\frac{m\pi}{L}w\right)dw\right) e^{-\left(\frac{m\pi}{L}\right)^2 kt}\sin\left(\frac{m\pi}{L}x\right) \\ &= \sum_{m=1}^{\infty}\frac{2}{L}e^{-\left(\frac{m\pi}{L}\right)^2 kt}\left\langle z_0(\cdot), \sin\left(\frac{m\pi}{L}\cdot\right)\right\rangle_{\mathbb{L}^2(0,L;\mathbb{R})}\sin\left(\frac{m\pi}{L}x\right), \end{aligned} \tag{9.14}$$

where $0 < x < L$ and $t > 0$.

If homogeneous Neumann BCs are used instead, the IBVP (9.11) becomes

$$\begin{cases} \frac{\partial}{\partial t} z(x,t) = k^2 \frac{\partial^2}{\partial x^2} z(x,t), \ 0 < x < L, \ t > 0, \\ z(x,0) = z_0(x), \ 0 < x < L, \\ \frac{\partial z}{\partial x}(0,t) = \frac{\partial z}{\partial x}(L,t) = 0, \ t > 0. \end{cases} \tag{9.15}$$

The solution of this IBVP is given by

$$z(x,t) = \sum_{m=0}^{\infty} \left(\frac{2}{L} \int_0^L z_0(w) \cos\left(\frac{m\pi}{L} w\right) dw \right) e^{-\left(\frac{m\pi}{L}\right)^2 kt} \cos\left(\frac{m\pi}{L} x\right). \tag{9.16}$$

We shall investigate various properties of these IBVPs and their solutions using MATLAB in the following **EXPLORE!** activities.

EXPLORE!

9.1 Parameter Play!

In this **EXPLORE!** you will investigate the effects of changing K, λ, and c used to define the diffusivity constant k^2. To do this, you will use the MATLAB Heat Equation 1D GUI to visualize the solutions to IBVP (9.11) and IBVP (9.15) for different choices of k^2.

MATLAB-Exercise 9.1.1. To begin, open **MATLAB** and in the command line, type:

MATLAB_Heat_Equation_1D

Use the GUI to answer the following questions. Make certain to click on the HELP button for instructions on how to use the GUI. A typical screenshot of the GUI can found in Figure 9.1.

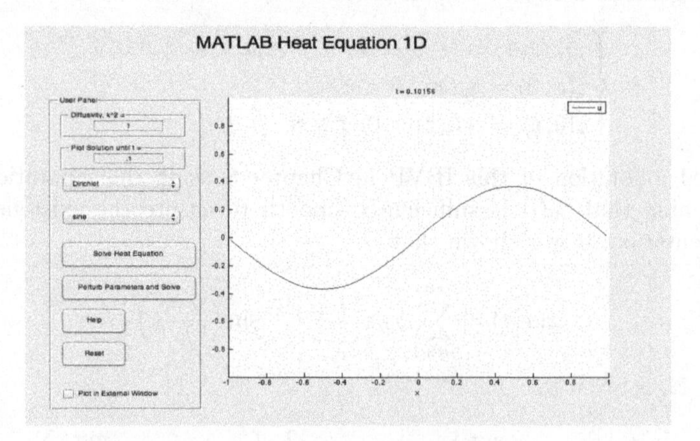

FIGURE 9.1: MATLAB Heat Equation 1D GUI

i.) Run the GUI with $k^2 = 1$, $t = 0.5$, and Dirichlet boundary conditions, and choose the sine option for the initial condition. The GUI will play a movie of the solution twice, and will produce a picture of the solution in time and space. The movie will give you a sense for the "speed" at which the temperature changes from its initial profile. The separate figure will be helpful when you compare solutions stemming from different diffusivity parameters.

ii.) Run the GUI again, but this time with $k^2 = 0.25$. To do this, simply click the box containing the diffusivity constant and replace the 1 with 0.25. Then, click *Solve.*

iii.) What did you observe in (ii)? Did the speed depicted in the movie seem faster or slower than in (i) with $k^2 = 1$? If you could not tell, compare the picture from the previous run of the GUI to the second picture just produced.

iv.) Make a conjecture for how the solution of IBVP (9.11) depends on the diffusivity constant.

v.) Test your conjecture, but this time use the exponential choice for the initial condition and use diffusivity constants both larger and smaller than 1.

vi.) Repeat questions (i)-(v), but this time choose Neumann boundary conditions and the cosine option for the initial condition. Go back and repeat the same questions with the initial condition option: 1 ($z_0(x) = 1$).

vii.) Summarize your findings by explicitly stating the solution's dependence on the diffusivity constant and on the physical parameters that comprise k^2. Also, point out any differences or commonalities you discovered based on the boundary conditions.

So far, our discussion of the heat equation has been restricted to the standard Dirichlet or Neumann boundary conditions. Another common boundary condition used in mathematical modeling is referred to as *periodic boundary conditions* . In the one-dimensional case, this amounts to assuming that the solution along the left boundary equals the solution along the right boundary. Slightly altering IBVP (9.11) leads us to the IBVP

$$\begin{cases} \frac{\partial}{\partial t} z(x,t) = k^2 \frac{\partial^2}{\partial x^2} z(x,t), \ 0 < x < L, \, t > 0, \\ z(x,0) = z_0(x), \ 0 < x < L, \\ z(0,t) = z(L,t), \ t > 0. \end{cases} \tag{9.17}$$

How will the new set of boundary conditions affect the solutions of the heat equation? It's time to find out.

viii.) Repeat questions (i)-(v), but this time choose periodic boundary conditions. Feel free to choose any of the initial condition options (even the step function, if you dare!). Remember your goal is to realize the effect that changing the diffusivity constant has on the solution.

■

EXPLORE!

9.2 Where Has All The Heat Gone?

In the previous **EXPLORE!** we looked at the effects of changing the diffusivity constant. Next, we investigate the effects of allowing the initial profile to diffuse for long periods of time. For example, in Figure 9.2 we see the initial profile quickly diffuses to the zero function, whereas in Figure 9.3 the solution still retains some features of the initial profile. But, what if we allowed the system corresponding to Figure 9.3 to diffuse longer? At some point, would the solution diffuse enough to look like zero function? Or, if not the zero function, how about some other function independent of time?

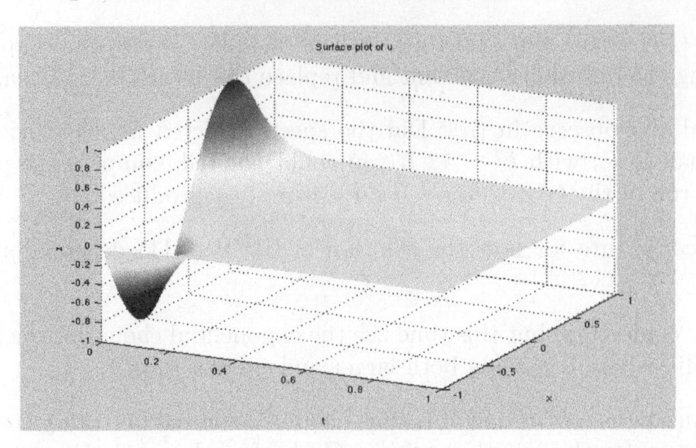

FIGURE 9.2: IBVP (9.11) with $k^2 = 1$, Dirichlet BC, initial profile: $\sin \pi x$

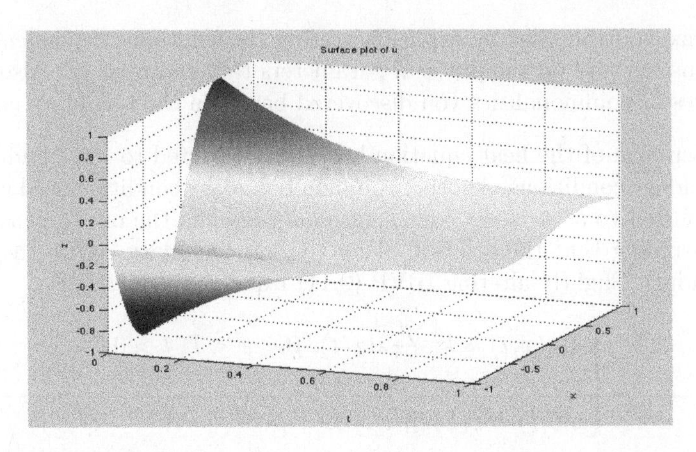

FIGURE 9.3: IBVP (9.11) with $k^2 = 0.25$, Dirichlet BC, initial profile: $\sin \pi x$

MATLAB-Exercise 9.1.2. Open **MATLAB** and in the command line, type:

MATLAB_Heat_Equation_1D

Use the GUI to answer the following questions.

i.) Does the solution to IBVP (9.11) when $k^2 = 0.25$ converge to *something* for the sine initial condition option? If so, to what function? Begin by trying $t = 2$. Keep in mind that computers cannot handle the infinite, so do not input $t = \infty$! Rather start with small times and then form a conclusion by taking larger times.

ii.) Repeat (i) with $k^2 = 0.1$. Did the solution converge? If so, did it converge to the same function as in (i)?

iii.) Now, let us investigate the long-term behavior of IBVP (9.15) using the cosine initial condition option. Explore whether or not the solution converges and if the convergence depends on the diffusivity parameter. (Hint: A good place to begin is to repeat (i) and (ii).)

iii.) Did the solution converge to the same function as in (ii)? Make a conjecture for why or why not.

iv.) In (i) and (iii), you were asked to investigate the long-term behavior of the heat equation for very 'nice' (i.e. smooth) initial profiles. But, what if the initial profile was not smooth? For example, what if it were the following step function:

$$z_0(x) = \begin{cases} 0, & -L \le x < \frac{-L}{2} \\ 1, & \frac{-L}{2} \le x \le \frac{L}{2} \\ 0, & \frac{L}{2} < x \le 1, \end{cases}$$

which is not differentiable on $[-L, L]$ and could possibly violate any patterns we observed in the previous questions. Use the GUI and the step function option for the initial condition to investigate the question, "*Where has all the heat gone*?" Explain your observations and be mindful that your answer should depend on the boundary condition you select.

v.) Summarize your findings from this numerical experiment. Point out not only the differences, but the similarities in solutions that are allowed to diffuse for long periods of time.

■

EXPLORE!

9.3 Continuous Dependence on Parameters

In practice, IBVPs (9.11), (9.15), and (9.17) depend on two measurements, the initial condition and the diffusivity parameter. In this **EXPLORE!**, we investigate the consequences of those two measurements being a little "off." We encountered this concept numerous times in Part I; in fact,

STOP! Reread Section 4.8 up to the **Disclaimer!** and review your work on **EXPLORE! Does (HCP) Depend Continuously on the Initial Data?**

In Part I, solutions to (HCP) were functions that mapped $[t_0, \infty)$ into $\mathbb{R}^N$. For example, in the $N = 1$ case, a solution $x(t)$ depended on only one variable, namely the time variable. This meant that to investigate continuous dependence we studied the difference between two solutions by observing

$$|x(t) - x^*(t)| \tag{9.18}$$

where $x^*(t)$ was the solution of (HCP) equipped with the perturbed parameters and/or initial conditions.

The basic idea of continuous dependence in Part I was that "small" perturbations in parameters or initial conditions should result in the difference in solutions being "small" for all times after the initial time. We can apply that same idea to IBVPs (9.11), (9.15), and (9.17) provided we first answer the question, "How do we study differences between two solutions when the solutions depend on time *and* another spatial variable?"

For concreteness, suppose $z(x,t)$ and $z^*(x,t)$ are solutions of IBVP (9.11) with slightly different parameters. Substituting these solutions into (9.18) yields the expression

$$|z(x,t) - z^*(x,t)| \tag{9.19}$$

which really is not good enough to capture the continuous dependence idea because we seek to track differences in time, not both time and space. To remove the dependence on the spatial variable we could focus solely on the largest difference in the spatial variable. That is, we can take the supremum of the difference over the spatial variable, namely

$$\sup_{x\in[-L,L]} |z(x,t) - z^*(x,t)| \tag{9.20}$$

STOP! Carefully explain how you compute (9.20). If you get stuck, refer to Definition 2.4.2.

STOP! Why does the function (9.20) only depend on t?

The quantity in (9.20) is commonly referred to as a *sup-norm*, defined as follows:

Definition 9.1.1. Let $u : [-L, L] \to \mathbb{R}$. The *sup-norm* of u, denoted by $\|u\|_\infty$, is given by

$$\|u\|_\infty = \sup_{x\in[-L,L]} |u(x)|$$

We are now in a position to explore the dependence on model parameters and/or initial conditions in IBVPs (9.11), (9.15), and (9.17). For the models to depend continuously on small changes in the parameters and/or initial changes, the sup-norm of the difference between the two solutions must be small for all times under consideration.

MATLAB-Exercise 9.1.3. Open **MATLAB** and in the command line, type:

MATLAB_Heat_Equation_1D

To explore continuous dependence using the GUI we must first solve the heat equation with a particular diffusivity constant and initial condition. Once a solution has been constructed, click the *Perturb Parameters and Solve* button to specify the perturbed diffusivity constant and a perturbation size for the initial condition. When prompted, select the sup-norm option to visualize the difference in the solutions. Use the GUI to answer the following questions.

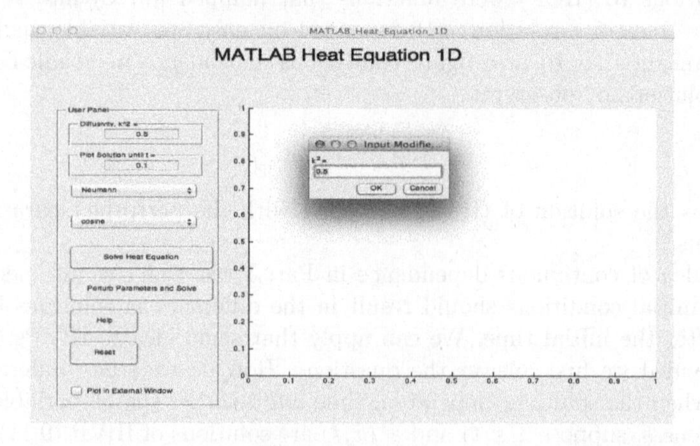

FIGURE 9.4: Perturbing Parameters and Initial Conditions with MATLAB Heat Equation 1D GUI

i.) Consider IBVP (9.11) with diffusivity constant 0.5. Explore whether or not the solution of this IBVP depends continuously on the sine option initial condition.

a.) Plot solutions until $t = 0.3$ and choose a perturbation size of 0.5. Did the difference become small?

b.) Repeat (a) with perturbation sizes of $0.25, 0.1, 0.01$, and 0.005.

c.) Make a conjecture regarding IBVP (9.11) continuous dependence on the initial condition.

d.) Repeat (a)-(c) using the exponential and step function options for the initial condition. Clearly articulate your conjectures by being precise with how you are checking "smallness."

ii.) Consider IBVP (9.15) using the cosine option for the initial condition and diffusivity constant 0.5. Explore whether or not the solution of this IBVP depends continuously on the diffusivity constant.

a.) Plot solutions until $t = 0.3$ and choose a perturbed diffusivity constant of 0.4. Did the difference become small?

b.) Repeat (a) with perturbed diffusivity constants of $0.45, 0.55, 0.48$, and 0.51.

c.) Make a conjecture regarding continuous dependence of the solution on the initial condition in IBVP (9.11).

d.) Repeat (a)-(c) using the exponential and step function options for the initial condition. Clearly articulate your conjectures by being precise with how you are checking "smallness."

iii.) Consider IBVP (9.17) using the exponential option for the initial condition and diffusivity constant 0.5. Explore whether or not the solution of this IBVP depends continuously on the diffusivity constant and on the initial condition.

a.) Plot solutions until $t = 0.3$ and choose a perturbed diffusivity constant of 0.4 and perturbation size 0.3. Did the difference become small?

b.) Repeat (a) with perturbed diffusivity constants of $0.45, 0.55, 0.48$, and 0.51 and perturbation sizes $0.15, 0.1, 0.01$, and 0.005.

c.) Make a conjecture regarding IBVP (9.11) continuous dependence on the initial condition.

d.) Repeat (a)-(c) with the cosine and step function options for the initial condition. Clearly articulate your conjectures by being precise with the way you are checking "smallness." In addition, if your conjecture states that the IBVP does *not* depend continuously on both the diffusivity constant and the initial condition, then determine if continuous dependence is possible for one of them individually.

iv.) In Part I, we observed that (HCP) depended continuously on the initial condition, but not necessarily on the model parameters. Was that the case here for IBVPs (9.11), (9.15), and (9.17)?

■

MATLAB-Exercise (9.1.3) relied heavily on the way we computed the norm to measure "smallness." The sup-norm essentially produces the worst spatial value when computing size. One could argue that this norm is overlooking some important features of a function. For example, consider the functions

$$v(x) = 1, \text{ where } x \in [0, L];$$

$$u(x) = \begin{cases} 1, & x \in [0, \frac{L}{2}) \cup (\frac{L}{2}, L] \\ 0, & x = 0 \end{cases}$$

Observe that $\|u - v\|_\infty = 1$ **(Why?)**, but the two functions are nearly identical! In fact, they agree at every point except one, $x = \frac{L}{2}$. If we want a norm that considers all points in the spatial domain we can use the so called $\mathbb{L}^2$-*norm*, which uses an integral to quantify size in a way that makes use of all points in the spatial domain.

Remark 9.1.1. The $\mathbb{L}^2$-norm was defined in Chapter 8 and can be viewed as an extension of the Euclidean norm (cf. Definition 2.6.3) where a finite sum has been replaced by an infinite sum (the integral).

Using a different norm could change the nature of the continuous dependence results. **(Why?)** Let's explore that notion below.

MATLAB-Exercise 9.1.4. Explore the dependence on model parameters and/or initial conditions using the L^2-norm in IBVPs (9.11), (9.15), and (9.17). When prompted to select a norm to visualize differences in the solution, select *All*.

i.) Repeat MATLAB-Exercise 9.1.3 now using the L^2-norm. Also for each question compare and contrast the different sizes generated by the different norms.

The previous questions have neglected an important piece of the continuous dependence puzzle, namely time. It is quite possible that if we plotted the solutions out to $t = 10$ we would have observed something different regarding continuous dependence. However, as seen in **EXPLORE!** 9.3, diffusion typically takes place quite fast, and solutions converge to a profile that looks flat and stays in the same place for all time. This should be reassuring regarding your observations and conjectures, except for those involving the step function. In this case, the solutions did not completely diffuse to a flat profile.

ii.) For each of the boundary conditions, investigate continuous dependence on the diffusivity constant and the step function initial condition. Use the diffusivity constant, say $k^2 = 0.5$, and plot solutions until $t = 2$ for Dirichlet and periodic BCs and $t = 5$ for Neumann. If you are unable to conjecture continuous dependence on both the diffusivity parameter and initial condition, state a conjecture regarding perturbing only the initial condition.

■

EXPLORE!
9.4 Nonhomogeneous BCs

Next, we consider IBVPs (9.11) and (9.15), equipped with nonhomogeneous BCs. To begin, consider the IBVP

$$\begin{cases} \frac{\partial}{\partial t} z(x,t) = k^2 \frac{\partial^2}{\partial x^2} z(x,t), \ 0 < x < L, \, t > 0, \\ z(x,0) = z_0(x), \ 0 < x < L, \\ z(0,t) = T_1, \ z(L,t) = T_2, \ t > 0, \end{cases} \tag{9.21}$$

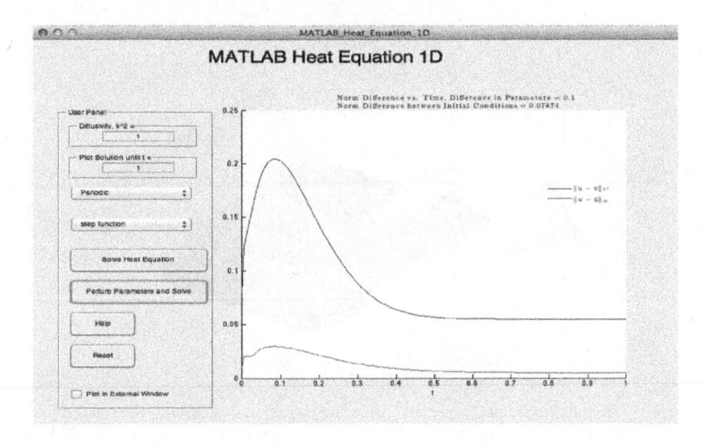

FIGURE 9.5: Visualizing Differences in Solutions with Different Norms

where T_1 and T_2 are nonzero constants. IBVP (9.21) is equipped with what are referred to as *nonhomogeneous Dirichlet* boundary conditions because $T_1$and T_2 are assumed to be nonzero. One can also equip (9.15) with *nonhomogeneous Neumann* boundary conditions to get

$$\begin{cases} \frac{\partial}{\partial t}z(x,t) = k^2 \frac{\partial^2}{\partial x^2} z(x,t), \ 0 < x < L, \, t > 0, \\ z(x,0) = z_0(x), \ 0 < x < L, \\ \frac{\partial z}{\partial x}(0,t) = T_1, \ \frac{\partial z}{\partial x}(L,t) = T_2, \ t > 0, \end{cases} \tag{9.22}$$

STOP! What would constitute a solution to (9.21) or to (9.22)? How will the nonhomogeneous BCs affect the shape of the solution?

MATLAB-Exercise 9.1.5. To begin, open **MATLAB** and in the command line, type:

MATLAB_Heat_NonhomogBC_1D

Use the GUI to answer the following questions. Make certain to click on the HELP button for instructions on how to use the GUI. Screenshots of outputs of the GUI can be found in Figure 9.6 and Figure 9.7.

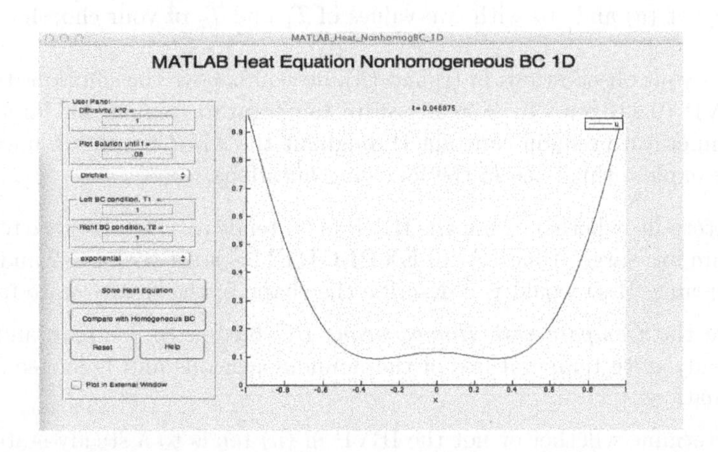

FIGURE 9.6: MATLAB Heat Equation Nonhomogeneous BC 1D GUI

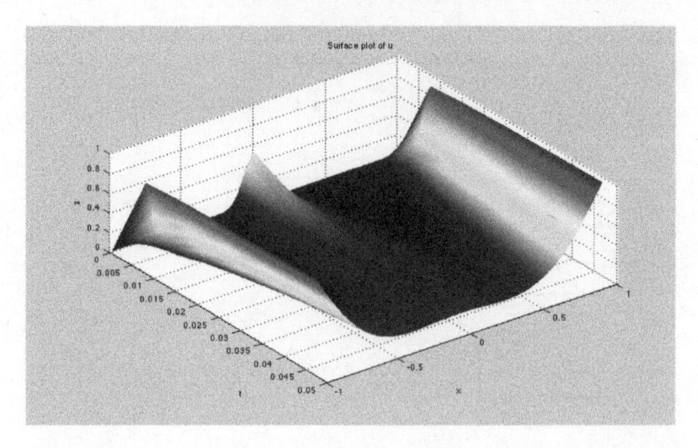

FIGURE 9.7: Solution to (9.21) with $T_1 = 1 = T_2$, $k^2 = 1$, and $z_0(x) = e^{-100x^2}$

i.) Solve IBVP 9.21 until $t = 1$ with the sine initial condition option, $k^2 = 0.75$ and $T_1 = 1 = T_2$.

  a.) What do you notice initially in the movie and for small times in the surface plot that is different than in the homogeneous boundary condition case? Why do you think this happens?

  b.) Click the *Compare with Homogeneous BC* button. Compare and contrast the two solutions.

  c.) Repeat (a) and (b) with two values of T_1 and T_2 of your choosing.

ii.) Solve IBVP 9.22 until $t = 1$ with the exponential initial condition option, $k^2 = 0.75$ and $T_1 = 1 = T_2$.

  a.) What do you notice initially in the movie and for small times in the surface plot that is so different than in the homogeneous boundary condition case? Why do you think this happens?

  b.) Click the *Compare with Homogeneous BC* button. Compare and contrast the two solutions.

  c.) Repeat (a) and (b) with two values of T_1 and T_2 of your choosing.

iii.) Based on your observations in (i) and (ii), do you believe the solutions to IBVP (9.21) and IBVP (9.22) tend to a steady-state temperature remain for large times? Does that temperature remain constant throughout the wire? If you are uncertain, that is OK. We explore this issue in the following questions.

  a.) Determine whether or not the IBVP in (i) tends to a steady-state temperature remain for "large" times. As in **EXPLORE!** 9.4 start with $t = 2$ and then increase the time. How would you describe the shape of the steady-state temperature?

  b.) Use the *Compare with Homogeneous BC* button to compare and contrast the steady-state temperatures of the nonhomogeneous and homogeneous BC heat equations.

  c.) Determine whether or not the IBVP in (ii) tends to a steady-state temperature for "large" times. Describe the shape of the steady-state temperature.

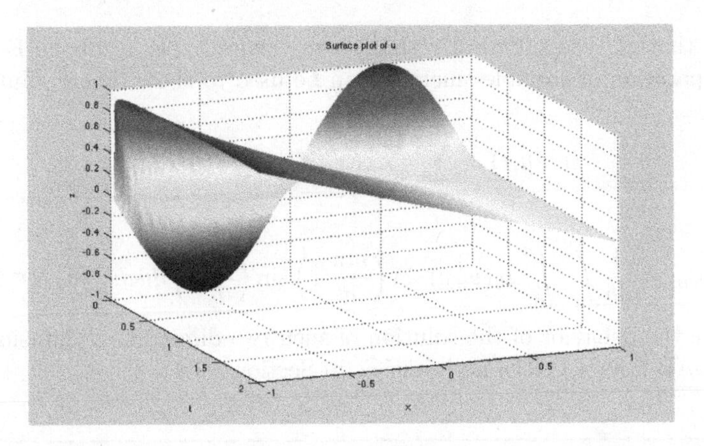

FIGURE 9.8: A Steady-State Temperature with Nonhomogeneous BCs

d.) Use the *Compare with Homogeneous BC* button to compare and contrast the steady-state temperatures of the nonhomogeneous and homogeoneous BC heat equations.

■

Next, we consider a similar model for heat conduction in a two-dimensional rectangular plate composed of an isotropic, uniform material, as pictured below:

9.1.2 Two-Dimensional Diffusion Equation

Assuming that the temperature is zero along the boundary of the rectangle, the IBVP describing the transient temperature at every point on the plate over time is given by

$$\begin{cases} \frac{\partial z}{\partial t}(x,y,t) = k^2\left(\frac{\partial^2 z}{\partial x^2}(x,y,t) + \frac{\partial^2 z}{\partial y^2}(x,y,t)\right), \; 0 < x < a, \; 0 < y < b, \; t > 0, \\ z(x,y,0) = z_0(x,y), \; 0 < x < a, \; 0 < y < b, \\ z(x,0,t) = 0 = z(x,b,t), \; 0 < x < a, \; t > 0, \\ z(0,y,t) = 0 = z(a,y,t), \; 0 < y < b, \; t > 0, \end{cases} \tag{9.23}$$

where $z(x,y,t)$ represents the temperature at the point (x,y) on the plate at time t.

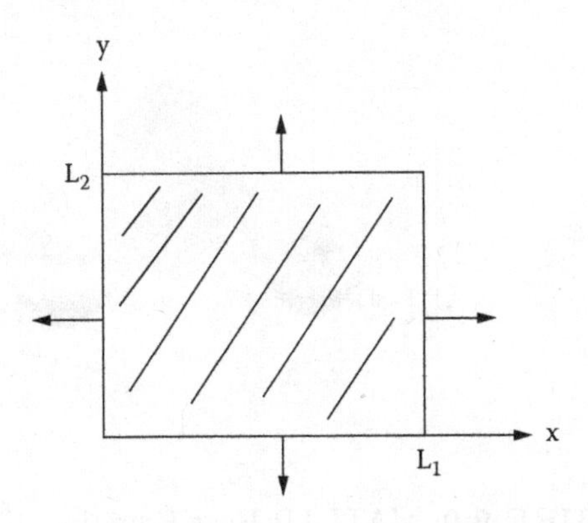

Assuming that $z_0(\cdot)$ is sufficiently smooth to ensure it has a unique Fourier representation, the separation of variables method can be used to show that a solution of (9.23) is given by

$$z(x,y,t) = \sum_{m=1}^{\infty}\sum_{n=1}^{\infty} b_{mn} X_m(x) Y_n(y) T_{mn}(t) \tag{9.24}$$

where

$$b_{mn} = \frac{4}{ab}\int_0^b \int_0^a z_0(w,v)\sin\left(\frac{m\pi}{a}w\right)\sin\left(\frac{n\pi}{b}v\right) dw dv, \quad m,\, n \in \mathbb{N}. \tag{9.25}$$

We investigate the behavior of the solution of such two-dimensional diffusion IBVPs using similar MATLAB **EXPLORE!** activities as in Section 9.1.1.

EXPLORE!

9.6 Effects of Parameters on Two-Dimensional Solutions

This **EXPLORE!** investigates the effects of changing K, λ, and c used to define the diffusivity constant k^2. The difference now when compared to **EXPLORE!** 9.1 is that the spatial variable has two components. You will use the MATLAB Heat Equation 2D GUI to visualize the solutions to IBVP (9.25) and its Neumann and periodic BCs counterparts for different choices of k^2 and determine whether or not your findings in one-dimension remain the same in two. For simplicity, the GUI solves the two-dimensional diffusion equation on a square.

MATLAB-Exercise 9.1.6. To begin, open **MATLAB** and in the command line, type:

MATLAB_Heat_Equation_2D

Use the GUI to answer the following questions. Make certain to click on the HELP button for instructions on how to use the GUI. Typical screenshots of the GUI can be found in Figure 9.9 and 9.10.

i.) Investigate the diffusivity constant's effect on IBVP (9.25) using the GUI.

a.) Run the GUI with $k^2 = 1$, $t = 0.5$, Dirichlet boundary conditions and choose the sine option for the initial condition. The GUI will play a movie of the solution

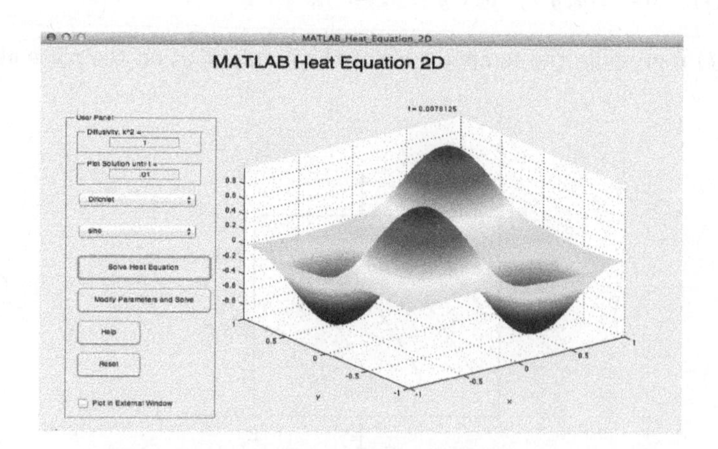

FIGURE 9.9: MATLAB Heat Equation 2D GUI

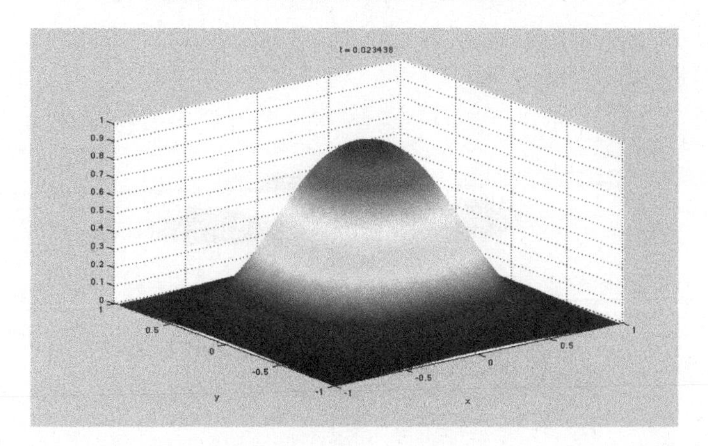

FIGURE 9.10: Two-Dimensional Heat Equation with $k^2 = 1$, periodic BC, initial profile: step function

twice, and will then output a picture of the solution at the final time. As before, the movie will give you a sense for the "speed" at which the temperature changes from its initial profile. The separate figure will be helpful when you compare solutions stemming from different diffusivity parameters.

b.) Run the GUI again, but this time with $k^2 = 0.25$. To do this, simply click the box containing the diffusivity constant and replace the 1 with 0.25. Then, click *Solve*.

c.) What did you observe in (b)? Did the speed exhibited in the movie seem faster or slower than in (a)? If you could not tell use the figure from the previous run of the GUI to the second figure just produced.

d.) Make a conjecture for how the solution of IBVP (9.21) depends on the diffusivity constant. Compare your conjecture with the one-dimensional case from **EXPLORE!** 9.1.

e.) Test your conjecture, but this time use the exponential choice for the initial condition and use diffusivity constants both larger and smaller than 1.

ii.) Modify IBVP (9.25) for periodic boundary conditions. Repeat questions (i) (a)-(e) using periodic boundary conditions. Choose any of the initial condition options. Remember, your goal is to realize the effect that changing the diffusivity constant has on the solution.

iii.) Repeat questions (i) (a)-(e), but this time choose Neumann boundary conditions and the cosine option for the initial condition. Go back and repeat with the initial condition option: 1.

iv.) Summarize your findings by explicitly stating the solution's dependence on the diffusivity constant and on the physical parameters that make-up k^2. Also, point out any differences or commonalities you discovered when using different types of boundary conditions. Compare and contrast your two-dimensional summary with the one-dimensional summary from **EXPLORE!** 9.1.

■

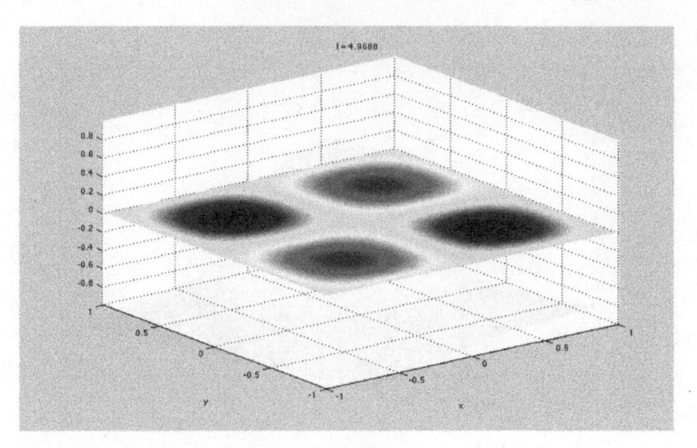

FIGURE 9.11: Two-Dimensional Heat Equation with $k^2 = .25$, Dirichlet BC, initial profile: $\sin(\pi x) sin(\pi y)$

EXPLORE!

9.7 Behavior of Solution As Time Goes On

Next, we investigate the effects of allowing a two-dimensional initial temperature profile to diffuse for long periods of time. This **EXPLORE!** activity is the two-dimensional version of **EXPLORE!** 9.2.

MATLAB-Exercise 9.1.7. Open **MATLAB** and in the command line, type:

MATLAB_Heat_Equation_2D

A particular example of the GUI's output can be found in Figure 9.11. Use the GUI to answer the following questions.

i.) Investigate the behavior of the solution of IBVP (9.21) for "large" times when $k^2 = 0.25$ using the sine initial condition option. Use the movie to describe the evolution of the solution, which is a surface, and to get an intuitive handle for happens as time gets larger. Begin by trying $t = 2$. Remember that computers cannot handle "infinity," so do not input $t = \infty$! Rather, start with small times and then convince yourself of the long term behavior of the the solution.

  a.) Repeat (i) with $k^2 = 0.1$. Did the solution converge? If so, did it converge to the same function as in (i).

ii.) Repeat (i) using the periodic BC and the cosine initial condition option. Explore whether or not the solution converges and if the convergence depends on the diffusivity parameter.

  a.) Did the solution using the periodic BC converge to the same function as in (i) and (ii)? Make a conjecture as to why or why not.

iii.) In **EXPLORE!** 9.2 we allowed a non-smooth initial temperature profile to diffuse for a long time. Repeat that exercise here by selecting the step function initial condition option in MATLAB_Heat_Equation_2D GUI. Describe your observations based on the different boundary conditions.

iv.) Summarize your observations found in the numerical experiment of (iii). Point out not only the differences, but also the similarities in solutions that are allowed to diffuse for long periods of time. Compare your summary here with the one you provided in **EXPLORE!** 9.2.

■

EXPLORE!

9.8 Continuous Dependence on Parameters and ICs

We now extend **EXPLORE!** 9.3 to the two-dimensional setting. The goal is to explore the effect that small perturbations of the diffusivity constant or initial condition have on solutions to IBVP (9.21) and its periodic and Neumann BCs counterparts. Doing this requires that we measure differences in solutions, so we need to define the sup-norm and the $\mathbb{L}^2$-norm for functions with two independent variables.

Definition 9.1.2. Let $u : [0, L] \times [0, L] \to \mathbb{R}$. The *sup-norm* and $\mathbb{L}^2$*-norm* of u on $[0, L] \times [0, L]$, denoted by $\|u\|_{\mathbb{C}([0,L]\times[0,L];\mathbb{R})}$ and $\|u\|_{\mathbb{L}^2([0,L]\times[0,L];\mathbb{R})}$, respectively, are given by

$$\|u\|_{\mathbb{C}([0,L]\times[0,L];\mathbb{R})} = \sup_{(x,y)\in[0,L]\times[0,L]} |u(x,y)|$$

$$\|u\|_{\mathbb{L}^2([0,L]\times[0,L];\mathbb{R})} = \left(\int_0^L \int_0^L u^2(x,y)\,dydx\right)^{\frac{1}{2}}$$

The following MATLAB exercise walks you through the same ideas as in the one-dimensional setting, but at a slightly quicker pace since both norms will be considered at the same time.

MATLAB-Exercise 9.1.8. Open **MATLAB** and in the command line, type:

MATLAB_Heat_Equation_2D

Exploring continuous dependence using the GUI requires us to first solve the heat equation with a particular diffusivity constant and initial condition. Once a solution has been constructed, click the *Perturb Parameters and Solve* button to specify the perturbed diffusivity constant and a perturbation size for the initial condition. A screenshot of inputting parameters into the GUI can be found in Figures 9.12 and others of a particular GUI outputs in Figure 9.13 and 9.14. When prompted to select a norm to visualize the difference in the solutions, please select *All*. Use the GUI to answer the following questions.

i.) Consider IBVP (9.21) with diffusivity constant 0.5. Explore whether or not the solution depends continuously on the sine option initial condition.

a.) Plot solutions until $t = 0.3$ and choose a perturbation size of 0.5. Did the difference become small?

b.) Repeat (a) with perturbation sizes of $0.25, 0.1, 0.01$, and 0.005.

c.) Make a conjecture regarding IBVP (9.21) continuous dependence on the diffusivity constant.

d.) Repeat (a)-(c) with the exponential and step function options for the initial condition. Clearly articulate your conjectures by precisely describing how you are checking "smallness."

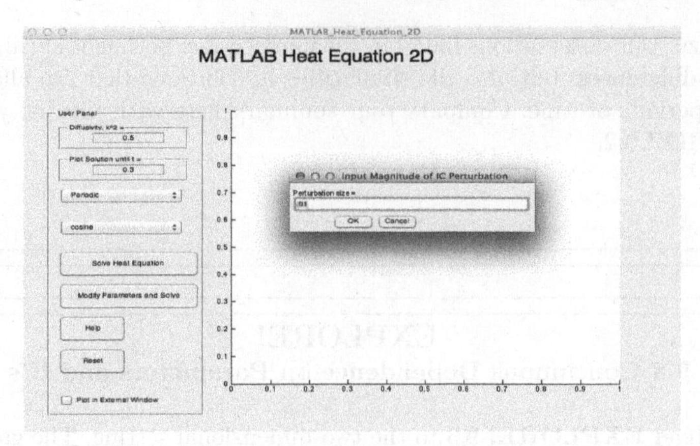

FIGURE 9.12: Perturbing Parameters and Initial Conditions with MATLAB Heat Equation 2D GUI

ii.) Consider the Neumann BC counterpart to IBVP (9.21) with the cosine option for the initial condition and diffusivity constant 0.5. Explore whether or not the solution of this IBVP depends continuously on the diffusivity constant.

  a.) Plot solutions until $t = 0.3$ and choose a perturbed diffusivity constant 0.4. Did the difference become small?

  b.) Repeat (a) with perturbed diffusivity constants of $0.45, 0.55, 0.48$, and 0.51.

  c.) Make a conjecture regarding continuous dependence of the solution on the diffusivity constant.

  d.) Repeat (a)-(c) with the exponential and step function options for the initial condition. Clearly articulate your conjectures by precisely describing how you are checking "smallness."

iii.) Consider the periodic BC counterpart to IBVP (9.21) with the exponential option for the initial condition and diffusivity constant 0.5. Explore whether or not the solution of this IBVP depends continuously on the diffusivity constant and on the initial condition.

  a.) Plot solutions until $t = 0.3$ and choose a perturbed diffusivity constant 0.4 and a perturbation size of 0.3. Did the difference become small?

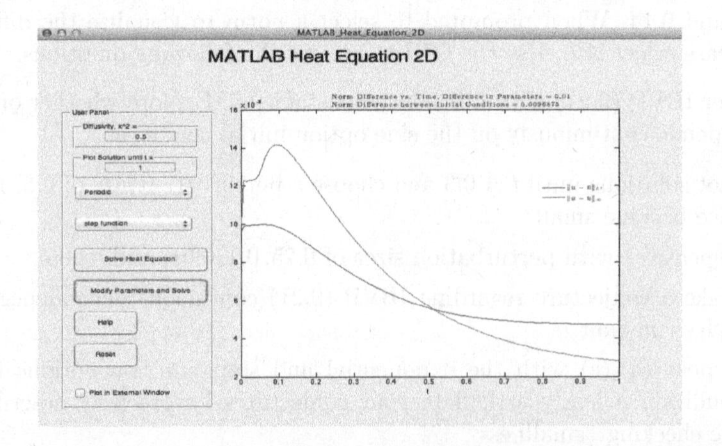

FIGURE 9.13: Visualizing Continuous Dependence with Different Norms

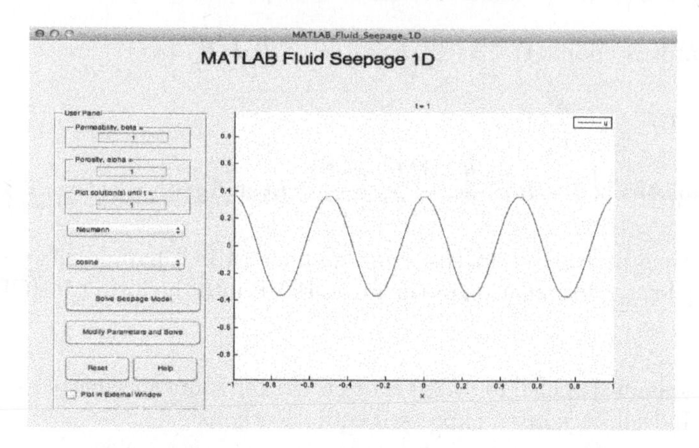

FIGURE 9.14: MATLAB Fluid Seepage 1D GUI

b.) Repeat (a) with perturbed diffusivity constants of $0.45, 0.55, 0.48$, and 0.51 and perturbation sizes $0.15, 0.1, 0.01$, and 0.005.

c.) Make a conjecture regarding continuous dependence of the solution on the initial condition and the diffusivity constant.

d.) Repeat (a)-(c) with the cosine and step function options for the initial condition. Clearly articulate your conjectures by precisely describing how you are checking "smallness." In addition, if your conjecture states the IBVP does *not* depend continuously on both the diffusivity constant and the initial condition, determine if continuous dependence is possible for one of them individually.

iv.) Summarize your findings and compare it with the observation you made in the one-dimensional case from **EXPLORE!** 9.3.

■

9.1.3 Abstraction Formulation

We focus on the first stage of the strategy outlined in Chapter 8, involved in reformulating the IBVPs arising in the models abstractly in the form (A-HCP), a form that resembles (HCP). Once we have become reasonably comfortable with what this entails, then we will move on to developing a theory that structurally resembles Part I. Along the way, we will encounter many rich differences between ODEs and PDEs which result in challenges that must be overcome. For now, though, we shall be content with formulating the IBVPs as abstract evolution equations.

We discussed how to formulate IBVP (9.11) as an abstract IVP that resembled (HCP) from Part I. We summarize the discussion here, for completeness.

Let $\mathcal{H} = \mathbb{L}^2(0, a; \mathbb{R})$, assume that $z_0(\cdot) \in \mathcal{H}$, and identify the solution and IC, respectively, by

$$u(t)[x] = z(x,t), \; 0 < x < L, \, t > 0, \tag{9.26}$$

$$u_0[x] = u(0)[x] = z(x,0) = z_0(x), \; 0 < x < L. \tag{9.27}$$

Define the operator $A : \text{dom}(A) \subset \mathbb{L}^2(0, L; \mathbb{R}) \to \mathbb{L}^2(0, L; \mathbb{R})$ by

$$A[f] = k^2 \frac{d^2}{dx^2}[f], \tag{9.28}$$

$$\text{dom}(A) = \left\{ f \in \mathcal{H} \,\middle|\, \exists \frac{df}{dx}, \frac{d^2 f}{dx^2}, \frac{d^2 f}{dx^2} \in \mathcal{H}, \text{ and } f(0) = f(L) = 0 \right\}.$$

Identifying the time derivatives in the same manner as in Chapter 8, we see that using (9.26) - (9.28) yields a reformulation of the given IBVP into the form (A-HCP) in the space $\mathbb{L}^2(0, L; \mathbb{R})$.

Remark. The domain specified in (9.28) is often written more succinctly using the Sobolev space $\mathbb{H}^2(0, L)$. Indeed, it is often expressed equivalently as

$$\text{dom}(A) = \left\{ f \in \mathbb{H}^2(0, L; \mathbb{R}) \,|\, f(0) = f(L) = 0 \right\}.$$

Exercise 9.1.1. Complete the following.

i.) Transform the IBVP (9.15) into an abstract evolution equation of the form (A-HCP). Clearly define all identifications.

ii.) Transform the IBVP (9.15) into an abstract evolution equation of the form (A-HCP). Clearly define all identifications.

9.2 Clay Consolidation and Seepage of Fluid Through Fissured Rocks

9.2.1 Seepage of Fluid Through Fissured Rocks

Consider a sizeable stack of rocks separated by a network of mini-cracks or fissures as the one pictured below. Liquid flows along the arteries of this network, but also through tiny pores in the rocks themselves.

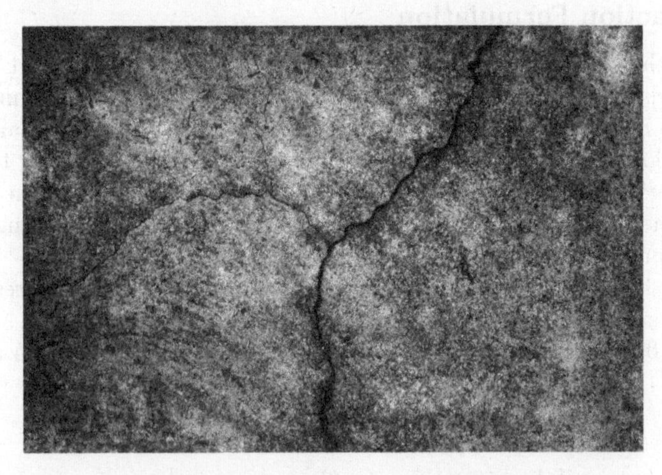

The modeling of such fluid flow in a bounded domain $\Omega \subset \mathbb{R}^3$ involves PDEs governing the pressure of the liquid in the fissures. We provide a sketch of the derivation of this model following the discussion provided by [3].

We shall make several assumptions, the first of which is a simplifcation of the underyling network with which we are dealing. We assume the width of the fissures exceeds the size of the pores, while the fissures themselves occupy a much smaller volume than the pores. For each point Q within the network, we assign the following quantities:

P_F= average pressure of liquid in the fissures near Q
P_p= average pressure of liquid in the pores near Q
$\overrightarrow{V_F}$= seepage velocity of liquid along the fissures in a particular direction near Q
$\overrightarrow{V_p}$= seepage velocity of liquid along the pores in a particular direction near Q

We impose the following assumptions:

1. Flow of liquid proceeds mainly along the fissures. So, the flow velocity through the blocks is very small as compared to the seepage along the fissures.
2. If the fissures are sufficiently narrow and the velocity is small enough, then Darcy's Law governs the motion of the liquid along the fissures. Precisely,

$$\overrightarrow{V_F} = -\frac{k_1}{\mu}\overrightarrow{\nabla} P_F, \tag{9.29}$$

where k_1 is the permeability of the system of fissures and μ is the viscosity of the liquid.
3. In accounting for the transfer of liquid between the blocks and fissures, we assume that there is a smooth change of pressure during the movement of the liquid in the fissures. As such, let

v= volume of liquid flow from the blocks into the fissures per unit time and unit volume of rock
α =rock characteristic
μ =viscosity

Then, a consequence of this assumption that we shall use going forward is

$$v = \frac{\alpha}{\mu}(P_F - P_p). \tag{9.30}$$

As such, the mass q of liquid that flows from the pores into the fissures per unit time per unit volume of rock is given by

$$\begin{aligned} q &= \text{density} \times \text{volume} \\ &= \rho \times v \\ &= \rho\left[\frac{\alpha}{\mu}(P_F - P_p)\right] \end{aligned} \tag{9.31}$$

The law of conservation of the mass of liquid present *in the fissures* yields

$$\frac{\partial}{\partial t}(m_F\rho) + \overrightarrow{\nabla}\cdot\left(\rho\overrightarrow{V_F}\right) - q = 0, \tag{9.32}$$

where m_F is the mass of liquid in the fissure. By (9.29), it follows that (9.32) is equivalent to

$$\frac{\partial}{\partial t}(m_F\rho) + \overrightarrow{\nabla}\cdot\left(-\rho\frac{k_1}{\mu}\overrightarrow{\nabla} P_F,\right) - q = 0. \tag{9.33}$$

Next, we would like to further simplify (9.33). To do so, we impose the following assumption:

4. The liquid is slightly compressible, so that

$$\rho = \rho_0 + \beta(\delta\rho), \tag{9.34}$$

where

ρ_0 =density of the liquid at the initial pressure measured within the network of the fissures

β = coefficient of compressibility

$\delta\rho$ =change in pressure relative to the standard pressure

Substituting (9.34) into (9.33) then yields

$$k_1 \triangle P_F + \alpha(P_p - P_F) = 0. \tag{9.35}$$

Next, we apply the law of conservation of mass of liquid present *in the pores* to obtain (similar to (9.32))

$$\frac{\partial}{\partial t}(m_p\rho) + \vec{\nabla}\cdot\left(\rho\vec{V_p}\right) + q = 0. \tag{9.36}$$

To simplify the PDEs slightly, we impose the following assumption:

5. The permeability of the blocks is low. This assumption enables us to drop the term $\vec{\nabla}\cdot\left(\rho\vec{V_p}\right)$ in (9.36) because this term measures the flux in the mass of liquid within the pores due to the inflow of liquid coming from the pores. As such, (9.36) simplifies to

$$\frac{\partial}{\partial t}(m_p\rho) + q = 0. \tag{9.37}$$

Under suitable conditions (as outlined in [3], it can be assumed that

$$dm_p = \beta_c dP_p, \tag{9.38}$$

where β_c is the coefficient of compressibility of the blocks. As such, using (9.31), (9.34), and (9.38) in (9.37) yields the equation

$$(\beta_c + m_0\beta)\frac{\partial P_p}{\partial t} + \frac{\alpha}{\mu}(P_p - P_F) = 0, \tag{9.39}$$

where m_0 is the magnitude of the porosity of the blocks at standard pressure.

Together, equations (9.35) and (9.39) describe the motion of the liquid in the network of fissures under assumptions (1) - (5). Eliminating P_p in these equations leads to the following PDE whose unknown is P_F:

$$\frac{\partial P_F}{\partial t} - \eta\frac{\partial}{\partial t}(\triangle P_F) = k\triangle P_F, \tag{9.40}$$

where $\eta = \frac{k_1}{\alpha}$ and $k = \frac{k_1}{\mu(\beta_c + m_0\beta)}$. Finally, (9.40) is equivalent to the following PDE, which is an example of a so-called *Sobolev equation*:

$$\frac{\partial}{\partial t}(P_F - \eta\triangle P_F) = k\triangle P_F. \tag{9.41}$$

This completes the derivation.

Equipping PDE (9.41) with an initial condition (describing the initial pressure in the fissures) and Neumann boundary conditions (which govern the pressure or pressure change along the boundary of the network of fissures) leads to the following IBVP:

$$\begin{cases} \frac{\partial}{\partial t}p(x,y,z,t) - \alpha\frac{\partial}{\partial t}\left(\Delta p(x,y,z,t)\right) = \beta\Delta p(x,y,z,t), \ (x,y,z) \in \Omega, \ t > 0, \\ p(x,y,z,0) = p_0(x,y,z), \ (x,y,z) \in \Omega, \\ \frac{\partial p}{\partial \mathbf{n}}(x,y,z,,t) = 0, \ (x,y,z) \in \partial\Omega, \ t > 0, \end{cases} \tag{9.42}$$

where $\Omega \subset \mathbb{R}^3$ is a bounded domain. The parameters α and β are dependent on the characteristics of the rocks (e.g., porosity and permeability), as you can tell by retracing through the derivation.

A simplified one-dimensional version of (9.42) on $[0,\pi]$ is given by

$$\begin{cases} \frac{\partial}{\partial t}\left(p(x,t) - \alpha\frac{\partial^2}{\partial x^2}p(x,t)\right) = \beta\frac{\partial^2}{\partial x^2}p(x,t), \ 0 < x < \pi, \ t > 0, \\ p(x,0) = p_0(x), \ 0 < x < \pi, \\ \frac{\partial p(0,t)}{\partial x} = \frac{\partial p(\pi,t)}{\partial x} = 0, \ t > 0. \end{cases} \tag{9.43}$$

where $\alpha, \beta > 0$. If one chooses periodic boundary conditions then (9.43) becomes

$$\begin{cases} \frac{\partial}{\partial t}\left(p(x,t) - \alpha\frac{\partial^2}{\partial x^2}p(x,t)\right) = \beta\frac{\partial^2}{\partial x^2}p(x,t), \ 0 < x < \pi, \ t > 0, \\ p(x,0) = p_0(x), \ 0 < x < \pi, \\ p(0,t) = p(\pi,t), \ t > 0. \end{cases} \tag{9.44}$$

The main difference between (9.43) and (9.15) is the presence of the term $-\alpha\frac{\partial^2 p}{\partial x^2}(x,t)$ within the time derivative.

Exercise 9.2.1. Verify that the solution of (9.43) (where $\alpha = \beta = 1$ for simplicity) is given by

$$\begin{aligned} p(x,t) &= S(t)[p_0](x) \\ &= \sum_{m=0}^{\infty} e^{-\left(\frac{m^2}{m^2+1}\right)t} \left\langle p_0(\cdot), \sqrt{\frac{2}{\pi}}\cos(m\cdot)\right\rangle_{\mathbb{L}^2(0,\pi;\mathbb{R})} \sqrt{\frac{2}{\pi}}\cos(mx), \end{aligned} \tag{9.45}$$

where $0 < x < \pi$, $t > 0$.

Exercise 9.2.2. Verify $S(t)[p_0]$ given in (9.45) satisfies the following semigroup properties:

i.) $S(0)[p_0](x_0) = p_0(x)$

ii.) $S(t_1+t_2)[p_0](x) = S(t_1)\left(S(t_2)[p_0]\right)(x)$

iii.) $S(t)[p_0+q_0](x) = S(t)[p_0](x) + S(t)[q_0](x)$

EXPLORE!
9.9 Parameter Play!

In this **EXPLORE!** activity, we visualize the effect that changing α and β has on solutions of the IBVPs (9.43) and (9.44), and their periodic counterpart. Before doing so, let us gain some intuition regarding the evolution of the solutions to IBVPs (9.43) and (9.44) by comparing them to the solutions of the heat equations (9.11) and (9.15), respectively.

Exercise 9.2.3. In this exercise, the term "model" refers to the solution of both IBVPs (9.43) and (9.44).

i.) Setting $\alpha = 0$ reduces the model to which IBVP we have seen before?

ii.) Based on (i) and **EXPLORE!** 9.1, how do expect the model's evolution to change if you increase β? decrease β?

iii.) Based on **EXPLORE!** 9.1, how do you expect the solution to evolve over time if β is very small?

MATLAB-Exercise 9.2.1. To begin, open **MATLAB** and in the command line, type:

MATLAB_Fluid_Seepage_1D

Use the GUI to answer the following questions. Make certain to click on the HELP button for instructions on how to use the GUI. A typical screenshot of the GUI can found in Figure 9.14.

i.) Solve (9.43) until $t = 0.5$ using $\alpha = 1$, $\beta = 2$, and the sine option for the initial condition. How would you compare the solution to the heat equation with the same initial and boundary conditions?

  a.) Set $\alpha = 0$ and keep everything else the same. Did the solution evolve in same fashion as the heat equation?

  b.) Keep increasing α by 0.5 until you can describe the effect that changing α has on the solution while keeping β fixed.

  c.) Set $\beta = 0$ and $\alpha = 1$. What did you observe after running the GUI? Is this what you expected?

  d.) Keep increasing β by 0.5 until you can describe the effect that changing β has on the solution while keeping α fixed.

  e.) Repeat (a)-(d) using the exponential initial condition option.

ii.) Solve (9.44) until $t = 0.5$ using $\alpha = 1$, $\beta = 2$, and the exponential option for the initial condition. How would you compare the solution to the heat equation with the same initial and boundary conditions?

  a.) Set $\alpha = 0$ and keep everything else the same. Did the solution evolve in same fashion as the heat equation?

  b.) Keep increasing α by 0.5 until you can describe the effect that changing α has on the solution while keeping β fixed. Does your answer differ from (i)-(b)?

  c.) Set $\beta = 0$ and $\alpha = 1$. What did you observe after running the GUI? Is this what you expected? Does your answer differ from (i)-(c)?

  d.) Keep increasing β by 0.5 until you can describe the effect that changing β has on the solution while keeping α fixed. Does your answer differ from (i)-(d)?

  e.) Repeat (a)-(d) using the cosine initial condition option.

iii.) Summarize your observations regarding the evolution of fluid seepage model and its dependence on model parameters. Compare your answer with **EXPLORE!** 9.1.

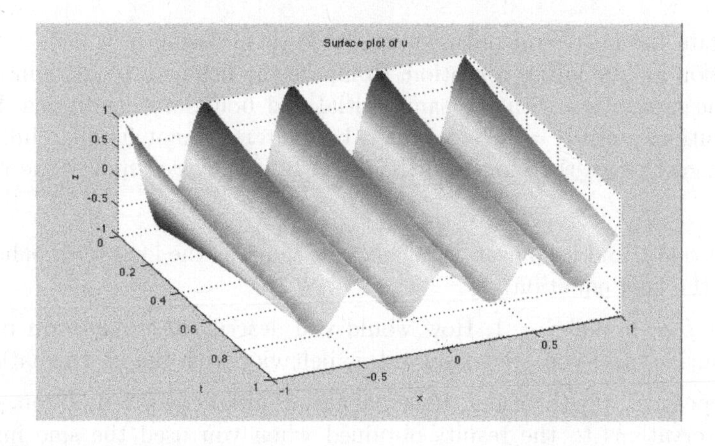

FIGURE 9.15: IBVP (9.43) with $\beta = 1$, $\alpha = 1$, initial profile: $\cos 4\pi x$.

iv.) Repeat either (i) or (ii) using the step function option for the initial condition. Do any of your observations differ from those presented in (iii)? How about when you compare your observations to solutions of the heat equation?

■

EXPLORE!

9.10 Long-Term Behavior

MATLAB-Exercise 9.2.2. To begin, open **MATLAB** and in the command line, type:

MATLAB_Fluid_Seepage_1D

Screenshots of particular GUI outputs can be found in Figures 9.15, 9.16, and 9.17. Use the GUI to answer the following questions.

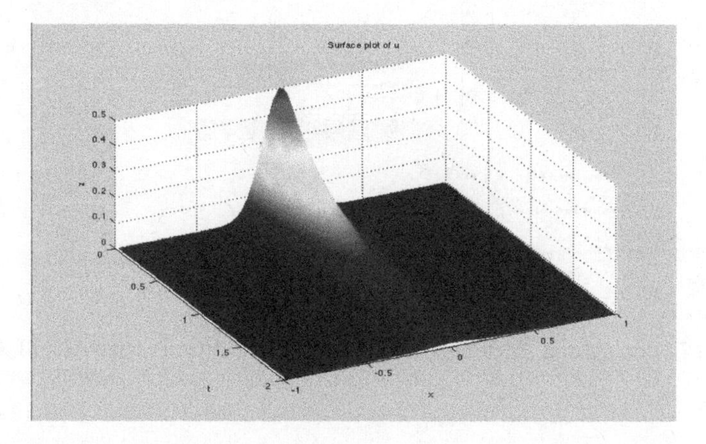

FIGURE 9.16: IBVP (9.43) with $\beta = 2$, $\alpha = 1$, initial profile: e^{-40x^2}

i.) Investigate the long-term behavior of IBVP (9.43) using $\alpha = 1$, $\beta = 2$, and the cosine option for the initial condition. Compare the behavior to the long-term behavior of the heat equation with the same initial and boundary conditions. Begin by plotting solutions until $t = 2$. Remember, computers cannot handle "infinity," so start with "small" times then increase them slowly if you need more time to describe the behavior.

a.) Set $\alpha = 0$ and keep everything else the same. Is the long-term behavior the same as the heat equation?

b.) Set $\beta = 0$ and $\alpha = 1$. How would you describe the long-term behavior of the solution? Did you ever observe this behavior with the heat equation?

c.) Repeat (i), (a)-(b) using the exponential initial condition option. Compare your observations to the results obtained when you used the sine initial condition option.

ii.) Investigate the long-term behavior of IBVP (9.44) using $\alpha = 1$, $\beta = 2$, and the cosine option for the initial condition. Compare the behavior to the long-term behavior of the heat equation with the same initial and boundary conditions.

a.) Set $\alpha = 0$ and keep everything else the same. Is the long-term behavior the same as the heat equation?

b.) Set $\beta = 0$ and $\alpha = 1$. How would you describe the long-term behavior of the solution? Did you ever observe this behavior with the heat equation?

c.) Repeat (i), (a)-(b) using the exponential initial condition option. Compare your observations to the results obtained when you used the sine initial condition option.

iii.) Summarize your observations regarding the evolution of fluid seepage model and its dependence on model parameters. Compare your answer with **EXPLORE!** 9.1.

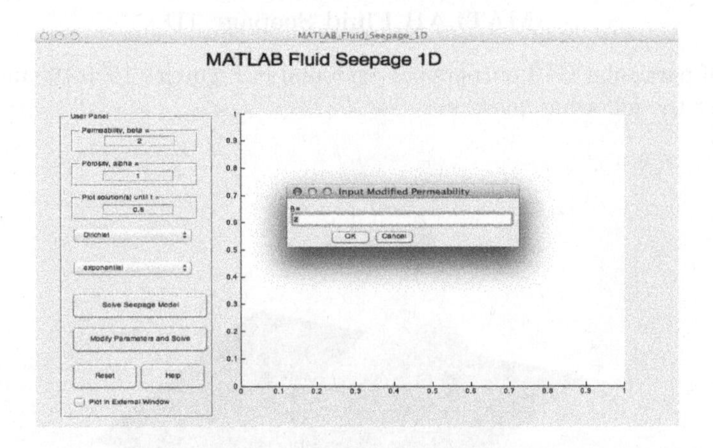

FIGURE 9.17: Perturbing Parameters and Initial Conditions with MATLAB Fluid Seepage 1D GUI

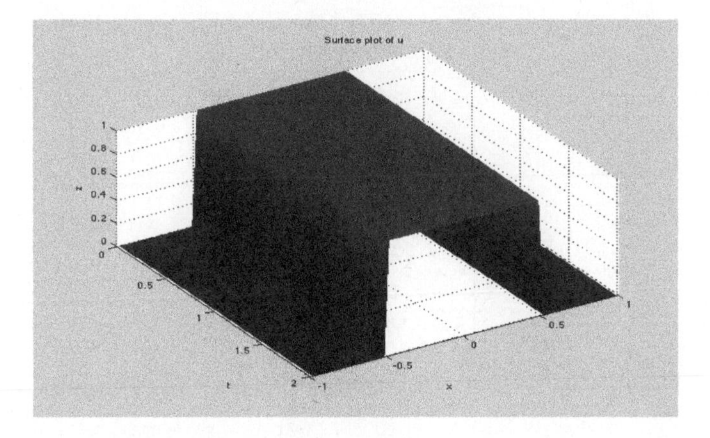

FIGURE 9.18: IBVP (9.44) with $\beta = 0.001$, $\alpha = 1$, initial profile: step function.

iv.) Repeat either (i) or (ii) using the step function option for the initial condition. Do any of your observations differ from those presented in (iii)? How about when you compare your observations to solutions of the heat equation?

■

EXPLORE!

9.11 Continuous Dependence on Parameters

MATLAB-Exercise 9.2.3. Open **MATLAB** and in the command line, type:

MATLAB_Fluid_Seepage_1D

To explore continuous dependence with the GUI we must first solve the IBVP with particular model parameters. Once a solution has been constructed, click the *Perturb Parameters and Solve* button to specify the perturbed parameters and a perturbation size for the initial condition. A screenshot of inputting parameters into the GUI can be found in Figure 9.18 and another of a particular GUI output in Figure 9.19. When prompted to select a norm to visualize the difference in the solutions, please select *All.* Use the GUI to answer the following questions.

i.) Consider IBVP (9.43) with model parameters, $\alpha = 1$ and $\beta = 2$. Explore whether or not the solution of this IBVP depends continuously on the cosine option initial condition.

  a.) Plot solutions until $t = 0.3$ and choose a perturbation size of 0.5. Did the difference become small? If so, in which norm?

  b.) Repeat (a) with perturbation sizes of $0.25, 0.1, 0.01$, and 0.005.

  c.) Make a conjecture regarding the continuous dependence of the solution of IBVP (9.43) on the initial condition.

  d.) Repeat (a)-(c) with the exponential and step function options for the initial condition. Clearly articulate your conjectures by precisely describing how you are checking "smallness."

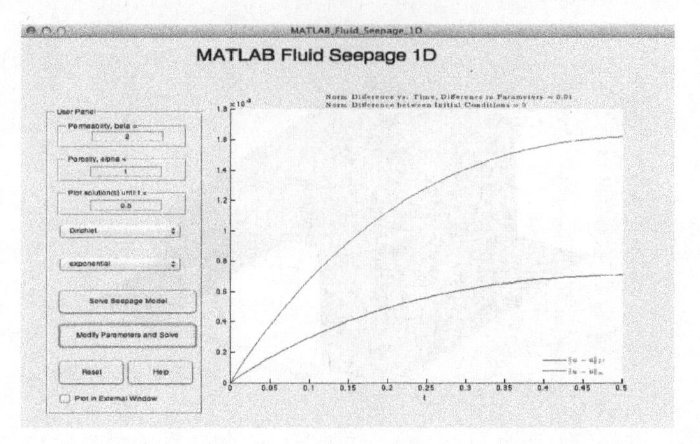

FIGURE 9.19: Visualizing Continuous Dependence with Different Norms

e.) Do your conjectures differ from those made in **EXPLORE!** 9.3 for the corresponding heat equations?

ii.) Consider IBVP (9.44) with the cosine initial condition option. Investigate whether or not the solution of the IBVP depends continuously on the model parameters near $\alpha = 1$ and $\beta = 2$.

a.) Plot solutions until $t = 0.3$. First, consider what happens when only α is perturbed. That is, keep $\beta = 2$ and the perturbation size for the initial condition should equal 0. Choose the perturbed porosity constant to equal 1.1. Did the difference become small?

b.) Repeat (a) with perturbed porosity constants of $0.9, 0.98$, and 1.01.

c.) Make a conjecture regarding continuous dependence of the solution on the porosity constant.

d.) Repeat (a)-(c) for perturbing β only. Choose perturbed permeability constants to equal $2.1, 1.9, 1.99$, and 2.01.

e.) Repeat (a)-(d) with the exponential and step function options for the initial condition. Clearly articulate your conjectures by precisely describing how you are checking "smallness."

f.) Investigate if the solution of IBVP (9.44) depends continuously on the model's parameters and the initial conditions by perturbing α and β in the same manner as in (a)-(e), but also by choosing initial condition perturbation sizes to equal $0.25, 0.1, 0.05$, and 0.01. Explain your observations and try to formulate a claim regarding the model's continuous dependence on the initial conditions and the parameters. Clearly describe any differences found using the two different norms.

iii.) Summarize your findings and compare them with those found in **EXPLORE!** 9.3.

iv.) As seen in the previous **EXPLORE!** 9.10, the fluid seepage model reduces to a completely different IBVP, namely the heat equation, when α is set equal to zero. This suggests that IBVPs (9.43) and (9.44) may be sensitive to changes when α is small.

a.) Choose $\alpha = 0.01, \beta = 2$, the exponential initial condition option, and plot solutions until $t = 0.75$. Choose a perturbed porosity constant equal to 0.005. Did the difference become small?

b.) Repeat (a) for perturbed porosity constants 0.001 and 0. Do you need to alter any of your conjectures from previous parts? If so, describe how.

c.) Repeat (a) and (b) for the step function initial condition option.

d.) If you answered *No* in either part (b) or (c), determine if perturbing the other model parameter and the initial condition makes any difference when α is small. The perturbation sizes are left to you, but if you need any guidance refer to parts (i) and (ii).

■

9.2.2 Two-Dimensional Seepage of Fluid Through Fissured Rocks

Now, let us consider the two-dimensional IBVPs (9.43) and (9.44) where instead of the spatial variable being confined to $[0, \pi]$ the ordered pair of spatial variables are confined to the rectangle, $[0, \pi] \times [0, \pi]$. The two-dimensional version of (9.43) is given by

$$\begin{cases} \frac{\partial}{\partial t}\left(p(x,y,t) - \alpha \Delta p(x,y,t)\right) = \beta \Delta p(x,y,t), \ 0 < x < \pi, 0 < y < \pi, \ t > 0, \\ p(x,y,0) = p_0(x,y), \ 0 < x < \pi, 0 < y < \pi \\ \frac{\partial p(0,y,t)}{\partial x} = 0 = \frac{\partial p(\pi,y,t)}{\partial x}, \ 0 < y < \pi, t > 0 \\ \frac{\partial p(x,0,t)}{\partial y} = 0 = \frac{\partial p(x,\pi,t)}{\partial y}, \ 0 < x < \pi, t > 0 \end{cases} \tag{9.46}$$

where

$$\Delta p(x,y,t) = \frac{\partial^2}{\partial x^2} p(x,y,t) + \frac{\partial^2}{\partial y^2} p(x,y,t).$$

Replacing the BCs in (9.46) with periodic boundary conditions yields

$$\begin{cases} \frac{\partial}{\partial t}\left(p(x,y,t) - \alpha \Delta p(x,y,t)\right) = \beta \Delta p(x,y,t), \ 0 < x < \pi, 0 < y < \pi, \ t > 0, \\ p(x,y,0) = p_0(x,y), \ 0 < x < \pi, 0 < y < \pi \\ p(0,y,t) = p(\pi,y,t), \ 0 < y < \pi, t > 0 \\ p(x,0,t) = p(x,\pi,t), \ 0 < x < \pi, t > 0 \end{cases} \tag{9.47}$$

EXPLORE!
9.12 Parameter Play!

In this **EXPLORE!**, we examine the effect of changing α and β in IBVPs (9.46) and (9.47) and compare these effects to those occurring in the one-dimensional case.

STOP! How did changing α and β in IBVPs (9.43) and (9.44) affect the evolution of the corresponding solutions?

MATLAB-Exercise 9.2.4. To begin, open **MATLAB** and in the command line, type:

MATLAB_Fluid_Seepage_2D

Use the GUI to answer the following questions. Make certain to click on the HELP button for instructions on how to use the GUI. A typical screenshot of the GUI can found in Figure 9.20.

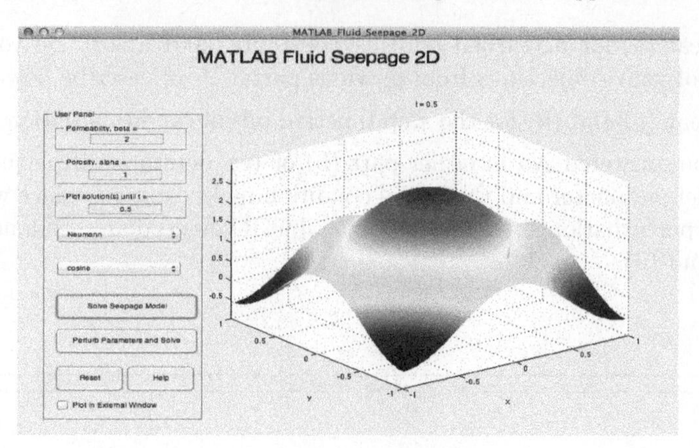

FIGURE 9.20: MATLAB Fluid Seepage 2D GUI

i.) Solve (9.46) until $t = 0.5$ using $\alpha = 1$, $\beta = 2$, and the cosine option for the initial condition. How would you compare the solution to the heat equation with the same initial and boundary conditions?

a.) Set $\alpha = 0$ and keep everything else the same. Did the solution evolve in same fashion as the heat equation?

b.) Keep increasing α by 0.5 until you can describe the effect that changing α has on the solution while keeping β fixed.

c.) Set $\beta = 0$ and $\alpha = 1$. What did you observe after running the GUI this time? Is this what you expected?

d.) Keep increasing β by 0.5 until you can describe the effect that changing β has on the solution while keeping α fixed.

e.) Repeat (a)-(d) using the exponential initial condition option and any positive model parameters.

ii.) Solve (9.47) until $t = 0.5$ using $\alpha = 1$, $\beta = 2$, and the cosine option for the initial condition. How would you compare the solution to the heat equation with the same initial and boundary conditions?

a.) Set $\alpha = 0$ and keep everything else the same. Did the solution evolve in same fashion as the heat equation?

b.) Keep increasing α by 0.5 until you can describe the effect that changing α has on the solution while keeping β fixed. Does your answer differ from (i)-(b)?

c.) Set $\beta = 0$ and $\alpha = 1$. What did you observe after running the GUI this time? Is this what you expected? Does your answer differ from (i)-(c)?

d.) Keep increasing β by 0.5 until you can describe the effect that changing β has on the solution while keeping α fixed. Does your answer differ from (i)-(d)?

e.) Repeat (a)-(d) using the exponential initial condition option and any positive model parameters.

iii.) Summarize your observations regarding the evolution of the solution of the fluid seepage model and its dependence on the model parameters. Compare your answer with **EXPLORE!** 9.9.

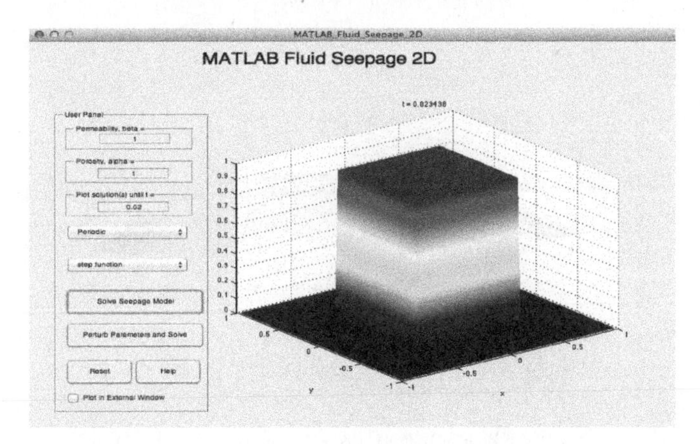

FIGURE 9.21: Fluid Seepage Model: periodic BCs, $\alpha = 1 = \beta$, initial profile: step function.

v.) Repeat either (i) or (ii) using the step function option for the initial condition. Do any of your observations differ from those presented in (iii)? How about when you compare your observations to solutions of the heat equation? See Figures 9.21 and 9.10 for one such comparison.

■

EXPLORE!

9.13 Long-Term Behavior

MATLAB-Exercise 9.2.5. To begin, open **MATLAB** and in the command line, type:

MATLAB_Fluid_Seepage_2D

Screenshots of particular GUI outputs can be found in Figures 9.22 and 9.23. Use the GUI to answer the following questions.

i.) Investigate the long-term behavior of IBVP (9.46) using $\alpha = 1$, $\beta = 2$, and the cosine option for the initial condition. Compare the behavior to the long-term behavior of the solution of the heat equation with the same initial and boundary conditions. Begin by plotting solutions until $t = 2$. Remember, computers cannot handle "infinity," so start with "small" times then increase them slowly if you need more time to describe the behavior. Figure 9.22 illustrates an example for which plotting solutions until $t = 2$, and even $t = 3$, does not fully capture the long-term behavior of the solution. However, watching the movies for these two time scenarios does strongly suggest what that behavior ought to be.

   a.) Set $\alpha = 0$ and keep everything else the same. Is the long-term behavior the same as the heat equation?

   b.) Set $\beta = 3$ and $\alpha = 1$. How would you describe the long-term behavior of the solution?

   c.) Set $\beta = 3$ and $\alpha = 0.5$. Compare the long-term behavior with the one described in (b).

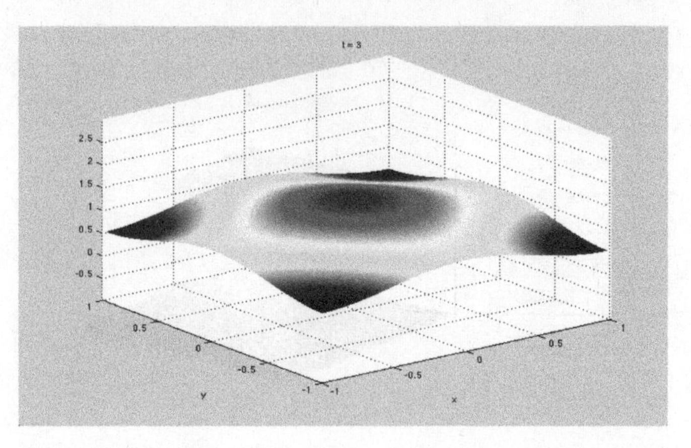

FIGURE 9.22: IBVP (9.47) with $\beta = 2$, $\alpha = 1$, initial profile: $\cos \pi x + \cos \pi y + 1$

d.) Repeat (i), (a)-(c) using the exponential initial condition option. Compare your observations to the results obtained when you used the sine initial condition option.

ii.) Investigate the long-term behavior of IBVP (9.44) using $\alpha = 1$, $\beta = 2$, and the cosine option for the initial condition. Compare the behavior to the long-term behavior of the heat equation with the same initial and boundary conditions.

a.) Set $\alpha = 0$ and keep everything else the same. Is the long-term behavior the same as the heat equation?

b.) Set $\beta = 0.001$ and $\alpha = 1$. How would you describe the long-term behavior of the solution?

c.) Set $\beta = 0.001$ and $\alpha = 0.5$. Compare the long-term behavior with the one described in (b).

d.) Repeat (i), (a)-(c) using the exponential initial condition option. Compare your observations to the results obtained when you used the sine initial condition option.

iii)) Summarize your observations regarding the long-term behavior of the solution of the fluid seepage model.

v.) Repeat either (i) or (ii) using the step function option for the initial condition. Do any of your observations differ from those presented in (iii)? How about when you compare your observations to solutions of the heat equation?

■

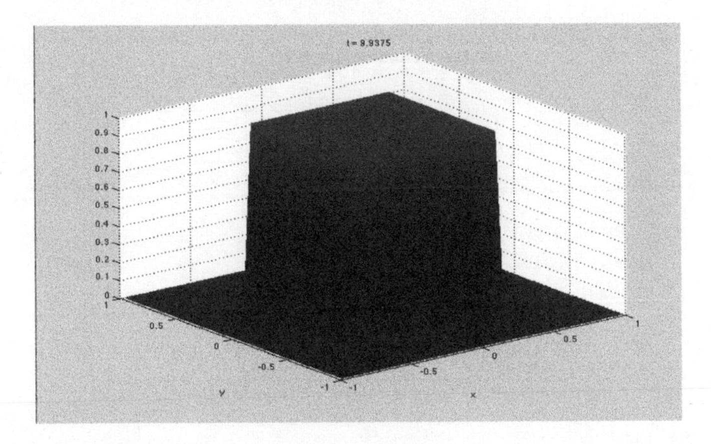

FIGURE 9.23: IBVP (9.47) with $\beta = .001$, $\alpha = 1$, initial profile: step function.

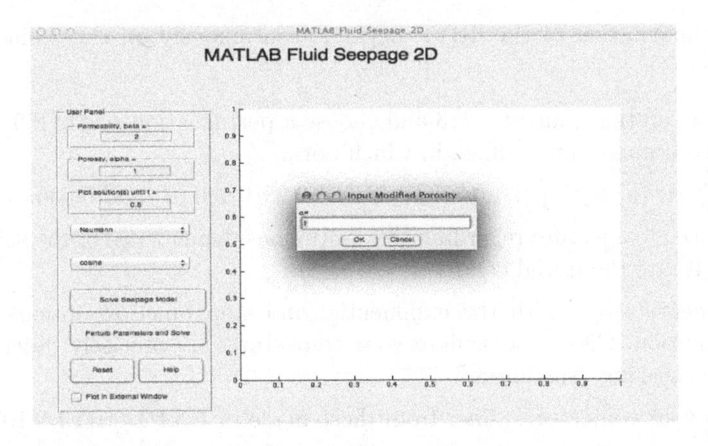

FIGURE 9.24: Perturbing Parameters and Initial Conditions with MATLAB Fluid Seepage 2D GUI

EXPLORE!

9.14 Continuous Dependence on Parameters

MATLAB-Exercise 9.2.6. Open **MATLAB** and in the command line, type:

MATLAB_Fluid_Seepage_2D

To explore continuous dependence using the GUI one must first solve the IBVP with particular model parameters. Once a solution has been constructed, click the *Perturb Parameters and Solve* button to specify the perturbed parameters and a perturbation size for the initial condition. A screenshot of inputting parameters into the GUI can be found in Figure 9.24 and another of a particular GUI output in Figure 9.25. When prompted to select a norm to visualize the difference in the solutions, please select *All.* Use the GUI to answer the following questions.

i.) Consider IBVP (9.46) with model parameters $\alpha = 1$ and $\beta = 2$. Explore whether

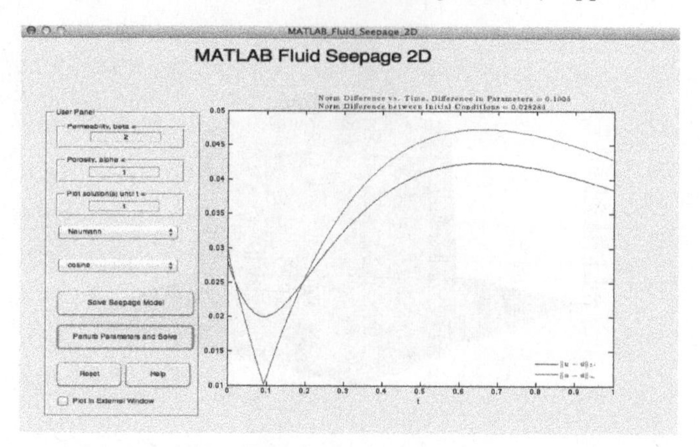

FIGURE 9.25: Differences Appear Small in Both Norms

or not the solution of this IBVP depends continuously on the cosine option initial condition.

a.) Plot solutions until $t = 0.3$ and choose a perturbation size of 0.5. Did the difference become small? If so, in which norm?

b.) Repeat (a) with perturbation sizes of $0.25, 0.1, 0.01$, and 0.005.

c.) Make a conjecture regarding the continuous dependence of the solution of IBVP (9.46) on the initial condition.

d.) Repeat (a)-(c) with the exponential and step function options for the initial condition. Clearly articulate your conjectures by precisely describing how you are checking "smallness."

e.) Do your conjectures differ from those made in **EXPLORE!** 9.10?

ii.) Consider IBVP (9.47) with the cosine initial condition option. Investigate whether or not the solution of the IBVP depends continuously on the model parameters near $\alpha = 1$ and $\beta = 2$.

a.) Plot solutions until $t = 0.3$. First consider what happens when only α is perturbed. That is, keep $\beta = 2$ and the perturbation size for the initial condition should equal 0. Choose the perturbed porosity constant to equal 1.1. Did the difference become small?

b.) Repeat (a) with perturbed porosity constants of $0.9, 0.98$, and 1.01.

c.) Make a conjecture regarding continuous dependence of the solution on the porosity constant.

d.) Repeat (a)-(c) for perturbing β only. Choose perturbed permeability constants to equal $2.1, 1.9, 1.99$, and 2.01.

e.) Repeat (a)-(d) with the exponential and step function options for the initial condition. Clearly articulate your conjectures by precisely describing how you are checking "smallness."

f.) Investigate if the solution of IBVP (9.47) depends continuously on the model's parameters and the initial conditions by perturbing α and β in the same manner as in (a)-(e), but also by choosing initial condition perturbation sizes to equal $0.25, 0.1, 0.05$, and 0.01. Explain your observations and try to formulate a claim regarding the model's continuous dependence on the initial conditions and the parameters. Clearly describe any differences in results when using the two different norms.

iii.) Summarize your findings and compare them to those found in **EXPLORE!** 9.10.

iv.) As seen in the previous **EXPLORE!**, the fluid seepage model reduces to a completely different IBVP, the heat equation, when α is set equal to zero. This suggests that IBVPs (9.46) and (9.47) may be sensitive to changes when α is small.

a.) Choose $\alpha = 0.01, \beta = 2$, the exponential initial condition option, and plot solutions until $t = 0.75$. Choose a perturbed porosity constant equal to 0.005. Did the difference become small?

b.) Repeat (a) for perturbed porosity constants 0.001 and 0. Do you need to alter any of your conjectures from previous parts? If so, describe how.

c.) Repeat (a) and (b) for the step function initial condition option.

d.) If you answered *No* in either part (b) or (c), determine if perturbing the other model parameter and the initial condition makes any difference when α is small. The perturbation sizes are left to you, but if you need any guidance refer to parts (i) and (ii).

■

9.2.3 Hypoplasticity

The erosion of beaches and grasslands is an ongoing environmental concern for various species of wildlife and human development. Avalanches occur due to the movement and changing of soil. Understanding such phenomena has important environmental ramifications. *Hypoplasticity* is an area of study that examines the behavior of granular solids, such as soil, sand, and clay. The IBVPs that arise in the modeling of this phenomena are complicated, mainly due to the presence of phase changes that the material undergoes.

We examine a particular form of a system of equations relating the fluid pressure and structural displacement, ignoring the physical meaning of the constants involved. Let $\Omega \subset \mathbb{R}^3$ be a bounded domain with smooth boundary $\partial\Omega$. The fluid pressure is denoted by $p(\mathbf{x}, t)$ and the (three-dimensional) structural displacement by $\mathbf{w}(\mathbf{x}, t)$ at position $\mathbf{x} = (x, y, z)$ in the soil at time t. Consider the following IBVP:

$$\begin{cases} -\beta_1 \nabla (\nabla \cdot \mathbf{w}(\mathbf{x}, t)) - \beta_2 \Delta \mathbf{w}(\mathbf{x}, t) + \beta_3 \nabla p(\mathbf{x}, t) = \mathbf{f}(\mathbf{x}, t), \ \mathbf{x} \in \Omega, \, t > 0, \\ \frac{\partial}{\partial t} (\beta_4 p(\mathbf{x}, t) + \beta_3 \nabla \cdot \mathbf{w}(\mathbf{x}, t)) - \nabla \cdot \beta_5 \nabla p(\mathbf{x}, t) = h(\mathbf{x}, t), \ \mathbf{x} \in \Omega, \, t > 0, \\ p(\mathbf{x}, 0) = p_0(\mathbf{x}), \ w(\mathbf{x}, 0) = w_0(\mathbf{x}), \ \mathbf{x} \in \Omega, \\ p(\mathbf{x}, t) = 0, \ \frac{\partial}{\partial \mathbf{n}} \mathbf{w}(\mathbf{x}, t) = 0, \ x \in \partial\Omega, \, t > 0. \end{cases} \tag{9.48}$$

It can be shown that solving the first PDE in (9.48) for $\mathbf{w}(\mathbf{x}, t)$ and then substituting this into the second PDE yields an abstract evolution equation in the space $\left(\mathbb{L}^2(\Omega)\right)^3$ of the form

$$\begin{cases} \frac{d}{dt} \left(\alpha p(t) - \nabla \cdot v^{-1} (\nabla p(t))\right) + \Delta p(t) = h(t), \ t > 0, \\ p(0) = p_0, \end{cases} \tag{9.49}$$

where

$$v(\mathbf{w}) = -\beta_1 \nabla (\nabla \cdot \mathbf{w})) - \beta_2 \Delta \mathbf{w}. \tag{9.50}$$

An overly simplified, yet comprehensible, one-dimensional version of this homogeneous evolution equation is given by

$$\begin{cases} \frac{\partial}{\partial t}\left(z(x,t) - \alpha \frac{\partial^2}{\partial x^2} z(x,t)\right) = \beta \frac{\partial^2}{\partial x^2} z(x,t),\ 0 < x < a,\ t > 0, \\ z(x,0) = z_0(x),\ 0 < x < a, \\ z(0,t) = z(a,t) = 0,\ t > 0, \end{cases} \tag{9.51}$$

where $\alpha, \beta > 0$. Or equipping (9.51) with periodic boundary conditions yields

$$\begin{cases} \frac{\partial}{\partial t}\left(z(x,t) - \alpha \frac{\partial^2}{\partial x^2} z(x,t)\right) = \beta \frac{\partial^2}{\partial x^2} z(x,t),\ 0 < x < a,\ t > 0, \\ z(x,0) = z_0(x),\ 0 < x < a, \\ z(0,t) = z(a,t),\ t > 0, \end{cases} \tag{9.52}$$

These IBVPs are very similar to the ones developed for the fluid seepage in a fissured rock model in the Subsection 9.2.1 **(What is different?)**, so the behavior should be quite similar.

EXPLORE!

9.15 Parameter Play!

In this **EXPLORE!**, we visualize the effects that changing α and β has on solutions to IBVPs (9.51) and (9.52). Furthermore, we compare solutions to the soil mechanics model to the fluid seepage model. Before doing so, let us make an informal argument for why solutions to the fluid seepage and the soil mechanics models may have similar behaviors. The only difference between the two models is the boundary conditions. As seen with the heat equation, boundary conditions play an essential role in the precise behavior of the solution, but certain qualitative features remain independent of the boundary condition. Thus, solutions to the two models may have similar qualitative features. In any case, we can use the following exercise to visualize the behavior of the IBVPs (9.51) and (9.52) and compare them to the fluid seepage model.

MATLAB-Exercise 9.2.7. To begin, open **MATLAB** and in the command line, type:

MATLAB_Soil_Mechanics_1D

Use the GUI to answer the following questions. Make certain to click on the HELP button for instructions on how to use the GUI. In the spirit of making comparisons more easily, the GUI will call α and β the porosity and permeability constants as in the Fluid Seepage GUIs. A typical screenshot of the GUI can found in Figure 9.26.

i.) Solve (9.51) until $t = 0.5$ using $\alpha = 1$, $\beta = 2$, and the sine option for the initial condition. How does this solution compare to the fluid seepage model with the same initial and boundary conditions?

   a.) Set $\alpha = 0.01$ and keep everything else the same. How would you describe the evolution of the solution as compared to the heat equation or fluid seepage model?

   b.) Keep increasing α by 0.5 until you can describe the effect that changing α has on the solution while keeping β fixed.

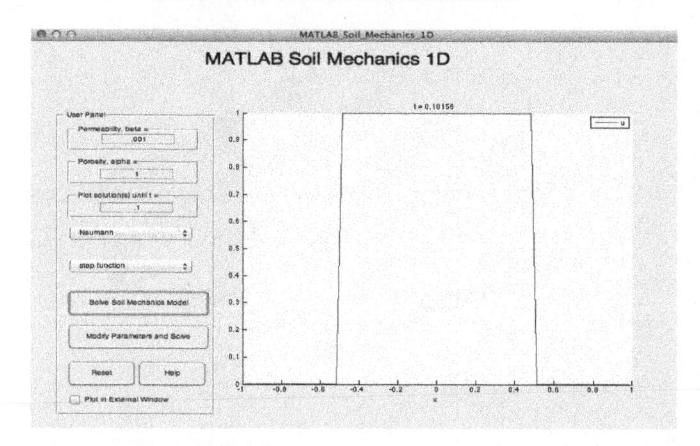

FIGURE 9.26: MATLAB Soil Mechanics 1D GUI

- c.) Set $\beta = 0.01$ and $\alpha = 1$. How would you describe the evolution of the solution as compared to the heat equation or fluid seepage model?
- d.) Keep increasing β by 0.5 until you can describe the effect that changing β has on the solution while keeping α fixed.
- e.) Repeat (a)-(d) using the exponential initial condition option.

ii.) Solve (9.52) until $t = 0.5$ using $\alpha = 1$, $\beta = 2$, and the sine option for the initial condition. How does this solution compare to the fluid seepage model with the same initial and boundary conditions?

- a.) Set $\alpha = 0.01$ and keep everything else the same. How would you describe the evolution of the solution as compared to the heat equation or fluid seepage model?
- b.) Keep increasing α by 0.5 until you can describe the effect that changing α has on the solution while keeping β fixed.
- c.) Set $\beta = 0.01$ and $\alpha = 1$. How would you describe the evolution of the solution as compared to the heat equation or fluid seepage model?
- d.) Keep increasing β by 0.5 until you can describe the effect that changing β has on the solution while keeping α fixed.
- e.) Repeat (a)-(d) using the exponential initial condition option.

iii.) Summarize your observations regarding the evolution of soil mechanics model and its dependence on model parameters. Compare your answer with **EXPLORE!** 9.9.

iv.) Repeat (i) and (ii) using the step function option for the initial condition. Do any of your observations differ from those presented in (iii)? How about when you compare your observations to solutions of the one-dimensional fluid seepage model?

■

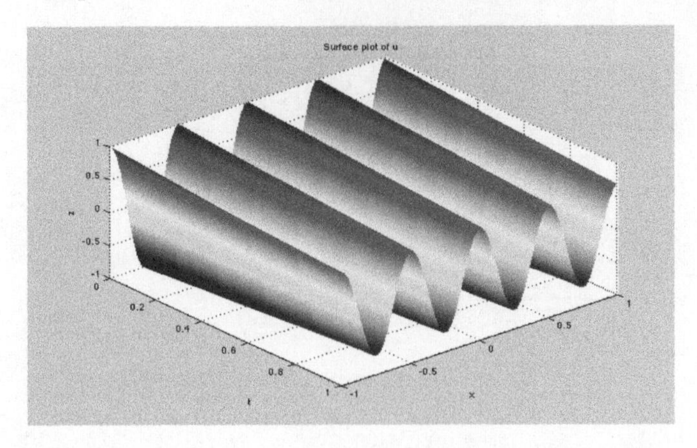

FIGURE 9.27: IBVP (9.51) with $\beta = 1$, $\alpha = 3$, initial profile: $\cos 4\pi x$

EXPLORE!

9.16 Long-Term Behavior

MATLAB-Exercise 9.2.8. To begin, open **MATLAB** and in the command line, type:

MATLAB_Soil_Mechanics_1D

Screenshots of particular GUI outputs can be found in Figures 9.27, 9.28, and 9.29. Use the GUI to answer the following questions.

i.) Investigate the long-term behavior of IBVP (9.51) using $\alpha = 1$, $\beta = 2$, and the sine option for the initial condition. Compare the behavior to the long-term behavior of the fluid seepage model with the same initial and boundary conditions. Begin by plotting solutions until $t = 2$. Remember, computers cannot handle "infinity," so start with "small" times then increase them slowly if you need more time to describe the behavior. Observe in Figure 9.29 that a clear pattern has emerged by $t = 3$. One could verify that pattern persists by plotting further out into time.

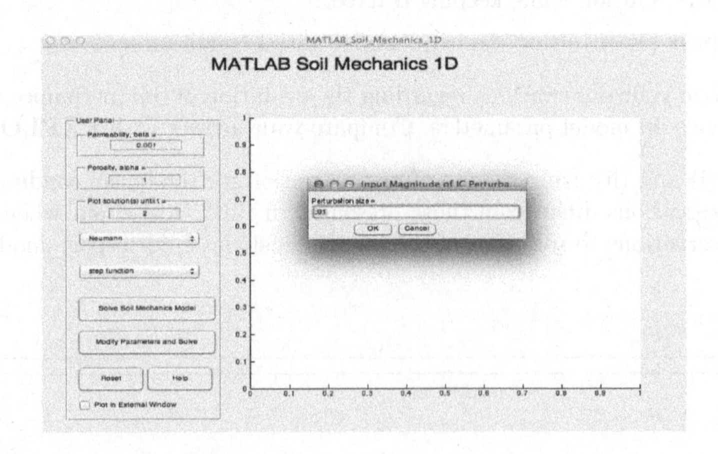

FIGURE 9.28: Perturbing Parameters and Initial Conditions with MATLAB Soil Mechanics 1D GUI

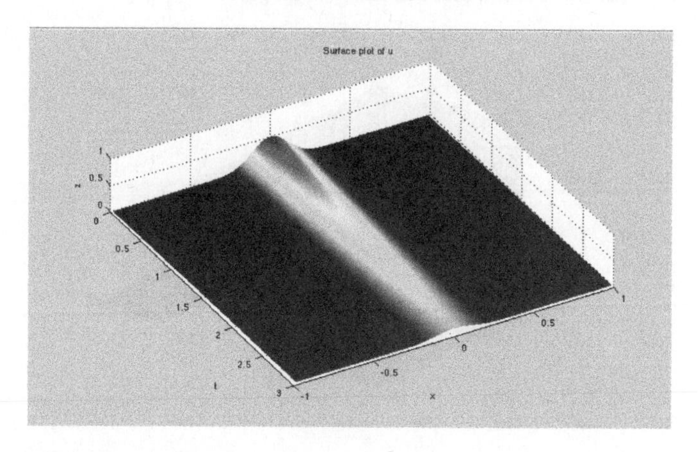

FIGURE 9.29: IBVP (9.51) with $\beta = 2$, $\alpha = 1$, initial profile: e^{-40x^2}

a.) Set $\alpha = 0.01$ and keep everything else the same. Is the long-term behavior similar to the heat equation and the fluid seepage model?

b.) Set $\beta = 0.01$ and $\alpha = 1$. How would you describe the long-term behavior of the solution?

c.) Repeat (i), (a)-(b) using the exponential initial condition option. Compare your observations to the results obtained when you used the sine initial condition option.

ii.) Investigate the long-term behavior of IBVP (9.52) using $\alpha = 1$, $\beta = 2$, and the sine option for the initial condition. Compare the behavior to the long-term behavior of the fluid seepage model with the same initial and boundary conditions.

a.) Set $\alpha = 0.01$ and keep everything else the same. Is the long-term behavior similar to the heat equation and the fluid seepage model? Explain.

b.) Set $\beta = 0.01$ and $\alpha = 1$. How would you describe the long-term behavior of the solution?

c.) Repeat (i), (a)-(b) using the exponential initial condition option. Compare your observations to the results obtained when you used the sine initial condition option.

iii.) Summarize your observations regarding the long-term behavior of IBVPs (9.51) and (9.52). Are you observations the same as those made in **EXPLORE!** 9.9? Explain.

iv.) Repeat either (i) and (ii) using the step function option for the initial condition. Do any of your observations differ from those presented in (iii)?

■

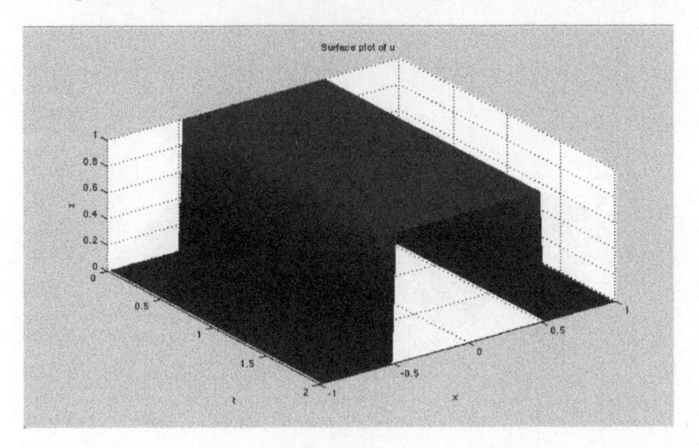

FIGURE 9.30: IBVP (9.51) with $\beta = 0.001$, $\alpha = 1$, initial profile: step function

---

EXPLORE!

9.17 Continuous Dependence on Parameters

MATLAB-Exercise 9.2.9. Open **MATLAB** and in the command line, type:

MATLAB_Soil_Mechanics_1D

To explore continuous dependence using the GUI we must first solve the IBVP with particular model parameters. Once a solution has been constructed, click the *Perturb Parameters and Solve* button to specify the perturbed parameters and a perturbation size for the initial condition. A screenshot of inputting parameters into the GUI can be found in Figure 9.30 and others of a particular GUI outputs in Figure 9.31. When prompted to select a norm to visualize the difference in the solutions, please select *All.* Use the GUI to answer the following questions.

i.) Consider IBVP (9.51) with model parameters, $\alpha = 1$ and $\beta = 2$. Explore whether or not the solution of this IBVP depends continuously on the sine option initial condition.

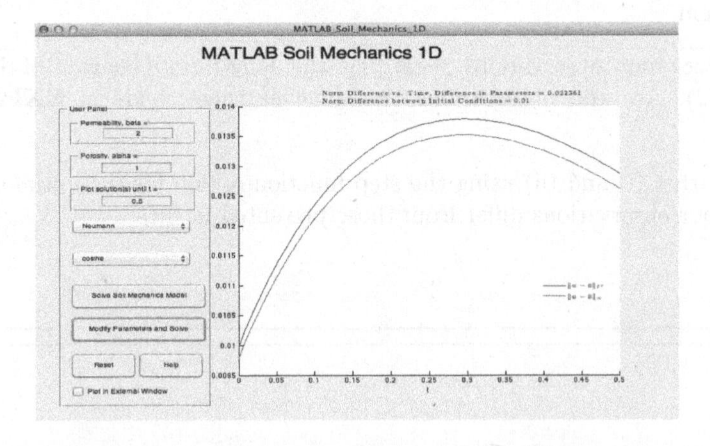

FIGURE 9.31: Differences in both norms tend to stay small and get even smaller over time.

a.) Plot solutions until $t = 0.3$ and choose a perturbation size of 0.5. Did the difference become small? If so, in which norm?

b.) Repeat (a) with perturbation sizes of $0.25, 0.1, 0.01$, and 0.005.

c.) Make a conjecture regarding the continuous dependence of the solution of IBVP (9.51) on the initial condition.

d.) Repeat (a)-(c) with the exponential and step function options for the initial condition. Clearly articulate your conjectures by precisely describing how you are checking "smallness."

e.) Do your conjectures differ from those made in **EXPLORE!** 9.11 for the corresponding fluid seepage model?

ii.) Consider IBVP (9.52) with the sine initial condition option. Investigate whether or not the solution of the IBVP depends continuously on the model parameters near $\alpha = 1$ and $\beta = 2$.

a.) Plot solutions until $t = 0.3$. First consider what happens when only α is perturbed. That is, keep $\beta = 2$ and the perturbation size for the initial condition should equal 0. Choose the perturbed porosity constant to equal 1.1. Did the difference become small?

b.) Repeat (a) with perturbed porosity constants of $0.9, 0.98$, and 1.01.

c.) Make a conjecture regarding continuous dependence of the solution on the porosity constant.

d.) Repeat (a)-c) for perturbing β only. Choose perturbed permeability constants to equal $2.1, 1.9, 1.99$, and 2.01.

e.) Repeat (a)-d) with the exponential and step function options for the initial condition. Clearly articulate your conjectures by precisely explaining how you are checking "smallness."

f.) Investigate if the solution of IBVP (9.52) depends continuously on the model's parameters and the initial conditions by perturbing α and β in the same manner as in (a)-(e), but also by choosing initial condition perturbation sizes to equal $0.25, 0.1, 0.05$ and 0.01. Explain your observations and try to formulate a claim regarding the model's continuous dependence on the initial conditions and the parameters. Clearly describe any differences that arise from using the two different norms.

iii.) Summarize your findings and compare them to those found in **EXPLORE!** 9.11.

iv.) As with the fluid seepage model, IBVP (9.51) and (9.52) reduces to a completely different IBVP, the heat equation, when α is set equal to zero. This suggests that IBVPs (9.51) and (9.52) may be sensitive to changes when α is small.

a.) Choose $\alpha = 0.01, \beta = 2$, the exponential initial condition option, and plot solutions until $t = 0.75$. Choose a perturbed porosity constant equal to 0.005. Did the difference become small?

b.) Repeat (a) for perturbed porosity constants 0.001 and 0. Do you need to alter any of your conjectures from previous parts? If so, describe how.

c.) Repeat (a) and (b) for the step function initial condition option.

d.) If you answered *No* in either part (b) or (c), determine if perturbing the other model parameter and the initial condition makes any difference when α is small. The perturbation sizes are left to you, but if you need any guidance refer to parts (i) and (ii).

■

9.2.4 Two-Dimensional Hypoplasticity

Next, let us consider two-dimensional versions of IBVPs (9.51) and (9.52) in which the spatial domain is the rectangle, $[0,a] \times [0,a]$. IBVP (9.43) becomes

$$\begin{cases} \frac{\partial}{\partial t}\left(p(x,y,t) - \alpha \Delta p(x,y,t)\right) + \beta \Delta p(x,y,t) = 0, \; 0 < x < a, 0 < y < a, \, t > 0, \\ p(x,y,0) = p_0(x,y), \; 0 < x < a, 0 < y < a \\ p(0,y,t) = 0 = p(a,y,t), \; 0 < y < a, t > 0 \\ p(x,0,t) = 0 = p(x,a,t), \; 0 < x < a, t > 0 \end{cases} \tag{9.53}$$

where

$$\Delta p(x,y,t) = \frac{\partial^2}{\partial x^2} p(x,y,t) + \frac{\partial^2}{\partial y^2} p(x,y,t)$$

If the BCs in (9.53) are replaced by periodic boundary conditions the IBVP becomes:

$$\begin{cases} \frac{\partial}{\partial t}\left(p(x,y,t) + \alpha \Delta p(x,y,t)\right) + \beta \Delta p(x,y,t) = 0, \; 0 < x < a, 0 < y < a t > 0, \\ p(x,y,0) = p_0(x,y), \; 0 < x < a, 0 < y < a \\ p(0,y,t) = p(\pi,y,t), \; 0 < y < a, t > 0 \\ p(x,0,t) = p(x,\pi,t), \; 0 < x < a, t > 0 \end{cases} \tag{9.54}$$

EXPLORE!
9.18 Parameter Play!

In this **EXPLORE!**, we examine the effect of changing α and β in IBVPs (9.53) and (9.54) and compare theses effects to the one-dimensional case.

STOP! How did changing α and β in IBVPs (9.51) and (9.52) affect the evolution of the solutions?

MATLAB-Exercise 9.2.10. To begin, open **MATLAB** and in the command line, type:

MATLAB_Soil_Mechanics_2D

Screenshots particular GUI outputs can be found in Figure 9.33 and 9.34. Use the GUI to answer the following questions. Make certain to click on the HELP button for instructions on how to use the GUI. A typical screenshot of the GUI can found in Figure 9.32.

i.) Solve (9.53) until $t = 0.5$ using $\alpha = 1$, $\beta = 2$, and the sine option for the initial condition. How would you compare the solution to the two-dimensional fluid seepage model with the same initial and boundary conditions?

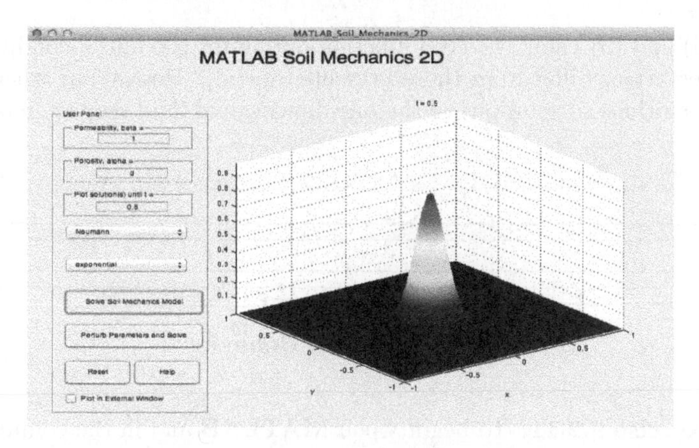

FIGURE 9.32: MATLAB Soil Mechanics 2D GUI

a.) Set $\alpha = 0.01$ and keep everything else the same. How would you describe the evolution of the solution as compared to the heat equation or fluid seepage model?

b.) Keep increasing α by 0.5 until you can describe the effect that changing α has on the solution while keeping β fixed.

c.) Set $\beta = 0.01$ and $\alpha = 1$. How would you describe the evolution of the solution as compared to the heat equation or fluid seepage model?

d.) Keep increasing β by 0.5 until you can describe the effect that changing β has on the solution while keeping α fixed.

e.) Repeat (a)-(d) using the exponential initial condition option.

ii.) Solve (9.54) until $t = 0.5$ using $\alpha = 1$, $\beta = 2$, and the sine option for the initial condition. How would you compare the solution to the two-dimensional fluid seepage model with the same initial and boundary conditions?

a.) Set $\alpha = 0.01$ and keep everything else the same. How would you describe the evolution of the solution as compared to the heat equation or fluid seepage model?

b.) Keep increasing α by 0.5 until you can describe the effect that changing α has on the solution while keeping β fixed.

c.) Set $\beta = 0.01$ and $\alpha = 1$. How would you describe the evolution of the solution as compared to the heat equation or fluid seepage model?

d.) Keep increasing β by 0.5 until you can describe the effect that changing β has on the solution while keeping α fixed.

e.) Repeat (a)-(d) using the exponential initial condition option.

iii.) Summarize your observations regarding the evolution of soil mechanics model and its dependence on model parameters. Compare your answer with **EXPLORE!** 9.9.

iv.) Repeat (i) and (ii) using the step function option for the initial condition. Do any of your observations differ from those presented in (iii)? How about when you compare your observations to solutions of the one-dimensional fluid seepage model?

■

EXPLORE!

9.19 Long-Term Behavior

MATLAB-Exercise 9.2.11. To begin, open **MATLAB** and in the command line, type:

MATLAB_Soil_Mechanics_2D

Screenshots particular GUI outputs can be found in Figure 9.33 and 9.34. Use the GUI to answer the following questions.

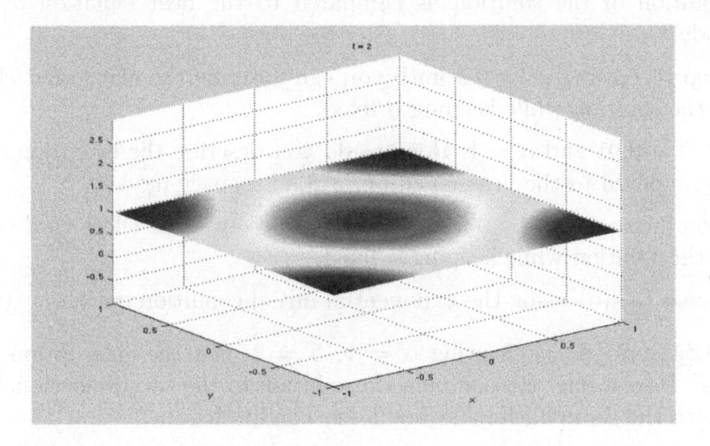

FIGURE 9.33: IBVP (9.47) with $\beta = 2$, $\alpha = 1$, initial profile: $\cos \pi x + \cos piy + 1$

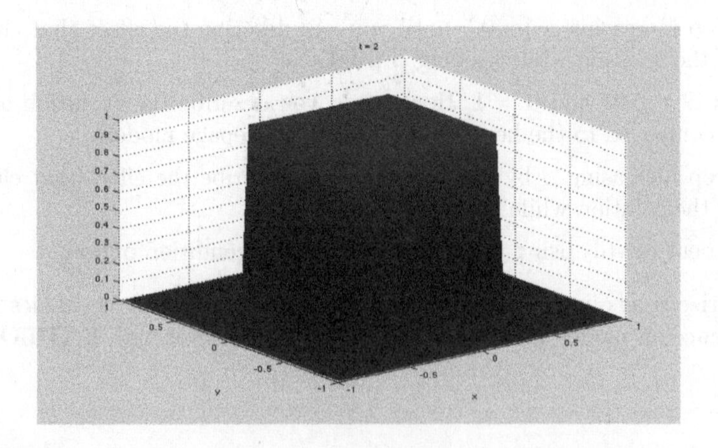

FIGURE 9.34: IBVP (9.53) with $\beta = .001$, $\alpha = 1$, initial profile: step function.

i.) Investigate the long-term behavior of IBVP (9.53) using $\alpha = 1$, $\beta = 2$, and the sine option for the initial condition. Compare the behavior to the long-term behavior of the heat equation with the same initial and boundary conditions. Begin by plotting solutions until $t = 2$. Remember, computers cannot handle "infinity," so start with "small" times then increase them slowly if you need more time to describe the behavior.

  a.) Set $\alpha = 0.01$ and keep everything else the same. Is the long-term behavior similar to the heat equation and the fluid seepage model?

  b.) Set $\beta = 0.01$ and $\alpha = 1$. How would you describe the long-term behavior of the solution?

  c.) Repeat (i), (a)-(b) using the exponential initial condition option. Compare your observations to the results obtained when you used the sine initial condition option.

ii.) Investigate the long-term behavior of IBVP (9.54) using $\alpha = 1$, $\beta = 2$, and the sine option for the initial condition. Compare the behavior to the long-term behavior of the fluid seepage model with the same initial and boundary conditions.

  a.) Set $\alpha = 0.01$ and keep everything else the same. Is the long-term behavior similar to the heat equation and the fluid seepage model?

  b.) Set $\beta = 0.01$ and $\alpha = 1$. How would you describe the long-term behavior of the solution?

  c.) Repeat (i), (a)-(b) using the exponential initial condition option. Compare your observations to the results obtained when you used the sine initial condition option.

iii.) Summarize your observations regarding the long-term behavior of IBVPs (9.53) and (9.54). Are you observations the same as those made in **EXPLORE!** 9.15? Explain.

iv.) Repeat either (i) and (ii) using the step function option for the initial condition. Do any of your observations differ from those presented in (iii)?

■

EXPLORE!
9.20 Continuous Dependence on Parameters

MATLAB-Exercise 9.2.12. Open **MATLAB** and in the command line, type:

MATLAB_Soil_Mechanics_2D

To explore continuous dependence with the GUI we must first solve the IBVP with particular model parameters. Once a solution has been constructed, click the *Perturb Parameters and Solve* button to specify the perturbed parameters and a perturbation size for the initial condition. A screenshot of inputting parameters into the GUI can be found in Figure 9.35 and another of a particular GUI output in Figure 9.36. When prompted to select a norm to visualize the difference in the solutions, please select *All*. Use the GUI to answer the following questions.

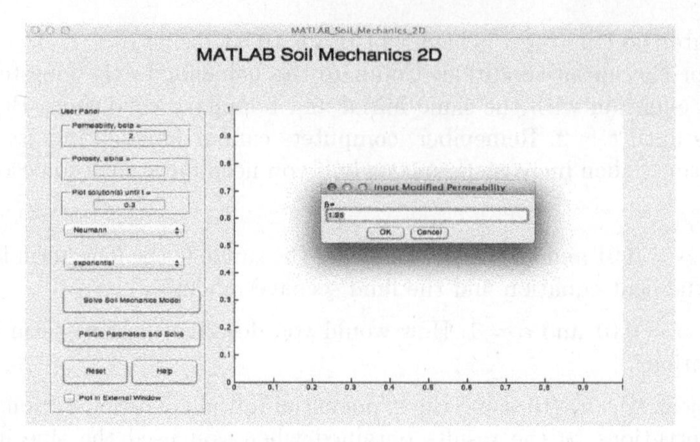

FIGURE 9.35: Perturbing Parameters and Initial Conditions with MATLAB Soil Mechanics 2D GUI

i.) Consider IBVP (9.53) with model parameters, $\alpha = 1$ and $\beta = 2$. Explore whether or not the solution of this IBVP depends continuously on the sine option initial condition.

  a.) Plot solutions until $t = 0.3$ and choose a perturbation size of 0.5. Did the difference become small? If so, in which norm?

  b.) Repeat (a) with perturbation sizes of $0.25, 0.1, 0.01$, and 0.005.

  c.) Make a conjecture regarding the continuous dependence of the solution IBVP (9.53) on the initial condition.

  d.) Repeat (a)-(c) with the exponential and step function options for the initial condition. Clearly articulate your conjectures by precisely describing how you are checking "smallness."

  e.) Do your conjectures differ from those made in **EXPLORE!** 9.16?

ii.) Consider IBVP (9.54) with the sine initial condition option. Investigate whether or not the solution of the IBVP depends continuously on the model parameters near $\alpha = 1$ and $\beta = 2$.

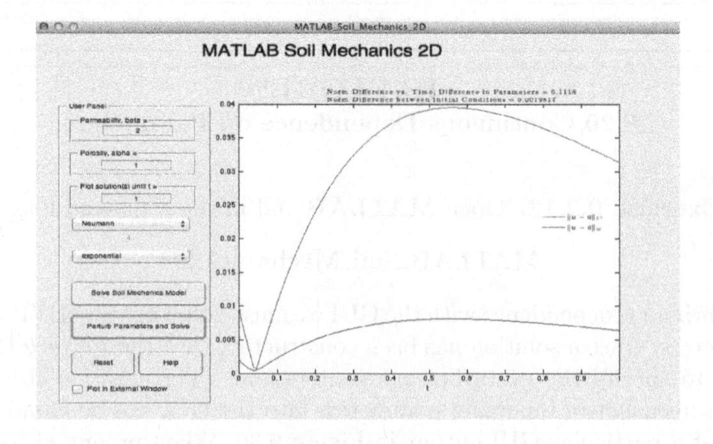

FIGURE 9.36: Moderate differences in model parameters and initial conditions for IBVP (9.53)

a.) Plot solutions until $t = 0.3$. First consider what happens when only α is perturbed. That is, keep $\beta = 2$ and the perturbation size for the initial condition should equal 0. Choose the perturbed porosity constant to equal 1.1. Did the difference become small?

b.) Repeat (a) with perturbed porosity constants of $0.9, 0.98$, and 1.01.

c.) Make a conjecture regarding continuous dependence of the solution on the porosity constant.

d.) Repeat (a)-(c) for perturbing β only. Choose perturbed permeability constants to equal $2.1, 1.9, 1.99$, and 2.01.

e.) Repeat (a)-(d) with the exponential and step function options for the initial condition. Clearly articulate your conjectures by precisely describing how you are checking "smallness."

f.) Investigate if the solution of IBVP (9.54) depends continuously on the model's parameters and the initial conditions by perturbing α and β in the same manner as in (a)-(e), but also by choosing initial condition perturbation sizes to equal $0.25, 0.1, 0.05$, and 0.01. Explain your observations and try to formulate a claim regarding the model's continuous dependence on the initial conditions and the parameters. Clearly describe any differences that arise from using the two different norms.

iii.) Summarize your findings and compare them to those found in **EXPLORE!** 9.16.

iv.) As with the fluid seepage model, IBVP (9.53) and (9.54) reduces to a completely different IBVP, namely the heat equation, when α is set equal to zero. This suggests that IBVPs (9.53) and (9.54) may be sensitive to changes when α is small.

a.) Choose $\alpha = 0.01, \beta = 2$, the exponential initial condition option, and plot solutions until $t = 0.75$. Choose a perturbed porosity constant equal to 0.005. Did the difference become small?

b.) Repeat (a) for perturbed porosity constants 0.001 and 0. Do you need to alter any of your conjectures from previous parts? If so, describe how.

c.) Repeat (a) and (b) for the step function initial condition option.

d.) If you answered *No* in either part (b) or (c), determine if perturbing the other model parameter and the initial condition makes any difference when α is small. The perturbation sizes are left to you, but if you need any guidance refer to parts (i) and (ii).

■

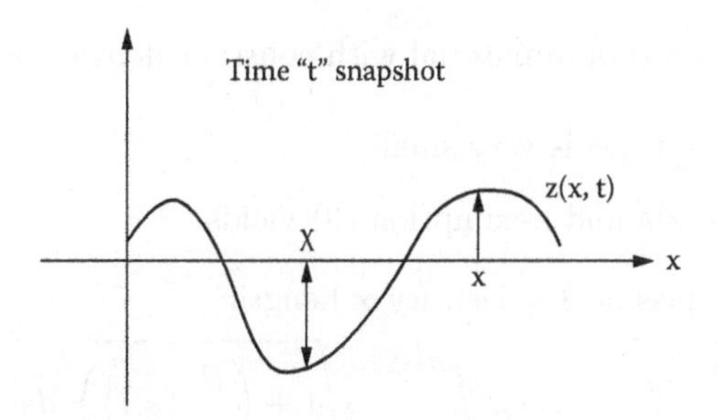

9.2.5 Abstract Formulation

The main difference between (9.43) and (9.11) is the presence of the term $-\frac{\partial^2 z}{\partial x^2}(x,t)$, which initially hinders our effort to transform (9.43) into the abstract form (A-HCP). As before, let $\mathcal{X} = \mathbb{L}^2(0,\pi;\mathbb{R})$ and define the operators $A : \text{dom}(A) \subset \mathcal{X} \to \mathcal{X}$ and $B : \text{dom}(B) \subset \mathcal{X} \to \mathcal{X}$ as follows:

$$\begin{aligned} A[f] &= f'', \ \text{dom}(A) = \left\{ f \in \mathbb{H}^2(0,\pi;\mathbb{R}) \mid f(0) = f(\pi) = 0 \right\}, \\ B[f] &= f - f'', \ \text{dom}(B) = \text{dom}(A). \end{aligned} \tag{9.55}$$

Making the identification $u(t)[x] = z(x,t)$ enables us to reformulate (9.43) as the following abstract evolution equation in $\mathbb{L}^2(0,\pi;\mathbb{R})$:

$$\begin{cases} (Bu)'(t) = Au(t), \ t > 0, \\ u(0) = u_0. \end{cases} \tag{9.56}$$

Exercise 9.2.4. Intuitively, what would be the natural thing to *try* to do in order to further express (9.56) in the form (A-HCP)? What conditions are needed to justify such a transformation?

We will study the intricacies of such problems in the next chapter. For the moment, let us just say A and B must be compatible in order to facilitate the further transition to the form (9.56).

9.3 The Classical Wave Equation and its Variants

The evolution over time of the vertical displacement of a vibrating string (rectangular membrane) of finite length (area) subject to small vibrations can be described by so-called *wave equations*. We introduce one- and two-dimensnional versions of this phenomena in this section.

9.3.1 One-Dimensional Wave Equation

Suppose that the vertical deflection (with respect to the x-axis) of a taut string of length L at position x along the string at time t is given by $z(x,t)$. Consider a small piece of string with endpoints at x and $x + \Delta x$. We shall follow the standard derivations (as presented by [39] and [26]) using Newton's Second Law of Motion. We impose the following assumptions:

1. The string remains in the plane as it vibrates, and moves only in the vertical direction.
2. The string is comprised of a material with constant density ρ.
3. The magnitude of $\frac{\partial z}{\partial x}(x,t)$ is very small.

Using the arc length formula and assumption (3) yields

$$\begin{aligned} \text{Mass of S} &= \text{Density} \times \text{Length} \\ &= \rho \int_x^{x+\Delta x} \sqrt{1 + \left(\frac{\partial z}{\partial x}(s,t) \right)^2} \, ds \\ &= \rho \Delta x. \end{aligned} \tag{9.57}$$

Ignoring external environmental effects, the only forces acting on the string are due to gravity (in the vertical downward direction) and due to the tension (or pull) of the string. We analyze each of these forces below. Refer to the following diagram for a visualization of these forces:

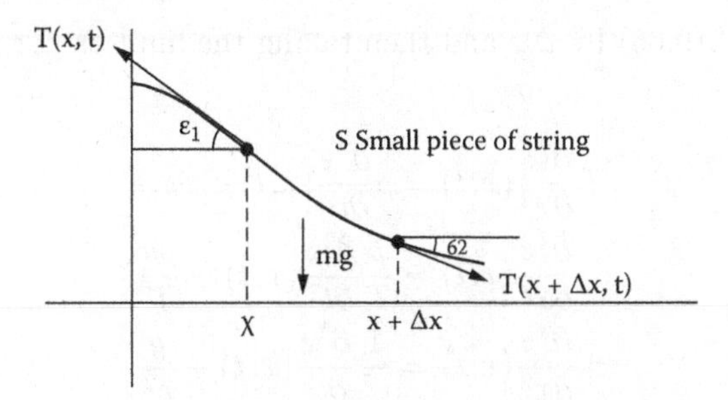

Horizontal Forces: The horizontal components of the force due to tension at x and $x+\Delta x$ are given by $T(x+\Delta x,t)\cos(\epsilon_1)$ and $-T(x,t)\cos(\epsilon_2)$.
Vertical Forces: The force due to gravity acts in the downward direction and is given by $-mg = -(\rho\Delta x)\,g$ by (9.57). The vertical components of the force due to tension at x and $x+\Delta x$ are given by $T(x+\Delta x,t)\sin(\epsilon_1)$ and $-T(x,t)\sin(\epsilon_2)$.

Since the motion only occurs in the vertical direction, the horizontal component of acceleration is zero. So, by Newton's Second Law of Motion, we have

$$T(x+\Delta x,t)\cos(\epsilon_1) - T(x,t)\cos(\epsilon_2) = 0$$
$$T(x+\Delta x,t)\cos(\epsilon_1) = T(x,t)\cos(\epsilon_2) \tag{9.58}$$

for all $0 < x < L$, $t > 0$. Since the string is taut, the tension can only change very slightly over time t. As such, (9.69) implies that the string has a constant tension T. That is,

$$T(x+\Delta x,t) = T(x,t) = T, \text{ for all } 0 < x < L, t > 0 \tag{9.59}$$

Next, using Newton's Second Law of Motion for the vertical motion yields

$$T(x+\Delta x,t)\sin(\epsilon_1) - T(x,t)\sin(\epsilon_2) - (\rho\Delta x)\,g = (\rho\Delta x)\frac{\partial^2}{\partial t^2}z(x,t). \tag{9.60}$$

Using (9.58) and (9.59) yields

$$T(x+\Delta x,t) = \frac{T}{\cos(\epsilon_1)} \tag{9.61}$$

$$T(x,t) = \frac{T}{\cos(\epsilon_2)} \tag{9.62}$$

Subsequently, substituting (9.61) and (9.62) into (9.60) yields

$$T\tan(\epsilon_1) - T\tan(\epsilon_2) - (\rho\Delta x)\,g = (\rho\Delta x)\frac{\partial^2}{\partial t^2}z(x,t). \tag{9.63}$$

Since a line segment making an angle θ with the $x-$axis has slope $\tan\theta$, it follows that

$$\tan(\epsilon_2) = \text{slope of the tangent line to } z(x,t) \text{ (in the } x-\text{direction)} = \frac{\partial z}{\partial x}(x,t), \tag{9.64}$$

$$\tan(\epsilon_1) = \text{slope of the tangent line to } z(x+\Delta x,t) \text{ (in the } x-\text{direction)} = \frac{\partial z}{\partial x}(x+\Delta x,t).$$

Substituting (9.64) into (9.63) yields

$$T\left(\frac{\partial z}{\partial x}(x+\Delta x,t)-\frac{\partial z}{\partial x}(x,t)\right)=(\rho\Delta x)\left(\frac{\partial^2 z}{\partial t^2}(x,t)+g\right). \tag{9.65}$$

Dividing both sides of (9.65) by Δx and then taking the limit as Δx goes to zero in (9.65) yields

$$\begin{aligned}T\frac{\partial^2 z}{\partial x^2}(x,t)&=\rho\frac{\partial^2 z}{\partial t^2}(x,t)+\rho g\\ \frac{\partial^2 z}{\partial x^2}(x,t)&=\frac{\rho}{T}\frac{\partial^2 z}{\partial t^2}(x,t)+\frac{\rho g}{T}\\ \frac{\partial^2 z}{\partial x^2}(x,t)&=\frac{1}{c^2}\frac{\partial^2 z}{\partial t^2}(x,t)+\frac{g}{c^2},\end{aligned} \tag{9.66}$$

where $c^2=\frac{T}{\rho}$.

Equipping (9.66) with Dirichlet BCs and standard ICs governing the initial profile and velocity over time yields the following IBVP:

$$\begin{cases}\frac{\partial^2}{\partial t^2}z(x,t)-c^2\frac{\partial^2}{\partial x^2}z(x,t) &=0,\ 0<x<L,\ t>0,\\ z(x,0)=z_0(x),\ \frac{\partial z}{\partial t}(x,0) &= z_1(x),\ 0<x<L,\\ z(0,t)=z(L,t) &=0,\ t>0.\end{cases} \tag{9.67}$$

The IBVP analogous to (9.67) when Neumann BCs are used instead of Dirichlet BCs is given by:

$$\begin{cases}\frac{\partial^2}{\partial t^2}z(x,t)-c^2\frac{\partial^2}{\partial x^2}z(x,t) &=0,\ 0<x<L,\ t>0,\\ z(x,0)=z_0(x),\ \frac{\partial z}{\partial t}(x,0) &= z_1(x),\ 0<x<L,\\ \frac{\partial z}{\partial x}(0,t)=\frac{\partial z}{\partial x}(L,t) &=0,\ t>0.\end{cases} \tag{9.68}$$

STOP! Interpret the meaning of Dirichlet BCs and Neumann BCs ($\frac{\partial z}{\partial x}(0,t)=\frac{\partial z}{\partial x}(L,t)$) in the context of vibrations of a string.

Exercise 9.3.1. Complete the following:

i.) Use the separation of variables technique to show that the solution of (9.67) is given by

$$\begin{aligned}z(x,t)&=\sum_{n=1}^{\infty}2\left[\langle z_0(\cdot),e_n(\cdot)\rangle_{\mathbb{L}^2}\cos(\lambda_n ct)+\frac{1}{L\lambda_n}\langle z_1(\cdot),e_n(\cdot)\rangle_{\mathbb{L}^2}\sin(\lambda_n ct)\right]\cdot\sin(\lambda_n x)\\ &=S(t)[z_0](x)\end{aligned} \tag{9.69}$$

where $e_n(\cdot)=\sin(\lambda_n\cdot)$ and $\lambda_n=\sqrt{\frac{n\pi}{L}},\ n\in\mathbb{N}$.

ii) Use separation of variables to solve (9.68). How does this compare to (9.69)?

Exercise 9.3.2. Verify $S(t)[z_0](x)$ given in (9.68) satisfies the semigroup properties

i.) $S(0)[z_0](x0)=z_0(x)$

ii.) $S(t_1+t_2)[z_0](x)=S(t_1)\left(S(t_2)[z_0]\right)(x)$

iii.) $S(t)[z_0 + z_0'](x) = S(t)[z_0](x) + S(t)[z_0'](x)$

EXPLORE!

9.21 Parameter Play!

In this **EXPLORE!** you will investigate the effects of changing the propagation constant c^2 in the one-dimensional wave equation. To do this, you will use the MATLAB Wave Equation 1D GUI to visualize the solutions to IBVP (9.67) and IBVP (9.68) for different choices of c^2.

MATLAB-Exercise 9.3.1. To begin, open **MATLAB** and in the command line, type:

MATLAB_Wave_Equation_1D

Use the GUI to answer the following questions. Make certain to click on the HELP button for instructions on how to use the GUI. A typical screenshot of the GUI can found in Figure 9.37.

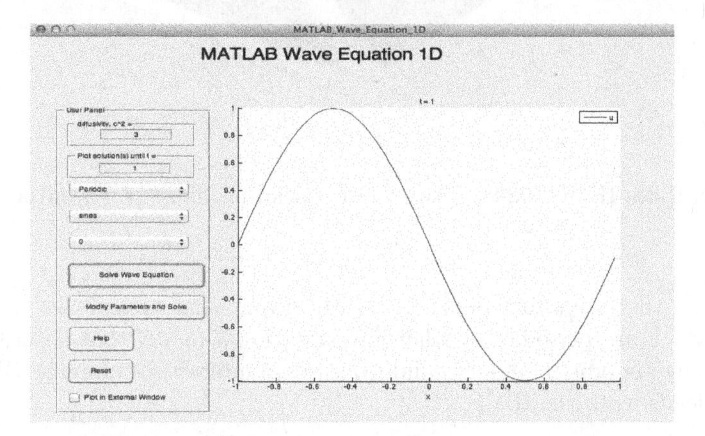

FIGURE 9.37: MATLAB Wave Equation 1D GUI

i.) Run the GUI with $c^2 = 2$, $t = 1$, Dirichlet boundary conditions and choose the sine option for the initial condition and the zero option for the initial velocity. The GUI will play a movie of the solution twice and then will produce a graph of the solution in time and space. The movie will give you a sense of the "speed" of the wave traveling from its initial profile. The separate figure will be helpful when you compare solutions stemming from different propagation constants.

   a.) Run the GUI again, but this time with $c^2 = 1$. To do this, simply click the box containing the diffusivity constant and replace 2 with 1. Then, click *Solve*.

   b.) What did you observe in (a)? Did the motion of the wave seem faster or slower? If you could not tell, compare the figure from the previous run of the GUI to the second figure just outputed.

   c.) Repeat (a) and (b), but this time change the initial velocity option to 1.

   d.) Based upon your observations, how does c^2 affect the motion of the waves? How did the non-zero initial condition affect the solution?

e.) Test your observations by using the exponential choice for the initial condition, propogation constants larger and smaller than 2, and different initial velocities. Articulate the evolution of the waves and how those surfaces were affected by the propogation constant. Try to qualitatively explain the changes in the solution when using non-zero initial velocities.

ii.) Repeat (i), but this time choose Neumann boundary conditions and the cosine option for the initial condition. Use initial velocity options of zero and one as in (i).

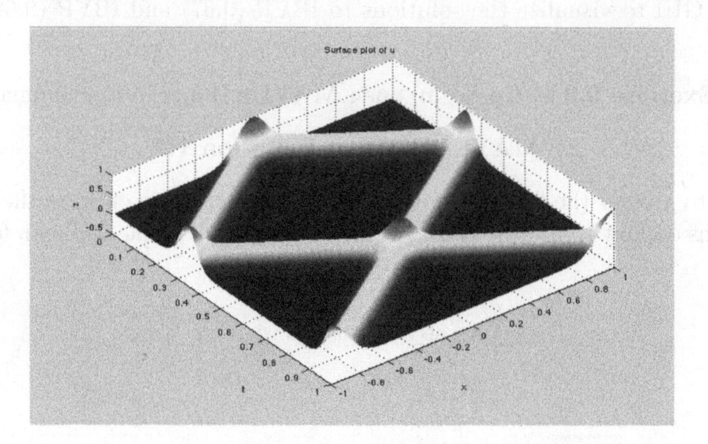

FIGURE 9.38: IBVP (9.68) with $c^2 = 1$, initial profile: e^{-40x^2}, initial velocity: 0

It is quite likely that your first encounter with graphs that looked like waves were sines and cosines, which are periodic, so when investigating a model that describes waves, one could argue using periodic boundary conditions are of interest. Equipping IBVP (9.67) with periodic BCs leads us to the IBVP

$$\begin{cases} \frac{\partial^2}{\partial t^2} z(x,t) - c^2 \frac{\partial^2}{\partial x^2} z(x,t) = 0, \; 0 < x < L, \, t > 0, \\ z(x,0) = z_0(x), \; \frac{\partial z}{\partial t}(x,0) = z_1(x), \; 0 < x < L, \\ z(0,t) = z(L,t), \, t > 0. \end{cases} \tag{9.70}$$

iii.) Repeat (i) using periodic boundary conditions. Feel free to choose any of the initial condition or velocity options. (If you choose the step function option, do not expect a very smooth solution.) Remember, your goal is to make observations regarding the effect of changing the propogation constant and the initial velocity.

iv.) How did solutions of the wave equation differ from those of the heat equation? A parameter was affixed to the second spatial derivative in both models, but their solutions looked quite different. Why do you think this is the case?

■

EXPLORE!

9.22 Long-Term Behavior

In this **EXPLORE!** we study long-term behavior of solutions to the wave equation. Figure 9.38 appears to exhibit a pattern, but we need to investigate more carefully.

MATLAB-Exercise 9.3.2. Open **MATLAB** and in the command line, type:

MATLAB_Wave_Equation_1D

Use the GUI to answer the following questions.

i.) Solve (9.67) up to $t = 3$ when $c^2 = 2$, sine is the initial condition option, and zero is the initial velocity option. Describe the evolution of the wave. Does a pattern emerge? If so, how would you describe that pattern? If not, allow the GUI to run until $t = 5$. Then, reassess.

  a.) What happens if you change the initial velocity option to 1? Does that evolution change? Does the pattern change?

  b.) Repeat (i) and (i)-(a) using the exponential initial condition option.

ii.) Investigate the long-term behavior of (9.67) by repeating (i) and all its parts beginning with the cosine initial condition option.

iii.) Investigate the long-term behavior of (9.70) by repeating (i) and all its parts beginning with the cosine initial condition option.

iv.) Compare and contrast the long-term behavior of the wave equation and the heat equation. What can you expect from each of the models in the long-term?

v.) Explore whether or not using the step function option for the initial condition changes any of the patterns you described above for each of the three boundary conditions. The choice of initial velocity is left up to you.

■

EXPLORE!

9.23 Continuous Dependence on Parameters

MATLAB-Exercise 9.3.3. To explore continuous dependence with the GUI we must first solve the wave equation with a particular propagation constant, initial condition, and initial velocity. Once a solution has been constructed, click the *Perturb Parameters and Solve* button to specify the perturbed propogation constant and a perturbation size for the initial position and velocity. For simplicity, the GUI assumes the perturbation size is the same for both the initial position and velocity. A screenshot of inputting parameters into the GUI can be found in Figure 9.40 and others of particular GUI outputs in Figures 9.41 and 9.42. Use the GUI to answer the following questions.

i.) Use the sup-norm and $\mathbb{L}^2$-norm to quantifiy size in your investigation of the dependence of the solution of the wave equation on the propogation constant, c^2, and the initial conditions. When prompted click on the *sup/L2* button.

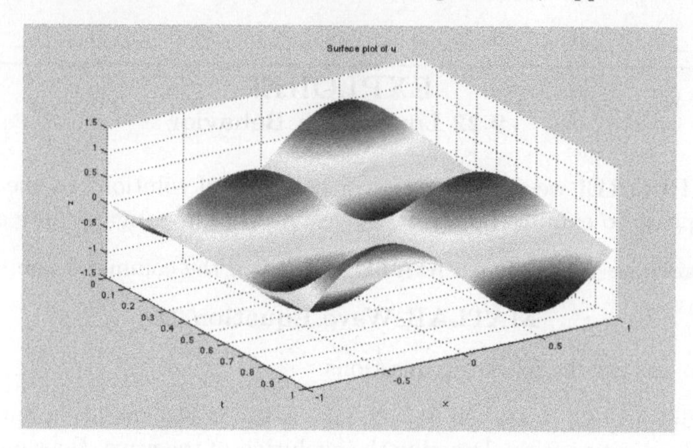

FIGURE 9.39: IBVP (9.70) with $c^2 = 3$, initial profile: $\sin \pi x$, initial velocity: 0

a.) Consider (9.67) and plot solutions until $t = 0.5$. Allow the GUI to run for $c^2 = 3$, initial position option set to sine, and initial velocity set to zero. Use the *Perturb Parameters and Solve* button to select a perturbation size of 0.5 and keep the propagation constant equal to 3. Were the differences 'small' ? Did the different norms produce different results?

b.) Repeat (a) with perturbation sizes of $0.25, 0.1, 0.01$, and 0.005. Make a conjecture regarding continuous dependence of the solution on the initial conditions.

c.) Test your conjecture by repeating (a) and (b) with a non-zero initial velocity and then again with the exponential initial position option. Do you need to alter your conjecture? If so, how?

d.) Consider (9.68) and plot solutions until $t = 0.5$. Allow the GUI to run for $c^2 = 3$, initial position option set to cosine, and initial velocity set to zero. Use the *Perturb Parameters and Solve* button to select a perturbed propagation constant of 3.1 and a perturbation size for the initial conditions equal to zero. Were the differences 'small'? Did the different norms produce different results?

e.) Repeat (d) with propagation constants of $2.9, 2.95, 2.99$, and 3.01. Make a con-

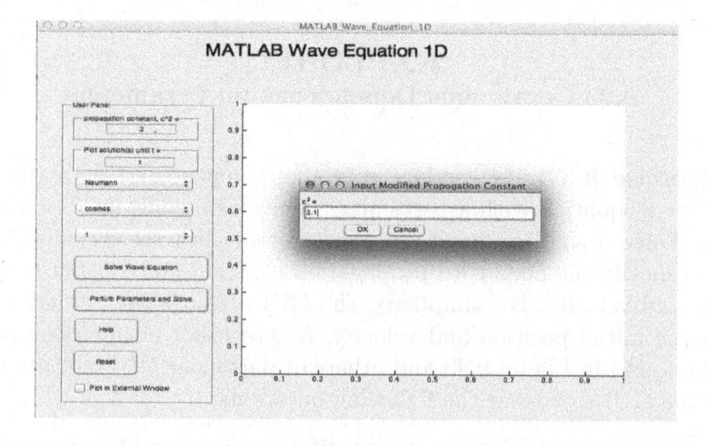

FIGURE 9.40: Perturbing Parameters and Initial Conditions with MATLAB Wave Equation 1D GUI

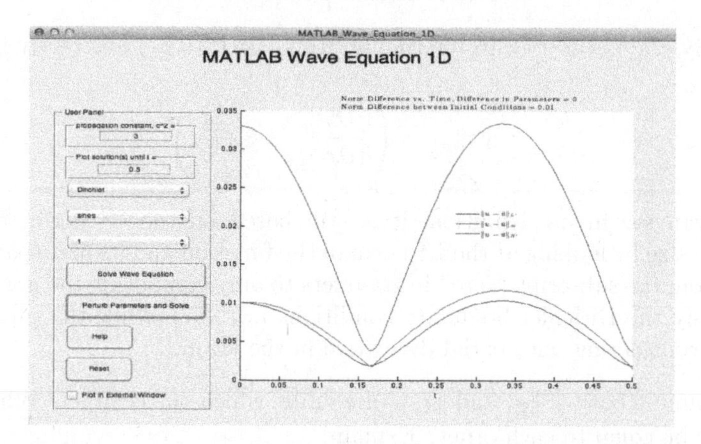

FIGURE 9.41: Perturbing the Initial Conditions only in (9.67)

jecture regarding continuous dependence of the solution on the propogation constants.

f.) Test your conjecture by repeating (d) and (e) with a non-zero initial velocity and then again with the exponential initial position option. Do you need to alter your conjecture? If so, how?

g.) Consider (9.70) and plot solutions until $t = 0.5$. Allow the GUI to run for $c^2 = 3$, initial position option set to cosine, and initial velocity set to zero. Use the *Perturb Parameters and Solve* button to select a perturbed propagation constant of 3.1 and a perturbation size for the initial conditions equal to 0.5. Were the differences 'small' ? Did the different norms produce different results?

h.) Repeat (g) with propagation constants of $2.95, 2.99$, and 3.01 and perturbation sizes of 0.1, 0.05, and 0.01. Make a conjecture regarding continuous dependence of the solution on the propagation constants and the initial conditions. Then, test your conjecture by repeating (g) and (h) with a non-zero initial velocity and then again with the exponential initial position option. Do you need to alter your conjecture? If so, how?

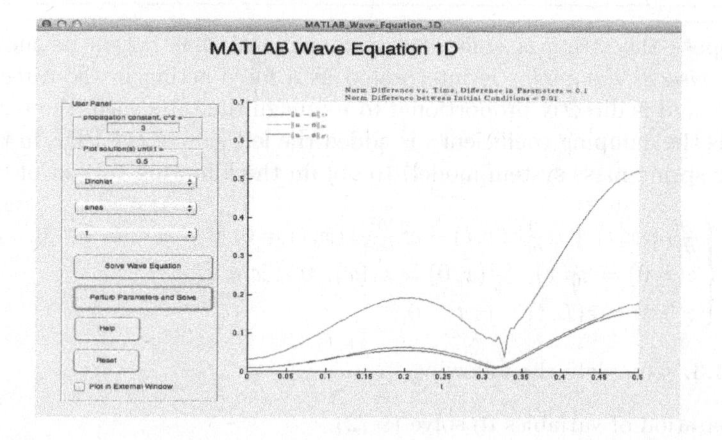

FIGURE 9.42: Perturbing both the Initial Conditions and the Propagation Constant in (9.67)

ii.) Ultimately, the norm used to reformulate this IBVP (9.67) abstractly is given by

$$\|z\|_{\mathbb{H}_0^1} = \left(\left\| \frac{\partial z}{\partial x} \right\|_{\mathbb{L}^2}^2 + \|z\|_{\mathbb{L}^2}^2 \right)^{\frac{1}{2}} \tag{9.71}$$

As you can see in its definition, it is still based on the $\mathbb{L}^2$ norm, but this norm quantifies size by looking at the $\mathbb{L}^2$-norm of the function and its first spatial derivative. In notation, the subscript "zero" in $\mathbb{H}_0^1$ refers to only computing the norm of functions that satisfy the Dirichlet boundary conditions in (9.67), while the superscript "one" refers to considering one spatial derivative in the norm.

a.) Assuming both $\|z\|_{\mathbb{H}_0^1}$ and $\|z\|_{\mathbb{L}^2}$ are finite, which one is larger? Why? Could they ever be equal to each other? Explain.

b.) Define a norm that quantifies size of functions that satisfy Neumann boundary conditions by considering the $\mathbb{L}^2$ norm of the function and its first spatial derivative. (Hint: It should look exactly the same as the right-side of (9.71).)

c.) Define a norm that quantifies the size of functions satisfying periodic boundary conditions by considering the $\mathbb{L}^2$ norm of the function and its first spatial derivative.

iii.) Explain why Figure 9.41 provides experimental evidence for continuous dependence of the solution on the initial conditions. Furthermore, why does Figure 9.42 suggest that the model does not depend continuously on small perturbations of the propagation constant?

iv.) Repeat (i) using the appropriate $\mathbb{H}^1$-norm defined in (ii). When prompted by the GUI to select a norm, choose *All* so you can compare the different norms in each part of (i). The GUI will automatically apply the appropriate $\mathbb{H}^1$ based on the boundary condition selected.

v.) Summarize your observations. Be careful to clearly specify which norm you are using when describing a phenomena.

■

Next, suppose the string is embodied in a material that resists its movement. Such resistance, or *viscous damping* , is interpreted as a force acting in the direction opposite to the velocity and is directly proportional to its magnitude. As such, the term $\alpha\frac{\partial}{\partial t}z(x,t)$, where $\alpha > 0$ is the damping coefficient , is added the left side of the PDE in the IBVPs (as we did for the spring-mass system model) to obtain the following variant of (9.67):

$$\begin{cases} \frac{\partial^2}{\partial t^2}z(x,t) + \alpha\frac{\partial}{\partial t}z(x,t) - c^2\frac{\partial^2}{\partial x^2}z(x,t) = 0, \ 0 < x < L, \, t > 0, \\ z(x,0) = z_0(x), \ \frac{\partial z}{\partial t}(x,0) = z_1(x), \ 0 < x < L, \\ z(0,t) = z(L,t) = 0, \, t > 0, \end{cases} \tag{9.72}$$

Exercise 9.3.3. Complete the following.

i.) Use separation of variables to solve (9.72).

ii.) Repeat (i) for the variant of (9.72) obtained by replacing Dirichlet BCs by Neumann BCs.

EXPLORE!

9.24 Parameter Play!

In this **EXPLORE!** you will investigate the effects of accounting for dampening in the wave model. To do this, you will use the GUI titled *MATLAB Wave Equation Damped 1D* to visualize solutions to IBVP (9.72) and its Neumann and periodic BCs counterparts for different choices of α.

MATLAB-Exercise 9.3.4. To begin, open **MATLAB** and in the command line, type:

MATLAB_Wave_Equation_Damp_1D

Use the GUI to answer the following questions. Make certain to click on the HELP button for instructions on how to use the GUI. A typical screenshot of the GUI can found in Figure 9.43.

i.) Run the GUI with $c^2 = 2$, $t = 1$, $\alpha = 5$, Dirichlet boundary conditions and choose the sine option for the initial condition and the zero option for the initial velocity. The GUI will play display a movie twice that compares the damped wave equation to the undamped wave equation. Once the movies end, a separate figure will display the surfaces of the two solutions in time and space.

   a.) Run the GUI again, but this time with $\alpha = 1$. To do this, simply click the box containing the dampening coefficient and replace 2 with 1. Then, click *Solve.*

   b.) What did you observe in (a)? How would describe the effect of the dampening term? Compare the dynamics of the damped solution to the undamped solution.

   c.) Repeat (a) and (b), but now change the initial velocity option to 1.

   d.) Based upon your observations, how does α affect the motion of the waves? How did the non-zero initial condition affect the solution?

   e.) Test your observations by using the exponential choice for the initial condition, propagation constants equal to 2, dampening coefficients larger and smaller than 5, and different initial velocities. Articulate the evolution of the waves and how those surfaces were affected by the dampening coefficient.

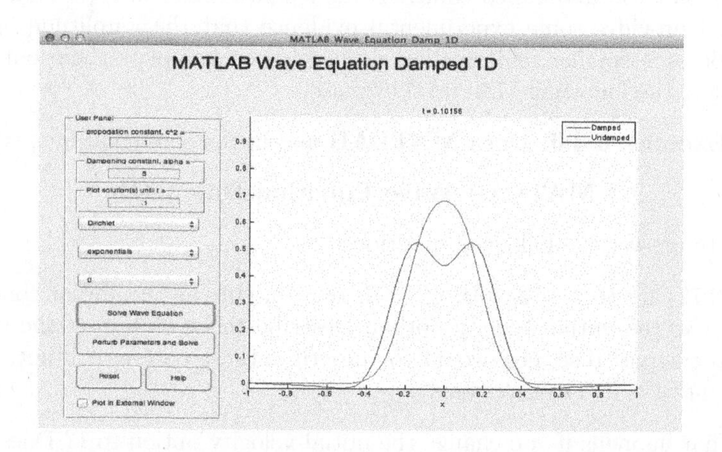

FIGURE 9.43: MATLAB Wave Equation Damp 1D GUI

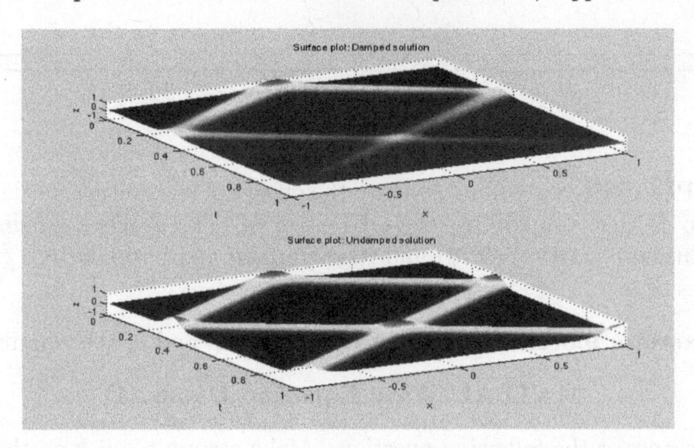

FIGURE 9.44: A Wave Equation Comparison, $c^2 = 3$, $\alpha = 5$, initial position: e^{-40x^2}, initial velocity: 0

ii.) Repeat (i), but this time choose Neumann boundary conditions and the cosine option for the initial condition. Use initial velocity options of zero and one as in (i).

iiii.) Repeat (i) using periodic boundary conditions. Feel free to choose any of the initial position or velocity options. (If you choose the step function option, do not expect a very smooth solution.) Remember, your goal is to make observations regarding the effect of changing the dampening coefficient.

■

EXPLORE!

9.25 Long-Term Behavior

Figure 9.44 suggests that the long-term behavior of the damped wave equation is completely different than the undamped equations investigated in **EXPLORE! 9.22**. There, we discovered that the undamped solutions tended to remain in a periodic motion. **EXPLORE!** 9.24 provides some experimental evidence that the amplitude of the periodic motion will become smaller and smaller, possibly even completely die out. To test this further complete the following MATLAB exercise.

MATLAB-Exercise 9.3.5. Open **MATLAB** and in the command line, type:

MATLAB_Wave_Equation_Damp_1D

Use the GUI to answer the following questions.

i.) Solve (9.72) up to $t = 3$ with $c^2 = 3$, $\alpha = 5$, sine as the initial condition option, and zero as the initial velocity option. Describe the evolution of the two waves. Do patterns emerge? If so, how would you describes those patterns? If not, allow the GUI to run until $t = 5$. Then, reassess.

a.) What happens if you change the initial velocity option to 1? Does the evolution change? More importantly, does the long-term pattern of the damped solution change? If necessary, plot solutions until $t = 5$ and then, reassess.

b.) Repeat (i) using the exponential initial condition option.

ii.) Investigate the long-term behavior of the damped wave equation with Neumann boundary conditions by repeating (i) and all its parts beginning with the cosine initial condition option.

iii.) Investigate the long-term behavior of the damped wave equation with periodic boundary conditions by repeating (i) and all its parts beginning with the cosine initial condition option.

iv.) Based on your observations in **EXPLORE!** 9.24 and above how will reducing the value of the dampening coefficient affect the long-term behavior of the damped wave equation?

v.) Test your answer in (iv) by repeating (i)-(iii) using α =1.

vi.) Explore whether or not the propogation constant effects the patterns observed in (i)-(v). (Hint: Follow the steps and ideas as in this **EXPLORE!** and those found in **EXPLORE!** 9.21.)

■

EXPLORE!
9.26 Continuous Dependence on Parameters

MATLAB-Exercise 9.3.6. As in previous continuous dependence **EXPLORE!s**, one must first solve the model with particular model parameters and initial conditions before using the *Perturb Parameters and Solve* button to specify the perturbed propogation constant and the perturbation size for the initial conditions. For simplicity, the GUI assumes the perturbation size is the same for both the initial position and velocity. A screenshot of inputting parameters into the GUI can be found in Figure 9.45 and another of a particular GUI outputs in Figure 9.46. Use the GUI to answer the following questions. When prompted to choose a norm select *All.*

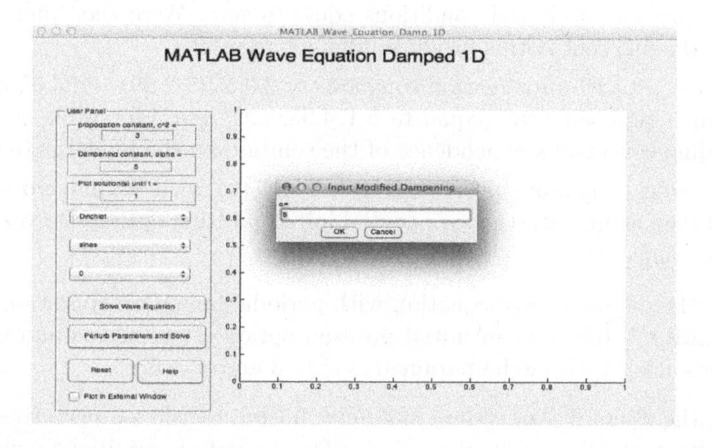

FIGURE 9.45: Perturbing Parameters and Initial Conditions with MATLAB Wave Equation Damped 1D GUI

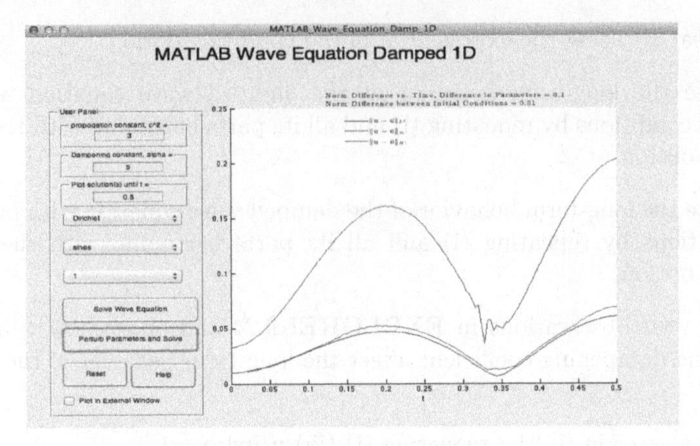

FIGURE 9.46: Perturbing Initial Conditions and Parameters

i.) Consider (9.72) and plot solutions until $t = 0.5$. Allow the GUI to run for $c^2 = 3$, $\alpha = 5$, initial position option set to sine, and initial velocity set to zero.

a.) Use the *Perturb Parameters and Solve* button to select a perturbation size of 0.5 and keep the propagation constant equal to 3 and the dampening coefficient equal to 5. Were the differences 'small' ? Did the different norms produce different results?

b.) Repeat (a) with perturbation sizes of $0.25, 0.1, 0.01$, and 0.005. Make a conjecture regarding continuous dependence of the solution on the initial conditions.

c.) Test your conjecture by repeating (a) and (b) with a non-zero initial velocity and then again with the exponential initial position option. Do you need to alter your conjecture? If so, how?

ii.) Consider the damped wave equation with Neumann boundary conditions and plot solutions until $t = 0.5$. Use an initial position option set to cosine and initial velocity set to zero along with model parameters $c^2 = 3$ and $\alpha = 5$.

a.) Use the *Perturb Parameters and Solve* button to select a perturbed propagation constant of 3.1, a perturbed dampening coefficient equal to 4.9, and a perturbation size for the initial conditions equal to zero. Were the differences 'small' ? Did the different norms produce different results?

b.) Repeat (e) with propagation constants of $2.9, 2.95, 2.99$, and 3.01 and perturbed dampening coefficients equal to $5.1, 4.95, 5.05$, and 4.99. Make a conjecture regarding continuous dependence of the solution on the model parameters.

c.) Test your conjecture by repeating (a) and (b) with a non-zero initial velocity and then again with the exponential initial position option. Do you need to alter your conjecture? If so, how?

iii.) Consider the damped wave equation with periodic boundary conditions and plot solutions until $t = 0.5$. Use an initial position option set to cosine and initial velocity set to zero along with model parameters $c^2 = 3$ and $\alpha = 5$.

a.) Use the *Perturb Parameters and Solve* button to select a perturbed propagation constant of 3.1, a perturbation size for the initial conditions equal to 0.5, and keep α equal to 5. Were the differences 'small' ? Did the different norms produce different results?

b.) Repeat (a) with propagation constants of 2.95, 2.99, and 3.01 and perturbation sizes of 0.1, 0.05, and 0.01. Make a conjecture regarding continuous dependence of the solution on the propagation constants and the initial conditions.

c.) Test your conjecture by repeating (a) and (b) with a non-zero initial velocity and then again with the exponential initial position option. Do you need to alter your conjecture? If so, how?

iv.) Compare and contrast your observations with those in **EXPLORE!** 9.23. Be careful to clearly specify which norm you are using when making any comparisons.

■

9.3.2 Two-Dimensional Wave Equations

Generalizing the discussion in Subsection 9.3.1 from a one-dimensional spatial domain to a bounded domain $\Omega \subset \mathbb{R}^N$ with smooth boundary $\partial\Omega$ is not difficult. Indeed, the resulting IBVP (9.67) is given by

$$\begin{cases} \frac{\partial^2}{\partial t^2} z(x,t) - c^2 \triangle z(x,t) = 0, \ x \in \Omega, \ t > 0, \\ z(x,0) = z_0(x), \ \frac{\partial z}{\partial t}(x,0) = z_1(x), \ x \in \Omega, \\ z(x,t) = 0, \ x \in \partial\Omega, \ t > 0. \end{cases} \tag{9.73}$$

In particular, if Ω is a two-dimensional bounded rectangular domain then IBVP (9.73) becomes

$$\begin{cases} \frac{\partial z}{\partial t^2}(x,y,t) = c^2 \left(\frac{\partial^2 z}{\partial x^2}(x,y,t) + \frac{\partial^2 z}{\partial y^2}(x,y,t) \right), 0 < x < a, \ 0 < y < b, \ t > 0, \\ z(x,y,0) = z_0(x,y), \ 0 < x < a, \ 0 < y < b, \\ z(x,0,t) = 0 = z(x,b,t), \ 0 < x < a, \ t > 0, \\ z(0,y,t) = 0 = z(a,y,t), \ 0 < y < b, \ t > 0, \end{cases} \tag{9.74}$$

Changing to Neumann BC in IBVP (9.74) leads to the following variant:

$$\begin{cases} \frac{\partial z}{\partial t^2}(x,y,t) = c^2 \left(\frac{\partial^2 z}{\partial x^2}(x,y,t) + \frac{\partial^2 z}{\partial y^2}(x,y,t) \right), 0 < x < a, \ 0 < y < b, \ t > 0, \\ z(x,y,0) = z_0(x,y), \ 0 < x < a, \ 0 < y < b, \\ \frac{\partial z(x,0,t)}{\partial y} = 0 = \frac{\partial z(x,b,t)}{\partial y}, \ 0 < x < a, \ t > 0, \\ \frac{\partial z(0,y,t)}{\partial x} = 0 = \frac{\partial z(a,y,t)}{\partial x}, \ 0 < y < b, \ t > 0, \end{cases} \tag{9.75}$$

STOP! Why do the IBVPs (9.74) and (9.75) essentially have four conditions along the boundary? Do they seem like natural extensions of the one-dimensional counterparts?

Exercise 9.3.4. Write down a version of the two-dimensional wave equation (9.75) on the rectangle this time equipped with periodic boundary conditions.

EXPLORE!

9.27 Two-Dimensional Parameter Play!

In this **EXPLORE!** you will investigate the effects of changing the parameter c^2 in the two-dimensional wave equation. To do this, you will use the MATLAB Wave Equation 2D GUI to visualize the solutions to IBVP (9.74) , IBVP (9.75), and their periodic boundary conditions counterpart for different choices of c^2.

MATLAB-Exercise 9.3.7. To begin, open **MATLAB** and in the command line, type:

MATLAB_Wave_Equation_2D

Use the GUI to answer the following questions. Make certain to click on the HELP button for instructions on how to use the GUI. A screenshot of inputting parameters into the GUI can be found in Figure 9.47 and another of a particular GUI output in Figure 9.48. Use the GUI to answer the following questions.

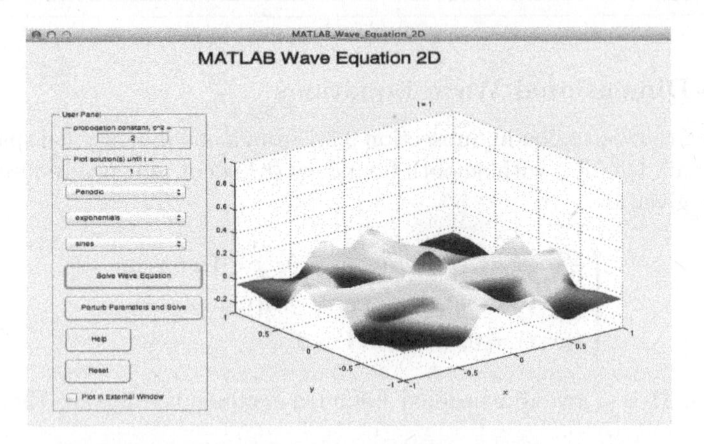

FIGURE 9.47: MATLAB Wave Equation 2D GUI

i.) Solve IBVP (9.74) when $c^2 = 2$, $t = 1$, choose the sine option for the initial condition and the zero option for the initial velocity. The GUI will play a movie of the solution twice then output a figure of the solution at the final time.

a.) Solve IBVP (9.74) again, but keep everything the same except change c^2 to equal 1. To do this, simply click the box containing the propogation constant and replace 2 with 1. Then, click *Solve.*

b.) What did you observe in (a)? Did the motion of the wave seem faster or slower as compared to $c^2 = 2$? Is your answer the same as in **EXPLORE!** 9.21 when you investigated the one-dimensional wave equation?

c.) Repeat (a) and (b), but change the initial velocity option to 1.

d.) Based upon your observations, how does c^2 affect the motion of the waves? How did the non-zero initial condition affect the solution? Compare your observations with those made in **EXPLORE!** 9.21.

e.) Test your observations by using the exponential choice for the initial condition, propagation constants larger and smaller than 2, and different initial velocities. Articulate the evolution of the waves and how those surfaces were affected by the propagation constant. Try to qualitatively explain the changes in the solution when using non-zero initial velocities.

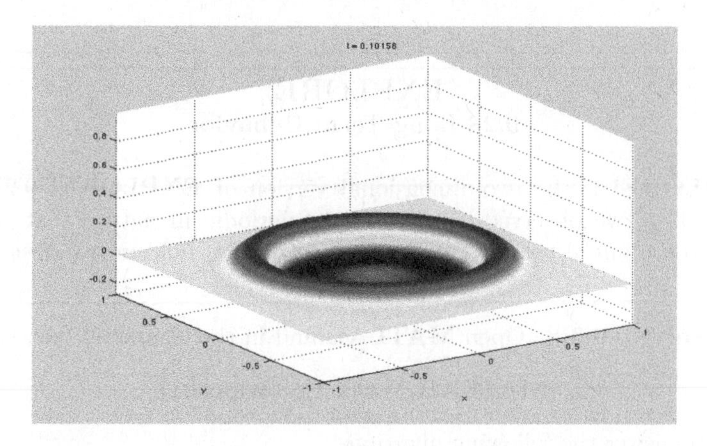

FIGURE 9.48: IBVP (9.74) with $c^2 = 3$, initial position: e^{-40x^2}, initial velocity: 0

ii.) Repeat (i) for IBVP (9.75), but begin with the cosine option for the initial condition. Use initial velocity options of zero and one as in (i).

iiii.) Repeat (i) using periodic boundary conditions. Feel free to choose any of the initial condition or velocity options. (If you choose the step function option, do not expect a very smooth solution, but it is quite beautiful. See Figure 9.49.) Remember, your goal is to make observations regarding the effect of changing the propagation constant and the initial velocity.

■

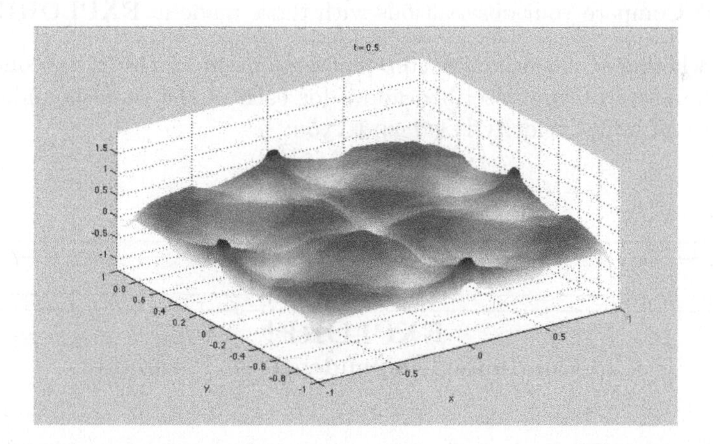

FIGURE 9.49: Two-Dimensional Wave Equation with periodic BCs: $c^2 = 5$; initial position: step function; initial velocity: 0

EXPLORE!

9.28 Long-Term Behavior

This **EXPLORE!** is the two-dimensional version of **EXPLORE!** 9.22. In the one-dimensional setting, we observed solutions were periodic in nature that would possibly translate due to the initial velocity. Will the same pattern hold in two-dimensions? Let us find out.

MATLAB-Exercise 9.3.8. Open **MATLAB** and in the command line, type:

MATLAB_Wave_Equation_2D

Use the GUI to answer the following questions.

i.) Solve IBVP (9.74) when $c^2 = 2$, $t = 3$, choose the sine option for the initial condition and the zero option for the initial velocity. Describe the evolution of the wave. Does a pattern emerge? If so, how would you describe that pattern? If not, allow the GUI to run until $t = 5$. Then, reassess.

  a.) What happens if you change the initial velocity option to *exponential*? Does that evolution change? Does the pattern change?

  b.) Repeat (i) and (i)-(a) using the exponential initial condition option.

ii.) Investigate the long-term behavior of IBVP (9.74) by repeating (i) and all its parts beginning with the cosine initial condition option.

iii.) Repeat (ii) using periodic boundary conditions.

iv.) Did the one-dimensional patterns present themselves for the two-dimensional wave equation? Compare your observations with those made in **EXPLORE!** 9.22.

v.) Explore whether or not using the step function option for the initial condition changes any of the patterns you described above for each of the three boundary conditions. The choice of initial velocity is left up to you.

■

EXPLORE!

9.23 Continuous Dependence on Parameters

MATLAB-Exercise 9.3.9. As in previous continuous dependence explorations, we must first solve the IBVP with particular model parameters and initial conditions. Once a solution has been constructed, click the *Perturb Parameters and Solve* button to specify the perturbed propagation constant and a perturbation size for the initial position and velocity. For simplicity, the GUI assumes the perturbation size is the same for both the initial position and initial velocity. A screenshot of inputting parameters into the GUI can be found in Figure 9.50 and others of particular GUI outputs in Figures 9.51 and 9.52. Use the GUI to answer the following question and when prompted to choose a norm to plot, select *All*.

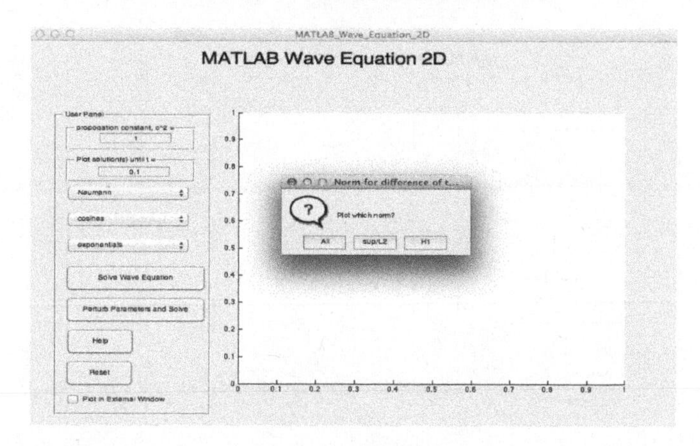

FIGURE 9.50: Perturbing Parameters and Initial Conditions with MATLAB Wave Equation 2D GUI

i.) Consider IBVP (9.74) with the propagation constant set equal to 3. Explore whether or not the solution of this IBVP depends continuously on the initial conditions up to $t = 0.5$.

   a.) Choose the initial position option sine and the initial velocity option zero. When prompted choose a perturbation size of 0.25. Is the difference between solutions "small"? Comment on the differences observed using the different norms.

   b.) Repeat (a) with perturbation sizes of $0.1, 0.01$, and 0.005. Make a conjecture regarding continuous dependence of the solution on the initial conditions.

   c.) Test your conjecture by repeating (a) and (b) with a non-zero initial velocity and then again with the exponential initial position option. Do you need to alter your conjecture? If so, how?

ii.) Consider IBVP (9.75) with the initial position option set to cosine and the initial velocity option set to zero. Explore whether or not the solution of this IBVP depends continuously on the propogation constant up to $t = 0.5$.

   a.) Choose $c^2 = 2$. When prompted select a perturbed propagation constant of 2.1.

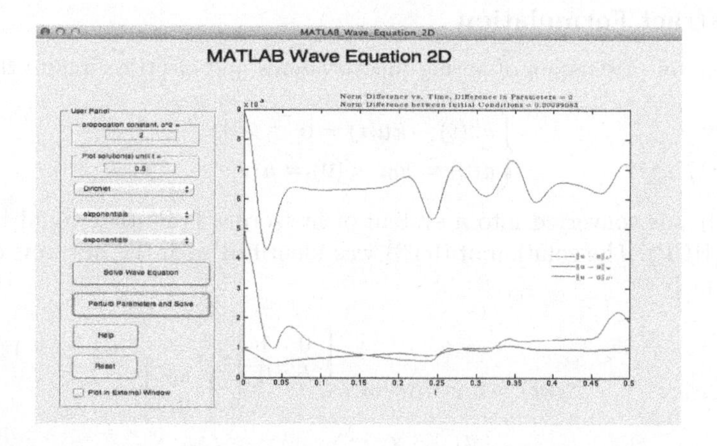

FIGURE 9.51: Perturbing the Initial Conditions only in (9.74)

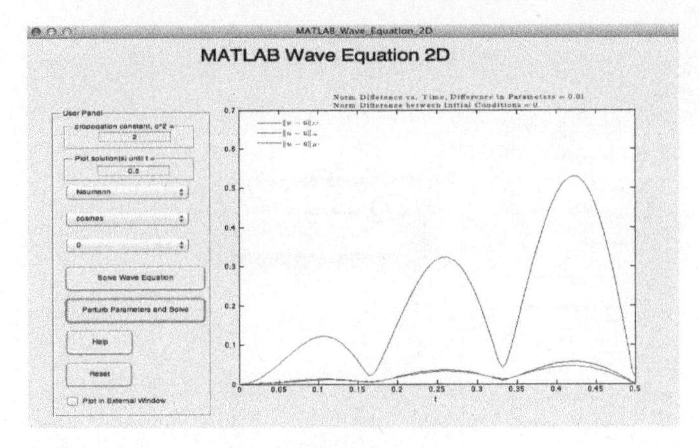

FIGURE 9.52: Perturbing both the Propagation Constant in (9.75)

How do the two solutions differ? Describe the differences based on the different norms.

b.) Repeat (a) with perturbed propagation constants of 1.9, 1.95, and 2.005. Make a conjecture regarding continuous dependence of the solution on the propagation constant.

c.) Test your conjecture by repeating (a) and (b) with a non-zero initial velocity and then again with the exponential initial position option. Do you need to alter your conjecture? If so, how?

iii.) Repeat (ii) with periodic boundary conditions.

iv.) Summarize your observations. Be careful to clearly specify which norm you are using when describing a phenomena. Compare your observations with those made for the one-dimensional wave equation.

■

9.3.3 Abstract Formulation

Recall from our discussion of spring-mass systems and electric circuits that the second order IVP

$$\begin{cases} u''(t) - ku(t) = 0, \ t > 0, \\ u(0) = u_0, \ u'(0) = u_1, \end{cases} \tag{9.76}$$

(where $k > 0$) was converted into a system of first-order IVPs that could be viewed as a special case (HCP). The solution of (9.68) was identified with the first row of the solution of (HCP), namely

$$\begin{aligned} u(t) &= \text{first row of } e^{\begin{bmatrix} 0 & 1 \\ k & 0 \end{bmatrix} t} \begin{bmatrix} u_0 \\ u_1 \end{bmatrix} \\ &= \frac{u_1}{\sqrt{k}} \sin\left(\sqrt{k}t\right) + u_0 \cos\left(\sqrt{k}t\right). \end{aligned} \tag{9.77}$$

Does the approach and ultimate form of the solution extend to handle (9.67)? We argue that (9.67) can be written as the abstract evolution equation of the form (A-HCP) by suitably choosing the state space $\mathcal{H}$ and the operator A. We do this by adapting the approach used in the finite-dimensional case. To this end, applying the change of variable

$$v_1 = z, v_2 = \frac{\partial z}{\partial t}$$
$$\frac{\partial v_1}{\partial t} = v_2, \frac{\partial v_2}{\partial t} = c^2 \frac{\partial^2 v_1}{\partial x^2} \tag{9.78}$$

enables us to express (9.67) as the equivalent system

$$\begin{cases} \frac{\partial}{\partial t}\begin{bmatrix} v_1 \\ v_2 \end{bmatrix}(x,t) = \begin{bmatrix} 0 & I \\ c^2 \frac{\partial^2}{\partial x^2} & 0 \end{bmatrix}\begin{bmatrix} v_1 \\ v_2 \end{bmatrix}(x,t), \ 0 < x < L, \ t > 0, \\ \begin{bmatrix} v_1 \\ v_2 \end{bmatrix}(x,0) = \begin{bmatrix} z_0 \\ z_1 \end{bmatrix}(x,0), \ 0 < x < L, \\ v_1(0,t) = v_1(L,t) = 0, \ t > 0. \end{cases} \tag{9.79}$$

We can deduce possible choices for the unknown u and operator A from this form, but the main difference this time is that the state space $\mathcal{H}$ must be a product space $K_1 \times K_2$ since the unknown is a vector consisting of two components. Looking back at the diffusion model, the choice of $\mathbb{L}^2(0,L;\mathbb{R})$ for the state space worked well from both the mathematical modeling and physical viewpoints when handling the term $-c^2\frac{\partial^2}{\partial x^2}$. As such, this space, or some closed subspace thereof, might be a reasonable choice for K_2. But, what about K_1?

At first glance, some natural choices for K_1 include $\mathbb{L}^2(0,L;\mathbb{R})$ (to mirror the first component) and $\mathbb{W}^{2,1}(0,L;\mathbb{R})$ (since we need at least one time derivative). However, it turns out that neither one of these is suitable for our present purposes. At this point, we must make use of other information inherent to the phenomenon being described; this time, we appeal to underlying physical principles and proceed as in [12]. Indeed, the total mechanical energy at time $t > 0$ given by

$$\frac{1}{2L}\int_0^L \left[\left|\frac{\partial z}{\partial x}(x,t)\right|^2 + |z(x,t)|^2 \right] dx = \frac{1}{2}\left[\left\|\frac{\partial z}{\partial x}(\cdot,t)\right\|_{\mathbb{L}^2}^2 + \|z(\cdot,t)\|_{\mathbb{L}^2}^2 \right] \tag{9.80}$$

is bounded above by the initial energy of the system (and is therefore finite). **(So what?)** As it turns out, the space $\mathbb{H}_0^1(0,L;\mathbb{R})$ equipped with the norm

$$\|z\|_{\mathbb{H}_0^1}^2 \equiv \left\|\frac{\partial z}{\partial x}\right\|_{\mathbb{L}^2}^2 + \|z\|_{\mathbb{L}^2}^2 \tag{9.81}$$

is a Hilbert space. Consequently, the following function space is meaningful:

$$\mathcal{H} = \mathbb{H}_0^1(0,L;\mathbb{R}) \times \mathbb{L}^2(0,L;\mathbb{R}) \tag{9.82}$$
$$\left\langle \begin{bmatrix} v_1 \\ v_2 \end{bmatrix}, \begin{bmatrix} v_1^\star \\ v_2^\star \end{bmatrix} \right\rangle_{\mathcal{H}} \equiv \int_0^L \left[\frac{\partial v_1}{\partial x}\frac{\partial v_1^\star}{\partial x} + v_2 v_2^\star \right] dx.$$

Now, we define the operator $A : \text{dom}(A) \subset \mathcal{H} \to \mathcal{H}$ by

$$A\begin{bmatrix} v_1 \\ v_2 \end{bmatrix} = \begin{bmatrix} 0 & I \\ c^2 \frac{\partial^2}{\partial x^2} & 0 \end{bmatrix}\begin{bmatrix} v_1 \\ v_2 \end{bmatrix} = \begin{bmatrix} v_1 \\ -c^2 \frac{\partial^2 v_2}{\partial x^2} \end{bmatrix} \tag{9.83}$$
$$\text{dom}(A) = \left(\mathbb{H}^2(0,L;\mathbb{R}) \cap \mathbb{H}_0^1(0,L;\mathbb{R})\right) \times \mathbb{H}_0^1(0,L;\mathbb{R}).$$

At this point, we comment that the BCs must be incorporated into the definition of dom(A) in a careful manner. The technical details can be more fully understood after taking a functional analysis course.

Exercise 9.3.5. Answer the following questions.

i.) How would the above discussion change if the Dirichlet BCs were replaced by Neumann BCs?

ii.) Work through the above discussion for (9.72). Show as many details as possible.

Transforming the multi-dimensional wave equation (9.73) into a system comparable to (9.79) amounts to using the more general matrix operator $\begin{bmatrix} 0 & I \\ c^2\triangle & 0 \end{bmatrix}$ and subsequently replacing every occurrence of the interval $(0, L)$ by Ω. The resulting function spaces are Hilbert spaces. Of course, the detail-checking becomes more involved, but the process closely resembles the one used in the one-dimensional setting. Summarizing, let $\Omega \subset \mathbb{R}^N$ be a bounded domain with smooth boundary $\partial\Omega$ and define the space $\mathcal{H} = \mathbb{H}_0^1(\Omega;\mathbb{R}) \times \mathbb{L}^2(\Omega;\mathbb{R})$. Then, we define the operator $A : \text{dom}(A) \subset \mathcal{H} \to \mathcal{H}$ by

$$\begin{aligned} A &= \begin{bmatrix} 0 & I \\ c^2\triangle & 0 \end{bmatrix} \\ \text{dom}(A) &= \left(\mathbb{H}^2(\Omega;\mathbb{R}) \cap \mathbb{H}_0^1(\Omega;\mathbb{R})\right) \times \mathbb{H}_0^1(\Omega;\mathbb{R}) \end{aligned} \tag{9.84}$$

and in so doing, are able to capture (9.73) as a special case of the same abstract evolution equation.

9.4 An Informal Recap: A First Step Toward Unification

Looking back through the models studied in this chapter reveals some commonalities among seemingly unrelated models. A similar observation was made in Part I and ultimately led to the formulation of an elegant theory for (HCP). Does the same happen here? It is too early to tell for certain, but we can make some important structural connections to Part I. Later, we will determine if the differences from Part I can be overcome.

First and foremost, the IBVPs discussed so far can be reformulated as an abstract evolution equation of the form

$$\begin{cases} (Bu)'(t) = Au(t), \ t > 0, \\ u(0) = u_0. \end{cases} \tag{9.85}$$

in some Hilbert space $\mathcal{H}$ by making judicious choices of the operators $A : \text{dom}(A) \subset \mathcal{H} \to \mathcal{H}$ and $B : \text{dom}(B) \subset \mathcal{H} \to \mathcal{H}$. In fact, for most part, we were able to choose the operator B to be the identity operator, thereby really making (9.85) resemble (HCP).

Arguably, making the judicious choice of the space $\mathcal{H}$ is the least intuitive part of the process, and unlike our experience in Part I, where the space was always $\mathbb{R}^N$, the initial selection of the space $\mathcal{H}$ changed to accomodate different spatial dimensions and smoothness requirements, and the actual choices of the operators A and B were defined on appropriate subspaces of $\mathcal{H}$ guided behind the scenes by the underlying physical setting of the problem.

Suffice it to say that the process of fitting these IBVPs into the abstract mold (9.85) is somewhat more involved than it was in Part I. This is expected, though, because we are converting a partial differential equation (in which the functions depend on more than one variable) into an ODE (in which there can **only be** one variable). So, the metamorphosis is more intricate.

Ultimately, we were able to make the IBVPs fit the same mold, which is an essential first step. Indeed, we can now embark on the second leg of the journey of formulating a theory for (9.85). For many of the IBVPs we considered, the separation of variables technique was able to be used to produce a rather complicated function (defined as an infinite series) that satisfied all components of the IBVP. In effect, this "showed" existence of a solution to each IBVP, but it speaks nothing to uniqueness of a solution. Admittedly, too, constructing these solutions was nowhere near as elegant as using Putzer's algorithm when constructing a solution of (HCP) in Part I. In fact, at first glance, we might be tempted to think that the solutions across models do not involve a quantity that plays the role of $e^{\mathbf{A}t}$. If this were true, this would be very bad news because the solution of (HCP) was given by $e^{\mathbf{A}t}u_0$, and we are trying to emulate the development of that theory in the present setting for (9.85). However, look back at the exercises in this chapter. At least for the most rudimentary versions of each of the IBVPs studied, the solution was written in the form $z(x,t) = S(t)\,[z_0]\,(x)$ satisfying the following properties:

1. $S(0)[v](x) = v(x)$
2. $S(t_1 + t_2)\,[v](x) = S(t_1)\,(S(t_2)\,[v])\,(x)$
3. $S(t)\,[v_1 + v_2](x) = S(t)\,[v_1](x) + S(t)\,[v_2](x)$

The first two of these properties resemble the following properties of $e^{\mathbf{A}t}$, when $\mathbf{A}$ was a square matrix:

1. $e^{\mathbf{A}(0)}\,[v] = v$
2. $e^{\mathbf{A}(t_1+t_2)}\,[v] = e^{\mathbf{A}t_1}\left(e^{\mathbf{A}t_2}\,[v]\right)$

So, it seems that there *is* some hidden quantity playing the role of $e^{\mathbf{A}t}$ here after all! This is good news, even though we have not yet nailed down the details. And, this begs more questions than it answers. Does the hidden quantity being identified as $S(t)$ behave in exactly the same manner as $e^{\mathbf{A}t}$? How, exactly, does $S(t)$ change when A, B, and $\mathcal{H}$ change? Can $S(t)$ be defined in the same way we defined $e^{\mathbf{A}t}$? Etc. This is a very rich body of material. Our first main goal is to establish a result analogous to Theorem 4.6.1, which, loosely put, says the following:

Pseudo-Theorem 1. As long as the initial condition is in the correct space and $\{e^{At} : t \geq 0\}$ exists, (9.85) has a unique classical solution given by $u(t) = e^{At}u_0$.

Here, we interpret $u(t)$ as an $\mathcal{H}$-valued function. Remember, $u(t)" = " \{z(x,t) : x \in \Omega\}$. So, we still have the connection back to the original PDE; it is simply buried in the space $\mathcal{H}$. We will address this more thoroughly later.

Being optimistic and assuming the above result holds, we next consider results concerning the continuous dependence of solutions on initial conditions and parameters. You conducted several MATLAB **EXPLORE!** activities focused on this issue. Ultimately, these results can be classified into two categories, depending on what pieces of data occurring in the IBVP are tweaked. Looking back at Section 4.8, we find that the first result involved characterizing the impact of tweaking only the initial condition in the IVPs. There, we observed that the solutions u and u_ϵ to the first-order IVPs

$$\begin{cases} u'(t) &= au(t), \\ u(0) &= u_0 \end{cases} \tag{9.86}$$

and

$$\begin{cases} u_\epsilon'(t) &= a u_\epsilon(t), \\ u_\epsilon(0) &= u_0 + \epsilon \end{cases} \tag{9.87}$$

were "close" when $|\epsilon|$ was small; that is,

$$\max\{|u(t) - u_\epsilon(t)| : 0 \le t \le T\} \to 0 \text{ as } |\epsilon| \to 0. \tag{9.88}$$

A more general result accounting for perturbations of both the parameter and IC was then established. Specifically, consider (9.86) together with the perturbed IVP

$$\begin{cases} u_{\delta\epsilon}'(t) &= (a+\delta)\, u_{\delta\epsilon}(t), \\ u_{\delta\epsilon}(0) &= u_0 + \epsilon \end{cases} \tag{9.89}$$

We argued that

$$\max\{|u(t) - u_{\delta\epsilon}(t)| : 0 \le t \le T\} \to 0 \text{ as } |\delta|, |\epsilon| \to 0. \tag{9.90}$$

Systems versions of these results that applied directly to (HCP) for any square matrix **A** were also formed, and the only significant change was replacing the norm used in (9.88) and (9.90) by the $\mathbb{R}^N$-norm.

Now, refer back to **EXPLORE!** 9.1, 9.6, and 9.8 for the heat equation, as well as the corresponding **EXPLORE!** activities for the other two models. We are really exploring a similar phenomena as above for IBVPs, right? Let us be a bit more deliberate about the similarity for a simple example. Consider the classical heat equation

$$\begin{cases} \frac{\partial}{\partial t} z(x,t) = k \frac{\partial^2}{\partial x^2} z(x,t), \ 0 < x < a, \ t > 0, \\ z(x,0) = z_0(x), \ 0 < x < a, \\ z(0,t) = z(a,t) = 0, \ t > 0 \end{cases} \tag{9.91}$$

Recall that $k = \frac{K}{\lambda c}$, where K, λ, c are physical constants obtained experimentally. Tweaking any of these three parameters will, in turn, result in a perturbation of the constant k appearing in (9.91). For instance, suppose we replace K by $K + \varepsilon$, for some $\varepsilon > 0$. The resulting constant then appearing in (9.91) would be $k_\varepsilon = \frac{K+\varepsilon}{\lambda c}$. Substituting this into (9.91) yields the following perturbed IBVP:

$$\begin{cases} \frac{\partial}{\partial t} z_\epsilon(x,t) = k_\varepsilon \frac{\partial^2}{\partial x^2} z_\epsilon(x,t), \ 0 < x < a, \ t > 0, \\ z_\epsilon(x,0) = z_0(x), \ 0 < x < a, \\ z_\epsilon(0,t) = z_\epsilon(a,t) = 0, \ t > 0 \end{cases} \tag{9.92}$$

We are interested in knowing how "close" the solutions of (9.91) and (9.92) are. Specifically, in the spirit of (9.88), we would like to show that

$$\max\{|z(x,t) - z_\epsilon(x,t)| : 0 \le t \le T, \ 0 \le x \le L\} \to 0 \text{ as } |\epsilon| \to 0. \tag{9.93}$$

For each individual IBVP, you studied the difference in (9.93) for different values of ε and concluded that, indeed, (9.93) held. This illustrates the fact that the benefit of establishing a more general result is that the moment an IBVP could appropriately be reformulated as (9.85) such that the result of Pseudo-Theorem 1 held, then we could draw the same conclusion without all the fuss. To this end, we need to understand how tweaking a parameter in a **PDE** of an IBVP impacts the transformation of that IBVP as an abstract evolution equation (9.85).

Recall that (9.91) was reformulated in the form of (9.85) by using the identity operator for B, and $A : \text{dom}(A) \subset \mathbb{L}^2(0,a;\mathbb{R}) \to \mathbb{L}^2(0,a;\mathbb{R})$ by

$$A[f] = k\frac{d^2}{dx^2}[f], \tag{9.94}$$
$$\text{dom}(A) = \left\{ f \in \mathbb{L}^2(0,a;\mathbb{R}) \,\middle|\, \exists \frac{df}{dx}, \frac{d^2 f}{dx^2} \in \mathbb{L}^2(0,L;\mathbb{R}), \text{ and } f(0) = f(a) = 0 \right\}.$$

But, how about IBVP (9.92)? It appears that the only difference between (9.91) and (9.92) is that the constant k is replaced by k_ε. In turn, this change in constant only impacts the definition of the operator A. **(Check this!)** As such, reformulating (9.92) in the form (9.85) simply requires that we use the following operator in place of A:

$$A_\varepsilon[f] = k_\varepsilon \frac{d^2}{dx^2}[f]$$

The same domain as the one used in (9.94) works for A_ε because constant multiples do not affect differentiability or integrability. So, (9.92) is reformulated as

$$\begin{cases} u_\varepsilon'(t) = A_\varepsilon u_\varepsilon(t), \ t > 0, \\ u_\varepsilon(0) = u_0. \end{cases} \tag{9.95}$$

in the same space $\mathcal{H}$ as (9.85). (This fact is essential since otherwise we could not study the difference between the two solutions since the norm of the difference would not be defined.)

This produces a very similar situation to the one studied in Part I. We ask the question, "How close are the solutions u and u_ε of these two abstract evolution equations?" It seems natural to simply use (9.88), but remember that $u(t)$ and $u_\varepsilon(t)$ are not real numbers this time; they are members of $\mathcal{H}$. This is similar to the hurdle we faced when extending the result from first-order IVPs to systems of IVPs in Part I. There, a simple change of norm to account for the change of setting was all that was needed. Is that all it takes here? Does simply replacing the absolute value by the $\mathcal{H}$-norm do the job? More precisely, is the correct formulation of our question, "Does $\max\{\|u(t) - u_\epsilon(t)\|_{\mathcal{H}} : 0 \le t \le T\} \to 0$ as $|\epsilon| \to 0$?"

You actually considered this question in **EXPLORE!** 9.3 for different choices of $\mathcal{H}$. **(Look back!)** A similar discussion when the initial condition is tweaked, and then when both the initial condition and parameters affecting the operator A are perturbed, holds. **(Convince yourself!)** It is reasonable to believe that this same abstract formulation should work for any of the IBVPs encountered thus far. From the vantage point of developing a coherent theory for (9.85), this is reassuring. However, we should not get lulled into a false sense of security. There is something to actually prove! Look back at the results in Section 4.8. They all involve estimates involving $e^{\mathbf{A}t}$, and in the current setting, we do not even know what plays the role of $e^{\mathbf{A}t}$, not to mention the properties it satisfies. So, even this part of the theory hinges on identifying and understanding the quantity $S(t)$.

Next, some of the **EXPLORE!** activities focused on long-term behavior of solutions of the IBVPs as $t \to \infty$. Visually, we focused on what happened to the solution "surface" as $t \to \infty$. Specifically, we would like to know how the level curves of the surface (in the time variable) evolve over time. For some of the models, we observed that the level curves approached zero uniformly throughout the spatial domain, while others exhibited periodic-like behavior over time. Looking back at Section 4.8, the long-term behavior seemed to, not surprisingly, depend on the nature of the operator A. This, in turn, impacted the nature of $e^{\mathbf{A}t}$ which itself, in turn, governs the evolution of the solution. Probing the result a bit deeper stills reveals that the nature of the eigenvalues of the matrix $\mathbf{A}$ dictated the long-term behavior of the solution. For the IBVPs now under consideration, the operators A are no longer matrices. So, what plays the role of the eigenvalues? Or, does some other feature of A play this role? We will investigate this more carefully later.

Finally, throughout the **EXPLORE!** activities, we studied all of the above results for different choices of boundary conditions. These arose as part of the modeling process of the various phenomena. The three main types of BCs used were of Dirichlet type, Neumann type, and periodic type. Always when reformulating the IBVPs as abstract evolution equations, the BCs were subsumed into the domain of the operator A. Technically, we must be careful when doing this because the domain of A must be a closed subspace of $\mathcal{H}$. Moreover, the class of functions from which we can choose the initial profile u_0 might be impacted by such a change. These are subtle details we will address later. Once done, such changes in the BCs do not interfere with the formulation of the above results.

Chapter 10

Formulating a Theory for (A-HCP)

Overview

We have encountered a large collection of IBVPs arising in the mathematical modeling of very different phenomena that can be all be reformulated as a single abstract evolution equation of the form (A-HCP), or a closely-related variant. Such commonality prompts us to determine if a theory analogous to Part I could be formulated to study all of these problems under the same theoretical umbrella. At the end of Chapter 9, we made some preliminary observations that gave us hope that such a theory could be developed. We make these observations more formal in this chapter.

10.1 Introducing (A-HCP)

As observed in Chapter 9, IBVPs consisting of PDEs often can be written in a more elegant form by identifying the unknown as a function-valued function $u(t)$, meaning that the outputs are elements of a Hilbert space of functions; suitably identifying operators $(A, \text{dom}(A))$ and $(B, \text{dom}(B))$to apply to $u(t)$ to capture the linear interactions amongst the terms in the PDEs; and identifying the initial condition. The nature of the abstract IVPs used to describe the scenarios and the questions posed are strikingly similar. Formally, we have the following form.

Definition 10.1.1. An abstract evolution equation of the form

$$(\mathbf{A-HCP}) \begin{cases} \frac{d}{dt}\left(u(t)\right) = A\left(u(t)\right),\ t > t_0, \\ u(t_0) = u_0, \end{cases} \tag{10.1}$$

in a Hilbert space $\mathcal{H}$, where $u : [t_0, \infty) \to \mathcal{H}$ is the unknown function, $A : \text{dom}(A) \subset \mathcal{H} \to \mathcal{H}$ is an appropriately-defined linear operator (like $\frac{d^2}{dx^2}$, $\triangle + I$, etc.), and $u_0 \in \mathcal{H}$ is a function describing the initial profile, is called an **abstract homogeneous Cauchy problem**, denoted (A-HCP).

What do we mean by a solution of (A-HCP)? This is actually somewhat of a loaded question. There is actually more than one way to define meaningful solutions to such a problem. The one with which we are most familiar resembles the definition of a classical solution in Part I, which basically says, "Plug in the function and if all parts are satisfied, then it is a solution." That is basically what we did in the PDEs studied in Chapter 9. Let us interpret this notion for (A-HCP) step-by-step.

If a function $u(t)$ were to satisfy (A-HCP) in a naïve sense, it would need to be differentiable on (t_0, ∞) (because $\frac{d}{dt}(u(t))$ occurs on the left-side of the equation); $u(t) \in \text{dom}(A)$, for all $t > t_0$ (because $Au(t)$ occurs on the right-side of the equation), and $u(t)$ would actually need to satisfy both equations in (A-HCP). These are all natural conditions, right? There is one small subtle addition, namely that we want the transition from $u(t_0)$ to $u(t)$ (for times t close to t_0) to be relatively smooth. Since we cannot have a two-sided derivative at t_0, we shall simply assume that $u(t)$ is right continuous at t_0.

Stop! Look back at Part I. Where did this issue arise for (HCP)?

Precisely, we have the following definition:

Definition 10.1.2. A *classical solution* of (A-HCP) is a function $u : [t_0, \infty) \to \mathcal{H}$ satisfying the following:

i.) $u(t)$ is continuous on $[t_0, \infty)$,
ii.) $u(t)$ is differentiable on (t_0, ∞),
iii.) $u(t) \in \text{dom}(A)$, for all $t > t_0$,
iv.) $u(t)$ satisfies both parts of (A-HCP).

10.2 Defining e^{At}

For some of the models in Chapters 9, when it was convenient to do so, we made a concerted effort to verify that the solution of the IBVP satisfied properties that resembled those similar to the matrix exponential from Part I. We did this even though we did not have a candidate for e^{At} in hand. The striking similarity in structure between (HCP) and (A-HCP) suggests that expecting a solution of (A-HCP) to be of the form $u(t) = e^{At}u_0$ is reasonable. But, what does e^{At} even mean in the present setting when A is an operator?

Recall that we presented two definitions of $e^{\mathbf{A}t}$ when $\mathbf{A} \in \mathbb{M}^N(\mathbb{R})$. The first one was formulated using a power series, namely

$$e^{\mathbf{A}t} = \sum_{k=0}^{\infty} \frac{(\mathbf{A}t)^k}{k!}.$$

Looking back through Section 4.3, specifically at the proof of Proposition 4.3.1, you will notice that we used the fact that $\sum_{k=0}^{\infty} \frac{\left(\|\mathbf{A}\|_{\mathbb{M}^N} t\right)^k}{k!}$ was a convergent series of real numbers. But, this only makes sense (from the so-called n^{th} *term test for divergence*) if $\lim_{k\to\infty} \frac{\left(\|\mathbf{A}\|_{\mathbb{M}^N} t\right)^k}{k!} = 0$, which in turn requires that $\|\mathbf{A}\|_{\mathbb{M}^N} < \infty$. This was not an issue when $\mathbf{A} \in \mathbb{M}^N(\mathbb{R})$ because the calculation of $\|\mathbf{A}\|_{\mathbb{M}^N}$ involved summing or taking the maximum of only finitely many positive real numbers. Contrarily, A is a more general operator (like $\frac{d^2}{dx^2}$, $\triangle + I$) in the present setting and unfortunately typically does not satisfy this type of condition. Consider the following **EXPLORE!** activity to see why.

EXPLORE!
Unbounded Operators

We use this **EXPLORE!** activity to formalize the observations made in MATLAB-Exercise 8.3.1 when we first encountered bounded and unbounded operators.

STOP! Return to Definition 8.3.12 and review the definitions of bounded and unbounded operators.

Consider the operator A introduced in Section 8.2 in (8.42) where, for simplicity, k and a are set equal to one namely $A : \mathrm{dom}(A) \subset \mathbb{L}^2(0,1;\mathbb{R}) \to \mathbb{L}^2(0,1;\mathbb{R})$ by

$$A[f] = \frac{d^2}{dx^2}[f],$$

$$\mathrm{dom}(A) = \left\{ f \in \mathbb{L}^2(0,1;\mathbb{R}) \,\middle|\, \frac{df}{dx}, \frac{d^2 f}{dx^2} \in \mathbb{L}^2(0,1;\mathbb{R}), \text{ and } f(0) = f(1) = 0 \right\}.$$

The notation

$$A : \mathrm{dom}(A) \subset \mathbb{L}^2(0,1;\mathbb{R}) \to \mathbb{L}^2(0,1;\mathbb{R})$$

means we are viewing A as a linear operator on $\mathbb{L}^2(0,1;\mathbb{R})$, but we restrict ourselves to selecting functions from the *dom*(A) before applying A. With this in mind, the definition of a *bounded* operator in this case is there exists $m \geq 0$ such that

$$\|A[f]\|_{\mathbb{L}^2(0,1;\mathbb{R})} \leq m \|f\|_{\mathbb{L}^2(0,1;\mathbb{R})}, \text{ for all } f \in dom(A). \tag{10.2}$$

Since the definition of an unbounded operator is not bounded, we must argue A does not satisfy (10.2). This amounts to showing for every fixed, yet arbitrary, $m \in \mathbb{R}$ there is a function $f \in dom(A) \subset \mathbb{L}^2(0,1;\mathbb{R})$ such that

$$\|A[f]\|_{\mathbb{L}^2(0,1;\mathbb{R})} > m \|f\|_{\mathbb{L}^2(0,1;\mathbb{R})},$$

or equivalently show there exists a sequence $(f_n)_{n \in \mathbb{N}} \subset dom(A) \subset \mathbb{L}^2(0,1;\mathbb{R})$ such that

$$\lim_{n\to\infty} \frac{\|A[f]\|_{\mathbb{L}^2(0,1;\mathbb{R})}}{\|f\|_{\mathbb{L}^2(0,1;\mathbb{R})}} = \infty \tag{10.3}$$

Here we use the sequence characterization (10.3) of an unbounded operator to verify that A is unbounded on $\mathbb{L}^2(0,1;\mathbb{R})$. The process has been broken down into several steps.

i.) Define

$$f_n(x) = \sin(2\pi n x),\ x \in [0,1],\ n \in \mathbb{N}$$

and show $(f_n)_{n \in \mathbb{N}} \subset dom(A) \subset \mathbb{L}^2(0,1;\mathbb{R})$ by completing the following steps for a fixed $n \in \mathbb{N}$.

a.) Show $f_n(0) = 0 = f_n(1)$.

b.) Show $\|f_n\|_{\mathbb{L}^2(0,1;\mathbb{R})} < \infty$

c.) Show $\left\| \frac{df_n}{dx} \right\|_{\mathbb{L}^2(0,1;\mathbb{R})} < \infty$

d.) Show $\left\| \frac{d^2 f_n}{dx^2} \right\|_{\mathbb{L}^2(0,1;\mathbb{R})} < \infty$

Therefore $(f_n)_{n\in\mathbb{N}} \subset dom(A) \subset \mathbb{L}^2(0,1;\mathbb{R})$.

ii.) Compute $\|f_n\|_{\mathbb{L}^2(0,1;\mathbb{R})}$.

iii.) Compute $\left\| \frac{d^2 f_n}{dx^2} \right\|_{\mathbb{L}^2(0,1;\mathbb{R})}$

iv.) Use (ii) and (iii) to show

$$\lim_{n\to\infty} \frac{\|A[f]\|_{\mathbb{L}^2(0,1;\mathbb{R})}}{\|f\|_{\mathbb{L}^2(0,1;\mathbb{R})}} = \infty$$

Therefore, A is unbounded on $\mathbb{L}^2(0,1;\mathbb{R})$.

v.) Repeat (i)-(iv) to show $A : \mathrm{dom}(A) \subset \mathbb{L}^2(0,1;\mathbb{R}) \to \mathbb{L}^2(0,1;\mathbb{R})$ given by

$$A[f] = \frac{d^2}{dx^2}[f],$$
$$\mathrm{dom}(A) = \left\{ f \in \mathbb{L}^2(0,1;\mathbb{R}) \middle| \frac{df}{dx}, \frac{d^2 f}{dx^2} \in \mathbb{L}^2(0,1;\mathbb{R}), \text{ and } \frac{df}{dx}(0) = \frac{df}{dx}(1) = 0 \right\}.$$

is unbounded on $\mathbb{L}^2(0,1;\mathbb{R})$. (Hint: Use the same sequence as before, but replace sine with cosine.)

vi.) Repeat (i)-(iv) using the sequence $f_n(x) = x^n$, $x \in [0,1]$, $n \in \mathbb{N}$ to show $A : \mathrm{dom}(A) \subset \mathbb{L}^2(0,1;\mathbb{R}) \to \mathbb{L}^2(0,1;\mathbb{R})$ given by

$$A[f] = \frac{d}{dx}[f],$$
$$\mathrm{dom}(A) = C^1([0,1])$$

is unbounded on $\mathbb{L}^2(0,1;\mathbb{R})$.

In effect, the conclusion we can draw from the above observation is that the operator A arising in (A-HCP), generally speaking, need not satisfy a condition of the sort $\|A\| < \infty$. As such, the power series approach will not provide a meaningful definition of e^{At} here. But, the situation is not hopeless; we have a second definition motivated by the limit definition of e^{At}, as described in Section 4.3.2.

The alternate definition provided when $\mathbf{A} \in \mathbb{M}^N(\mathbb{R})$ was

$$e^{\mathbf{A}t} = \lim_{n\to\infty} \left[\left(\mathbf{I} - \frac{t}{n}\mathbf{A} \right)^{-1} \right]^n.$$

In order for this definition to be meaningful, the existence of the quantity $\left(\mathbf{I} - \frac{t}{n}\mathbf{A}\right)^{-1}$ for $\frac{t}{n} \in (0,\infty)$ was essential, and was the fact that its integer powers $\left[\left(\mathbf{I} - \frac{t}{n}\mathbf{A}\right)^{-1}\right]^n$ remain bounded (to ensure the limit was finite). We need to interpret this in our new context when A is an operator and so, we need to understand what it means for $\left(I - \frac{t}{n}A\right)^{-1}$ to exist. For convenience, let us rename the quantity as follows:

$$r_A(\omega) = (I - \omega A)^{-1}, \; \omega > 0.$$

(Note that we identified the expression $\frac{t}{n}$ as ω.)

STOP! Suppose that $\mathbf{A} \in \mathbb{M}^N(\mathbb{R})$ (so that $\mathbf{A} : \mathbb{R}^N \to \mathbb{R}^N$) and assume $r_{\mathbf{A}}(\omega)$ exists. What does this mean?

i.) Fill in the blanks: Given y in the space ________, we can find x in the space ________ such that $(I - \omega A)x = y$.
ii.) How many such x's can there be? Why?

Now, how does this interpretation change if we replace this nice situation $\mathbf{A} : \mathbb{R}^N \to \mathbb{R}^N$ by an operator $A : \text{dom}(A) \subset \mathcal{H} \to \mathcal{H}$? Mimicking the above definition, saying $r_A(\omega)$ exists means that given $g \in \mathcal{H}$, there exists a unique $f \in \text{dom}(A)$ such that $(I - \omega A)f = g$. Is this reasonable? Let us consider the following two examples to find out.

Example 10.2.1. Let $\mathcal{X} = \mathbb{C}([0,\infty);\mathbb{R})$ and define $A : \text{dom}(A) \subset \mathcal{X} \to \mathcal{X}$ by

$$\begin{cases} Af &= -\frac{df}{dx} \\ \text{dom}(A) &= \left\{ f \in \mathcal{X} : \frac{df}{dx} \in \mathcal{X} \text{and } f(0) = 0 \right\} \end{cases}$$

Let $\omega > 0$ and let $g \in \mathcal{X}$. We must solve the equation $(I - \omega A)f = g$ for f. Observe that

$$\begin{aligned} (I - \omega A)f &= I(f) - \omega A(f) \\ &= f - \omega\left(-\frac{df}{dx}\right) \\ &= f + \omega\frac{df}{dx} \end{aligned}$$

As such, the equation $(I - \omega A)f = g$ becomes the IVP

$$\begin{cases} f(x) + \omega\frac{df}{dx}(x) &= g(x),\ x > 0, \\ f(0) &= 0 \end{cases}$$

Note that the initial condition arises because of that exact condition being specified in the domain of A. This is a first-order linear ODE that can be solved using variation of parameters to conclude that

$$f(x) = e^{-\frac{1}{\omega}x} \int_0^x \frac{e^{\frac{1}{\omega}s} g(s)}{\omega} ds.$$

(Verify this!) Clearly, $f(0) = 0$ **(Why?)** and it follows from the Fundamental Theorem of Calculus that $f \in \mathcal{X}$. **(Tell why.)** Hence, we conclude that $f \in \text{dom}(A)$ and so, we have identified a solution to the IBVP. The uniqueness of this solution follows from the uniqueness theorem for (Non-CP). **(Tell why.)** Consequently, $r_A(\omega)$ exists for all $\omega > 0$.

The next example is a little more involved, yet illustrates the same point.

Example 10.2.2. Let $\mathcal{X} = \mathbb{L}^2([0,L];\mathbb{R})$ and define $A : \text{dom}(A) \subset \mathcal{X} \to \mathcal{X}$ by

$$\begin{cases} Af &= c\frac{d^2 f}{dx^2} \\ \text{dom}(A) &= \left\{ f \in \mathcal{X} : \frac{df}{dx}, \frac{d^2 f}{dx^2} \in \mathcal{X} \text{and } f(0) = f(L) = 0 \right\} \end{cases}$$

Let $\omega > 0$ and let $g \in \mathcal{X}$. We must solve the equation $(I - \omega A) f = g$ for f. Observe that

$$\begin{aligned}(I - \omega A) f &= I(f) - \omega A(f)\\ &= f - \omega \left(c \frac{d^2 f}{dx^2} \right)\\ &= f - \omega c \frac{d^2 f}{dx^2}\end{aligned}$$

As such, the equation $(I - \omega A) f = g$ becomes the second-order IVP

$$\begin{cases} f(z) - \omega c \frac{d^2 f}{dz^2}(z) = g(z), \; z > 0, \\ f(0) = f(L) = 0 \end{cases} \tag{10.4}$$

Look back at the discussion of spring-mass systems in Part I. It seems reasonable to expect a solution of the form

$$f(z) = \sum_{k=0}^{\infty} a_k \sin \left(\frac{2\pi k}{L} z \right).$$

Substituting this expression into (10.4) yields

$$\sum_{k=0}^{\infty} a_k \sin \left(\frac{2\pi k}{L} z \right) - \omega c \sum_{k=0}^{\infty} (-a_k) \left(\frac{2\pi k}{L} \right)^2 \sin \left(\frac{2\pi k}{L} z \right) = g(z)$$

and so,

$$\sum_{k=0}^{\infty} \left[1 + \omega c \left(\frac{2\pi k}{L} \right)^2 \right] a_k \sin \left(\frac{2\pi k}{L} z \right) = g(z). \tag{10.5}$$

Since $g \in \mathbb{L}^2([0, L]; \mathbb{R})$, it follows that g has a unique Fourier representation given by

$$g(z) = \sum_{k=0}^{\infty} b_k \sin \left(\frac{2\pi k}{L} z \right), \tag{10.6}$$

where

$$b_k = \left\langle g(\cdot), \sin \left(\frac{2\pi k}{L} \cdot \right) \right\rangle_{\mathbb{L}^2([0,L];\mathbb{R})}, \; k = 0, 1, 2, ...$$

Equating (10.5) and (10.6) amounts to equating corresponding coefficients, namely

$$\left[1 + \omega c \left(\frac{2\pi k}{L} \right)^2 \right] a_k = b_k$$

so that

$$a_k = \frac{b_k}{1 + \omega c \left(\frac{2\pi k}{L} \right)^2}, \; k = 0, 1, 2, ...$$

Hence,

$$f(z) = \sum_{k=0}^{\infty} \frac{b_k}{1 + \omega c \left(\frac{2\pi k}{L} \right)^2} \sin \left(\frac{2\pi k}{L} z \right).$$

Observe that $f(0) = f(L) = 0$ **(Check this!)**. The uniqueness of f follows again from the theory for (Non-CP) and the uniqueness of the Fourier coefficients.

Is $f \in \text{dom}(A)$? This would require that f, $\frac{df}{dx}$, $\frac{d^2f}{dx^2} \in \mathbb{L}^2([0,L];\mathbb{R})$. For illustrative purposes we will prove $\frac{d^2f}{dx^2} \in \mathbb{L}^2([0,L];\mathbb{R})$. To this end, observe that

$$\frac{df}{dz}(z) = \sum_{k=0}^{\infty} \frac{b_k\left(\frac{2\pi k}{L}\right)}{1+\omega c\left(\frac{2\pi k}{L}\right)^2} \cos\left(\frac{2\pi k}{L}z\right)$$

$$\frac{d^2f}{dz^2}(z) = \sum_{k=0}^{\infty} \frac{-b_k\left(\frac{2\pi k}{L}\right)^2}{1+\omega c\left(\frac{2\pi k}{L}\right)^2} \sin\left(\frac{2\pi k}{L}z\right)$$

Observe that

$$\begin{aligned}
\int_0^L \left[\frac{d^2f}{dz^2}(z)\right]^2 dz &= \int_0^L \left[\sum_{k=0}^{\infty} \frac{-b_k\left(\frac{2\pi k}{L}\right)^2}{1+\omega c\left(\frac{2\pi k}{L}\right)^2} \sin\left(\frac{2\pi k}{L}z\right)\right]^2 dz \\
&=\leq \int_0^L \left[\sum_{k=0}^{\infty} \frac{|b_k|\left(\frac{2\pi k}{L}\right)^2}{1+\omega c\left(\frac{2\pi k}{L}\right)^2}\right]^2 dz \\
&=\leq \int_0^L \left[\frac{1}{\omega c}\sum_{k=0}^{\infty} |b_k|\right]^2 dz \\
&=\leq \frac{L}{\omega c}M^2 \\
&< \infty
\end{aligned}$$

The bound on $\sum_{k=0}^{\infty} |b_k|$ follows from the convergence of the Fourier series for g.

Mimicking this argument, one can show that f, $\frac{df}{dx} \in \mathbb{L}^2([0,L];\mathbb{R})$. **(Try it!)** Upon doing so, we conclude that $f \in \text{dom}(A)$. Consequently, $r_A(\omega)$ exists for all $\omega > 0$.

These two examples suggest that requiring that $r_A(\omega)$ exists for all $\omega > 0$ when A is an operator is not unreasonable. But, this does not imply that $r_A(\omega)$ is bounded. And, if it turns out to be unbounded, then we will run into the same problem we did with the power series approach. So, we are prompted to now determine if $r_A(\omega)$ is **always** bounded. Unfortunately, the answer is NO!

This is one of the critical differences between the present setting and that of Part I. In Part I, $e^{\mathbf{A}t}$ exists for all choices of $\mathbf{A} : \mathbb{R}^N \to \mathbb{R}^N$, but when we transition to the present more complicated setting of PDEs, we must be more selective with what operators A we are allowed to use. Consequently, not every IBVP, even if it can be symbolically captured using the form (A-HCP), will fall under the parlance of the theory develop in this chapter.

We need to identify a verifiable sufficient condition that can be imposed on the operator A that, when it holds **and** $r_A(\omega)$ exists for all $\omega > 0$, then $r_A(\omega)$ is a bounded operator. The answer to this question does exist, but the reason why it works is a consequence of a rather technical development beyond the scope of this text. At this level of our discussion, it is important simply to realize that there are conditions that can be imposed on A such that, when satisfied, we can define e^{At} using the limit definition approach. Furthermore, we were careful to choose models in Chapters 9 and 12 in such a way that all resulting IBVPs would fall within the scope of the theory being developed. A complete theoretical discussion can be found in [33].

The following pseudo-theorem summarizes the situation:

Pseudo-Theorem 2. Let $A : \text{dom}(A) \subset \mathcal{X} \to \mathcal{X}$ be a "nice" operator for which there exist positive real numbers a and M such that

$$\|r_A(\omega)f\|_{\mathcal{H}} \leq M(I - \omega a)^{-1}\|f\|_{\mathcal{H}},$$

for all $\omega > 0$ and for all $f \in \mathcal{H}$. Then,

$$S(t)f = \lim_{n\to\infty}\left[\left(1 - \frac{t}{n}A\right)^{-1}\right]^n f$$

exists and maps $[0, T]$ into $\text{dom}(A)$.

The following activity is intended to illustrate this theorem for some of the IBVPs you encountered in Chapter 9. While it does not constitute a proof of the result, it might at least help to convince you of its validity for our models.

EXPLORE!

The Emergence of e^{At}

We know the solution of (A-HCP) is given abstractly by e^{At}, but what is it? How do we get our hands on it? So far, we have been successful in identifying e^{At} from the solution of PDEs using separation of variables. However, separation of variables is very problem-dependent and does not always work. As such, finding another way to compute e^{At} will be of great benefit. This is where Pseudo-Theorem 2 comes to the rescue. In this **EXPLORE!** project, we use MATLAB to visualize the norm difference

$$\left\| e^{At}f - \left[\left(I - \frac{t}{n}A\right)^{-1}\right]^n f \right\|_{\mathbb{L}^2(-1,1;\mathbb{R})}, \ t \in [0, T]$$

for different choices of functions, f. As n gets larger we hope MATLAB will illustrate smaller and smaller norm differences. This specifically will at least intuitively imply

$$e^{At}f = \lim_{n\to\infty}\left[\left(I - \frac{t}{n}A\right)^{-1}\right]^n f \tag{10.7}$$

To begin, open **MATLAB** and in the command line, type:

MATLAB_Emergence_eAt

Use the GUI to answer the following questions. Make certain to click on the HELP button for instructions on how to use the GUI. A screenshot from the GUI can found in Figure 10.1. The GUI assumes a model, boundary condition, function, final time (T), and tolerance, labeled epsilon, are specified before clicking *Solve*. The GUI will automatically increase n until the norm difference is less than the specified tolerance.

i.) Consider the one-dimensional Neumann BC heat equation

$$\begin{cases} \frac{\partial}{\partial t}z(x,t) = k^2\frac{\partial^2}{\partial x^2}z(x,t), \ -1 < x < 1, \, t > 0, \\ z(x,0) = z_0(x), \ -1 < x < 1, \\ \frac{\partial z}{\partial x}(-1,t) = \frac{\partial z}{\partial x}(1,t) = 0, \ t > 0. \end{cases}$$

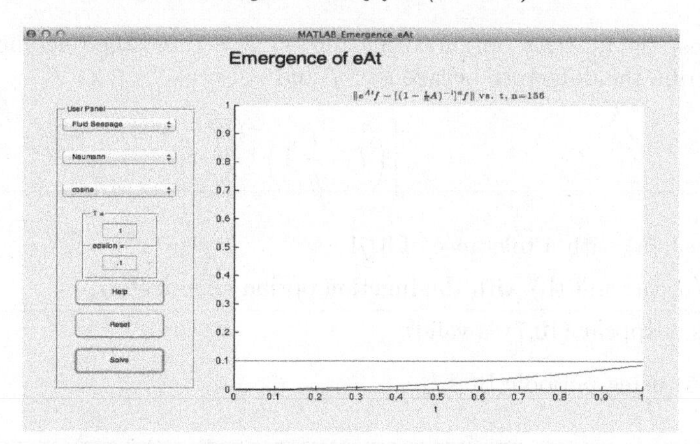

FIGURE 10.1: MATLAB Emergence eAt GUI

a.) Choose the function option cosine and set $T = 1$ and the tolerance equal to 0.1. Describe the difference between $e^{At}f$ and

$$\left[\left(I - \frac{t}{n}A\right)^{-1}\right]^n f.$$

b.) Repeat (a) with a tolerance of 0.01.

c.) Repeat (a) and (b) with the function option exponential.

d.) Does it appear (10.7) is valid?

ii.) Repeat (i) using periodic BCs.

iii.) Consider the periodic BC Fluid Seepage/Soil Mechanics model

$$\begin{cases} \frac{\partial}{\partial t}\left(p(x,t) - \alpha \frac{\partial^2}{\partial x^2} p(x,t)\right) = \beta \frac{\partial^2}{\partial x^2} p(x,t), \ -1 < x < 1, \ t > 0, \\ p(x,0) = p_0(x), \ \ -1 < x < 1, \\ p(-1,t) = p(1,t), \ \ t > 0. \end{cases}$$

a.) Choose the function option sine and set $T = 1$ and the tolerance equal to 0.1. Describe the difference between $e^{At}f$ and

$$\left[\left(I - \frac{t}{n}A\right)^{-1}\right]^n f.$$

b.) Repeat (a) with a tolerance of 0.01.

c.) Repeat (a) and (b) with the function option exponential.

d.) Does it appear (10.7) is valid?

iv.) Repeat (iii) using periodic BCs.

v.) Consider Neumann BC wave equation

$$\begin{cases} \frac{\partial^2}{\partial t^2} z(x,t) - c^2 \frac{\partial^2}{\partial x^2} z(x,t) & = 0, \ -1 < x < 1, \ t > 0, \\ z(x,0) = z_0(x), \ \frac{\partial z}{\partial t}(x,0) & = 0, \ -1 < x < 1, \\ \frac{\partial z}{\partial x}(-1,t) = \frac{\partial z}{\partial x}(1,t) & = 0, \ t > 0. \end{cases}$$

(Note: For simplicity, we assume the initial speed of the wave is 0.)

a.) Choose the function option cosine and set $T = 1$ and the tolerance equal to 0.1. Describe the difference between $e^{At}f$ and

$$\left[\left(I - \frac{t}{n}A\right)^{-1}\right]^{n} f.$$

b.) Repeat (a) with a tolerance of 0.01.

c.) Repeat (a) and (b) with the function option exponential.

d.) Does it appear (10.7) is valid?

vi.) Repeat (iii) using periodic BCs.

While this activity does not constitute a proof, hopefully it is reassuring that the solution of (A-HCP) can be expressed using $S(t)f$ defined in Pseudo-Theorem 2. Moreover, given that $S(t)u_0$ appears to be equal to the solution of the IBVP in each case and we showed that the solution satisfied exponential-like properties, it is not difficult to believe (or show) that $S(t)$ itself would satisfy these properties as well. As such, the following definition is natural and constitutes the analog of the matrix exponential $\{e^{\mathbf{A}t} : t \geq 0\}$ associated with (A-HCP):

Definition 10.2.1. A collection of operators $\{S(t) : \mathcal{H} \to \mathcal{H} \,|t \geq 0\}$ satisfying

i.) $S(t)$ is a bounded, linear operator, for every $t \geq 0$,

ii.) $S(0) = I$, where I is the identity operator on $\mathcal{H}$,

iii.) $S(t+s) = S(t)S(s)$, for all $t, s \geq 0$,

iv.) $\lim_{t\to 0^+} \|S(t)f - f\|_{\mathcal{H}} = 0$, for every $f \in \mathcal{H}$,

is called a *semigroup of bounded linear operators on* $\mathcal{H}$.

We write e^{At} in place of $S(t)$ to highlight the fact the semigroup behaves like the old matrix exponential. Note that condition (iv) above is built into the definition to ensure the right-continuity of $S(t) = e^{At}$ at $t = 0$, and is reasonable since we want the transition from initial profile to state at a close by time to be smooth.

Example 10.2.3. The following family of operators arises in the solution of the heat equation equipped with Neumann BCs. For each $t \geq 0$, define the operator $S(t) : \mathbb{L}^2(0, a; \mathbb{R}) \to \mathbb{L}^2(0, a; \mathbb{R})$ by

$$S(t)[f][x] = \sum_{m=0}^{\infty} \left(\frac{2}{a} \int_0^a f(w) \cos\left(\frac{m\pi}{a} w\right) dw \right) e^{-\left(\frac{m\pi}{a}\right)^2 kt} \cos\left(\frac{m\pi}{a} x\right).$$

We claim that $\{S(t) : t \geq 0\}$ is a semigroup on $\mathbb{L}^2(0, a; \mathbb{R})$.

First, we show that $S(t)$ is a bounded, linear operator. Let $t \geq 0$. For any $f, g \in \mathbb{L}^2(0, a; \mathbb{R})$, applying the linearity of the integral and convergent series immediately yields

$$S(t)[f+g][x] = S(t)[f][x] + S(t)[g][x], \; 0 < x < a.$$

This proves linearity. As for boundedness, let $f \in \mathbb{L}^2(0, a; \mathbb{R})$. Using standard inequalities and properties of convergent series with the fact that $\sup\left\{\left|\cos\left(\frac{m\pi}{a}w\right)\right| : w \in [0,a]\right\} \leq 1$ yields

$$\begin{aligned}
\|S(t)[f]\|^2_{\mathbb{L}^2(0,a;\mathbb{R})} &=\leq 2\int_0^a \left[\frac{2}{a}\|f\|^2_{\mathbb{L}^2(0,a;\mathbb{R})} + \left(\frac{2}{a}\right)^2 \|f\|^2_{\mathbb{L}^2(0,a;\mathbb{R})} M \sum_{m=1}^{\infty} e^{-\left(\frac{m\pi}{a}\right)^2 2kt}\right] dx \\
&=\leq \overline{M}\,\|f\|^2_{\mathbb{L}^2(0,a;\mathbb{R})} < \infty,
\end{aligned}$$

for some positive constants M and $\overline{M}$ depending on $a, k, t,$ and the convergent series $\sum_{m=1}^{\infty} e^{-\left(\frac{m\pi}{a}\right)^2 2kt}$. This proves boundedness.

Next, let $f \in \mathbb{L}^2(0,a;\mathbb{R})$. We use the fact that $\left\{\sqrt{\frac{2}{a}}\cos\left(\frac{m\pi}{a}\cdot\right) : m \in \mathbb{N}\cup\{0\}\right\}$ is an orthonormal basis of $\mathbb{L}^2(0,a;\mathbb{R})$ to see that

$$\begin{aligned}
S(0)[f][x] &= \sum_{m=0}^{\infty}\left(\frac{2}{a}\int_0^a f(w)\cos\left(\frac{m\pi}{a}w\right) dw\right)\cos\left(\frac{m\pi}{a}x\right) \\
&= \sum_{m=0}^{\infty}\left\langle f(\cdot), \sqrt{\frac{2}{a}}\cos\left(\frac{m\pi}{a}\cdot\right)\right\rangle_{\mathbb{L}^2(0,a;\mathbb{R})} \sqrt{\frac{2}{a}}\cos\left(\frac{m\pi}{a}x\right) \\
&= f(x).
\end{aligned}$$

Hence, $S(0) = I$, where I is the identity operator on $\mathbb{L}^2(0,a;\mathbb{R})$.

Finally, using the fact that $e^{-\left(\frac{m\pi}{a}\right)^2 k(t_1+t_2)} = e^{-\left(\frac{m\pi}{a}\right)^2 kt_1} \cdot e^{-\left(\frac{m\pi}{a}\right)^2 kt_2}$, it follows immediately that $S(t_1+t_2)[f][x] = S(t_2)(S(t_1)[f][x]),\ \forall f \in \mathbb{L}^2(0,a;\mathbb{R})$. This establishes the semigroup property. Thus, we have shown that $\{S(t) : t \geq 0\}$ is a semigroup on $\mathbb{L}^2(0,a;\mathbb{R})$.□

Exercise 10.2.1. Verify that the operators arising in the solution of the Sobolev-type IBVP arising in Section 8.2 constitute a semigroup of operators on $\mathbb{L}^2(0,a;\mathbb{R})$.

10.3 Properties of e^{At}

Now that we have defined e^{At} in a manner analogous to Part I, the next step is to verify that the salient characteristics of the matrix exponential (from Section 4.5) carry over to the present setting. Bearing in mind that the familiar concepts of limit, derivative, etc. all extend to the new setting, it can be shown that the properties listed in the following theorem hold.

Theorem 10.3.1. Properties of e^{At}
Let $\{e^{At} : \mathcal{H} \to \mathcal{H} \,|\, t \geq 0\}$ be a semigroup of bounded linear operators on $\mathcal{H}$ and that $A : \mathrm{dom}(A) \subset \mathcal{H} \to \mathcal{H}$ is a sufficiently nice operator. Then,

i.) There exist constants $\omega \in \mathbb{R}$ and $M \geq 1$ such that

$$\left\|e^{tA}\right\| \leq Me^{\omega t}, \textit{ for all } t \geq 0.$$

ii.) For every $x_0 \in \mathcal{H}$, the function $g\colon [0,\infty) \to \mathcal{H}$ defined by $g(t) = e^{At}x_0$ is continuous.
iii.) For every $t_0 \geq 0$, the function $h : \mathcal{H} \to \mathcal{H}$ defined by $h(x) = e^{At_0}x$ is continuous.

iv.) For every $x \in \text{dom}(A)$ and $t \geq 0$, $e^{At}x \in \text{dom}(A)$. Moreover, $\forall x \in \text{dom}(A)$, the function $g: (0, \infty) \to \mathcal{H}$ defined by $g(t) = e^{At}x_0$ is continuously differentiable and satisfies

$$\frac{d}{dt}e^{At}x_0 = \underbrace{Ae^{At}x_0 = e^{At}Ax_0}_{\text{i.e., } A \text{ and } e^{At} \text{ commute}}.$$

v.) For every $x_0 \in \mathcal{H}$ and $t_0 \geq 0$,

$$A\int_0^{t_0} e^{As}x_0 ds = e^{At_0}x_0 - x_0.$$

Proof. Verifying (ii) - (v) follows the same line of reasoning as in Section 4.5 with a simple norm change. **(Convince yourself.)** We prove (i) below. To do so, we show there exist $\delta_0 > 0$ and $M > 1$ such that

$$\sup_{0 \leq t \leq \delta_0} \left\| e^{At} \right\| \leq M. \tag{10.8}$$

First, suppose, by way of contradiction, that (10.8) does not hold. Then, $\forall n \in \mathbb{N}$, $\exists t_n \in \left[0, \frac{1}{n}\right]$ such that $\left\| e^{At_n} \right\| \geq n$. **(Why?)** As such, we know that there exists at least one $z_0 \in \mathcal{H}$ for which the sequence $\left\{ \left\| e^{At_n}z_0 \right\|_{\mathcal{H}} : n \in \mathbb{N} \right\}$ is unbounded. **(Why?)** This results in a contradiction since the continuity of $\left\{ e^{At} : t \geq 0 \right\}$ implies that $\lim_{n \to \infty} e^{At_n}z_0 = z_0$. **(Tell how.)**

Now, let $m \in \mathbb{N}$, and $0 \leq \tau \leq \delta_0$. Define $t = m\delta_0 + \tau$. Observe that

$$\begin{aligned}
\left\| e^{At} \right\| &= \left\| e^{A(m\delta_0 + \tau)} \right\| \\
&= \left\| \underbrace{e^{A\delta_0} \cdots e^{A\delta_0}}_{m \text{ times}} e^{A\tau} \right\| \\
&=\leq \left\| e^{A\delta_0} \right\|^m \left\| e^{A\tau} \right\| \leq M^{m+1}.
\end{aligned}$$

Since $t \geq m\delta_0$,

$$M^{m+1} \leq M \cdot M^{\frac{t}{\delta_0}} = Me^{\omega t},$$

where $\omega = \delta_0^{-1} \log M$. **(Why?)** This completes the proof of (i).

10.4 The Abstract Homogeneous Cauchy Problem: Well-posedness

The initial-value problem (A-HCP) is an abstract evolution equation within a Hilbert space setting. We seek a solution in the sense of Definition 11.1.2. For simplicity, we begin by considering (A-HCP) with $t_0 = 0$. The properties of e^{At} listed in Section 11.3 enable us to establish the following theorem.

Theorem 10.4.1. If $A : \text{dom}(A) \subset \mathcal{H} \to \mathcal{H}$ satisfies the conditions of Pseudo-Theorem 2, then for every $u_0 \in \text{dom}(A)$, (A-HCP) (with $t_0 = 0$) has a unique classical solution in the sense of Definition 11.1.2 given by $u(t) = e^{At}u_0$.

Exercise 10.4.1. Follow the proof of Theorem 4.6.1 in Part I line-by-line to prove this result. Indicate any modification necessary and why each line then follows from the properties introduced in this chapter.

Remark 10.4.1. Notice that we only asserted that the conclusion of Theorem 11.4.1 holds provided that $u_0 \in \text{dom}(A)$. But, why can we not take u_0 to be any element of the underlying space $\mathcal{H}$? It turns out that $u(t) = e^{At}u_0$ need not be differentiable if $u_0 \in \mathcal{H}\backslash\text{dom}(A)$ and so, for such u_0, the resulting function $u(t)$ would not satisfy the first equation of (A-HCP).

As in Part I, prescribing the initial condition at $t = 0$ is not necessary. The following corollary is an immediate consequence of the above theorem.

Corollary 10.4.1. If $A : \text{dom}(A) \subset \mathcal{H} \to \mathcal{H}$ satisfies the conditions of Pseudo-Theorem 2, then for every $u_0 \in \text{dom}(A)$, (A-HCP) has a unique classical solution in the sense of Definition 11.1.2 given by $u(t) = e^{A(t-t_0)}u_0$.

All IBVPs investigated in Chapters 9 and 12 can be reformulated as (A-HCP) in some appropriate Hilbert space such that the operators satisfy the conditions of Pseudo-Theorem 2, so that these two results guarantee the existence and uniqueness of classical solutions of all of them!

The ease with which we were able to show the previous theorem extends to the continuous dependence result as well. For instance, consider (A-HCP) and for each $n \in \mathbb{N}$, consider the IVP

$$\begin{cases} \frac{d}{dt}\left(u_n(t)\right) = A\left(u_n(t)\right), \ t > t_0, \\ u_n(t_0) = \left(u_0\right)_n, \end{cases} \tag{10.9}$$

in $\mathcal{H}$, where $\left(u_0\right)_n \in \text{dom}(A)$. Note that the only change occurs with the initial condition. Assuming that $\left\|\left(u_0\right)_n - u_0\right\|_{\mathcal{H}} \to 0$ as $n \to \infty$, we would like for the corresponding sequence of classical solutions of (10.9) to get closer to the solution of (A-HCP). Observe that the classical solution of (10.9) is $u_n(t) = e^{A(t-t_0)}\left(u_0\right)_n$ and so,

$$\begin{aligned} &\lim_{n\to\infty} \max\left\{\left\|u_n(t) - u(t)\right\|_{\mathcal{H}} : t_0 \le t \le t_0 + T\right\} = \\ &\lim_{n\to\infty} \max\left\{\left\|e^{A(t-t_0)}\left(u_0\right)_n - e^{A(t-t_0)}u_0\right\|_{\mathcal{H}} : t_0 \le t \le t_0 + T\right\} \le \\ &\underbrace{\max\left\{\left\|e^{At}\right\|_{\mathcal{H}} : t_0 \le t \le t_0 + T\right\}}_{\text{Bounded}} \cdot \lim_{n\to\infty} \left\|\left(u_0\right)_n - u_0\right\|_{\mathcal{H}} = 0 \end{aligned}$$

STOP! Interpret the meaning of this result. Also, compare it to the other way we formulated a continuous dependence result in Chapter 8 (using u_0 and $u_0 + \varepsilon$).

APPLY IT!!

To put the continuous dependence result above in context for the models listed below, complete the following:

I.) List the physical parameters, in the context of the model, for which the theory guarantees a continuous dependence result.

II.) Compare and contrast your answer in (I) with your observations from the continuous dependence **EXPLORE!** activities in Chapter 9.

i.) The Neumann BC heat equation

$$\begin{cases} \frac{\partial}{\partial t}z(x,t) = k^2\frac{\partial^2}{\partial x^2}z(x,t), \ 0 < x < L, \ t > 0, \\ z(x,0) = z_0(x), \ 0 < x < L, \\ \frac{\partial z}{\partial x}(0,t) = \frac{\partial z}{\partial x}(L,t) = 0, \ t > 0. \end{cases}$$

ii.) The two-dimensional Dirichlet BC heat equation

$$\begin{cases} \frac{\partial z}{\partial t}(x,y,t) = k^2\left(\frac{\partial^2 z}{\partial x^2}(x,y,t) + \frac{\partial^2 z}{\partial y^2}(x,y,t)\right), \ 0<x<a, \ 0<y<b, \ t>0, \\ z(x,y,0) = z_0(x,y), \ 0<x<a, \ 0<y<b, \\ z(x,0,t) = 0 = z(x,b,t), \ 0<x<a, \ t>0, \\ z(0,y,t) = 0 = z(a,y,t), \ 0<y<b, \ t>0, \end{cases}$$

iii.) The periodic BC Fluid Seepage/Soil Mechanics model

$$\begin{cases} \frac{\partial}{\partial t}\left(p(x,t) - \alpha\frac{\partial^2}{\partial x^2}p(x,t)\right) = \beta\frac{\partial^2}{\partial x^2}p(x,t), \ 0<x<\pi, \ t>0, \\ p(x,0) = p_0(x), \ 0<x<\pi, \\ p(0,t) = p(\pi,t), \ t>0. \end{cases}$$

iv.) The Neumann BC wave equation

$$\begin{cases} \frac{\partial^2}{\partial t^2}z(x,t) - c^2\frac{\partial^2}{\partial x^2}z(x,t) &= 0, \ 0<x<L, \ t>0, \\ z(x,0) = z_0(x), \ \frac{\partial z}{\partial t}(x,0) &= z_1(x), \ 0<x<L, \\ \frac{\partial z}{\partial x}(0,t) = \frac{\partial z}{\partial x}(L,t) &= 0, \ t>0. \end{cases}$$

v.) The damped wave equation

$$\begin{cases} \frac{\partial^2}{\partial t^2}z(x,t) + \alpha\frac{\partial}{\partial t}z(x,t) - c^2\frac{\partial^2}{\partial x^2}z(x,t) = 0, \ 0<x<L, \ t>0, \\ z(x,0) = z_0(x), \ \frac{\partial z}{\partial t}(x,0) = z_1(x), \ 0<x<L, \\ z(0,t) = z(L,t) = 0, \ t>0, \end{cases}$$

What about the other "continuous dependence" **EXPLORE!** activities? Those involved tweaking the operator A itself. If you look back at Section 4.8 in Part I, you will notice this result was more involved and was formulated under restrictions on $\{e^{At}: \mathcal{H} \to \mathcal{H} \,|t \geq 0\}$ itself. Similar, and worse, technical difficulties arise here, but the underlying intuition and end results are the same. Indeed, here is a formal result in this spirit.

Suppose that an operator A is sufficiently nice so that $\{e^{At}: \mathcal{H} \to \mathcal{H} \,|t \geq 0\}$ is well-defined, and that a "sufficiently well-behaved" operator B is added to A. Intuitively, as long as $A+B$ is "relatively close" to A and is not too badly behaved, the semigroup $\{e^{(A+B)t}: \mathcal{H} \to \mathcal{H} \,|t \geq 0\}$ should be well-defined. This, in turn, would suggest that the corresponding eveolution equation

$$\begin{cases} \frac{d}{dt}(u(t)) = (A+B)(u(t)), \ t>t_0, \\ u(t_0) = u_0, \end{cases}$$

should have a unique classical solution in the spirit of the above corollary. But, is this true, and how "nice" must the operator B really be?

A partial answer to this question is given by the following proposition.

Proposition 10.4.1. (A Perturbation Result)
If $A: \mathrm{dom}(A) \subset \mathcal{H} \to \mathcal{H}$ is such that $\{e^{At}: t \geq 0\}$ is well-defined on $\mathcal{H}$ and B is a bounded linear operator, then $A+B$ is also nice enough to ensure that $\{e^{(A+B)t}: t \geq 0\}$ is well-defined on $\mathcal{H}$. A formula is given by $e^{(A+B)t} = \sum_{n=0}^{\infty} u_n(t)$, where $\{u_n\}$ is defined recursively by

$$\begin{cases} u_n(t) = \int_0^t e^{A(t-s)}Bu_{n-1}(s)ds, \\ u_0(t) = e^{At}. \end{cases}$$

Exercise 10.4.2. Let us consider an example. Let $\mathbf{A} = \begin{bmatrix} \alpha & 0 \\ 0 & \beta \end{bmatrix}$, where $0 > \alpha > \beta$, and $\mathbf{B} = \begin{bmatrix} 2 & 0 \\ 0 & 3 \end{bmatrix}$.

i.) Determine an explicit formula for $e^{(\mathbf{A}+\mathbf{B})t}$.

ii.) What is a more efficient way to compute $e^{(\mathbf{A}+\mathbf{B})t}$ in this case? Verify that you get the same result as in (i).

It follows that the IVP

$$\begin{cases} u'(t) = (A+B)\,u(t), \ t > 0, \\ u(0) = u_0, \end{cases} \tag{10.10}$$

has a unique classical solution given by $u(t) = e^{(A+B)t}u_0, \ \forall t \geq 0$.

As an illustrative example, consider the following perturbed heat equation:

$$\begin{cases} \frac{\partial}{\partial t} z(x,t) = k\frac{\partial^2}{\partial x^2} z(x,t) + z(x,t), \ 0 < x < a, \ t > 0, \\ z(x,0) = z_0(x), \ 0 < x < a, \\ z(t,0) = z(t,a) = 0. \end{cases} \tag{10.11}$$

This IBVP can be written in the abstract form (10.10), where A and x_0 are defined as in our discussion of the heat equation, and we identify $B : \mathbb{L}^2(0,a;\mathbb{R}) \to \mathbb{L}^2(0,a;\mathbb{R})$ as the identity operator, which is clearly bounded and linear. Hence, (10.11) has a unique classical solution on $\mathbb{L}^2(0,a;\mathbb{R})$.

APPLY IT!!

We are equipped with enough pseudo-theoretical power to be able to address all of the models studied in Chapter 9. For each of the IBVPs listed below complete the following:

I.) Reformulate the IBVP in the abstract form (A-HCP).

II.) In the context of the IBVP, state what *would* need to be shown in order to guarantee a unique classical solution exists.

i.) The Neumann BC heat equation

$$\begin{cases} \frac{\partial}{\partial t} z(x,t) = k^2\frac{\partial^2}{\partial x^2} z(x,t), \ 0 < x < L, \ t > 0, \\ z(x,0) = z_0(x), \ 0 < x < L, \\ \frac{\partial z}{\partial x}(0,t) = \frac{\partial z}{\partial x}(L,t) = 0, \ t > 0. \end{cases} \tag{10.12}$$

ii.) The periodic BC heat equation

$$\begin{cases} \frac{\partial}{\partial t} z(x,t) = k^2\frac{\partial^2}{\partial x^2} z(x,t), \ 0 < x < L, \ t > 0, \\ z(x,0) = z_0(x), \ 0 < x < L, \\ z(0,t) = z(L,t), \ t > 0. \end{cases} \tag{10.13}$$

iii.) The two-dimensional Dirichlet BC heat equation

$$\begin{cases} \frac{\partial z}{\partial t}(x,y,t) = k^2\left(\frac{\partial^2 z}{\partial x^2}(x,y,t) + \frac{\partial^2 z}{\partial y^2}(x,y,t)\right), 0 < x < a,\, 0 < y < b,\ t > 0, \\ z(x,y,0) = z_0(x,y),\ 0 < x < a,\, 0 < y < b, \\ z(x,0,t) = 0 = z(x,b,t),\ 0 < x < a,\, t > 0, \\ z(0,y,t) = 0 = z(a,y,t),\ 0 < y < b,\, t > 0, \end{cases} \tag{10.14}$$

iv.) The periodic BC Fluid Seepage/Soil Mechanics model

$$\begin{cases} \frac{\partial}{\partial t}\left(p(x,t) - \alpha\frac{\partial^2}{\partial x^2}p(x,t)\right) = \beta\frac{\partial^2}{\partial x^2}p(x,t),\ 0 < x < \pi,\, t > 0, \\ p(x,0) = p_0(x),\ 0 < x < \pi, \\ p(0,t) = p(\pi,t),\ t > 0. \end{cases} \tag{10.15}$$

v.) The Neumann BC wave equation

$$\begin{cases} \frac{\partial^2}{\partial t^2}z(x,t) - c^2\frac{\partial^2}{\partial x^2}z(x,t) & = 0,\ 0 < x < L,\, t > 0, \\ z(x,0) = z_0(x),\ \frac{\partial z}{\partial t}(x,0) & = z_1(x),\ 0 < x < L, \\ \frac{\partial z}{\partial x}(0,t) = \frac{\partial z}{\partial x}(L,t) & = 0,\, t > 0. \end{cases} \tag{10.16}$$

vi.) The damped wave equation

$$\begin{cases} \frac{\partial^2}{\partial t^2}z(x,t) + \alpha\frac{\partial}{\partial t}z(x,t) - c^2\frac{\partial^2}{\partial x^2}z(x,t) = 0,\ 0 < x < L,\, t > 0, \\ z(x,0) = z_0(x),\ \frac{\partial z}{\partial t}(x,0) = z_1(x),\ 0 < x < L, \\ z(0,t) = z(L,t) = 0,\, t > 0, \end{cases} \tag{10.17}$$

vii.) The two-dimensional Dirichlet BC wave equation

$$\begin{cases} \frac{\partial z}{\partial t^2}(x,y,t) = c^2\left(\frac{\partial^2 z}{\partial x^2}(x,y,t) + \frac{\partial^2 z}{\partial y^2}(x,y,t)\right), 0 < x < a,\, 0 < y < b,\ t > 0, \\ z(x,y,0) = z_0(x,y),\ 0 < x < a,\, 0 < y < b, \\ z(x,0,t) = 0 = z(x,b,t),\ 0 < x < a,\, t > 0, \\ z(0,y,t) = 0 = z(a,y,t),\ 0 < y < b,\, t > 0, \end{cases} \tag{10.18}$$

10.5 A Brief Glimpse of Long-Term Behavior

Understanding the long-term behavior (as $t \to \infty$) of the classical solution of (A-HCP) is important to attaining a more complete picture of the evolution of the solution of the IBVPs expressed in this form. We investigated such behavior in the finite-dimensional case and discovered a very strong link between the long-term behavior of the solution and the eigenvalues of **A**. The existence of such a link is not particularly surprising since the eigenvalues of **A** played a crucial role in the structure of the matrix exponential, and the classical solution of (HCP) was, in turn, expressed uniquely in terms of this matrix exponential. Specifically, the classical solution of (HCP) was shown to be exponentially stable if and only if the real part of all eigenvalues of $\mathbf{A}$ were negative.

Can we establish such a result for the abstract IVP (A-HCP) in a Hilbert space $\mathcal{H}$? In response to this question, there is a substantive theory devoted to describing the relationships among the so-called *spectral properties* of A, the structure of its semigroup $\{e^{At} \mid t \geq 0\}$, and the stability of this semigroup (and, in turn, the classical solution of (A-HCP)). A complete discussion of this elegant theory would take us far afield of our present goals.

The following notions of stability are analogous to those introduced in Part I.

Definition 10.5.1. A semigroup $\{e^{At} : t \geq 0\}$ on $\mathcal{H}$ is
i.) *uniformly stable* if $\lim_{t\to\infty} \|e^{At}\| = 0$;
ii.) *exponentially stable* if $\exists M \geq 1$ and $\alpha > 0$ such that $\|e^{At}\| \leq Me^{-\alpha t}$, $\forall t \geq 0$;
iii.) *strongly stable* if $\lim_{t\to\infty} \|e^{At}z\|_{\mathcal{H}} = 0$, $\forall z \in \mathcal{H}$.

APPLY IT!!

Use the "long-term behavior" **EXPLORE!** activities in Chapter 9 to make a conjecture regarding the stability properties of the operators A arising in the reformulation of IBVPs (10.12)-(10.17).

10.6 Looking Ahead

By way of preparation for Chapter 12, we consider some modified versions of the classical heat equation, now with increased complexity.

Suppose an external space-dependent term is added to the right-side of the PDE in the classical heat equation IBVP to account for an external heat source. This leads to the modified IBVP

$$\begin{cases} \frac{\partial}{\partial t}z(x,t) = \frac{\partial^2}{\partial x^2}z(x,t) + f(x), \ 0 < x < a, \, t > 0, \\ z(x,0) = 0, \ 0 < x < a, \\ z(0,t) = z(a,t) = 0, \ t > 0, \end{cases} \tag{10.19}$$

where the forcing term $f : [0,a] \to \mathbb{R}$ can be given by, for instance,

$$f(x) = \begin{cases} x, & 0 \leq x \leq \frac{a}{2}, \\ a - x, & \frac{a}{2} \leq x \leq a. \end{cases} \tag{10.20}$$

Exercise 10.6.1. Complete the following:

i.) Verify directly that a classical solution of (10.19) is given by

$$z(x,t) = \frac{4}{a}\sum_{n=1}^{\infty} \frac{(-1)^{n-1}}{(2n-1)^4}\left(1 - e^{-(2n-1)^2 t}\right)\sin\left((2n-1)x\right). \tag{10.21}$$

ii.) Does there exist $\lim_{t\to\infty} |z(x,t)|$? Explain.

iii.) Comment on the differences between (10.21) and the homogeneous heat equation (when $f = 0$).

We would like to formulate such an IBVP abstractly in a Hilbert space and then express its classical solution using a suitable semigroup. Recall that the unique classical solution of the classical heat equation IBVP coupled with the homogeneous Neumann BCs is $u(t) = e^{At}[u_0]$, $t \geq 0$, where the operators $e^{At} : \mathbb{L}^2(0,a;\mathbb{R}) \to \mathbb{L}^2(0,a;\mathbb{R})$ are defined by

$$e^{At}(g)[x] = \sum_{m=0}^{\infty} \left(\frac{2}{a} \int_0^a g(w) \cos\left(\frac{m\pi}{a} w\right) dw \right) e^{-\left(\frac{m\pi}{a}\right)^2 kt} \cos\left(\frac{m\pi}{a} x\right). \tag{10.22}$$

We now incorporate a forcing term $f : [0,a] \times [0,\infty) \to \mathbb{R}$ into the heat conduction model to obtain the following more general IBVP:

$$\begin{cases} \frac{\partial}{\partial t} z(x,t) = k \frac{\partial^2}{\partial x^2} z(x,t) + f(x,t), \; 0 < x < a, \; t > 0, \\ z(x,0) = z_0(x), \; 0 < x < a, \\ \frac{\partial z}{\partial x}(0,t) = \frac{\partial z}{\partial x}(a,t) = 0, \; t > 0. \end{cases} \tag{10.23}$$

Exercise 10.6.2. Try to rewrite (10.23) as an abstract evolution equation. What new difficulties do you encounter?

Assuming that f is sufficiently smooth, it can be expressed uniquely in the form

$$f(x,t) = \sum_{m=1}^{\infty} f_m(t) \sin\left(\frac{m\pi}{a} x\right), \; 0 < x < a, \; t > 0, \tag{10.24}$$

where

$$f_m(t) = \frac{2}{a} \int_0^a f(w,t) \sin\left(\frac{m\pi}{a} w\right) dw, \; t > 0. \tag{10.25}$$

Assuming that the solution of (10.23) is of the form

$$z(x,t) = \sum_{m=1}^{\infty} z_m(t) \sin\left(\frac{m\pi}{a} x\right), \; 0 < x < a, \; t > 0, \tag{10.26}$$

complete the following exercise.

Exercise 10.6.3. Complete the following:

i.) Substitute (10.26) into (10.23) to show that $\forall m \in \mathbb{N}$, $z_m(t)$ satisfies the IVP

$$\begin{cases} z_m'(t) + \frac{m^2\pi^2}{a^2} z_m(t) = f_m(t), \; t > 0, \\ z(0) = z_m^0. \end{cases} \tag{10.27}$$

ii.) Use the variation of parameters technique to show that the solution of (10.27) is given by

$$z_m(t) = z_m^0 e^{-\left(\frac{m^2\pi^2}{a^2}\right)t} + \int_0^t e^{-\left(\frac{m^2\pi^2}{a^2}\right)(t-s)} f_m(s) ds, \; t > 0. \tag{10.28}$$

iii.) Use (10.22) and (10.25) in (10.28) to show that the solution of (10.27) can be simplified to

$$z(x,t) = e^{At}\left(z_0\right)[x] + \int_0^t \int_0^a \frac{2}{a} \sum_{m=1}^{\infty} e^{-\left(\frac{m^2\pi^2}{a^2}\right)(t-s)} \sin\left(\frac{m\pi}{a}w\right) \sin\left(\frac{m\pi}{a}x\right) f(w,s)dwds. \tag{10.29}$$

iv.) Finally, use (10.29) to further express the solution of (10.23) in the form

$$z(\cdot,t) = e^{At}[z_0][\cdot] + \int_0^t e^{A(t-s)} f(s,\cdot)ds, \ t > 0. \tag{10.30}$$

Equation (10.30) is reminiscent of the familiar variation of parameters formula from Part I, and suggests that if we identified the terms correctly between (10.23) and the abstract evolution equation

$$\begin{cases} \frac{du}{dt}(t) = Au(t) + F(t), \ t > 0, \\ u(0) = u_0, \end{cases} \tag{10.31}$$

then a viable representation formula for the classical solution might be

$$u(t) = e^{At}[u_0] + \int_0^t e^{A(t-s)} F(s)ds, \ t > 0.$$

Does such a formula make sense? If so, what conditions can be imposed on F to ensure that u given by this formula actually satisfies (10.31)? Such questions are the focus of our investigation in the next chapter.

Chapter 11

The Next Wave of Mathematical Models—With Forcing

Overview

As in the ODE models explored in Part I, nonnegligible external forces are often present and not negligible and so they need to be incorporated, when forming a mathematical model of a phenomenon. The nature of such external forces can be rather complex and their description can be very "nonlinear" in the sense of depending not only on time and position, but also on the solution itself and its partial derivatives. As before, we start off relatively tame and consider only forcing terms that are independent of the unknown function we seek to describe. More precisely, we only consider forcing terms of the form $f(x,t)$.

For the **EXPLORE!** projects to follow, you shall use each of the following forcing terms:

$$f_1(x,t) = A, \text{ where } A \in \mathbb{R}\setminus\{0\} \tag{11.1}$$

$$f_2(x,t) = Ax + Bt, \text{ where } A, B \in \mathbb{R}\setminus\{0\} \tag{11.2}$$

$$f_3(x,t) = \cos(\omega t), \text{ where } \omega > 0 \tag{11.3}$$

$$f_4(x,t) = e^{-at}, \text{ where } a > 0 \tag{11.4}$$

$$f_5(x,t) = \begin{cases} A, & a \le x \le b,\ t > 0 \\ 0 & \text{otherwise} \end{cases}, \text{ where } a < b \tag{11.5}$$

In the case of a two-dimensional model, (11.2) will be replaced by

$$f_2(x,y,t) = A(x+y) + Bt, \text{ where } A, B \in \mathbb{R}\setminus\{0\} \tag{11.6}$$

and (11.5) by

$$f_5(x,y,t) = \begin{cases} A, & a \le x \le b,\ c \le y \le d,\ t > 0 \\ 0 & \text{otherwise} \end{cases}, \text{ where } a < b. \tag{11.7}$$

The other forcing functions are independent of the spatial variabes and remain the same in two-dimensions.

11.1 Turning Up the Heat—Variants of the Heat Equation

11.1.1 One-Dimensional Diffusion with Forcing

Suppose an external space-dependent term is added to the right-side of the PDE in (9.15) to account for an external heat source. We now incorporate a forcing term $f : [0,a] \times [0,\infty) \to \mathbb{R}$ into the heat conduction model to obtain the following more general IBVP:

$$\begin{cases} \frac{\partial}{\partial t} z(x,t) = k^2 \frac{\partial^2}{\partial x^2} z(x,t) + f(x,t),\ 0 < x < a,\ t > 0, \\ z(x,0) = z_0(x),\ 0 < x < a, \\ \frac{\partial z}{\partial x}(0,t) = \frac{\partial z}{\partial x}(a,t) = 0,\ t > 0. \end{cases} \tag{11.8}$$

Exercise 11.1.1. Try to rewrite (11.8) as an abstract evolution equation. What new difficulties do you encounter?

Assuming that f is sufficiently smooth, it can be expressed uniquely in the form

$$f(x,t) = \sum_{m=1}^{\infty} f_m(t) \sin\left(\frac{m\pi}{a} x\right),\ 0 < x < a,\ t > 0, \tag{11.9}$$

where the coefficients $f_m(t)$ are defined by

$$f_m(t) = \frac{2}{a} \int_0^a f(w,t) \sin\left(\frac{m\pi}{a} w\right) dw,\ t > 0. \tag{11.10}$$

(This follows from an extension of the Fourier series discussion. See [11] for more details.)

Assuming that the solution of (11.8) is of the form

$$z(x,t) = \sum_{m=1}^{\infty} z_m(t) \sin\left(\frac{m\pi}{a} x\right),\ 0 < x < a,\ t > 0, \tag{11.11}$$

complete the following exercise.

Exercise 11.1.2. Complete the following:

i.) Substitute (11.11) into (11.8) to show that $\forall m \in \mathbb{N}$, $z_m(t)$ satisfies the IVP

$$\begin{cases} z_m'(t) + \frac{m^2\pi^2}{a^2} z_m(t) = f_m(t),\ t > 0, \\ z(0) = z_m^0. \end{cases} \tag{11.12}$$

ii.) Use the variation of parameters technique to show that the solution of (11.12) is given by

$$z_m(t) = z_m^0 e^{-\left(\frac{m^2\pi^2}{a^2}\right)t} + \int_0^t e^{-\left(\frac{m^2\pi^2}{a^2}\right)(t-s)} f_m(s)ds,\ t > 0. \tag{11.13}$$

iii.) Use the associated semigroup $e^{At} : \mathbb{L}^2(0, a; \mathbb{R}) \to \mathbb{L}^2(0, a; \mathbb{R})$ linked to the homogeneous version of IBVP (11.8) given by

$$e^{At}(g)[x] = \sum_{m=0}^{\infty} \left(\frac{2}{a} \int_0^a g(w) \cos\left(\frac{m\pi}{a} w\right) dw \right) e^{-\left(\frac{m\pi}{a}\right)^2 kt} \cos\left(\frac{m\pi}{a} x\right)$$

and (11.10) in (11.13) to show that the solution of (11.12) can be simplified to

$$\begin{aligned} z(x,t) &= e^{At}(z_0)[x] \\ &+ \int_0^t \int_0^a \frac{2}{a} \sum_{m=1}^{\infty} e^{-\left(\frac{m^2\pi^2}{a^2}\right)(t-s)} \sin\left(\frac{m\pi}{a} w\right) \sin\left(\frac{m\pi}{a} x\right) f(w,s)dwds. \end{aligned} \tag{11.14}$$

iv.) Finally, use (11.14) to further express the solution of (11.8) in the form

$$z(\cdot,t) = e^{At}[z_0][\cdot] + \int_0^t e^{A(t-s)} f(s,\cdot)ds,\ t > 0. \tag{11.15}$$

Exercise 11.1.3. As a direct application of the process outlined in Exercise 11.1.2, consider the following IBVP equipped with homogeneous Dirichlet

$$\begin{cases} \frac{\partial}{\partial t} z(x,t) = \frac{\partial^2}{\partial x^2} z(x,t) + f(x,t),\ 0 < x < a,\ t > 0, \\ z(x,0) = 0,\ 0 < x < a, \\ z(0,t) = z(a,t) = 0,\ t > 0, \end{cases} \tag{11.16}$$

where the forcing term $f : [0,a] \to \mathbb{R}$ is defined by

$$f(x) = \begin{cases} x, & 0 \le x \le \frac{a}{2}, \\ a - x, & \frac{a}{2} \le x \le a. \end{cases} \tag{11.17}$$

i.) Verify directly that a mild solution of (11.16) is given by

$$z(x,t) = \frac{4}{a} \sum_{n=1}^{\infty} \frac{(-1)^{n-1}}{(2n-1)^4} \left(1 - e^{-(2n-1)^2 t}\right) \sin\left((2n-1)x\right). \tag{11.18}$$

ii.) Comment on the differences between (11.18) and (9.14) (when $f = 0$).

Next, consider the following several **EXPLORE!** activities analogous to what you completed previously for the homogeneous diffusion equation. The goal of these exercises is two-fold. One, you should compare what is happening in the current setting as a result of incorporating a term governing external forcing into the PDE portion of the IBVP to what happened for the homogeneous versions of the IBVPs. Two, pay attention to the new results that we are trying to develop analogous to the results established for nonhomogeneous ODEs in Part I.

EXPLORE!

11.1 Parameter Play!

In this **EXPLORE!** you will investigate the effects of changing K, λ, and c used to define the diffusivity constant k^2. To do this, you will use the MATLAB Heat Equation Nonhomogeneous 1D GUI to visualize the solutions to IBVP (11.8) and IBVP (11.16) for different choices of k^2.

MATLAB-Exercise 11.1.1. To begin, open **MATLAB** and in the command line, type:

MATLAB_Heat_Equation_Non_1D

Use the GUI to answer the following questions. Make certain to click on the HELP button for instructions on how to use the GUI. A typical screenshot of the GUI can found in Figure 11.1 and a particular GUI outputs in Figure 11.2, 11.3, and 11.4.

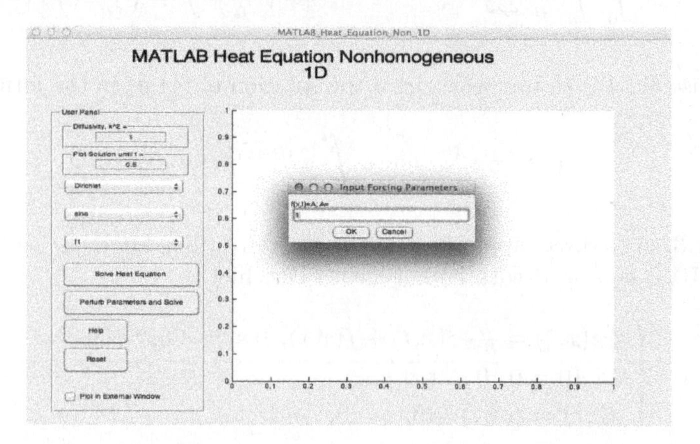

FIGURE 11.1: MATLAB Heat Equation Nonhomogeneous 1D GUI

i.) Consider IBVP (11.16) with the sine initial condition option and the forcing function option f1.

a.) Use $k^2 = 1$, plot solutions until $t = 0.5$ and when prompted set A in the definition of $f_1(x,t)$ equal to 1. Describe the evolution of the solution. How did it differ from the solution of the corresponding homogeneous heat equation?

b.) Based on your intuition gained in Chapter 9 regarding the heat equation, how do you expect the solution of (11.16) to evolve when $k^2 = 2$, as compared to (a)? Check your intuition by using the GUI to visualize this situation.

c.) What do you expect to happen if $k^2 = 0.25$? Test your hypothesis with the GUI.

d.) Repeat (a) using $A = 0.5$. How would you compare this situation with the one in (a) and the homogeneous heat equation solution?

e.) Repeat (d) using $A = 0.25, 0.1$, and 0.001. In addition, make a conjecture for whether or not these nonhomogeneous solutions converge to the homogeneous solution as $f_1(x,t)$ converges to the zero function.

f.) What if A gets larger? Explore this by choosing a sequence of increasingly larger values for A. Summarize your observations.

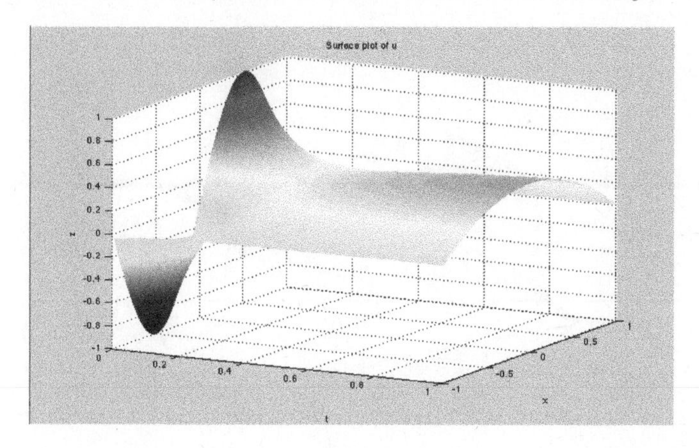

FIGURE 11.2: IBVP (11.16) with $k^2 = 1$, $f(x,t) = 1$, initial profile: $\sin \pi x$.

g.) Repeat (a)-(f) using the exponential initial condition option. Did you observe any behavior contradictory to your conjectures in (b), (c), and (e)? If so, how?

h.) Repeat (a)-(f) using the step function initial condition option. Did you observe any behavior contradictory to your conjectures in (b), (c), and (e)? If so, how?

ii.) Repeat (i) with the forcing function option f5.

iii.) Consider IBVP (11.8) with the cosine initial condition option and the forcing function option f2.

a.) Use $k^2 = 1$, plot solutions until $t = 0.5$, and when prompted, set A and B in the definition of $f_2(x,t)$, equal to 1 and -1, respectively . Describe the evolution of the solution. How did it differ from the solution of the corresponding homogeneous heat equation?

b.) Based on your intuition gained in Chapter 9 regarding the heat equation, how do you expect the solution of (11.16) to evolve when $k^2 = 2$, as compared to (a)? Check your intuition by using the GUI to visualize this situation.

c.) What do you expect to happen if $k^2 = 0.25$? Test your hypothesis with the GUI.

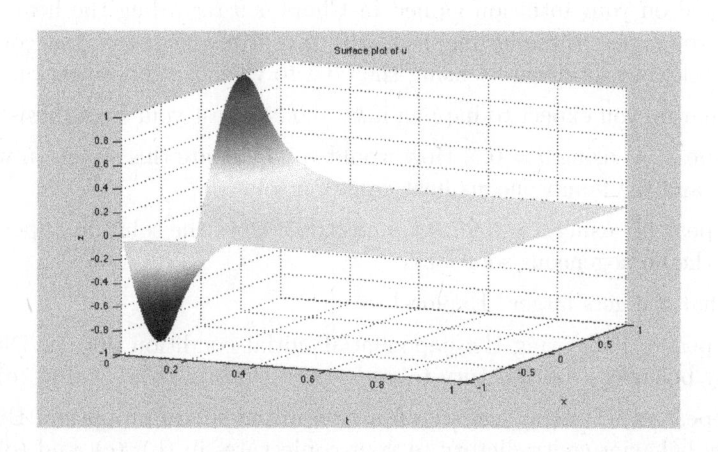

FIGURE 11.3: IBVP (11.16) with $k^2 = 1$, $f(x,t) = 0.01$, initial profile: $\sin \pi x$.

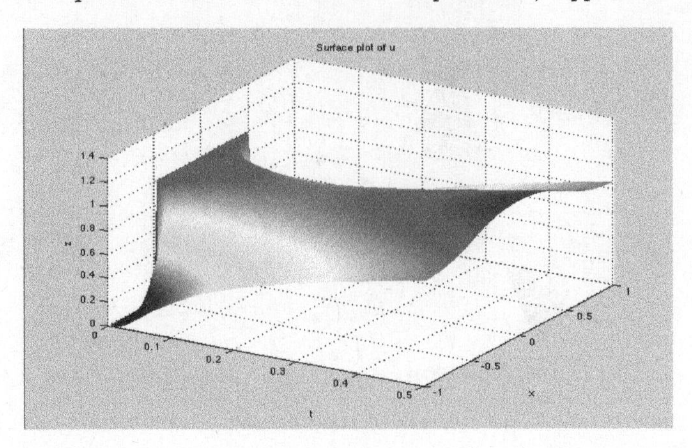

FIGURE 11.4: IBVP (11.8) with $k^2 = 1$, forcing function and initial condition: step function.

d.) Repeat (a) using $A = 0.5$ and $B = -0.5$. How would you compare this situation with the one in (a) and the homogeneous heat equation solution?

e.) Repeat (d) using $A = 0.25, 0.1$, and 0.001 and $B = -0.25, -0.1$, and -0.001. In addition, make a conjecture for whether or not these nonhomogenous solutions converge to the homogeneous solution as $f_2(x,t)$ converges to the zero function.

f.) What if A gets larger and B more negative? Explore!

g.) Repeat (a)-(f) using the exponential initial condition option. Did you observe any behavior contradictory to your conjectures in (b), (c), and (e)? If so, how?

h.) Repeat (a)-(f) using the step function initial condition option. Did you observe any behavior contradictory to your conjectures in (b), (c), and (e)? If so, how?

iv.) Consider IBVP (11.8) with the cosine initial condition option and the forcing function option f4.

a.) Use $k^2 = 1$, plot solutions until $t = 0.5$, and when prompted, set a in the definition of $f_4(x,t)$ equal to 1. Describe the evolution of the solution. How did it differ from the solution of the corresponding homogeneous heat equation?

b.) Based on your intuition gained in Chapter 9 regarding the heat equation, how do you expect the solution of (11.16) to evolve when $k^2 = 2$, as compared to (a)? Check your intuition by using the GUI to visualize this situation.

c.) What do you expect to happen if $k^2 = 0.25$? Test your hypothesis with the GUI.

d.) Repeat (a) using $a = 0.5$. How would you compare this situation with the one in (a) and the homogeneous heat equation solution?

e.) Repeat (d) using $a = 0.25, 0.1$, and 0.001. Does the solution appear to get closer to the homogeneous solution?

f.) What if a gets larger? Explore!

g.) Repeat (a)-(f) using the exponential initial condition option. Did you observe any behavior contradictory to your conjectures in (b), (c), and (e)? If so, how?

h.) Repeat (a)-(f) using the step function initial condition option. Did you observe any behavior contradictory to your conjectures in (b), (c), and (e)? If so, how?

v.) Provide an IBVP for (11.8) equipped with periodic boundary condition.

vi.) In the GUI select periodic boundary conditions, the cosine initial condition option, and the forcing function option f3.

a.) Use $k^2 = 1$, plot solutions until $t = 0.5$, and when prompted, set ω in the definition of $f_3(x,t)$ equal to 1. Describe the evolution of the solution. How did it differ from the solution of the corresponding homogeneous heat equation?

b.) Based on your intuition gained in Chapter 9 regarding the heat equation, how do you expect the solution of (11.16) to evolve when $k^2 = 2$, as compared to (a)? Check your intuition by using the GUI to visualize this situation.

c.) What do you expect to happen if $k^2 = 0.25$? Test your hypothesis with the GUI.

d.) Repeat (a) using $a = 0.5$. How would you compare this situation with the one in (a) and the homogeneous heat equation solution?

e.) Repeat (d) using $a = 0.25, 0.1$, and 0.001. Explain why the solution does **not** tend to the homogeneous solution.

f.) What if a gets larger? Explore!

g.) Repeat (a)-(f) using the exponential initial condition option. Did you observe any behavior contradictory to your conjectures in (b), (c), and (e)? If so, how?

h.) Repeat (a)-(f) using the step function initial condition option. Did you observe any behavior contradictory to your conjectures in (b), (c), and (e)? If so, how?

■

EXPLORE!
11.2 Where Has All The Heat Gone?

MATLAB-Exercise 11.1.2. To begin, open **MATLAB** and in the command line, type:

MATLAB_Heat_Equation_Non_1D

Use the GUI to answer the following questions. Make certain to click on the HELP button for instructions on how to use the GUI. Examples of GUI outputs can be found in Figures 11.5, 11.6, 11.7, and 11.8.

i.) Consider IBVP (11.16) with the sine initial condition option and the forcing function option f1.

a.) Investigate the behavior of the solution of IBVP (11.16) for large times. Use $k^2 = 2$ and when prompted set A in the definition of $f_1(x,t)$ equal to 1. Begin by plotting solutions until $t = 2$. If necessary, slowly increase the plotting time until you are able to describe the behavior of the solution. Remember, a computer cannot handle the infinite! Compare the behavior here with the homogeneous heat equation.

b.) Repeat (a) using $A = 0.5$. How would you compare this situation with the one in (a) and the homogeneous heat equation solution?

c.) Repeat (b) using $A = 0.25, 0.02$, and 0.001. How is the forcing function affecting the long-term behavior of the solution?

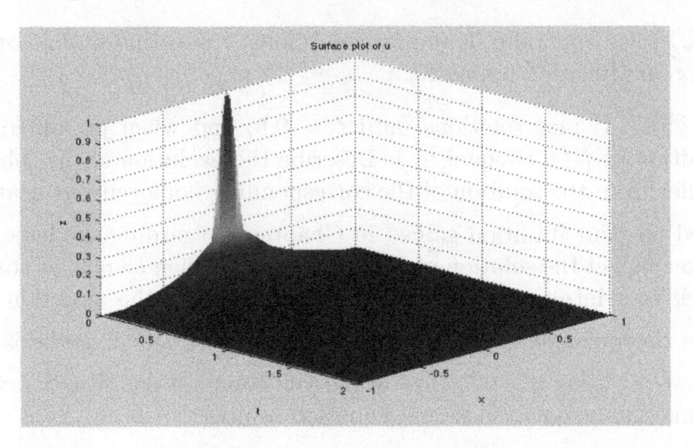

FIGURE 11.5: IBVP (11.16) with $k^2 = 1$, forcing function: step function with $A = 0.001$, initial condition: e^{-100x^2}.

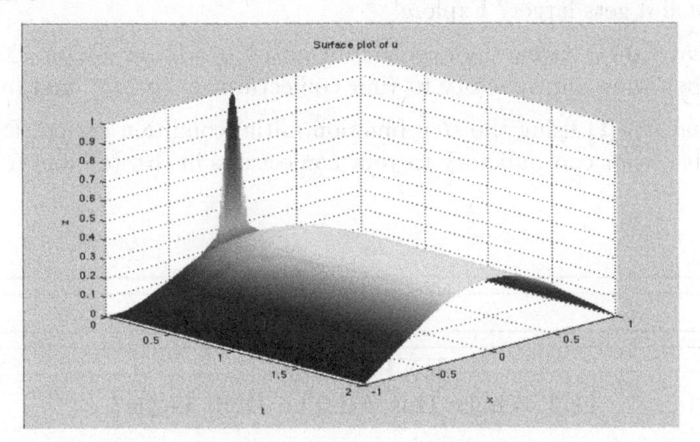

FIGURE 11.6: IBVP (11.16) with $k^2 = 1$, forcing function: step function with $A = 2$, initial condition: e^{-100x^2}.

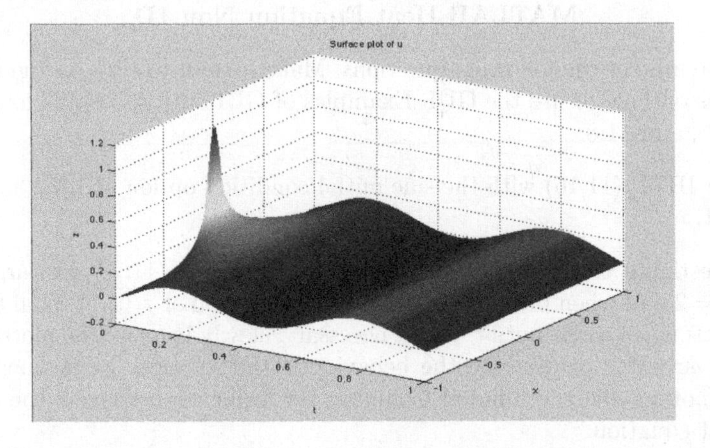

FIGURE 11.7: IBVP (11.8) with $k^2 = 1$, forcing function: $f(x,t) = \cos(10t)$ and initial condition: e^{-100x^2}.

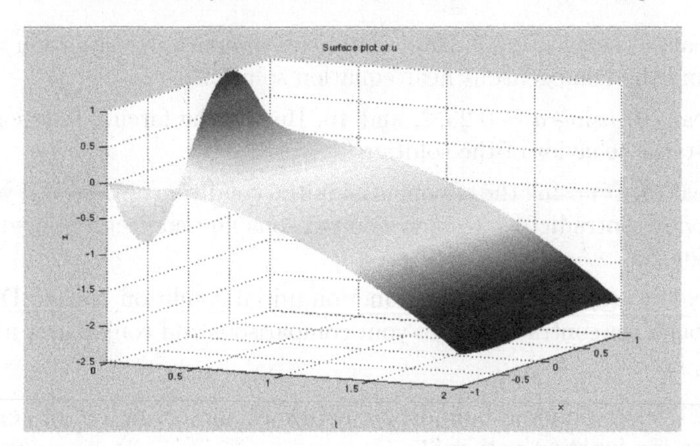

FIGURE 11.8: IBVP (11.8) with Periodic BCs, $k^2 = 1$, forcing function: $f(x,t) = x - t$ and initial condition: $\sin \pi x$.

d.) Repeat (a)-(c) using the exponential initial condition option. Did you observe any behavior contradictory to your comparisons and conjectures made in (a)-(c)? If so, how?

e.) Repeat (a)-(c) using the step function initial condition option. Did you observe any behavior contradictory to your comparisons and conjectures made in (a)-(c)? If so, how?

ii.) Repeat (i) with the forcing function option f5.

iii.) Consider IBVP (11.8) with the cosine initial condition option and the forcing function option f2.

a.) Investigate the behavior of the solution of IBVP (11.8) for large times. Use $k^2 = 2$ and when prompted, set A and B in the definition of $f_2(x,t)$, equal to 1 and -1. Begin by plotting solutions until $t = 2$. If necessary, slowly increase the plotting time until you are able to describe the behavior of the solution. Compare the behavior here with the homogeneous heat equation.

b.) Repeat (a) using $A = 0.5$. How would you compare this situation with the one in (a) and the homogeneous heat equation solution?

c.) Repeat (b) using $A = 2, 0.25$, and 0.001 and $B = -2, -0.25$, and -0.001. How is the forcing function affecting the long-term behavior of the solution?

d.) Repeat (a)-(c) using the exponential initial condition option. Did you observe any behavior contradictory to your comparisons and conjectures made in (a)-(c)? If so, how?

e.) Repeat (a)-(c) using the step function initial condition option. Did you observe any behavior contradictory to your comparisons and conjectures made in (a)-(c)? If so, how?

iv.) Consider IBVP (11.8) with the cosine initial condition option and the forcing function option f4.

a.) Investigate the behavior of the solution of IBVP (11.8) for large times. Use $k^2 = 2$ and when prompted, set a in the definition of $f_4(x,t)$ equal to 1. Begin by plotting solutions until $t = 2$. If necessary, slowly increase the plotting time until you are able to describe the behavior of the solution. Compare the behavior here with the homogeneous heat equation.

b.) Repeat (a) using $a = 0.5$. How would you compare this situation with the one in (a) and the homogeneous heat equation solution?

c.) Repeat (b) using $a = 0.25, 2$, and 10. How is the forcing function affecting the long-term behavior of the solution?

d.) Repeat (a)-(c) using the exponential initial condition option. Did you observe any behavior contradictory to your comparisons and conjectures made in (a)-(c)? If so, how?

e.) Repeat (a)-(c) using the step function initial condition option. Did you observe any behavior contradictory to your comparisons and conjectures made in (a)-(c)? If so, how?

v.) In the GUI select periodic boundary conditions, the cosine initial condition option, and the forcing function option f3.

a.) Investigate the IBVP solution's behavior for large times. Use $k^2 = 2$ and when prompted, set ω in the definition of $f_3(x,t)$ equal to 1. Begin by plotting solutions until $t = 2$. If necessary, slowly increase the plotting time until you are able to describe the behavior of the solution. Compare the behavior here with the homogeneous heat equation.

b.) Repeat (a) using $a = 0.5$. How would you compare this situation with the one in (a) and the homogeneous heat equation solution?

c.) Repeat (b) using $a = 0.25, 2$, and 10. How is the forcing function affecting the long-term behavior of the solution?

d.) Repeat (a)-(c) using the exponential initial condition option. Did you observe any behavior contradictory to your comparisons and conjectures made in (a)-(c)? If so, how?

e.) Repeat (a)-(c) using the step function initial condition option. Did you observe any behavior contradictory to your comparisons and conjectures made in (a)-(c)? If so, how?

■

EXPLORE!

11.3 Continuous Dependence on Parameters

MATLAB-Exercise 11.1.3. To explore continuous dependence with the GUI one must first solve the nonhomogeneous heat equation with a particular diffusivity constant, initial condition, and forcing function. Once a solution has been constructed, click the *Perturb Parameters and Solve* button to specify the perturbed diffusivity constant and a perturbation size for the initial position. The GUI will also allow for a perturbed forcing function parameter. When prompted to select a norm choose *All* to investigate dependence using both norms at the same time. To begin, open **MATLAB** and in the command line, type:

MATLAB_Heat_Equation_Non_1D

Use the GUI to answer the following questions. Make certain to click on the HELP button for instructions on how to use the GUI. A screenshot of inputting parameters into the GUI can be found in Figure 11.9 and others of particular GUI outputs in Figures 11.10–11.14.

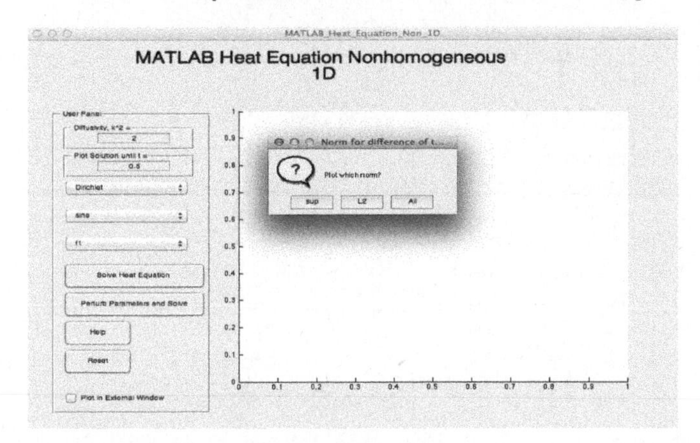

FIGURE 11.9: Perturbing parameters, initial conditions, and forcing function parameters with MATLAB Heat Equation Nonhomogeneous 1D GUI

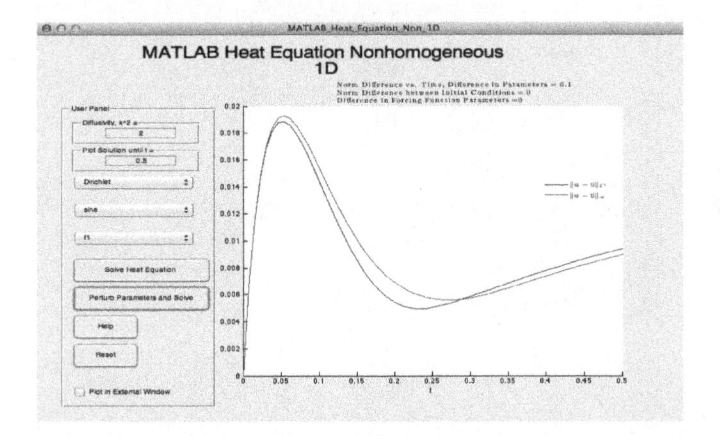

FIGURE 11.10: Perturbing the diffusivity constant in IBVP (11.16). Forcing function, $f_1(x,t) = 1$.

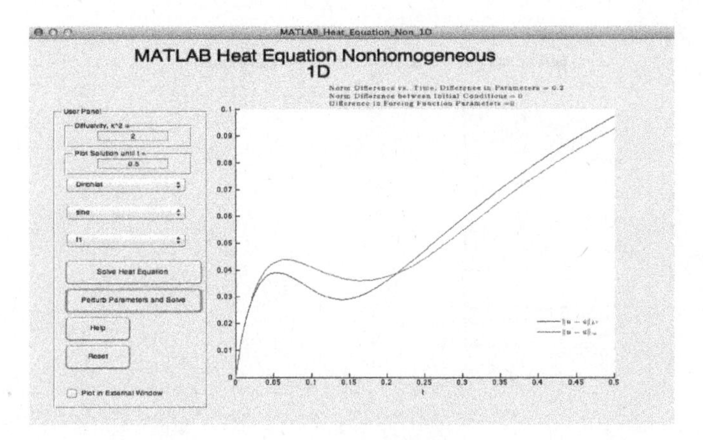

FIGURE 11.11: Perturbing the diffusivity constant in IBVP (11.16). Forcing function, $f_1(x,t) = 5$.

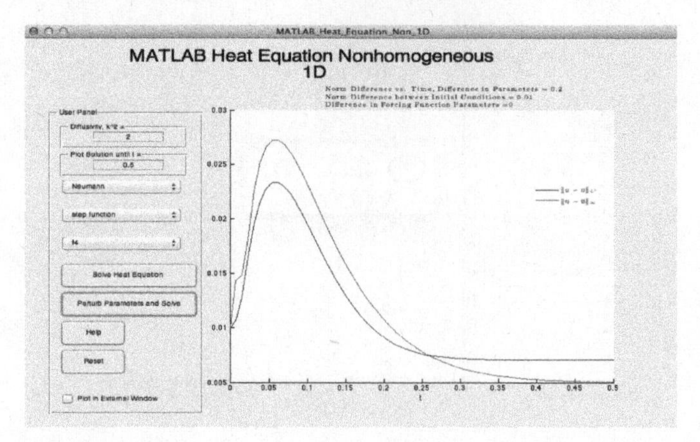

FIGURE 11.12: Perturbing the diffusivity constant and initial condition in IBVP (11.8). Forcing function, $f_4(x,t) = e^{-5t}$.

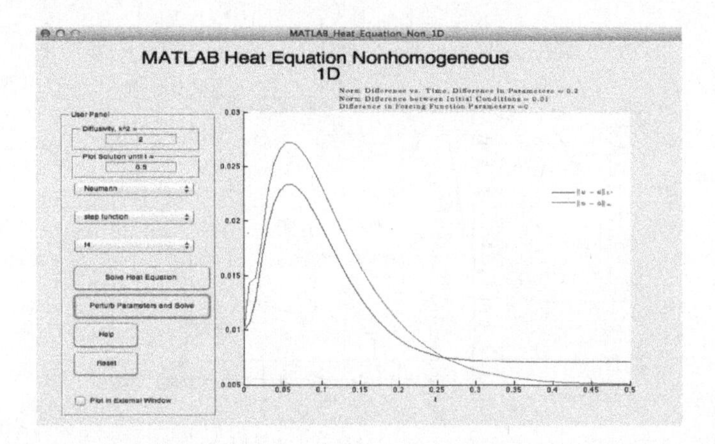

FIGURE 11.13: Perturbing the diffusivity constant and initial condition in IBVP (11.8). Forcing function, $f_4(x,t) = e^{-0.5t}$.

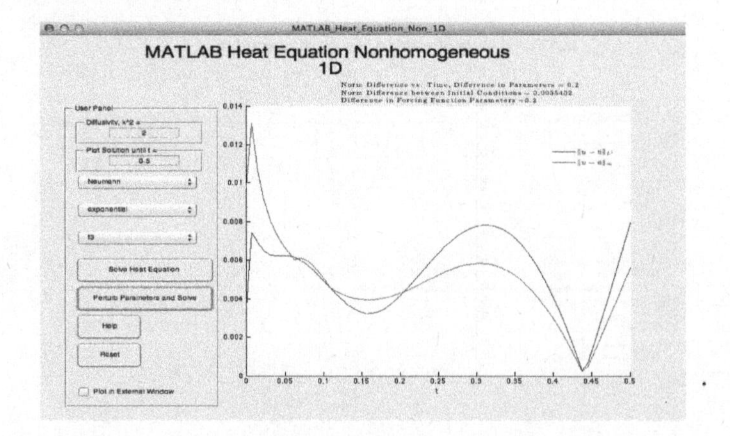

FIGURE 11.14: Perturbing the diffusivity constant, the initial condition, and forcing function parameter in IBVP (11.8). Forcing function, $f_3(x,t) = \cos 10t$.

i.) Consider IBVP (11.16) with the sine initial condition option and the forcing function option f1.

a.) Investigate the continuous dependence of the solution of IBVP (11.16) on the diffusivity constant. Use $k^2 = 2$ and when prompted set A in the definition of $f_1(x,t)$ equal to 1. Plot solutions until $t = 0.5$. Once the movie has finished playing, click the *Perturb Parameters and Solve* button and choose a perturbed diffusivity constant $k^2 = 1.8$. Keep all other parameters the same as in the original run of the GUI. Describe the differences in the solutions. Compare and contrast the results using the different norms.

b.) Choose $k^2 = 1.9, 2.1, 1.99,$ and 2.01. Record your observations regarding the difference in the solutions. Make a conjecture regarding continuous dependence on the diffusivity constant.

c.) Repeat (a) and (b) using $A = 5$ in the definition of $f_1(x,t)$.

d.) Repeat (a) and (b) using $A = 0.5$ in the definition of $f_1(x,t)$

e.) How did the forcing function influence the differences in the solution? (Hint: See Figures 11.10 and 11.11 for one example.)

d.) Repeat (a)-(e) using the exponential initial condition option. Did you observe any behavior contradictory to your comparisons and conjectures made in (a)-(e)? If so, how?

e.) Repeat (a)-(e) using the step function initial condition option. Did you observe any behavior contradictory to your comparisons and conjectures made in (a)-(e)? If so, how?

ii.) Repeat (i) with the forcing function option f5.

iii.) Consider IBVP (11.8) with the cosine initial condition option and the forcing function option f2.

a.) Investigate the continuous dependence of the solution of IBVP (11.16) on the initial condition. Use $k^2 = 2$ and when prompted, set A and B in the definition of $f_2(x,t)$, equal to 1 and -1, respectively . Plot solutions until $t = 0.5$. Once the movie has finished playing, click the *Perturb Parameters and Solve* button and choose a perturbation value equal to 0.1. Keep all other parameters the same as in the original run of the GUI. Describe the differences in the solutions. Compare and contrast the results using the different norms.

b.) Choose perturbation values equal to $0.05, 0.01,$ and 0.001. Record your observations regarding the difference in the solutions. Make a conjecture regarding continuous dependence on the initial condition.

c.) Repeat (a) and (b) using $A = 5$ and $B = -5$ in the definition of $f_2(x,t)$.

d.) Repeat (a) and (b) using $A = 0.5$ and $B = -0.5$ in the definition of $f_2(x,t)$

e.) How did the forcing function influence the differences in the solution?

f.) Repeat (a)-(e) using the exponential initial condition option. Did you observe any behavior contradictory to your comparisons and conjectures made in (a)-(e)? If so, how?

g.) Repeat (a)-(e) using the step function initial condition option. Did you observe any behavior contradictory to your comparisons and conjectures made in (a)-(e)? If so, how?

iv.) Consider IBVP (11.8) with the cosine initial condition option and the forcing function option f4.

a.) Investigate the continuous dependence of the solution of IBVP (11.16) on the initial condition and diffusivity constant. Use $k^2 = 2$ and when prompted, set a in the definition of $f_4(x, t)$ equal to 1. Plot solutions until $t = 0.5$. Once the movie has finished playing, click the *Perturb Parameters and Solve* button and choose a perturbed diffusivity constant equal to 1.8 and perturbation value equal to 0.1. Keep all other parameters the same as in the original run of the GUI. Describe the differences in the solutions. Compare and contrast the results using the different norms.

b.) Choose perturbed diffusivity constants equal to $2.1, 1.9$, and 2.001 and perturbation values equal to $0.05, 0.01$, and 0.001. Record your observations regarding the difference in the solutions. Make a conjecture regarding continuous dependence on the initial condition and the diffusivity constant.

c.) Repeat (a) and (b) using $a = 5$ in the definition of $f_4(x, t)$.

d.) Repeat (a) and (b) using $a = 0.5$ in the definition of $f_4(x, t)$

e.) How did the forcing function influence the differences in the solution? (Hint: See Figures 11.12 and 11.13 for one example.)

f.) Repeat (a)-(e) using the exponential initial condition option. Did you observe any behavior contradictory to your comparisons and conjectures made in (a)-(e)? If so, how?

g.) Repeat (a)-(e) using the step function initial condition option. Did you observe any behavior contradictory to your comparisons and conjectures made in (a)-(e)? If so, how?

v.) In the GUI select periodic boundary conditions, the cosine initial condition option, and the forcing function option f3.

a.) Investigate the continuous dependence of the solution of IBVP (11.8) on the initial condition, the diffusivity constant, and the forcing function parameter. Use $k^2 = 2$ and when prompted, set ω in the definition of $f_3(x, t)$ equal to 1. Plot solutions until $t = 0.5$. Once the movie has finished playing, click the *Perturb Parameters and Solve* button and choose a perturbed diffusivity constant equal to 1.8, perturbation value equal to 0.1, and forcing function parameter equal to 1.1. Describe the differences in the solutions. Compare and contrast the results using the different norms.

b.) Choose perturbed diffusivity constants equal to $2.1, 1.9$, and 2.001, perturbation values equal to $0.05, 0.01$, and 0.001, and forcing function parameters equal to $0.9, 0.95$, and 1.01. Record your observations regarding the difference in the solutions. Make a conjecture regarding continuous dependence on the initial condition and the diffusivity constant.

c.) Repeat (a)-(b) using the exponential initial condition option. Did you observe any behavior contradictory to your comparisons and conjectures made in (a)-(b)? If so, how?

d.) Repeat (a)-(b) using the step function initial condition option. Did you observe any behavior contradictory to your comparisons and conjectures made in (a)-(b)? If so, how?

vi.) Summarize your observations regarding continuous dependence. Make sure you address the potential issue of using different norms and the influence the forcing function had on the differences in the solutions.

■

Next, we consider the same IBVPs, but now investigate what happens if the BCs are not homogeneous.

EXPLORE!
11.4 Nonhomogeneous BCs

To begin, consider the IBVP

$$\begin{cases} \frac{\partial}{\partial t}z(x,t) = k^2\frac{\partial^2}{\partial x^2}z(x,t) + f(x,t),\ 0 < x < L,\ t > 0, \\ z(x,0) = z_0(x),\ 0 < x < L, \\ z(0,t) = T_1,\ z(L,t) = T_2,\ t > 0, \end{cases} \tag{11.19}$$

where $T_1$and T_2 are nonzero constants. This is the analogue to IBVP (9.21); here we have simply added a forcing term. If we equip this the IBVP instead with Neumann BCs, we get

$$\begin{cases} \frac{\partial}{\partial t}z(x,t) = k^2\frac{\partial^2}{\partial x^2}z(x,t) + f(x,t),\ 0 < x < L,\ t > 0, \\ z(x,0) = z_0(x),\ 0 < x < L, \\ \frac{\partial z(0,t)}{\partial x} = T_1,\ \frac{\partial z(L,t)}{\partial x} = T_2,\ t > 0, \end{cases} \tag{11.20}$$

In the following exercise, we visualize the effect nonhomogeneous BCs have on the nonhomogeneous heat equation, and furthermore, compare the solution to the case of homogeneous BCs.

MATLAB-Exercise 11.1.4. To begin, open **MATLAB** and in the command line, type:

MATLAB_Heat_Non_NonhomogBC_1D

Use the GUI to answer the following questions. Make certain to click on the HELP button for instructions on how to use the GUI. Screenshots of inputting parametres and GUI outputs can be found in Figures 11.15, 11.16, and 11.17.

i.) Solve IBVP (11.19) until $t = 1$ with the sine initial condition option, forcing function option f1, $k^2 = 2$ and $T_1 = 1 = T_2$. When prompted, choose A in the definition of $f_1(x,t)$ to equal 1.

a.) What do you notice initially in the movie and for small times in the surface plot that is so different than in the homogeneous boundary condition case? Why do you think this happens? Does the solution appear to stabilize to a steady-state temperature? If you need more time, increase the plotting time in increments of one until a pattern emerges.

b.) Click the *Compare with Homogeneous BC* button. Compare and contrast the two solutions.

c.) Repeat (a) and (b) with two values of T_1 and T_2 of your choosing.

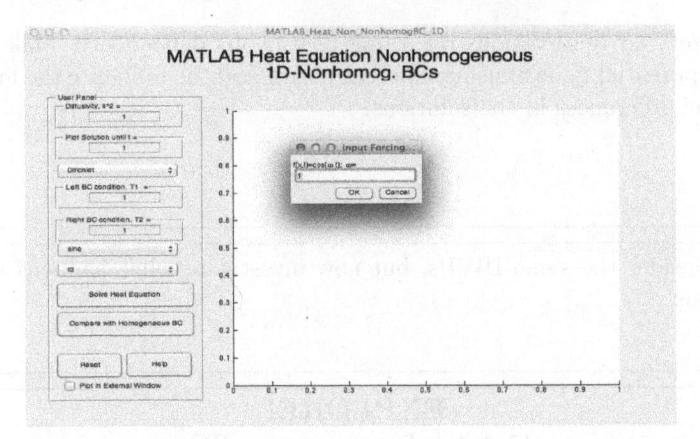

FIGURE 11.15: MATLAB Heat Equation Nonhomogeneous 1D-Nonhomg. BCs GUI

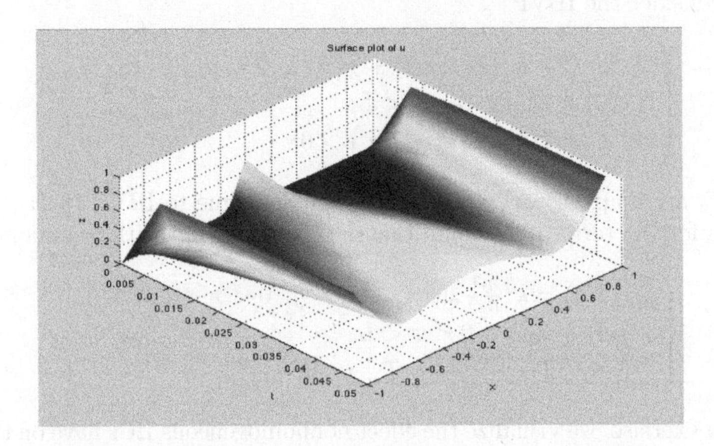

FIGURE 11.16: Solution to (11.19) with $T_1 = 1 = T_2$, $k^2 = 1$, initial condition: e^{-100x^2} and forcing function: $f_5(x,t)$ with $A = 5$.

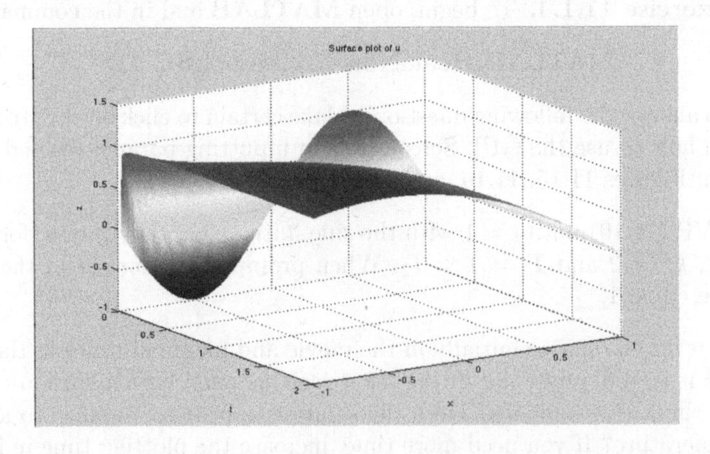

FIGURE 11.17: Same IBVP as in Figure 9.8 with the addition of the forcing term $f_1(x,t) = 1$.

d.) Repeat (a)-(c) using $A = 5$. How did increasing A in the definition of $f_1(x,t)$ affect the solution to IBVP 11.19?

e.) Repeat (a)-(c) usin $A = 0.01$. How did decreasing A to almost zero affect the solution to IBVP (11.19)?

f.) Repeat (a)-(e) using the exponential option for the initial condition.

g.) Repeat (a)-(e) using the step function option for the initial condition.

h.) How did changing the initial conditions change the evolution of the solution? Did the effect of changing the forcing parameters remain the same for each initial condition?

ii.) Repeat (i) with forcing function option f5.

iii.) Solve IBVP 11.20 until $t = 1$ with the cosine initial condition option, forcing function option f2, $k^2 = 2$ and $T_1 = -1 = T_2$. When prompted, choose A and B in the definition of $f_2(x,t)$ to equal 1 and -1, respectively.

a.) Describe the evolution of the solution. Does the solution appear to stabilize to a steady-state temperature? If you need more time, increase the plotting time in increments of one until a pattern emerges.

b.) Click the *Compare with Homogeneous BC* button. Compare and contrast the two solutions.

c.) Repeat (a) and (b) with two values of T_1 and T_2 of your choosing.

d.) Repeat (a)-(c) using $A = 5$ and $B = -1$. How did increasing A in the definition of $f_2(x,t)$ affect the solution to IBVP 11.20?

e.) Repeat (a)-(c) using $A = 0.01 = B$. How did decreasing A and B to almost zero affect the solution to IBVP 11.20?

f.) Repeat (a)-(e) using the exponential option for the initial condition.

g.) Repeat (a)-(e) using the step function option for the initial condition.

h.) How did changing the initial conditions change the evolution of the solution? Did the effect of changing the forcing parameters remain the same for each initial condition?

iv.) Solve IBVP 11.20 until $t = 1$ with the cosine initial condition option, forcing function option f3, $k^2 = 2$ and $T_1 = -1 = T_2$. When prompted, choose ω in the definition of $f_3(x,t)$ to equal 5.

a.) Describe the evolution of the solution. Does the solution appear to stabilize to a steady-state temperature? If you need more time, increase the plotting time in increments of one until a pattern emerges.

b.) Click the *Compare with Homogeneous BC* button. Compare and contrast the two solutions.

c.) Repeat (a) and (b) with two values of T_1 and T_2 of your choosing.

d.) Repeat (a)-(c) using $\omega = 10$. How did increasing ω in the definition of $f_3(x,t)$ affect the solution to IBVP 11.20?

e.) Repeat (a)-(c) using $\omega = 0$.

f.) Repeat (a)-(e) using the exponential option for the initial condition.

g.) Repeat (a)-(e) using the step function option for the initial condition.

h.) How did changing the initial conditions change the evolution of the solution? Did the effect of changing the forcing parameters remain the same for each initial condition?

■

11.1.2 Two-Dimensional Diffusion with Forcing

We now consider a similar model for heat conduction with external forcing in a two-dimensional rectangular plate composed of an isotropic, uniform material. Suppose that a heat source is positioned in proximity to one edge of a rectangular slab of material for which we are monitoring the temperature over time. Assuming that its intensity increases with time, a possible generalization of IBVP (9.23) describing this scenario is given by

$$\begin{cases} \frac{\partial}{\partial t}z(x,y,t) &= k^2\Delta z(x,y,t) + f(x,y,t),\ 0<x<a,\ 0<y<b,\ t>0, \\ z(x,y,0) &= z_0(x,y),\ 0<x<a,\ 0<y<b, \\ z(x,0,t) &= 0 = z(x,b,t),\ 0<x<a,\ t>0, \\ z(0,y,t) &= 0 = z(a,y,t),\ 0<y<b,\ t>0, \end{cases} \tag{11.21}$$

where $z(x,y,t)$ represents the temperature at the point (x,y) on the plate at time t and f and z_0 are sufficiently smooth as to both have a unique Fourier-type representation of the type introduced in (11.9). For example,

$$\begin{cases} \frac{\partial}{\partial t}z(x,y,t) &= k^2\Delta z(x,y,t) + t^2 + x + 2y,\ 0<x<a,\ 0<y<b,\ t>0, \\ z(x,y,0) &= \sin 2x + \cos 2y,\ 0<x<a,\ 0<y<b, \\ z(x,0,t) &= 0 = z(x,b,t),\ 0<x<a,\ t>0, \\ z(0,y,t) &= 0 = z(a,y,t),\ 0<y<b,\ t>0, \end{cases} \tag{11.22}$$

would describe the temperature at the point (x,y) for an initial temperature profile of $\sin 2x + \cos 2y$ and forcing function $t^2 + x + 2y$. Notice that the forcing function affects the system both in time and space.

Returning to (11.21), the separation of variables method can be used to show that the form of a solution of (9.23) (the homogeneous version of (11.21)) is given by

$$z(x,y,t) = \sum_{m=1}^{\infty}\sum_{n=1}^{\infty} b_{mn}X_m(x)Y_n(y)T_{mn}(t). \tag{11.23}$$

Emulating the process introduced in Exercise 12.2, we assume that f can be expressed uniquely in the form

$$f(x,y,t) = \sum_{m=1}^{\infty}\sum_{n=1}^{\infty} f_{mn}(t)X_m(x)Y_n(y),\ 0<x<a,\ 0<y<b,\ t>0, \tag{11.24}$$

where the coefficients $f_{mn}(t)$ are defined by

$$f_{mn}(t) = \frac{4}{ab}\int_0^a\int_0^b f(v,w,t)X\left(\frac{m\pi}{a}v\right)Y\left(\frac{n\pi}{b}v\right)dwdv,\ t>0. \tag{11.25}$$

We assume that the solution of (11.21) is of the form

$$z(x,y,t) = \sum_{m=1}^{\infty}\sum_{n=1}^{\infty} z_{mn}(t) X\left(\frac{m\pi}{a}x\right) Y\left(\frac{n\pi}{b}y\right) \quad 0 < x < a,\ t > 0. \tag{11.26}$$

As before, for each $m, n \in \mathbb{N}$, $z_{mn}(t)$ satisfies the IVP

$$\begin{cases} z'_{mn}(t) + \left(\frac{m^2\pi^2}{a^2} + \frac{n^2\pi^2}{b^2}\right) z_{mn}(t) = f_{mn}(t),\ t > 0, \\ z_{mn}(0) = z^0_{mn}. \end{cases} \tag{11.27}$$

This is a linear ODE which can be solved using variation of parameters. Once done and you have substituted these functions into (11.26), it can be shown that the solution is given by

$$z(\cdot,t) = e^{At}[z_0][\cdot] + \int_0^t e^{A(t-s)} f(s,\cdot) ds,\ t > 0, \tag{11.28}$$

where the semigroup is now the one obtained when studying the homogeneous version of (11.21).

As in previous sections, one can equip the heat equation with Neumann or periodic boundary conditions, which alters the evolution of the solution as compared to the Dirichlet case. With Neumann boundary conditions, (11.21) becomes

$$\begin{cases} \frac{\partial}{\partial t} z(x,y,t) & = k^2 \Delta z(x,y,t) + f(x,y,t),\ 0 < x < a,\ 0 < y < b,\ t > 0, \\ z(x,y,0) & = z_0(x,y),\ 0 < x < a,\ 0 < y < b, \\ \frac{\partial z(x,0,t)}{\partial y} & = 0 = \frac{\partial z(x,b,t)}{\partial y},\ 0 < x < a,\ t > 0, \\ \frac{\partial z(0,y,t)}{\partial x} & = 0 = \frac{\partial z(a,y,t)}{\partial x},\ 0 < y < b,\ t > 0, \end{cases} \tag{11.29}$$

whereas when using periodic boundary conditions, (11.21) becomes

$$\begin{cases} \frac{\partial}{\partial t} z(x,y,t) & = k^2 \Delta z(x,y,t) + f(x,y,t),\ 0 < x < a,\ 0 < y < b,\ t > 0, \\ z(x,y,0) & = z_0(x,y),\ 0 < x < a,\ 0 < y < b, \\ z(x,0,t) & = z(x,b,t),\ 0 < x < a,\ t > 0, \\ z(0,y,t) & = z(a,y,t),\ 0 < y < b,\ t > 0. \end{cases} \tag{11.30}$$

We investigate the behavior of the solution of such two-dimensional diffusion IBVPs using MATLAB as in the previous section.

EXPLORE!

11.5 Parameter Play!

In this **EXPLORE!** you will investigate the effects of changing K, λ, and c used to define the diffusivity constant k^2. To do this, you will use the MATLAB Heat Equation Nonhomogeneous 2D GUI to visualize the solutions to IBVP (11.8) and IBVP (11.16) for different choices of k^2.

MATLAB-Exercise 11.1.5. To begin, open **MATLAB** and in the command line, type:

MATLAB_Heat_Equation_Non_2D

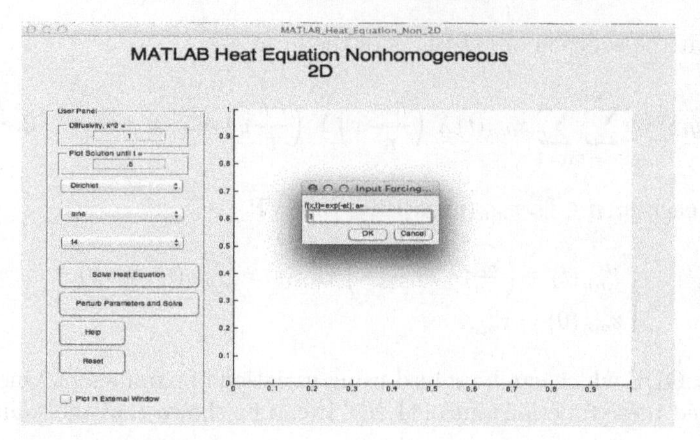

FIGURE 11.18: MATLAB Heat Equation Nonhomogeneous 2D GUI

Use the GUI to answer the following questions. Make certain to click on the HELP button for instructions on how to use the GUI. Screenshots of inputting parameters and GUI output can be found in Figure 11.18 and Figure 11.19.

i.) Consider IBVP (11.21) with the sine initial condition option and the forcing function option f1.

a.) Use $k^2 = 1$, plot solutions until $t = 0.5$ and when prompted set A in the definition of $f_1(x, t)$ equal to 1. Describe the evolution of the solution. How did it differ from the corresponding one-dimensional nonhomogeneous heat equation?

b.) Based on your intuition gained in Chapter 9 regarding the heat equation, how do you expect the solution of (11.21) to evolve when $k^2 = 2$, as compared to (a)? Check your intuition by using the GUI to visualize this situation.

c.) What do you expect to happen if $k^2 = 0.25$? Test your hypothesis with the GUI.

d.) Repeat (a) using $A = 0.5$. How would you compare this situation with the one in (a) and the homogeneous heat equation solution?

e.) Repeat (d) using $A = 0.25, 0.1$, and 0.001. In addition, make a conjecture for

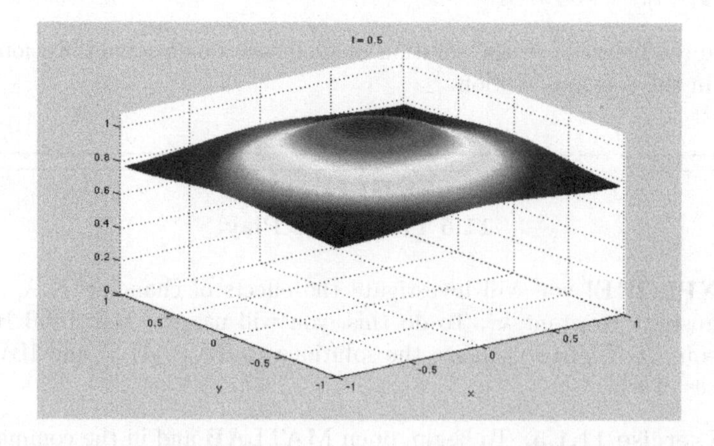

FIGURE 11.19: IBVP (11.29) with $k^2 = 1$, forcing function and initial condition: step function.

whether or not these nonhomogeneous solutions converge to the homogeneous solution as $f_1(x,t)$ converges to the zero function.

f.) What if A gets larger? Explore this by choosing a sequence of increasingly larger values for A. Summarize your observations.

g.) Repeat (a)-(f) using the exponential initial condition option. Did you observe any behavior contradictory to your conjectures in (b), (c), and (e)? If so, how?

h.) Repeat (a)-(f) using the step function initial condition option. Did you observe any behavior contradictory to your conjectures in (b), (c), and (e)? If so, how?

ii.) Repeat (i) with the forcing function option f5.

iii.) Consider IBVP (11.29) with the cosine initial condition option and the forcing function option f2.

a.) Use $k^2 = 1$, plot solutions until $t = 0.5$, and when prompted, set A and B in the definition of $f_2(x,t)$, equal to 1 and -1, respectively . Describe the evolution of the solution. How did it differ from the corresponding one-dimensional nonhomogeneous heat equation?

b.) Based on your intuition gained in Chapter 9 regarding the heat equation, how do you expect the solution of (11.29) to evolve when $k^2 = 2$, as compared to (a)? Check your intuition by using the GUI to visualize this situation.

c.) What do you expect to happen if $k^2 = 0.25$? Test your hypothesis with the GUI.

d.) Repeat (a) using $A = 0.5$ and $B = -0.5$. How would you compare this situation with the one in (a) and the homogeneous heat equation solution?

e.) Repeat (d) using $A = 0.25, 0.1$, and 0.001 and $B = -0.25, -0.1$, and -0.001. In addition, make a conjecture for whether or not these nonhomogenous solutions converge to the homogeneous solution as $f_2(x,t)$ converges to the zero function.

f.) What if A gets larger and B more negative? Explore!

g.) Repeat (a)-(f) using the exponential initial condition option. Did you observe any behavior contradictory to your conjectures in (b), (c), and (e)? If so, how?

h.) Repeat (a)-(f) using the step function initial condition option. Did you observe any behavior contradictory to your conjectures in (b), (c), and (e)? If so, how?

iv.) Consider IBVP (11.29) with the cosine initial condition option and the forcing function option f4.

a.) Use $k^2 = 1$, plot solutions until $t = 0.5$, and when prompted, set a in the definition of $f_4(x,t)$ equal to 1. Describe the evolution of the solution. How did it differ from the corresponding one-dimensional nonhomogeneous heat equation?

b.) Based on your intuition gained in Chapter 9 regarding the heat equation, how do you expect the solution of (11.29) to evolve when $k^2 = 2$, as compared to (a)? Check your intuition by using the GUI to visualize this situation.

c.) What do you expect to happen if $k^2 = 0.25$? Test your hypothesis with the GUI.

d.) Repeat (a) using $a = 0.5$. How would you compare this situation with the one in (a) and the homogeneous heat equation solution?

e.) Repeat (d) using $a = 0.25, 0.1$, and 0.001. Does the solution appear to get closer to the homogeneous solution?

f.) What if a gets larger? Explore!

g.) Repeat (a)-(f) using the exponential initial condition option. Did you observe any behavior contradictory to your conjectures in (b), (c), and (e)? If so, how?

h.) Repeat (a)-(f) using the step function initial condition option. Did you observe any behavior contradictory to your conjectures in (b), (c), and (e)? If so, how?

v.) Consider IBVP (11.30) with $k^2 = 1$, the cosine initial condition option, and the forcing function option f3.

a.) Use $k^2 = 1$, plot solutions until $t = 0.5$, and when prompted, set ω in the definition of $f_3(x,t)$ equal to 1. Describe the evolution of the solution. How did it differ from the corresponding one-dimensional nonhomogeneous heat equation?

b.) Based on your intuition gained in Chapter 9 regarding the heat equation, how do you expect the solution of (11.30) to evolve when $k^2 = 2$, as compared to (a)? Check your intuition by using the GUI to visualize this situation.

c.) What do you expect to happen if $k^2 = 0.25$? Test your hypothesis with the GUI.

d.) Repeat (a) using $a = 0.5$. How would you compare this situation with the one in (a) and the homogeneous heat equation solution?

e.) Repeat (d) using $a = 0.25, 0.1$, and 0.001. Explain why the solution does **not** tend to the homogeneous solution.

f.) What if a gets larger? Explore!

g.) Repeat (a)-(f) using the exponential initial condition option. Did you observe any behavior contradictory to your conjectures in (b), (c), and (e)? If so, how?

h.) Repeat (a)-(f) using the step function initial condition option. Did you observe any behavior contradictory to your conjectures in (b), (c), and (e)? If so, how?

■

EXPLORE!

11.6 Where Has All The Heat Gone?

MATLAB-Exercise 11.1.6. To begin, open **MATLAB** and in the command line, type:

MATLAB_Heat_Equation_Non_2D

Use the GUI to answer the following questions. Make certain to click on the HELP button for instructions on how to use the GUI. Screenshots of GUI output can be found in Figures 11.20 and 11.21.

i.) Consider IBVP (11.21) with the sine initial condition option and the forcing function option f1.

a.) Investigate the behavior of the solution of IBVP (11.21) for large times. Use $k^2 = 2$ and when prompted set A in the definition of $f_1(x,t)$ equal to 1. Begin by plotting solutions until $t = 2$. If necessary, slowly increase the plotting time until you are able to describe the behavior of the solution. Remember, a computer cannot handle the infinite! Compare the behavior here with the homogeneous heat equation.

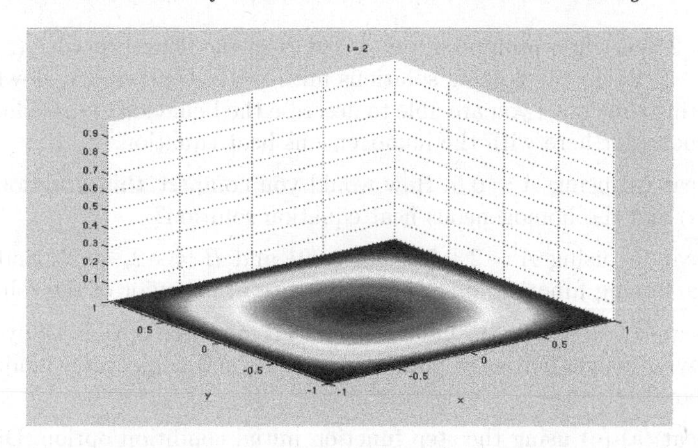

FIGURE 11.20: IBVP (11.21) with $k^2 = 1$, forcing function: step function with $A = 0.001$, initial condition: e^{-40x^2}.

b.) Repeat (a) using $A = 0.5$. How would you compare this situation with the one in (a) and the homogeneous heat equation solution?

c.) Repeat (b) using $A = 2, 0.25$, and 0.001. How is the forcing function affecting the long-term behavior of the solution?

d.) Repeat (a)-(c) using the exponential initial condition option. Did you observe any behavior contradictory to your comparisons and conjectures made in (a)-(c)? If so, how?

e.) Repeat (a)-(c) using the step function initial condition option. Did you observe any behavior contradictory to your comparisons and conjectures made in (a)-(c)? If so, how?

ii.) Repeat (i) with the forcing function option f5.

iii.) Consider IBVP (11.29) with the cosine initial condition and the forcing function option f2.

a.) Investigate the behavior of the solution of IBVP (11.29) for large times. Use

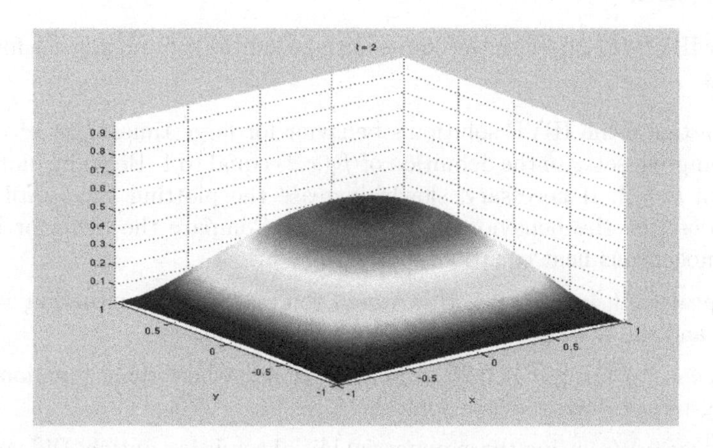

FIGURE 11.21: IBVP (11.21) with $k^2 = 1$, forcing function: step function with $A = 2$, initial condition: e^{-40x^2}.

$k^2 = 2$ and when prompted, set A and B in the definition of $f_2(x,t)$, equal to 1 and -1. Begin by plotting solutions until $t = 2$. If necessary, slowly increase the plotting time until you are able to describe the behavior of the solution. Compare the behavior here with the homogeneous heat equation.

b.) Repeat (a) using $A = 0.5$. How would you compare this situation with the one in (a) and the homogeneous heat equation solution?

c.) Repeat (b) using $A = 2, 0.25$, and 0.001 and $B = -2, -0.25$, and -0.001. How is the forcing function affecting the long-term behavior of the solution?

d.) Repeat (a)-(c) using the exponential initial condition option. Did you observe any behavior contradictory to your comparisons and conjectures made in (a)-(c)? If so, how?

e.) Repeat (a)-(c) using the step function initial condition option. Did you observe any behavior contradictory to your comparisons and conjectures made in (a)-(c)? If so, how?

iv.) Consider IBVP (11.29) with the cosine initial condition option and the forcing function option f4.

a.) Investigate the behavior of the solution of IBVP (11.29) for large times. Use $k^2 = 2$ and when prompted, set a in the definition of $f_4(x,t)$ equal to 1. Begin by plotting solutions until $t = 2$. If necessary, slowly increase the plotting time until you are able to describe the behavior of the solution. Compare the behavior here with the homogeneous heat equation.

b.) Repeat (a) using $a = 0.5$. How would you compare this situation with the one in (a) and the homogeneous heat equation solution?

c.) Repeat (b) using $a = 0.25, 2$, and 10. How is the forcing function affecting the long-term behavior of the solution?

d.) Repeat (a)-(c) using the exponential initial condition option. Did you observe any behavior contradictory to your comparisons and conjectures made in (a)-(c)? If so, how?

e.) Repeat (a)-(c) using the step function initial condition option. Did you observe any behavior contradictory to your comparisons and conjectures made in (a)-(c)? If so, how?

v.) Consider IBVP (11.30) with the cosine initial condition option and the forcing function option f3.

a.) Investigate the IBVP solution's behavior for large times. Use $k^2 = 2$ and when prompted, set ω in the definition of $f_3(x,t)$ equal to 1. Begin by plotting solutions until $t = 2$. If necessary, slowly increase the plotting time until you are able to describe the behavior of the solution. Compare the behavior here with the homogeneous heat equation.

b.) Repeat (a) using $a = 0.5$. How would you compare this situation with the one in (a) and the homogeneous heat equation solution?

c.) Repeat (b) using $a = 0.25, 2$, and 10. How is the forcing function affecting the long-term behavior of the solution?

d.) Repeat (a)-(c) using the exponential initial condition option. Did you observe any behavior contradictory to your comparisons and conjectures made in (a)-(c)? If so, how?

e.) Repeat (a)-(c) using the step function initial condition option. Did you observe any behavior contradictory to your comparisons and conjectures made in (a)-(c)? If so, how?

■

EXPLORE!

11.7 Continuous Dependence on Parameters

MATLAB-Exercise 11.1.7. To explore continuous dependence with the GUI one must first solve the nonhomogeneous heat equation with a particular diffusivity constant, initial condition, and forcing function. Once a solution has been constructed, click the *Perturb Parameters and Solve* button to specify the perturbed diffusivity constant and a perturbation size for the initial position. The GUI will also allow for a perturbed forcing function parameter. When prompted to select a norm choose *All* to investigate dependence using both norms at the same time. To begin, open **MATLAB** and in the command line, type:

MATLAB_Heat_Equation_Non_2D

Use the GUI to answer the following questions. Make certain to click on the HELP button for instructions on how to use the GUI. Screenshots of inputting parameters and GUI output can be found in Figures 11.22–11.27.

i.) Consider IBVP (11.21) with the sine initial condition option and the forcing function option f1.

a.) Investigate the continuous dependence of the solution on the diffusivity constant for IBVP (11.21). Use $k^2 = 2$ and when prompted set A in the definition of $f_1(x,t)$ equal to 1. Plot solutions until $t = 0.5$. Once the movie has finished playing, click the *Perturb Parameters and Solve* button and choose a perturbed diffusivity constant $k^2 = 1.8$. Keep all other parameters the same as in the original run of the GUI. Describe the differences in the solutions. Compare and contrast the results using the different norms.

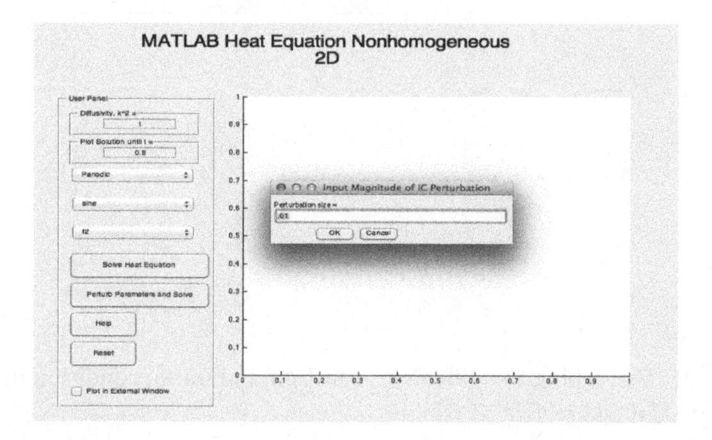

FIGURE 11.22: Perturbing parameters, initial conditions, and forcing function parameters with MATLAB Heat Equation Nonhomogeneous 2D GUI

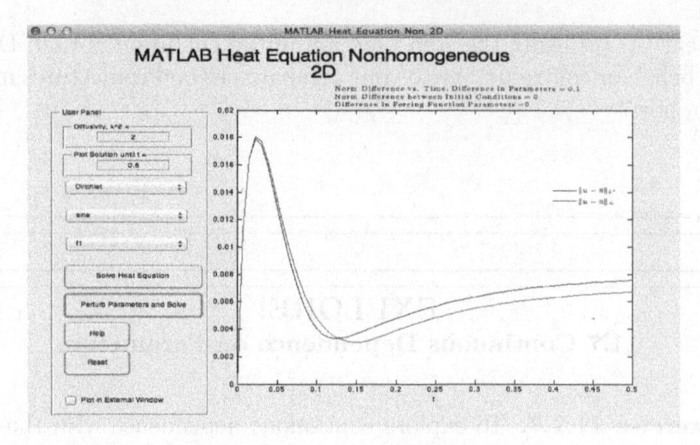

FIGURE 11.23: Perturbing the diffusivity constant in IBVP (11.21). Forcing function, $f_1(x,t) = 1$.

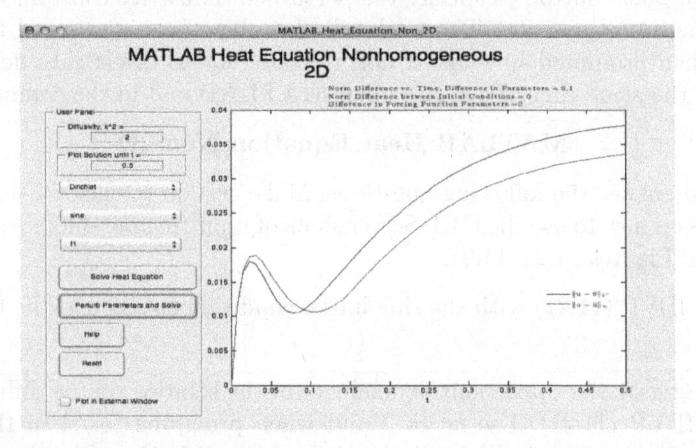

FIGURE 11.24: Perturbing the diffusivity constant in IBVP (11.21). Forcing function, $f_1(x,t) = 5$.

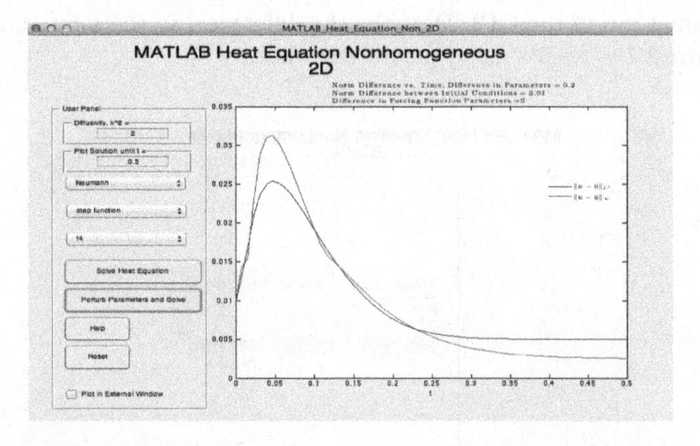

FIGURE 11.25: Perturbing the diffusivity constant and initial condition in IBVP (11.29). Forcing function, $f_4(x,t) = e^{-5t}$.

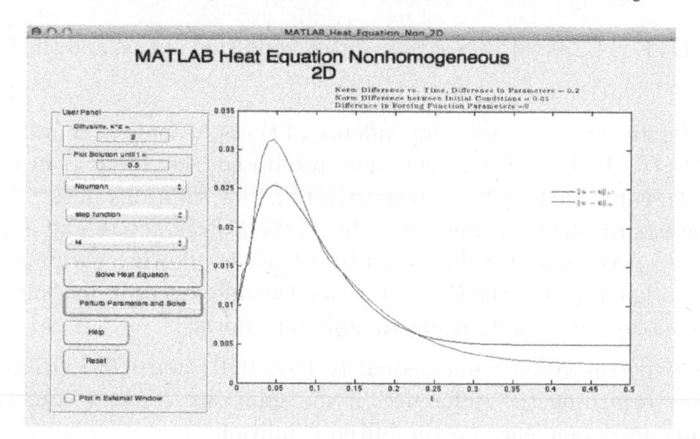

FIGURE 11.26: Perturbing the diffusivity constant and initial condition in IBVP (11.29). Forcing function, $f_4(x,t) = e^{-0.5t}$.

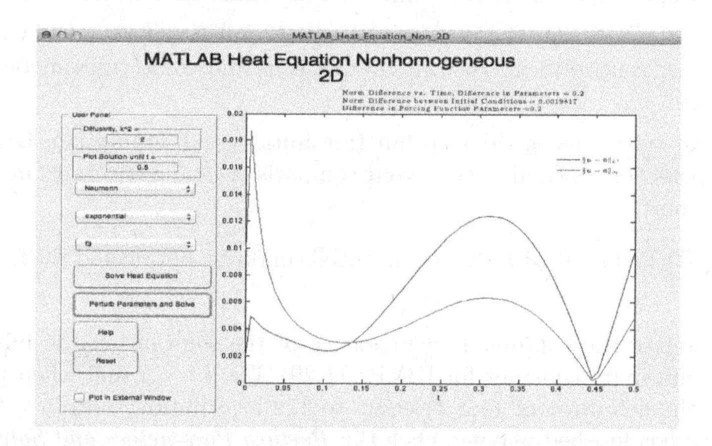

FIGURE 11.27: Perturbing the diffusivity constant, the initial condition, and forcing function parameter in IBVP (11.29). Forcing function, $f_3(x,t) = \cos 10t$.

b.) Choose $k^2 = 1.9, 2.1, 1.99$, and 2.01. Record your observations regarding the difference in the solutions. Make a conjecture regarding continuous dependence on the diffusivity constant.

c.) Repeat (a) and (b) using $A = 5$ in the definition of $f_1(x,t)$.

d.) Repeat (a) and (b) using $A = 0.5$ in the definition of $f_1(x,t)$.

e.) How did the forcing function influence the differences in the solution? (Hint: See Figures 11.10 and 11.11 for one example.)

d.) Repeat (a)-(e) using the exponential initial condition option. Did you observe any behavior contradictory to your comparisons and conjectures made in (a)-(e)? If so, how?

e.) Repeat (a)-(e) using the step function initial condition option. Did you observe any behavior contradictory to your comparisons and conjectures made in (a)-(e)? If so, how?

ii.) Repeat (i) with the forcing function option f5.

iii.) Consider IBVP (11.29) with the cosine initial condition option and the forcing function option f2.

a.) Investigate the continuous dependence of the solution on the initial condition for IBVP (11.29). Use $k^2 = 2$ and when prompted, set A and B in the definition of $f_2(x,t)$, equal to 1 and -1, respectively . Plot solutions until $t = 0.5$. Once the movie has finished playing, click the *Perturb Parameters and Solve* button and choose a perturbation value equal to 0.1. Keep all other parameters the same as in the original run of the GUI. Describe the differences in the solutions. Compare and contrast the results using the different norms.

b.) Choose perturbation values equal to $0.05, 0.01$, and 0.001. Record your observations regarding the difference in the solutions. Make a conjecture regarding continuous dependence on the initial condition.

c.) Repeat (a) and (b) using $A = 5$ and $B = -5$ in the definition of $f_2(x,t)$.

d.) Repeat (a) and (b) using $A = 0.5$ and $B = -0.5$ in the definition of $f_2(x,t)$.

e.) How did the forcing function influence the differences in the solution?

d.) Repeat (a)-(e) using the exponential initial condition option. Did you observe any behavior contradictory to your comparisons and conjectures made in (a)-(e)? If so, how?

e.) Repeat (a)-(e) using the step function initial condition option. Did you observe any behavior contradictory to your comparisons and conjectures made in (a)-(e)? If so, how?

iv.) Consider IBVP (11.29) with the cosine initial condition option and the forcing function option f4.

a.) Investigate the continuous dependence of the solution on the initial condition and diffusivity constant for IBVP (11.29). Use $k^2 = 2$ and when prompted, set a in the definition of $f_4(x,t)$ equal to 1. Plot solutions until $t = 0.5$. Once the movie has finished playing, click the *Perturb Parameters and Solve* button and choose a perturbed diffusivity constant equal to 1.8 and perturbation value equal to 0.1. Keep all other parameters the same as in the original run of the GUI. Describe the differences in the solutions. Compare and contrast the results using the different norms.

b.) Choose perturbed diffusivity constants equal to $2.1, 1.9$, and 2.001 and perturbation values equal to $0.05, 0.01$, and 0.001. Record your observations regarding the difference in the solutions. Make a conjecture regarding continuous dependence on the initial condition and the diffusivity constant.

c.) Repeat (a) and (b) using $a = 5$ in the definition of $f_4(x,t)$.

d.) Repeat (a) and (b) using $a = 0.5$ in the definition of $f_4(x,t)$

e.) How did the forcing function influence the differences in the solution? (Hint: See Figures 11.12 and 11.13 for one example.)

d.) Repeat (a)-(e) using the exponential initial condition option. Did you observe any behavior contradictory to your comparisons and conjectures made in (a)-(e)? If so, how?

e.) Repeat (a)-(e) using the step function initial condition option. Did you observe any behavior contradictory to your comparisons and conjectures made in (a)-(e)? If so, how?

v.) Consider IBVP (11.29) with the cosine initial condition option and the forcing function option f3.

a.) Investigate the IBVP continuous dependence on the initial condition, the diffusivity constant, and the forcing function parameter. Use $k^2 = 2$ and when prompted, set ω in the definition of $f_3(x,t)$ equal to 1. Plot solutions until $t = 0.5$. Once the movie has finished playing, click the *Perturb Parameters and Solve* button and choose a perturbed diffusivity constant equal to 1.8, perturbation value equal to 0.1, and forcing function parameter equal to 1.1. Describe the differences in the solutions. Compare and contrast the results using the different norms.

b.) Choose perturbed diffusivity constants equal to $2.1, 1.9$, and 2.001, perturbation values equal to $0.05, 0.01$, and 0.001, and forcing function parameters equal to $0.9, 0.95$, and 1.01. Record your observations regarding the difference in the solutions. Make a conjecture regarding continuous dependence on the initial condition and the diffusivity constant.

c.) Repeat (a)-(b) using the exponential initial condition option. Did you observe any behavior contradictory to your comparisons and conjectures made in (a)-(b)? If so, how?

d.) Repeat (a)-(b) using the step function initial condition option. Did you observe any behavior contradictory to your comparisons and conjectures made in (a)-(b)? If so, how?

vi.) Summarize your observations regarding continuous dependence. Make sure you address the potential issue of using different norms and the influence the forcing function had on the differences in the solutions.

■

11.1.3 Abstract Formulation

The structure of the IBVPs studied in the previous two subsections, together with the structure of their solutions, is reminiscent of the familiar variation of parameters formula for nonhomogeneous linear ODES studied in Part I, and suggests that if we identified the terms correctly between the IBVPs and the abstract evolution equation

$$\begin{cases} u'(t) = Au(t) + F(t), \ t > 0, \\ u(0) = u_0, \end{cases} \tag{11.31}$$

then a viable representation formula for the solution might very well be

$$u(t) = e^{At}[u_0] + \int_0^t e^{A(t-s)} F(s) ds, \ t > 0. \tag{11.32}$$

The only missing step *per se* is formulating the external forcing term $f(x,t)$ or $f(x,y,t)$ as a Hilbert-space valued function $F(t)$. This is actually quite natural and is similar to how we interpret the solution function itself in this manner. Symbolically, we would define $F : [0,T] \to \mathcal{H}$ by

$$F(t)[x] = f(x,t), \ 0 < x < L, 0 < t < T. \tag{11.33}$$

Exercise 11.1.4. Consider IBVP (11.8) with each of the five forcing functions (11.1) - (11.5). For each, reformulate the IBVP as an abstract evolution equation in an appropriate Hilbert space in the form (11.31).

11.2 Seepage of Fluid Through Fissured Rocks

11.2.1 One-Dimensional Fissured Rock Model with Forcing

As with the heat model, external factors that were originally ignored in the formulation of the mathematical model for a network of fissured rocks can now be taken into account if they can be expressed in the form $f(x,t)$. Doing so leads to an IBVP of the form :

$$\begin{cases} \frac{\partial}{\partial t}\left(p(x,t) - \alpha\frac{\partial^2}{\partial x^2}p(x,t)\right) = \beta\frac{\partial^2}{\partial x^2}p(x,t) + f(x,t),\ 0 < x < \pi,\ t > 0, \\ p(x,0) = p_0(x),\ 0 < x < \pi, \\ \frac{\partial p(0,t)}{\partial x} = \frac{\partial p(\pi,t)}{\partial x} = 0,\ t > 0. \end{cases} \tag{11.34}$$

If one wanted to include external factors for fluid seepage within a network of fissured rocks where periodic boundary conditions are appropriate, the model becomes

$$\begin{cases} \frac{\partial}{\partial t}\left(p(x,t) - \alpha\frac{\partial^2}{\partial x^2}p(x,t)\right) = \beta\frac{\partial^2}{\partial x^2}p(x,t) + f(x,t),\ 0 < x < \pi,\ t > 0, \\ p(x,0) = p_0(x),\ 0 < x < \pi, \\ p(0,t) = p(\pi,t),\ t > 0. \end{cases} \tag{11.35}$$

As shown in Chapter 9, the fluid seepage and the soil mechanics models are the same equation, yet coupled with different boundary conditions. The one-dimensional soil mechanics model

$$\begin{cases} \frac{\partial}{\partial t}\left(p(x,t) - \alpha\frac{\partial^2}{\partial x^2}p(x,t)\right) = \beta\frac{\partial^2}{\partial x^2}p(x,t) + f(x,t),\ 0 < x < \pi,\ t > 0, \\ p(x,0) = p_0(x),\ 0 < x < \pi, \\ p(0,t) = p(\pi,t) = 0,\ t > 0. \end{cases} \tag{11.36}$$

differs only from the fluid seepage model in that it uses Dirichlet BCs. For brevity, we now combine both models by introducing a Dirichlet BCs option to the fluid seepage GUI and corresponding exercises.

EXPLORE!

11.8 Parameter Play!

In this **EXPLORE!**, we visualize the effect changing α and β have on solutions to IBVPs (11.36)-(11.35) when a forcing function is present. The **EXPLORE!** is a direct extension of the homogeneous case studied in **EXPLORE!** 9.9.

MATLAB-Exercise 11.2.1. To begin, open **MATLAB** and in the command line, type:

MATLAB_Fluid_Seepage_Non_1D

Use the GUI to answer the following questions. Make certain to click on the HELP button for instructions on how to use the GUI. A screenshot of inputting parameters and GUI output can be found in Figure 11.28 and Figure 11.29.

i.) Consider IBVP (11.36) with the sine initial condition option and the forcing function option f1.

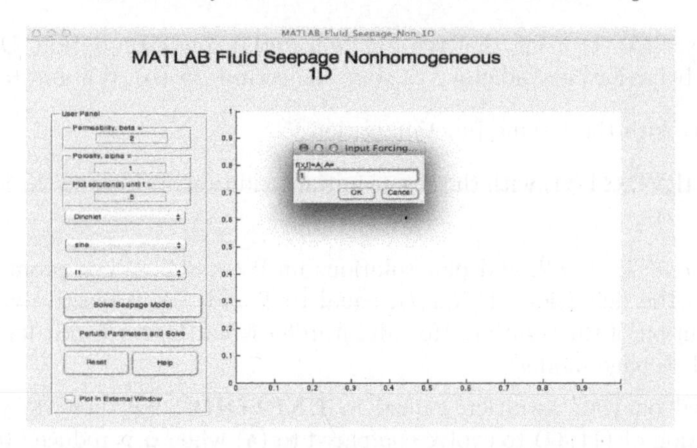

FIGURE 11.28: MATLAB Fluid Seepage Nonhomogeneous 1D GUI

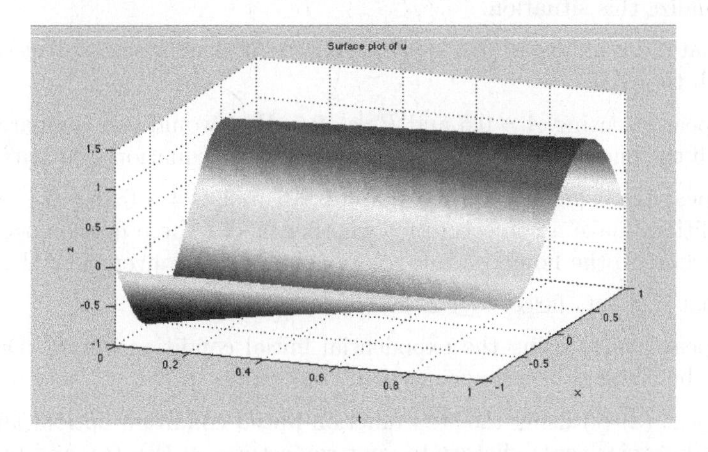

FIGURE 11.29: IBVP (11.36) with $\beta = 1$, $\alpha = 3$, $f(x,t) = 5$, initial profile: $\sin \pi x$.

a.) Use $\alpha = 1$, $\beta = 2$, and plot solutions until $t = 0.5$. When prompted set A, in the definition of $f_1(x,t)$ equal to 1. Describe the evolution of the solution. How did it differ from the corresponding homogeneous fluid seepage model?

b.) Based on your intuition gained in **EXPLORE!** 9.9, how do you expect the solution of (11.36) to evolve compared to (a) when β is reduced to 0.5, keeping everything the else the same as in (a)? Check your intuition by using the GUI to visualize this situation.

c.) What if β is increased to 5 keeping everything else the same? Test your hypothesis with the GUI.

d.) Repeat (a) using $A = 0.5$. How would you compare this situation with the one in (a) and the homogeneous heat equation solution?

e.) Repeat (a) using $A = 0.25, 0.1$, and 0.001. In addition, make a conjecture for whether or not these nonhomogeneous solutions converge to the homogeneous solution as $f_1(x,t)$ converges to the zero function.

f.) What if A gets larger? Explore this by choosing a sequence of increasingly larger values for A. Summarize your observations.

g.) Repeat (a)-(f) using the exponential initial condition option. Did you observe any behavior contradictory to your conjectures in (b), (c), and (e)? If so, how?

h.) Repeat (a)-(f) using the step function initial condition option. Did you observe any behavior contradictory to your conjectures in (b), (c), and (e)? If so, how?

ii.) Repeat (i) with the forcing function option f5.

iii.) Consider IBVP (11.34) with the cosine initial condition option and the forcing function option f2.

a.) Use $\alpha = 1$, $\beta = 2$, and plot solutions until $t = 0.5$. When prompted set A and B in the definition of $f_2(x, t)$, equal to 1 and -1, respectively . Describe the evolution of the solution. How did it differ from the corresponding homogeneous fluid seepage model?

b.) Based on your intuition gained in **EXPLORE!** 9.9, how do you expect the solution of (11.34) to evolve compared to (a) when α is reduced to 0.25, keeping everything else the same as in (a)? Check you intuition by using the GUI to visualize this situation.

c.) What if α is increased to 5 keeping everything else the same? Test your hypothesis with the GUI.

d.) Repeat (a) using $A = 0.5$ and $B = -0.5$. How would you compare this situation with the one in (a) and the homogeneous heat equation solution?

e.) Repeat (a) using $A = 0.25, 0.1$, and 0.001 and $B = -0.25, -0.1$, and -0.001. In addition, make a conjecture for whether or not these nonhomogenous solutions converge to the homogeneous solution as $f_2(x, t)$ converges to the zero function.

f.) What if A gets larger and B more negative. Explore!

g.) Repeat (a)-(f) using the exponential initial condition option. Did you observe any behavior contradictory to your conjectures in (b), (c), and (e)? If so, how?

h.) Repeat (a)-(f) using the step function initial condition option. Did you observe any behavior contradictory to your conjectures in (b), (c), and (e)? If so, how?

iv.) Consider IBVP (11.34) with the cosine initial condition option and the forcing function option f4.

a.) Use $\alpha = 1$, $\beta = 2$, and plot solutions until $t = 0.5$. When prompted set a in the definition of $f_4(x, t)$ equal to 1. Describe the evolution of the solution. How did it differ from the corresponding homogeneous fluid seepage model?

b.) Based on your intuition gained in **EXPLORE!** 9.9, how do you expect the solution of (11.34) to evolve compared to (a) when β is reduced to 0.01, keeping everything the else the same as in (a)?

c.) What do you expect to happen if $\beta = 0.001$? Test your hypothesis with the GUI.

d.) Repeat (a) using $a = 0.5$. How would you compare this situation with the one in (a) and the homogeneous fluid seepage model solution?

e.) Repeat (d) using $a = 0.25, 0.1$, and 0.001. Does the solution appear to get closer to the homogeneous solution?

f.) What if a gets larger? Explore!

g.) Repeat (a)-(f) using the exponential initial condition option. Did you observe any behavior contradictory to your conjectures in (b), (c), and (e)? If so, how?

h.) Repeat (a)-(f) using the step function initial condition option. Did you observe any behavior contradictory to your conjectures in (b), (c), and (e)? If so, how?

v.) Consider IBVP (11.35) with the cosine initial condition option and the forcing function option f3.

a.) Use $\alpha = 1$, $\beta = 2$, and plot solutions until $t = 0.5$. When prompted set ω, in the definition of $f_3(x, t)$, equal to 1. Describe the evolution of the solution. How did it differ from the solution of the corresponding homogeneous heat equation?

b.) Based on your intuition gained in **EXPLORE!** 9.9, how do you expect the solution of (11.34) to evolve compared to (a) when α is reduced to 0.01, keeping everything the else the same as in (a)?

c.) What do you expect to happen if $\alpha = 0.001$? Test your hypothesis with the GUI.

d.) Repeat (a) using $\omega = 0.5$. How would you compare this situation with the one in (a) and the homogeneous fluid seepage solution?

e.) Repeat (d) using $\omega = 0.25, 0.1$, and 0.001. Explain why the solution does **not** tend to the homogeneous solution.

f.) What if ω gets larger? Explore!

g.) Repeat (a)-(f) using the exponential initial condition option. Did you observe any behavior contradictory to your conjectures in (b), (c), and (e)? If so, how?

h.) Repeat (a)-(f) using the step function initial condition option. Did you observe any behavior contradictory to your conjectures in (b), (c), and (e)? If so, how?

■

EXPLORE!

11.9 Long Term Behavior

MATLAB-Exercise 11.2.2. To begin, open **MATLAB** and in the command line, type:

MATLAB_Fluid_Seepage_Non_1D

Use the GUI to answer the following questions. Make certain to click on the HELP button for instructions on how to use the GUI. Screenshots of GUI output can be found in Figure 11.30 and Figure 11.31.

i.) Consider IBVP (11.36) with the sine initial condition option and the forcing function option f1.

a.) Investigate the behavior of the solution of IBVP (11.36) for large times. Use $\alpha = 1$, $\beta = 2$, and when prompted set A in the definition of $f_1(x, t)$ equal to 1. Begin by plotting solutions until $t = 2$. If necessary, slowly increase the plotting time until you are able to describe the long-term behavior of the solution. Remember, a computer cannot handle the infinite. Compare the nonhomogeneous long-term behavior with the corresponding long-term behavior of the homogeneous fluid seepage model.

b.) Repeat (a) using $A = 0.5$. How would you compare this situation with the one in (a) and the homogeneous fluid seepage model?

c.) Repeat (a) using $A = 0.25, 0.1$, and 0.001. How is the forcing function affecting the long-term behavior of the solution?

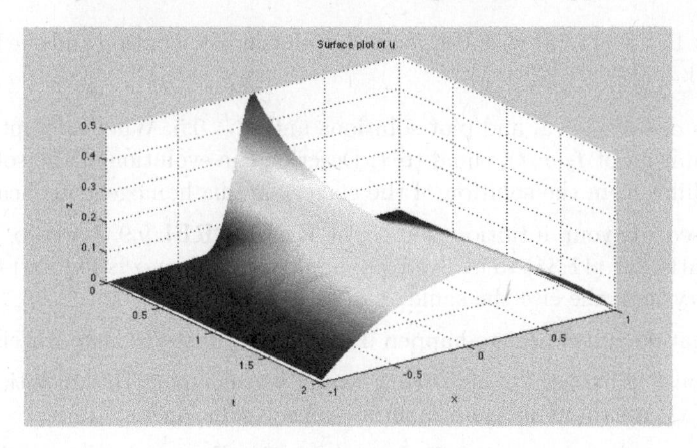

FIGURE 11.30: IBVP (9.49) with $\beta = 2$, $\alpha = 1$, initial profile: e^{-40x^2}, forcing function $f_1(x,t) = 1$.

d.) Repeat (a)-(c) using the exponential initial condition option. Did you observe any behavior contradictory to your conjectures in (a)-(c)? If so, how?

e.) Repeat (a)-(c) using the step function initial condition option. Did you observe any behavior contradictory to your conjectures in (a)-(c)? If so, how?

ii.) Repeat (i) with the forcing function option f5.

iii.) Consider IBVP (11.34) with the cosine initial condition option and the forcing function option f2.

a.) Investigate the behavior of the solution of IBVP (11.34) for large times. Use $\alpha = 1$, $\beta = 2$, and when prompted, set A and B in the definition of $f_2(x,t)$, equal to 1 and -1, respectively . Begin by plotting solutions until $t = 2$. If necessary, slowly increase the plotting time until you are able to describe the long-term behavior of the solution. Compare the nonhomogeneous long-term behavior with the corresponding long-term behavior of the homogeneous fluid seepage model.

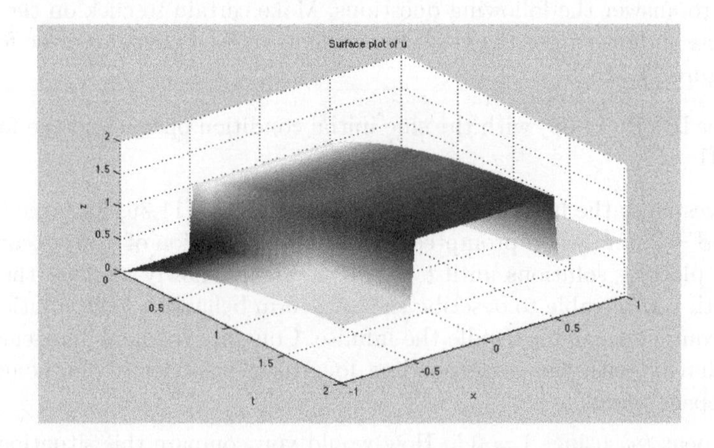

FIGURE 11.31: IBVP (11.34) with $\beta = .001$, $\alpha = 1$, initial profile: step function, forcing function: $f_4(x,t) = e^{-t}$.

b.) Repeat (a) using $A = 0.5$ and $B = -0.5$. How would you compare this situation with the one in (a) and the homogeneous fluid seepage model?

c.) Repeat (a) using $A = 0.25, 0.1$, and 0.001 and $B = -0.25, -0.1$, and -0.001. How is the forcing function affecting the long-term behavior of the solution?

d.) Repeat (a)-(c) using the exponential initial condition option. Did you observe any behavior contradictory to your conjectures in (a)-(c)? If so, how?

e.) Repeat (a)-(c) using the step function initial condition option. Did you observe any behavior contradictory to your conjectures in (a)-(c)? If so, how?

iv.) Consider IBVP (11.34) with the cosine initial condition option and the forcing function option f4.

a.) Investigate the behavior of the solution of IBVP (11.34) for large times. Use $\alpha = 1$, $\beta = 2$, and when prompted, set a in the definition of $f_4(x,t)$ equal to 1. Begin by plotting solutions until $t = 2$. If necessary, slowly increase the plotting time until you are able to describe the long-term behavior of the solution. Compare the nonhomogeneous long-term behavior with the corresponding long-term behavior of the homogeneous fluid seepage model.

b.) Repeat (a) using $a = 0.5$. How would you compare this situation with the one in (a) and the homogeneous fluid seepage model?

c.) Repeat (a) using $a = 0.25, 0.1$, and 0.001. How is the forcing function affecting the long-term behavior of the solution?

d.) Repeat (a)-(c) using the exponential initial condition option. Did you observe any behavior contradictory to your conjectures in (a)-(c)? If so, how?

e.) Repeat (a)-(c) using the step function initial condition option. Did you observe any behavior contradictory to your conjectures in (a)-(c)? If so, how?

v.) Consider IBVP (11.35) with the cosine initial condition option and the forcing function option f3.

a.) Investigate the behavior of the solution of IBVP (11.34) for large times. Use $\alpha = 1$, $\beta = 2$, and when prompted, set ω in the definition of $f_3(x,t)$ equal to 1. Begin by plotting solutions until $t = 2$. If necessary, slowly increase the plotting time until you are able to describe the long-term behavior of the solution. Compare the nonhomogeneous long-term behavior with the corresponding long-term behavior of the homogeneous fluid seepage model.

b.) Repeat (a) using $\omega = 0.5$. How would you compare this situation with the one in (a) and the homogeneous fluid seepage model?

c.) Repeat (a) using $\omega = 0.25, 1$, and 10. How is the forcing function affecting the long-term behavior of the solution?

d.) Repeat (a)-(c) using the exponential initial condition option. Did you observe any behavior contradictory to your conjectures in (a)-(c)? If so, how?

e.) Repeat (a)-(c) using the step function initial condition option. Did you observe any behavior contradictory to your conjectures in (a)-(c)? If so, how?

■

EXPLORE!

11.10 Continuous Dependence on Parameters

MATLAB-Exercise 11.2.3. To explore continuous dependence with the GUI one must first solve the nonhomogeneous fluid seepage model with particular model parameters, initial condition, and forcing function. Once a solution has been constructed, click the *Perturb Parameters and Solve* button to specify the perturbed model parameters and a perturbation size for the initial position. The GUI will also allow for a perturbed forcing function parameter. When prompted to select a norm choose *All* to investigate dependence using both norms at the same time. To begin, open **MATLAB** and in the command line, type:

MATLAB_Fluid_Seepage_Non_1D

Use the GUI to answer the following questions. Make certain to click on the HELP button for instructions on how to use the GUI. Screenshots of inputting parameters and GUI output can be found in Figures 11.32–11.37.

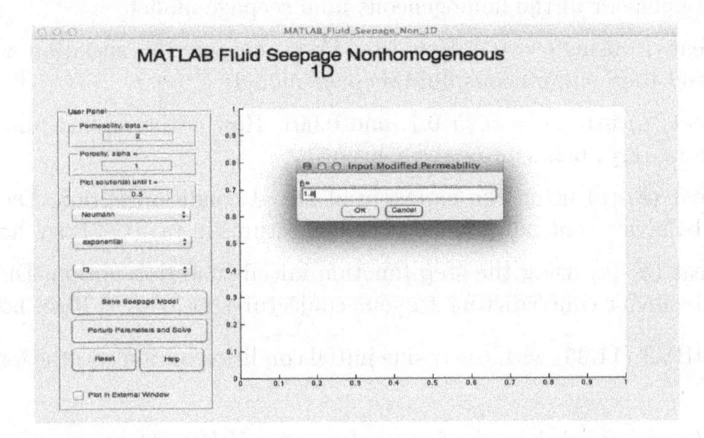

FIGURE 11.32: Perturbing parameters, initial conditions, and forcing function parameters with MATLAB Fluid Seepage Nonhomogeneous 1D GUI

i.) Consider IBVP (11.36) with the sine initial condition option and the forcing function option f1.

a.) Investigate the continuous dependence of the solution on the model parameters for IBVP (11.36). Use $k^2 = 1$, $\alpha = 1$, $\beta = 2$, and when prompted set A in the definition of $f_1(x,t)$ equal to 1. Plot solutions until $t = 0.5$. Once the movie has finished playing, click the *Perturb Parameters and Solve* button and choose a perturb β by setting it equal to 1.8 and α by setting equal to 1.1. Keep all other parameters the same as in the original run of the GUI. Describe the differences in the solutions. Compare and contrast the results using the different norms.

b.) Choose $\beta = 1.9, 1.99$, and 2.01 and $\alpha = 0.9, 1.01$, and 0.99. Record your observations regarding the difference in the solutions. Make a conjecture regarding continuous dependence on the model parameters.

c.) Repeat (a) and (b) using $A = 5$ in the definition of $f_1(x,t)$.

d.) Repeat (a) and (b) using $A = 0.5$ in the definition of $f_1(x,t)$.

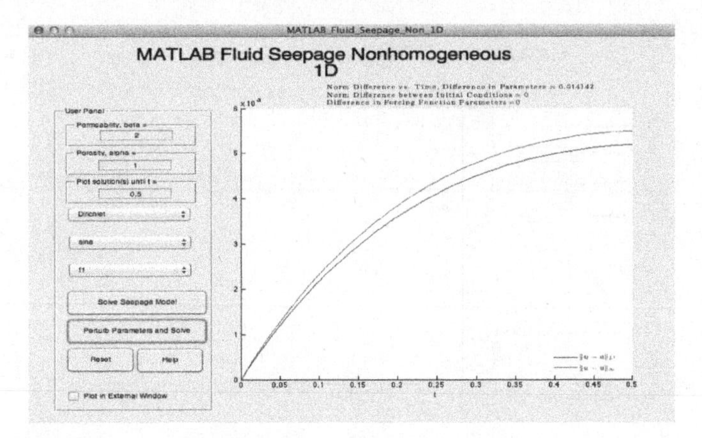

FIGURE 11.33: Perturbing the model parameters in IBVP (11.36). Forcing function, $f_5(x,t)$ with $A = 1$.

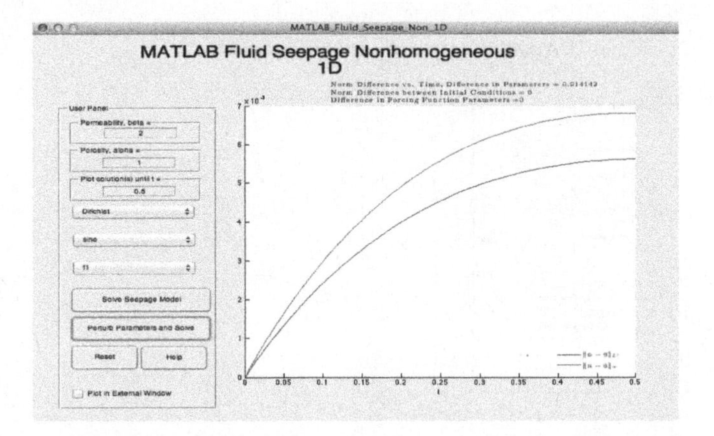

FIGURE 11.34: Perturbing the model parameters in IBVP (11.36). Forcing function, $f_5(x,t)$ with $A = 5$.

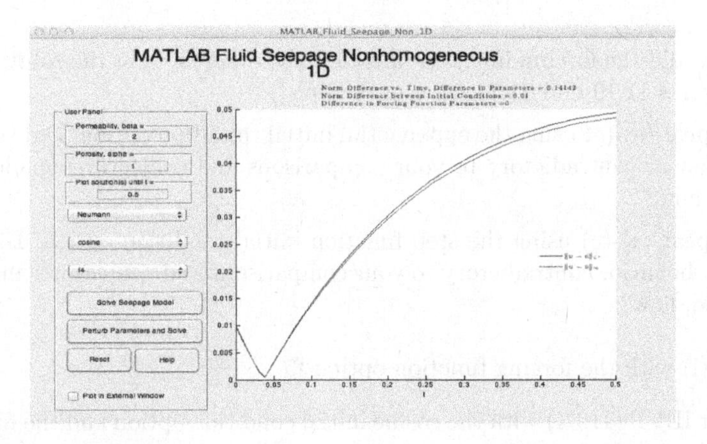

FIGURE 11.35: Perturbing the model parameters and initial condition in IBVP (11.34). Forcing function, $f_4(x,t) = e^{-5t}$.

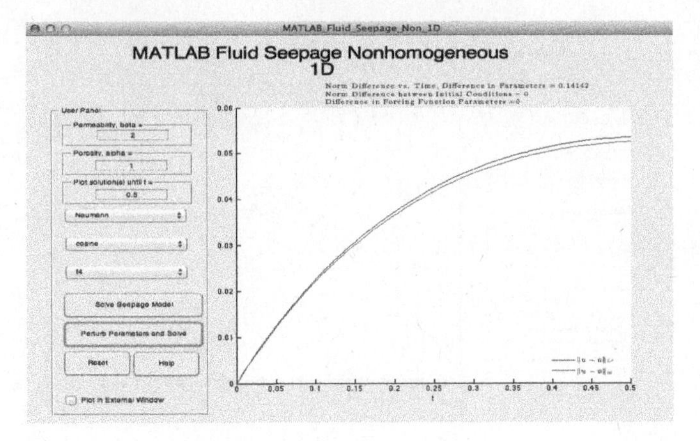

FIGURE 11.36: Perturbing model parameters and initial condition in IBVP (11.34). Forcing function, $f_4(x,t) = e^{-0.5t}$.

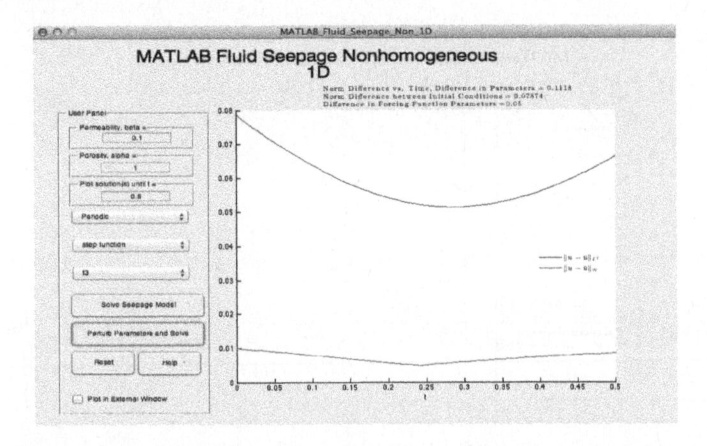

FIGURE 11.37: Perturbing the diffusivity constant, the initial condition, and forcing function parameter in IBVP (11.35). Forcing function, $f_3(x,t) = \cos t$.

e.) How did the forcing function influence the differences in the solution? (Hint: See Figures 11.10 and 11.11 for one example.)

d.) Repeat (a)-(e) using the exponential initial condition option. Did you observe any behavior contradictory to your comparisons and conjectures made in (a)-(e)? If so, how?

e.) Repeat (a)-(e) using the step function initial condition option. Did you observe any behavior contradictory to your comparisons and conjectures made in (a)-(e)? If so, how?

ii.) Repeat (i) with the forcing function option f5.

iii.) Consider IBVP (11.34) with the cosine initial condition option and the forcing function option f2.

a.) Investigate the continuous dependence of the solution on the initial condition for IBVP (11.34). Use $k^2 = 1$, $\alpha = 1$, $\beta = 2$, and when prompted, set A and B in the definition of $f_2(x,t)$, equal to 1 and -1, respectively . Plot solutions until $t = 0.5$. Once the movie has finished playing, click the *Perturb Parameters and Solve* button and choose a perturbation value equal to 0.1. Keep all other parameters the same as in the original run of the GUI. Describe the differences in the solutions. Compare and contrast the results using the different norms.

b.) Choose perturbation values equal to $0.05, 0.01$, and 0.001. Record your observations regarding the difference in the solutions. Make a conjecture regarding continuous dependence on the initial condition.

c.) Repeat (a) and (b) using $A = 5$ and $B = -5$ in the definition of $f_2(x,t)$.

d.) Repeat (a) and (b) using $A = 0.5$ and $B = -0.5$ in the definition of $f_2(x,t)$.

e.) How did the forcing function influence the differences in the solution?

d.) Repeat (a)-(e) using the exponential initial condition option. Did you observe any behavior contradictory to your comparisons and conjectures made in (a)-(e)? If so, how?

e.) Repeat (a)-(e) using the step function initial condition option. Did you observe any behavior contradictory to your comparisons and conjectures made in (a)-(e)? If so, how?

iv.) Consider IBVP (11.34) with the cosine initial condition option and the forcing function option f4.

a.) Investigate the continuous dependence of the solution on the initial condition and the model parameters for IBVP (11.34). Use $k^2 = 1$, $\alpha = 1$, $\beta = 2$, and when prompted, set a in the definition of $f_4(x,t)$ equal to 1. Plot solutions until $t = 0.5$. Once the movie has finished playing, click the *Perturb Parameters and Solve* button and choose a perturbed permeability constant equal to 1.8, porsity to equal 0.8, and perturbation value equal to 0.1. Keep all other parameters the same as in the original run of the GUI. Describe the differences in the solutions. Compare and contrast the results using the different norms.

b.) Choose perturbed permeability constants to equal $2.1, 1.9$, and 2.001, porosity constants to equal $1.1, 0.9$, and 1.01, and perturbation values equal to $0.05, 0.01$, and 0.001. Record your observations regarding the difference in the solutions. Make a conjecture regarding continuous dependence on the initial condition and the model parameters.

c.) Repeat (a) and (b) using $a = 5$ in the definition of $f_4(x,t)$.

d.) Repeat (a) and (b) using $a = 0.5$ in the definition of $f_4(x,t)$.

e.) How did the forcing function influence the differences in the solution? (Hint: See Figures 11.12 and 11.13 for one example.)

f.) Repeat (a)-(e) using the exponential initial condition option. Did you observe any behavior contradictory to your comparisons and conjectures made in (a)-(e)? If so, how?

g.) Repeat (a)-(e) using the step function initial condition option. Did you observe any behavior contradictory to your comparisons and conjectures made in (a)-(e)? If so, how?

v.) Consider IBVP (11.35) with the cosine initial condition option, and the forcing function option f3.

a.) Investigate the continuous dependence of the solution on the initial condition, the model parameters, and the forcing function parameter for IBVP (11.35). Use $k^2 = 1$, $\alpha = 1$, $\beta = 0.1$, and when prompted, set ω in the definition of $f_3(x,t)$ equal to 1. Plot solutions until $t = 0.5$. Once the movie has finished playing, click the *Perturb Parameters and Solve* button and choose a perturbed permeability constant equal to 1.8, porsity to equal 0.8, perturbation value equal to 0.1, and forcing function parameter equal to 1.1. Describe the differences in the solutions. Compare and contrast the results using the different norms.

b.) Choose perturbed permeability constants to equal 0.01, 0.05, and 0.11, porosity constants to equal 1.1, 0.9, and 1.01, perturbation values equal to 0.05, 0.01, and 0.001, and forcing function parameters equal to 0.9, 0.95, and 1.01. Record your observations regarding the difference in the solutions. Make a conjecture regarding continuous dependence on the initial condition, model parameters, and the forcing function parameter.

c.) Repeat (a)-(b) using the exponential initial condition option. Did you observe any behavior contradictory to your comparisons and conjectures made in (a)-(b)? If so, how?

d.) Repeat (a)-(b) using the step function initial condition option. Did you observe any behavior contradictory to your comparisons and conjectures made in (a)-(b)? If so, how?

vi.) Summarize your observations regarding continuous dependence. Make sure you address the potential issue of using different norms and the influence the forcing function had on the differences in the solutions.

■

11.2.2 Higher-Dimensional Fissured Rock Model with Forcing

Consider the two-dimensional models with forcing described by the IBVPs

$$\begin{cases} \frac{\partial}{\partial t}p(x,y,t) & -\alpha\frac{\partial}{\partial t}\left(\Delta p(x,y,t)\right) = \beta\Delta p(x,y,t) + f(x,y,t),\ 0 < x < \pi,\ 0 < y < \pi,\ t > 0, \\ z(x,y,0) & = 0 = z_0(x,y),\ 0 < x < \pi,\ 0 < y < \pi, \\ z(x,0,t) & = 0 = z(x,b,t),\ 0 < x < \pi,\ t > 0, \\ z(0,y,t) & = 0 = z(a,y,t),\ 0 < y < \pi,\ t > 0, \end{cases} \tag{11.37}$$

$$\begin{cases} \frac{\partial}{\partial t}p(x,y,t) & -\alpha\frac{\partial}{\partial t}\left(\Delta p(x,y,t)\right) = \beta\Delta p(x,y,t) + f(x,y,t),\ 0 < x < \pi,\ 0 < y < \pi,\ t > 0, \\ z(x,y,0) & = z_0(x,y),\ 0 < x < \pi,\ 0 < y < \pi, \\ \frac{\partial z(x,0,t)}{\partial y} & = 0 = \frac{\partial z(x,b,t)}{\partial y},\ 0 < x < \pi,\ t > 0, \\ \frac{\partial z(0,y,t)}{\partial x} & = 0 = \frac{\partial z(a,y,t)}{\partial x},\ 0 < y < \pi,\ t > 0, \end{cases} \tag{11.38}$$

$$\begin{cases} \frac{\partial}{\partial t}p(x,y,t) & -\alpha\frac{\partial}{\partial t}\left(\Delta p(x,y,t)\right) = \beta\Delta p(x,y,t) + f(x,y,t),\ 0 < x < \pi,\ 0 < y < \pi,\ t > 0, \\ z(x,y,0) & = z_0(x,y),\ 0 < x < \pi,\ 0 < y < \pi, \\ z(x,0,t) & = z(x,b,t),\ 0 < x < \pi,\ t > 0, \\ z(0,y,t) & = z(a,y,t),\ 0 < y < \pi,\ t > 0, \end{cases} \tag{11.39}$$

IBVP (11.37) is equipped with Dirichlet boundary conditions, (11.38) with Neumann boundary conditions, and (11.39) with periodic boundary conditions. As in the one-dimensional case, we explore the effects that the parameters have on the evolution of the solution, long-term behavior, as well as continuous dependence on the model parameters and initial condition.

EXPLORE!

11.11 Parameter Play!

In this **EXPLORE!**, we visualize the effect changing α and β have on solutions to IBVPs (11.37)-(11.39) when a forcing function is present.

MATLAB-Exercise 11.2.4. To begin, open **MATLAB** and in the command line, type:

MATLAB_Fluid_Seepage_Non_2D

Use the GUI to answer the following questions. Make certain to click on the HELP button for instructions on how to use the GUI. A typical screenshot of the GUI can found in Figure 11.38

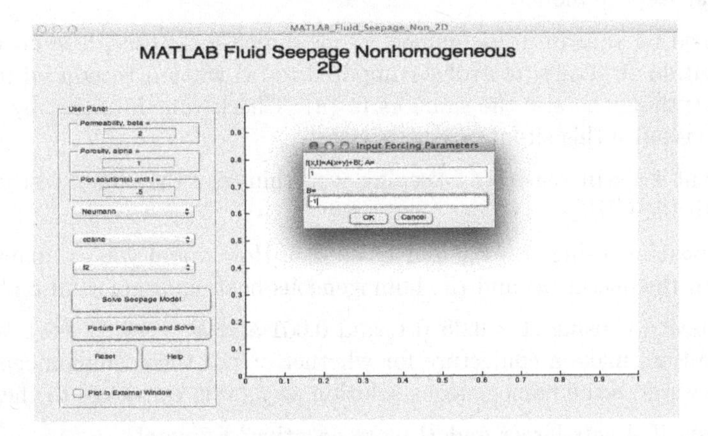

FIGURE 11.38: MATLAB Fluid Seepage Nonhomogeneous 2D GUI

i.) Consider IBVP (11.37) with the sine initial condition option and the forcing function option f1.

a.) Use $\alpha = 1$, $\beta = 2$, and plot solutions until $t = 0.5$. When prompted set A, in the definition of $f_1(x, t)$ equal to 1. Describe the evolution of the solution. How did it differ from the corresponding homogeneous fluid seepage model?

b.) Based on your intuition gained in one-dimensional case, how do you expect the solution of (11.37) to evolve compared to (a) when β is reduced to 0.5, keeping everything the else the same as in (a)? Check your intuition by using the GUI to visualize this situation.

c.) What if β is increased to 5 keeping everything else the same? Test your hypothesis with the GUI.

d.) Repeat (a) using $A = 0.5$. How would you compare this situation with the one in (a) and the homogeneous heat equation solution?

e.) Repeat (a) using $A = 0.25, 0.1$, and 0.001. In addition, make a conjecture for whether or not these nonhomogeneous solutions converge to the homogeneous solution as $f_1(x,t)$ converges to the zero function.

f.) What if A gets larger? Explore this by choosing a sequence of increasingly larger values for A. Summarize your observations.

g.) Repeat (a)-(f) using the exponential initial condition option. Did you observe any behavior contradictory to your conjectures in (b), (c), and (e)? If so, how?

h.) Repeat (a)-(f) using the step function initial condition option. Did you observe any behavior contradictory to your conjectures in (b), (c), and (e)? If so, how?

ii.) Repeat (i) with the forcing function option f5.

iii.) Consider IBVP (11.38) with the cosine initial condition option and the forcing function option f2.

a.) Use $\alpha = 1$, $\beta = 2$, and plot solutions until $t = 0.5$. When prompted set A and B in the definition of $f_2(x,t)$, equal to 1 and -1, respectively . Describe the evolution of the solution. How did it differ from the corresponding homogeneous fluid seepage model?

b.) Based on your intuition gained in one-dimensional case, how do you expect the solution of (11.38) to evolve compared to (a) when α is reduced to 0.25, keeping everything the else the same as in (a)? Check your intuition by using the GUI to visualize this situation.

c.) What if α is increased to 5 keeping everything else the same? Test your hypothesis with the GUI.

d.) Repeat (a) using $A = 0.5$ and $B = -0.5$. How would you compare this situation with the one in (a) and the homogeneous heat equation solution?

e.) Repeat (a) using $A = 0.25, 0.1$, and 0.001 and $B = -0.25, -0.1$, and -0.001. In addition, make a conjecture for whether or not these nonhomogenous solutions converge to the homogeneous solution as $f_2(x,t)$ converges to the zero function.

f.) What if A gets larger and B more negative? Explore!

g.) Repeat (a)-(f) using the exponential initial condition option. Did you observe any behavior contradictory to your conjectures in (b), (c), and (e)? If so, how?

h.) Repeat (a)-(f) using the step function initial condition option. Did you observe any behavior contradictory to your conjectures in (b), (c), and (e)? If so, how?

iv.) Consider IBVP (11.38) with the cosine initial condition option and the forcing function option f4.

a.) Use $\alpha = 1$, $\beta = 2$, and plot solutions until $t = 0.5$. When prompted set a in the definition of $f_4(x,t)$ equal to 1. Describe the evolution of the solution. How did it differ from the corresponding homogeneous fluid seepage model?

b.) Based on your intuition gained in the one-dimensional case, how do you expect the solution of (11.34) to evolve compared to (a) when β is reduced to 0.01, keeping everything the else the same as in (a)?

c.) What do you expect to happen if $\beta = 0.001$? Test your hypothesis with the GUI.

d.) Repeat (a) using $a = 0.5$. How would you compare this situation with the one in (a) and the homogeneous fluid seepage model solution?

e.) Repeat (d) using $a = 0.25, 0.1$, and 0.001. Does the solution appear to get closer to the homogeneous solution?

f.) What if a gets larger? Explore!

g.) Repeat (a)-(f) using the exponential initial condition option. Did you observe any behavior contradictory to your conjectures in (b), (c), and (e)? If so, how?

h.) Repeat (a)-(f) using the step function initial condition option. Did you observe any behavior contradictory to your conjectures in (b), (c), and (e)? If so, how?

v.) Consider IBVP (11.39) with the cosine initial condition option and the forcing function option f3.

a.) Use $\alpha = 1$, $\beta = 2$, and plot solutions until $t = 0.5$. When prompted set ω, in the definition of $f_3(x,t)$, equal to 1. Describe the evolution of the solution. How did it differ from the solution of the corresponding homogeneous heat equation?

b.) Based on your intuition gained in the one-dimensional case, how do you expect the solution of (11.39) to evolve compared to (a) when α is reduced to 0.01, keeping everything the else the same as in (a)?

c.) What do you expect to happen if $\alpha = 0.001$? Test your hypothesis with the GUI.

d.) Repeat (a) using $\omega = 0.5$. How would you compare this situation with the one in (a) and the homogeneous fluid seepage solution?

e.) Repeat (d) using $\omega = 0.25, 0.1$, and 0.001. Explain why the solution does **not** tend to the homogeneous solution.

f.) What if ω gets larger? Explore!

g.) Repeat (a)-(f) using the exponential initial condition option. Did you observe any behavior contradictory to your conjectures in (b), (c), and (e)? If so, how?

h.) Repeat (a)-(f) using the step function initial condition option. Did you observe any behavior contradictory to your conjectures in (b), (c), and (e)? If so, how?

■

EXPLORE!

11.12 Long Term Behavior

MATLAB-Exercise 11.2.5. To begin, open **MATLAB** and in the command line, type:

MATLAB_Fluid_Seepage_Non_2D

Use the GUI to answer the following questions. Make certain to click on the HELP button for instructions on how to use the GUI. A typical screenshot of GUI output can be found in Figure 11.39.

i.) Consider IBVP (11.37) with the sine initial condition option and the forcing function option f1.

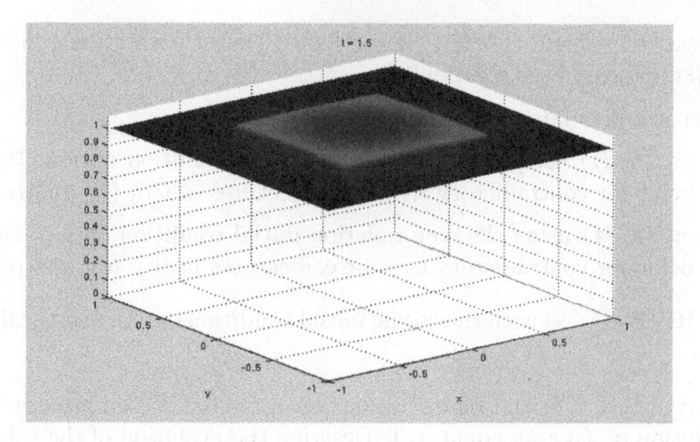

FIGURE 11.39: IBVP (11.38) with $\beta = 2$, $\alpha = 1$, initial profile: step function, forcing function $f_4(x,t) = e^{-t}$. Solution appears to become more flat over time, yet continues to translate along the positive z-axis.

a.) Investigate the behavior of the solution of IBVP (11.37) for large times. Use $\alpha = 1$, $\beta = 2$, and when prompted set A in the definition of $f_1(x,t)$ equal to 1. Begin by plotting solutions until $t = 2$. If necessary, slowly increase the plotting time until you are able to describe the long-term behavior of the solution. Remember, a computer cannot handle the infinite. Compare the nonhomogeneous long-term behavior with the corresponding long-term behavior of the homogeneous fluid seepage model.

b.) Repeat (a) using $A = 0.5$. How would you compare this situation with the one in (a) and the homogeneous fluid seepage model?

c.) Repeat (a) using $A = 0.25, 0.1$, and 0.001. How is the forcing function affecting the long-term behavior of the solution?

d.) Repeat (a)-(c) using the exponential initial condition option. Did you observe any behavior contradictory to your conjectures in (a)-(c)? If so, how?

e.) Repeat (a)-(c) using the step function initial condition option. Did you observe any behavior contradictory to your conjectures in (a)-(c)? If so, how?

ii.) Repeat (i) with the forcing function option f5.

iii.) Consider IBVP (11.38) with the cosine initial condition option and the forcing function option f2.

a.) Investigate the behavior of the solution of IBVP (11.38) for large times. Use $\alpha = 1$, $\beta = 2$, and when prompted, set A and B in the definition of $f_2(x,t)$, equal to 1 and -1, respectively . Begin by plotting solutions until $t = 2$. If necessary, slowly increase the plotting time until you are able to describe the long-term behavior of the solution. Compare the nonhomogeneous long-term behavior with the corresponding long-term behavior of the homogeneous fluid seepage model.

b.) Repeat (a) using $A = 0.5$ and $B = -0.5$. How would you compare this situation with the one in (a) and the homogeneous fluid seepage model?

c.) Repeat (a) using $A = 0.25, 0.1$, and 0.001 and $B = -0.25, -0.1$, and -0.001. How is the forcing function affecting the long-term behavior of the solution?

d.) Repeat (a)-(c) using the exponential initial condition option. Did you observe any behavior contradictory to your conjectures in (a)-(c)? If so, how?

e.) Repeat (a)-(c) using the step function initial condition option. Did you observe any behavior contradictory to your conjectures in (a)-(c)? If so, how?

iv.) Consider IBVP (11.38) with the cosine initial condition option and the forcing function option f4.

a.) Investigate the behavior of the solution of IBVP (11.38) for large times. Use $\alpha = 1$, $\beta = 2$, and when prompted, set a in the definition of $f_4(x,t)$ equal to 1. Begin by plotting solutions until $t = 2$. If necessary, slowly increase the plotting time until you are able to describe the long-term behavior of the solution. Compare the nonhomogeneous long-term behavior with the corresponding long-term behavior of the homogeneous fluid seepage model.

b.) Repeat (a) using $a = 0.5$. How would you compare this situation with the one in (a) and the homogeneous fluid seepage model?

c.) Repeat (a) using $a = 0.25, 0.1$, and 0.001. How is the forcing function affecting the long-term behavior of the solution?

d.) Repeat (a)-(c) using the exponential initial condition option. Did you observe any behavior contradictory to your conjectures in (a)-(c)? If so, how?

e.) Repeat (a)-(c) using the step function initial condition option. Did you observe any behavior contradictory to your conjectures in (a)-(c)? If so, how?

v.) Consider IBVP (11.39) with the cosine initial condition option and the forcing function option f3.

a.) Investigate the behavior of the solution of IBVP (11.38) for large times. Use $\alpha = 1$, $\beta = 2$, and when prompted, set ω in the definition of $f_3(x,t)$ equal to 1. Begin by plotting solutions until $t = 2$. If necessary, slowly increase the plotting time until you are able to describe the long-term behavior of the solution. Compare the nonhomogeneous long-term behavior with the corresponding long-term behavior of the homogeneous fluid seepage model.

b.) Repeat (a) using $\omega = 0.5$. How would you compare this situation with the one in (a) and the homogeneous fluid seepage model?

c.) Repeat (a) using $\omega = 0.25, 1$, and 10. How is the forcing function affecting the long-term behavior of the solution?

d.) Repeat (a)-(c) using the exponential initial condition option. Did you observe any behavior contradictory to your conjectures in (a)-(c)? If so, how?

e.) Repeat (a)-(c) using the step function initial condition option. Did you observe any behavior contradictory to your conjectures in (a)-(c)? If so, how?

■

EXPLORE!

11.13 Continuous Dependence on Parameters

MATLAB-Exercise 11.2.6. To explore continuous dependence with the GUI one must first solve the nonhomogeneous fluid seepage model with particular model parameters, initial condition, and forcing function. Once a solution has been constructed, click the *Perturb Parameters and Solve* button to specify the perturbed model parameters and a perturbation size for the initial position. The GUI will also allow for a perturbed forcing function parameter. When prompted to select a norm choose *All* to investigate dependence using both norms at the same time. To begin, open **MATLAB** and in the command line, type:

MATLAB_Fluid_Seepage_Non_2D

Use the GUI to answer the following questions. Make certain to click on the HELP button for instructions on how to use the GUI. Screenshots of inputting parameters and GUI output can be found in Figures 11.40–11.44.

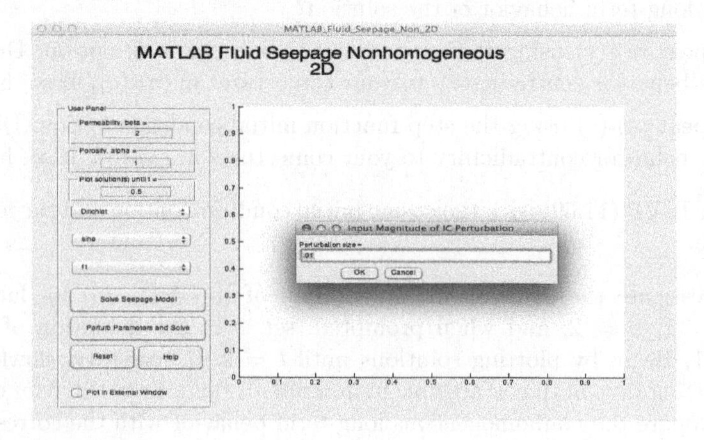

FIGURE 11.40: Perturbing parameters, initial conditions, and forcing function parameters with MATLAB Fluid Seepage Nonhomogeneous 2D GUI

i.) Consider IBVP (11.37) with the sine initial condition option and the forcing function option f1.

a.) Investigate the continuous dependence of the solution on the model paramters for IBVP (11.37). Use $k^2 = 1$, $\alpha = 1$, $\beta = 2$, and when prompted set A in the definition of $f_1(x,t)$ equal to 1. Plot solutions until $t = 0.5$. Once the movie has finished playing, click the *Perturb Parameters and Solve* button and choose a peturb β by setting it equal to 1.8 and α by setting equal to 1.1. Keep all other parameters the same as in the original run of the GUI. Describe the differences in the solutions. Compare and contrast the results using the different norms.

b.) Choose $\beta = 1.9, 1.99$, and 2.01 and $\alpha = 0.9, 1.01$, and 0.99. Record your observations regarding the difference in the solutions. Make a conjecture regarding continuous dependence on the model parameters.

c.) Repeat (a) and (b) using $A = 5$ in the definition of $f_1(x,t)$.

d.) Repeat (a) and (b) using $A = 0.5$ in the definition of $f_1(x,t)$.

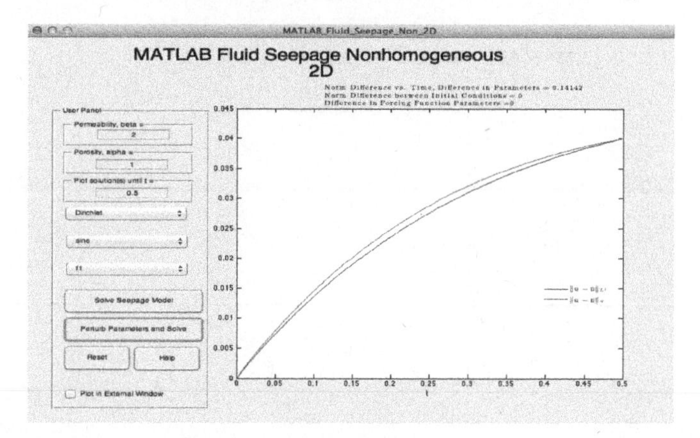

FIGURE 11.41: Perturbing the model parameters in IBVP (11.37). Forcing function, $f_5(x,t)$ with $A = 5$.

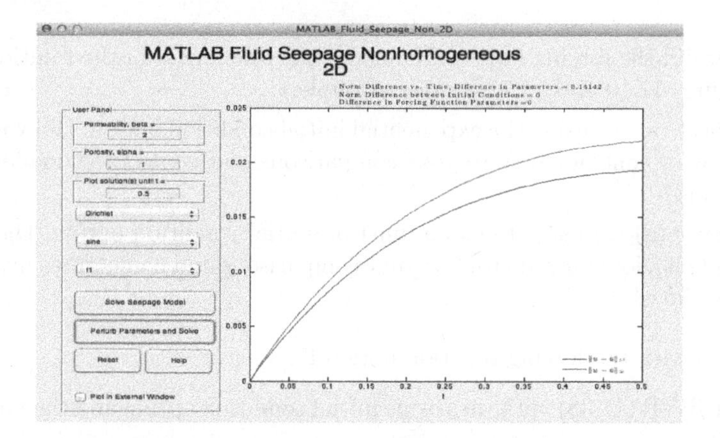

FIGURE 11.42: Perturbing the model parameters in IBVP (11.37). Forcing function, $f_5(x,t)$ with $A = 1$.

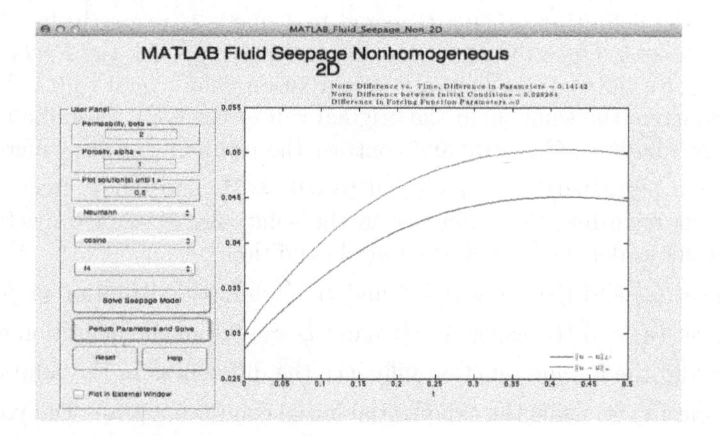

FIGURE 11.43: Perturbing the model parameters and initial condition in IBVP (11.38). Forcing function, $f_4(x,t) = e^{-5t}$.

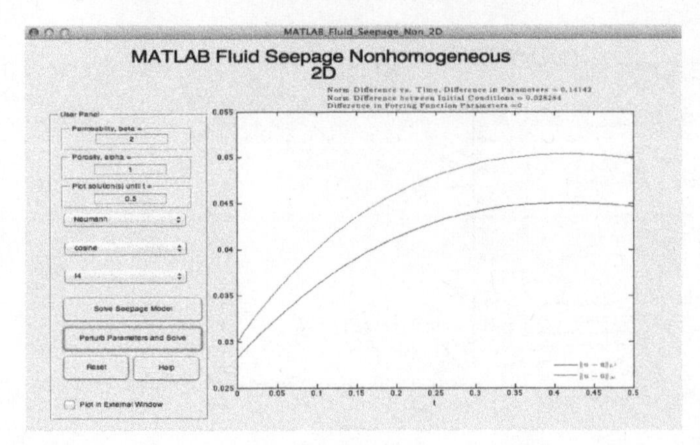

FIGURE 11.44: Perturbing model parameters and initial condition in IBVP (11.38). Forcing function, $f_4(x,t) = e^{-0.5t}$.

e.) How did the forcing function influence the differences in the solution? (Hint: See Figures 11.10 and 11.11 for one example.)

d.) Repeat (a)-(e) using the exponential initial condition option. Did you observe any behavior contradictory to your comparisons and conjectures made in (a)-(e)? If so, how?

e.) Repeat (a)-(e) using the step function initial condition option. Did you observe any behavior contradictory to your comparisons and conjectures made in (a)-(e)? If so, how?

ii.) Repeat (i) with the forcing function option f5.

iii.) Consider IBVP (11.38) with the cosine initial condition option and the forcing function option f2.

a.) Investigate the continuous dependence of the solution on the initial condition for IBVP (11.38). Use $k^2 = 1$, $\alpha = 1$, $\beta = 2$, and when prompted, set A and B in the definition of $f_2(x,t)$, equal to 1 and -1, respectively . Plot solutions until $t = 0.5$. Once the movie has finished playing, click the *Perturb Parameters and Solve* button and choose a perturbation value equal to 0.1. Keep all other parameters the same as in the original run of the GUI. Describe the differences in the solutions. Compare and contrast the results using the different norms.

b.) Choose perturbation values equal to $0.05, 0.01$, and 0.001. Record your observations regarding the difference in the solutions. Make a conjecture regarding continuous dependence on the initial condition.

c.) Repeat (a) and (b) using $A = 5$ and $B = -5$ in the definition of $f_2(x,t)$.

d.) Repeat (a) and (b) using $A = 0.5$ and $B = -0.5$ in the definition of $f_2(x,t)$.

e.) How did the forcing function influence the differences in the solution?

f.) Repeat (a)-(e) using the exponential initial condition option. Did you observe any behavior contradictory to your comparisons and conjectures made in (a)-(e)? If so, how?

g.) Repeat (a)-(e) using the step function initial condition option. Did you observe any behavior contradictory to your comparisons and conjectures made in (a)-(e)? If so, how?

iv.) Consider IBVP (11.38) with the cosine initial condition and the forcing function option f4.

a.) Investigate the continuous dependence of the solution on the initial condition and the model parameters for IBVP (11.38) . Use $k^2 = 1$, $\alpha = 1$, $\beta = 2$, and when prompted, set a in the definition of $f_4(x,t)$ equal to 1. Plot solutions until $t = 0.5$. Once the movie has finished playing, click the *Perturb Parameters and Solve* button and choose a perturbed permeability constant equal to 1.8, porosity to equal 0.8, and perturbation value equal to 0.1. Keep all other parameters the same as in the original run of the GUI. Describe the differences in the solutions. Compare and contrast the results using the different norms.

b.) Choose perturbed permeability constants equal to $2.1, 1.9$, and 2.001, porosity constants equal to $1.1, 0.9$, and 1.01, and perturbation values equal to $0.05, 0.01$, and 0.001. Record your observations regarding the difference in the solutions. Make a conjecture regarding continuous dependence on the initial condition and the model parameters.

c.) Repeat (a) and (b) using $a = 5$ in the definition of $f_4(x,t)$.

d.) Repeat (a) and (b) using $a = 0.5$ in the definition of $f_4(x,t)$.

e.) How did the forcing function influence the differences in the solution? (Hint: See Figures 11.12 and 11.13 for one example.)

d.) Repeat (a)-(e) using the exponential initial condition option. Did you observe any behavior contradictory to your comparisons and conjectures made in (a)-(e)? If so, how?

e.) Repeat (a)-(e) using the step function initial condition option. Did you observe any behavior contradictory to your comparisons and conjectures made in (a)-(e)? If so, how?

v.) Consider IBVP (11.39) with the cosine initial condition option, and the forcing function option f3.

a.) Investigate the continuous dependence of the solution on the initial condition, the model parameters, and the forcing function parameter for IBVP (11.39). Use $k^2 = 1$, $\alpha = 1$, $\beta = 0.1$, and when prompted, set ω in the definition of $f_3(x,t)$ equal to 1. Plot solutions until $t = 0.5$. Once the movie has finished playing, click the *Perturb Parameters and Solve* button and choose a perturbed permeability constant equal to 1.8, porosity equal to 0.8, perturbation value equal to 0.1, and forcing function parameter equal to 1.1. Describe the differences in the solutions. Compare and contrast the results using the different norms.

b.) Choose perturbed permeability constants equal to $0.01, 0.05$, and 0.11, porosity constants equal to $1.1, 0.9$, and 1.01, perturbation values equal to $0.05, 0.01$, and 0.001, and forcing function parameters equal to $0.9, 0.95$, and 1.01. Record your observations regarding the difference in the solutions. Make a conjecture regarding continuous dependence on the initial condition, model parameters, and the forcing function parameter.

c.) Repeat (a)-(b) using the exponential initial condition option. Did you observe any behavior contradictory to your comparisons and conjectures made in (a)-(b)? If so, how?

d.) Repeat (a)-(b) using the step function initial condition option. Did you observe any behavior contradictory to your comparisons and conjectures made in (a)-(b)? If so, how?

vi.) Summarize your observations regarding continuous dependence. Make sure you address the potential issue of using different norms and the influence the forcing function had on the differences in the solutions.

■

11.2.3 Abstract Formulation

Exercise 11.2.1. Consider the IBVP

$$\begin{cases} \frac{\partial}{\partial t}\left(z(x,t) - \frac{\partial^2}{\partial x^2}z(x,t)\right) = \frac{\partial^2}{\partial x^2}z(x,t) + 1 + x, \ 0 < x < \pi, \ t > 0, \\ z(x,0) = 1 + x^3, \ 0 < x < \pi, \\ z(0,t) = z(\pi,t) = 0, \ t > 0. \end{cases} \tag{11.40}$$

i.) Following the discussion in Section 9.2.5, express (11.40) as an abstract evolution equation of the form

$$\begin{cases} (Bu)'(t) = Au(t) + F(t), \ t > 0, \\ u(0) = u_0. \end{cases} \tag{11.41}$$

ii.) What is the form of the variation of parameters solution here?

11.3 The Classical Wave Equation and its Variants

11.3.1 One-Dimensional Wave Equation with Source Effects

Incorporating a forcing term (also known as a *source function*) into the IBVPs (9.67) and (9.68) accounts for the effect of the source of the wave due to the surrounding medium and which serve to drive the wave. Ignoring damping factors for the moment, the resulting IBVPs are of the form

$$\begin{cases} \frac{\partial^2}{\partial t^2}z(x,t) - c^2\frac{\partial^2}{\partial x^2}z(x,t) = f(x,t), \ 0 < x < L, \ t > 0, \\ z(x,0) = z_0(x), \ \frac{\partial z}{\partial t}(x,0) = z_1(x), \ 0 < x < L, \\ z(0,t) = z(L,t) = 0, \ t > 0. \end{cases} \tag{11.42}$$

The analogous IBVP to (11.42) when Neumann BCs are used instead of Dirichlet BCs is given by:

$$\begin{cases} \frac{\partial^2}{\partial t^2}z(x,t) - c^2\frac{\partial^2}{\partial x^2}z(x,t) = f(x,t), \ 0 < x < L, \ t > 0, \\ z(x,0) = z_0(x), \ \frac{\partial z}{\partial t}(x,0) = z_1(x), \ 0 < x < L, \\ \frac{\partial z}{\partial x}(0,t) = \frac{\partial z}{\partial x}(L,t) = 0, \ t > 0. \end{cases} \tag{11.43}$$

Lastly, if the IBVP is equipped with periodic BCs, the new IBVP becomes

$$\begin{cases} \frac{\partial^2}{\partial t^2}z(x,t) - c^2\frac{\partial^2}{\partial x^2}z(x,t) = f(x,t), \ 0 < x < L, \ t > 0, \\ z(x,0) = z_0(x), \ \frac{\partial z}{\partial t}(x,0) = z_1(x), \ 0 < x < L, \\ z(0,t) = z(L,t), \ t > 0. \end{cases} \tag{11.44}$$

Assuming that f is sufficiently smooth, it can be expressed uniquely in the form

$$f(x,t) = \sum_{m=1}^{\infty} \left[f_m^1(t) \sin\left(\frac{m\pi}{L}x\right) + f_m^2(t) \cos\left(\frac{m\pi}{L}x\right) \right], \; 0 < x < L, \; t > 0, \tag{11.45}$$

where the coefficients are defined by

$$f_m^1(t) = \frac{2}{L} \int_0^L f(w,t) \sin\left(\frac{m\pi}{L}w\right) dw, \; t > 0, \tag{11.46}$$

$$f_m^2(t) = \frac{2}{L} \int_0^L f(w,t) \cos\left(\frac{m\pi}{L}w\right) dw, \; t > 0. \tag{11.47}$$

Exercise 11.3.1. Use (11.45) and follow the process for solving the nonhomogeneous diffusion equation to solve the IBVP (11.44).

EXPLORE!
11.14 Parameter Play!

In this **EXPLORE!** you will investigate the effects of changing the parameter c^2 in IBVPs (11.42)-(11.44). To do this, you will use the MATLAB Wave Equation Nonhomogeneous 1D GUI to visualize the solutions for different choices of c^2.

MATLAB-Exercise 11.3.1. To begin, open **MATLAB** and in the command line, type:

MATLAB_Wave_Equation_Non_1D

Use the GUI to answer the following questions. Make certain to click on the HELP button for instructions on how to use the GUI. Screenshots of inputting parameters and GUI output can be found in Figures 11.45–11.49.

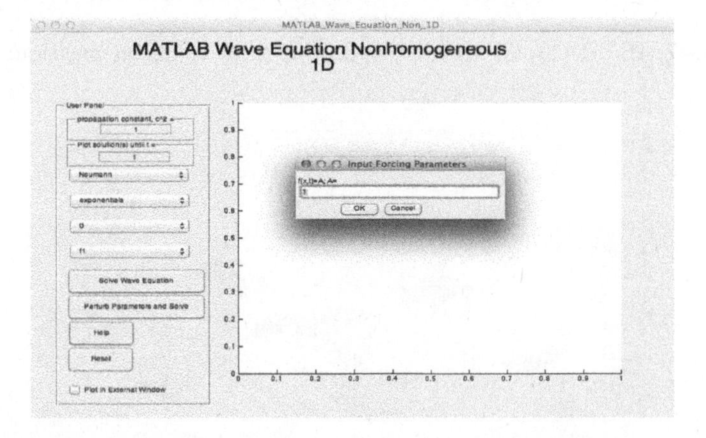

FIGURE 11.45: MATLAB Wave Equation Nonhomogeneous 1D GUI

i.) Consider IBVP (11.42) with the sine initial position option, the zero initial velocity option, and the forcing function option f1.

  a.) Use $c^2 = 1$, plot solutions until $t = 1$ and when prompted set A in the definition of $f_1(x,t)$ equal to 1. Describe the evolution of the solution. How did it differ from the corresponding homogeneous wave equation?

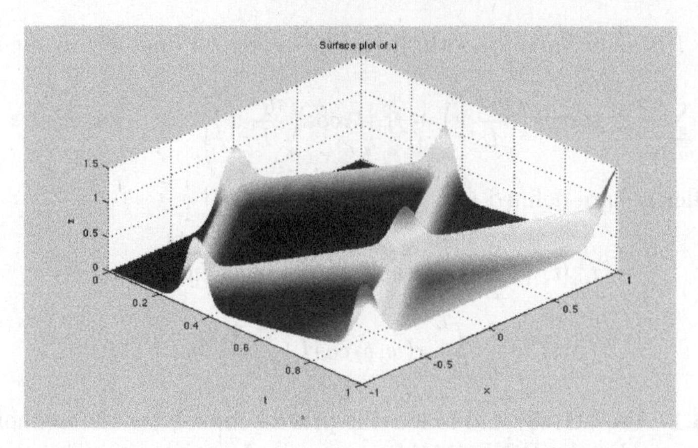

FIGURE 11.46: IBVP (11.43) with $c^2 = 3$, $f(x,t) = 0.01$, initial position: e^{-40x^2}, initial velocity: 0.

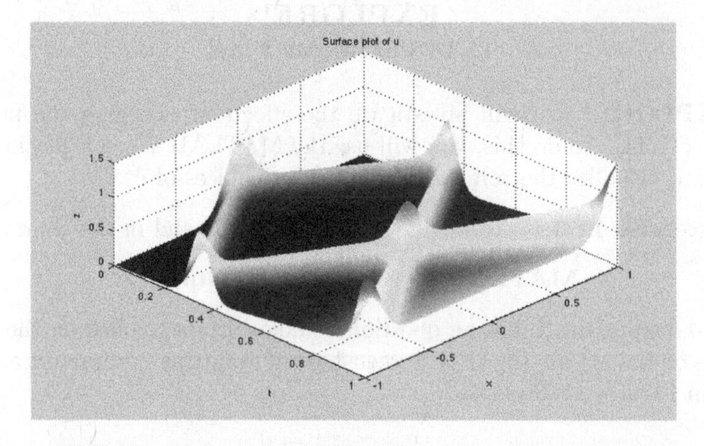

FIGURE 11.47: IBVP (11.43) with $c^2 = 3$, $f(x,t) = 1$, initial position: e^{-40x^2}, initial velocity: 0.

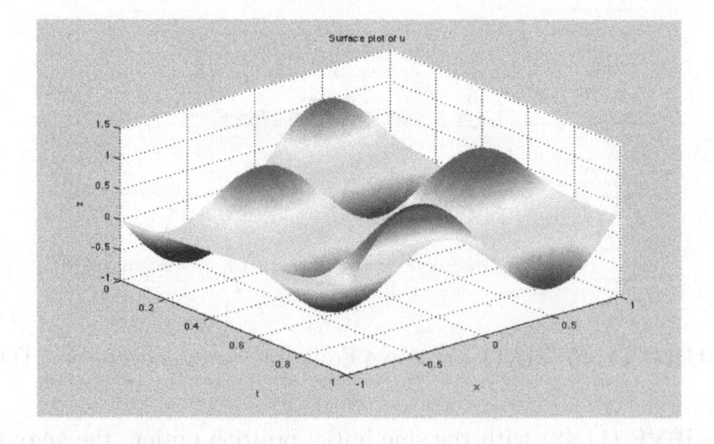

FIGURE 11.48: IBVP (11.44) with $c^2 = 1$, $f_3(x,t) = \cos t$, initial position: $\sin \pi x$, initial velocity: 0.

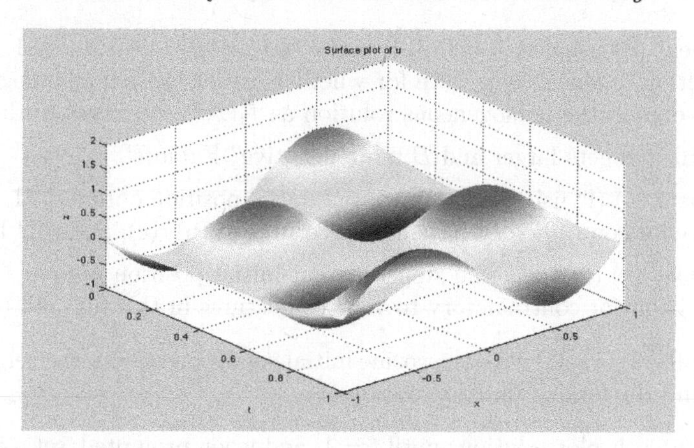

FIGURE 11.49: IBVP (11.44) with $c^2 = 1$, $f_3(x,t) = \cos 0.01t$, initial position: $\sin \pi x$, initial velocity: 0.

b.) Based on your intuition gained in Chapter 9 regarding the wave equation, how do you expect the solution of (11.42) to evolve when $c^2 = 2$, as compared to (a)? Check your intuition by using the GUI to visualize this situation.

c.) What do you expect to happen if $c^2 = 0.25$? Test your hypothesis with the GUI.

d.) Repeat (a) using $A = 0.5$. How would you compare this situation with the one in (a) and the homogeneous heat equation solution?

e.) Repeat (d) using $A = 0.25, 0.1$, and 0.001. In addition, make a conjecture for whether or not these nonhomogeneous solutions converge to the homogeneous solution as $f_1(x,t)$ converges to the zero function.

f.) What if A gets larger? Explore this by choosing a sequence of increasingly larger values for A. Summarize your observations.

g.) Repeat (a)-(f) using the exponential initial position option. Did you observe any behavior contradictory to your conjectures in (b), (c), and (e)? If so, how?

h.) Repeat (a)-(f) using the step function initial position option. Did you observe any behavior contradictory to your conjectures in (b), (c), and (e)? If so, how?

ii.) Repeat (i) with the forcing function option f5.

iii.) Consider IBVP (11.43) with the cosine initial position option, the cosine initial velocity option, and the forcing function option f2.

a.) Use $c^2 = 1$, plot solutions until $t = 1$, and when prompted, set A and B in the definition of $f_2(x,t)$, equal to 1 and -1, respectively . Describe the evolution of the solution. How did it differ from the corresponding homogeneous wave equation?

b.) Based on your intuition gained in Chapter 9 regarding the heat equation, how do you expect the solution of (11.43) to evolve when $c^2 = 2$, as compared to (a)? Check your intuition by using the GUI to visualize this situation.

c.) What do you expect to happen if $c^2 = 0.25$? Test your hypothesis with the GUI.

d.) Repeat (a) using $A = 0.5$ and $B = -0.5$. How would you compare this situation with the one in (a) and the homogeneous wave equation solution?

e.) Repeat (d) using $A = 0.25, 0.1$, and 0.001 and $B = -0.25, -0.1$, and -0.001. In addition, make a conjecture for whether or not these nonhomogenous solutions converge to the homogeneous solution as $f_2(x, t)$ converges to the zero function.

f.) What if A gets larger and B more negative? Explore!

g.) Repeat (a)-(f) using the exponential initial position option. Did you observe any behavior contradictory to your conjectures in (b), (c), and (e)? If so, how?

h.) Repeat (a)-(f) using the step function initial position option. Did you observe any behavior contradictory to your conjectures in (b), (c), and (e)? If so, how?

iv.) Consider IBVP (11.43) with the cosine initial condition option, the zero initial velocity option, and the forcing function option f4.

a.) Use $c^2 = 1$, plot solutions until $t = 1$, and when prompted, set a in the definition of $f_4(x, t)$ equal to 1. Describe the evolution of the solution. How did it differ from the corresponding homogeneous wave equation?

b.) Based on your intuition gained in Chapter 9 regarding the heat equation, how do you expect the solution of (11.43) to evolve when $c^2 = 2$, as compared to (a)? Check your intuition by using the GUI to visualize this situation.

c.) What do you expect to happen if $c^2 = 0.25$? Test your hypothesis with the GUI.

d.) Repeat (a) using $a = 0.5$. How would you compare this situation with the one in (a) and the homogeneous wave equation solution?

e.) Repeat (d) using $a = 0.25, 0.1$, and 0.001. Does the solution appear to get closer to the homogeneous solution?

f.) What if a gets larger? Explore!

g.) Repeat (a)-(f) using the exponential initial position option. Did you observe any behavior contradictory to your conjectures in (b), (c), and (e)? If so, how?

h.) Repeat (a)-(f) using the step function initial position option. Did you observe any behavior contradictory to your conjectures in (b), (c), and (e)? If so, how?

v.) Consider IBVP (11.44) with the cosine initial condition option, any choice for the initial velocity, and the fourth forcing function option, f3.

a.) Use $c^2 = 1$, plot solutions until $t = 1$, and when prompted, set ω in the definition of $f_3(x, t)$ equal to 1. Describe the evolution of the solution. How did it differ from the corresponding homogeneous wave equation?

b.) Based on your intuition gained in Chapter 9 regarding the wave equation, how do you expect the solution of (11.44) to evolve when $c^2 = 2$, as compared to (a)? Check your intuition by using the GUI to visualize this situation.

c.) What do you expect to happen if $c^2 = 0.25$? Test your hypothesis with the GUI.

d.) Repeat (a) using $a = 0.5$. How would you compare this situation with the one in (a) and the homogeneous wave equation solution?

e.) Repeat (d) using $a = 0.25, 0.1$, and 0.001. Explain why the solution should **not** tend to the homogeneous solution. Note: Unlike the previous two models, the periodic nature of the wave equation along with the periodic forcing function are making it difficult to compare with the homogeneous case. It's there...watch the boundaries!

f.) What if a gets larger? Explore!

g.) Repeat (a)-(f) using the exponential initial position option. Did you observe any behavior contradictory to your conjectures in (b), (c), and (e)? If so, how?

h.) Repeat (a)-(f) using the step function initial position option. Did you observe any behavior contradictory to your conjectures in (b), (c), and (e)? If so, how?

■

EXPLORE!

11.15 Long Term Behavior?

MATLAB-Exercise 11.3.2. To begin, open **MATLAB** and in the command line, type:

MATLAB_Wave_Equation_Non_1D

Use the GUI to answer the following questions. Make certain to click on the HELP button for instructions on how to use the GUI. Screenshots of GUI output can be found in Figure 11.50 and 11.51.

i.) Consider IBVP (11.42) with the sine initial position option, zero as the initial velocity option, and the forcing function option f1.

a.) Investigate the behavior of the solution of IBVP (11.42) for large times. Use $c^2 = 3$ and when prompted set A in the definition of $f_1(x,t)$ equal to 1. Begin by plotting solutions until $t = 2$. If necessary, slowly increase the plotting time until you are able to describe the behavior of the solution. Remember, a computer cannot handle the infinite! Compare the behavior here with the homogeneous wave equation.

b.) Repeat (a) using $A = 0.5$. How would you compare this situation with the one in (a) and the homogeneous heat equation solution?

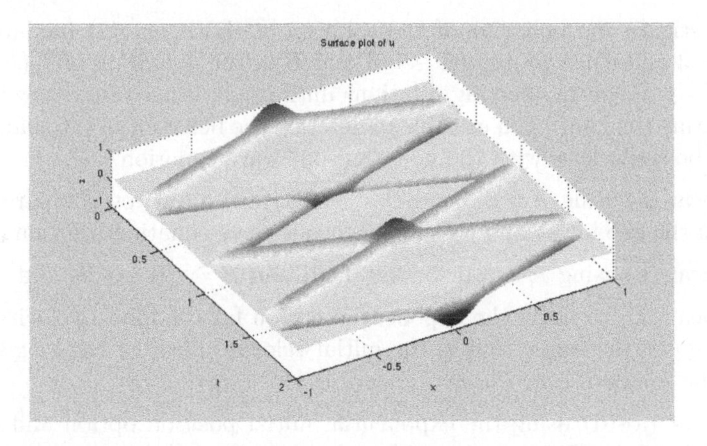

FIGURE 11.50: IBVP (11.42) with $c^2 = 3$, forcing function:$f_1(x,t) = 1$, initial position: e^{-40x^2}, initial velocity: 0.

c.) Repeat (b) using $A = 5, 0.25$, and 0.001. How is the forcing function affecting the long-term behavior of the solution?

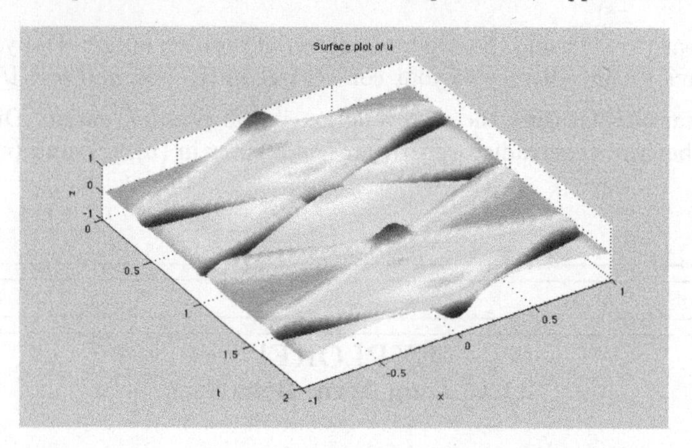

FIGURE 11.51: IBVP (11.42) with $c^2 = 3$, forcing function: $f_1(x,t) = 5$, initial position: e^{-40x^2}, initial velocity: 0.

d.) Repeat (a)-(c) using the exponential option for the initial velocity. Keep everything else the same. How is the initial velocity affecting the long-term behavior of the solution?

e.) Repeat (a)-(d) using the exponential initial position option and for the initial velocity, zero. Did you observe any behavior contradictory to your comparisons and conjectures made in (a)-(d)? If so, how?

f.) Repeat (a)-(d) using the step function initial position option and for the initial velocity, zero. Did you observe any behavior contradictory to your comparisons and conjectures made in (a)-(d)? If so, how?

ii.) Repeat (i) with the forcing function option f5.

iii.) Consider IBVP (11.43) with the cosine initial position option, zero as the initial velocity, and the forcing function option f2.

a.) Investigate the behavior of the solution of IBVP (11.43) for large times. Use $c^2 = 3$ and when prompted, set A and B in the definition of $f_2(x,t)$, equal to 1 and -1. Begin by plotting solutions until $t = 2$. If necessary, slowly increase the plotting time until you are able to describe the behavior of the solution. Compare the behavior here with the homogeneous wave equation.

b.) Repeat (a) using $A = 0.5$ and $B = -0.5$. How would you compare this situation with the one in (a) and the homogeneous wave equation solution?

c.) Repeat (b) using $A = 5, 0.25$, and 0.001 and $B = -5, -0.25$, and -0.001.

d.) Repeat (a)-(c) using the exponential option for the initial velocity. Keep everything else the same. How is the initial velocity affecting the long-term behavior of the solution?

e.) Repeat (a)-(d) using the exponential initial position option and for the initial velocity, zero. Did you observe any behavior contradictory to your comparisons and conjectures made in (a)-(d)? If so, how?

f.) Repeat (a)-(d) using the step function initial position option and for the initial velocity, zero. Did you observe any behavior contradictory to your comparisons and conjectures made in (a)-(d)? If so, how?

iv.) Consider IBVP (11.43) with the cosine initial condition option, zero as the initial velocity, and the forcing function option f4.

a.) Investigate the behavior of the solution of IBVP (11.43) for large times. Use $c^2 = 3$ and when prompted, set a in the definition of $f_4(x, t)$ equal to 1. Begin by plotting solutions until $t = 2$. If necessary, slowly increase the plotting time until you are able to describe the behavior of the solution. Compare the behavior here with the homogeneous wave equation.

b.) Repeat (a) using $a = 0.5$. How would you compare this situation with the one in (a) and the homogeneous wave equation solution?

c.) Repeat (b) using $a = 0.25, 2$, and 10. How is the forcing function affecting the long-term behavior of the solution?

d.) Repeat (a)-(c) using the exponential option for the initial velocity. Keep everything else the same. How is the initial velocity affecting the long-term behavior of the solution?

e.) Repeat (a)-(d) using the exponential initial position option and for the initial velocity, zero. Did you observe any behavior contradictory to your comparisons and conjectures made in (a)-(d)? If so, how?

f.) Repeat (a)-(d) using the step function initial position option and for the initial velocity, zero. Did you observe any behavior contradictory to your comparisons and conjectures made in (a)-(d)? If so, how?

v.) Consider IBVP (11.43) with the cosine initial condition option, zero as the initial velocity, and the fourth forcing function option, f3.

a.) Investigate the IBVP solution's behavior for large times. Use $c^2 = 3$ and when prompted, set ω in the definition of $f_3(x, t)$ equal to 1. Begin by plotting solutions until $t = 2$. If necessary, slowly increase the plotting time until you are able to describe the behavior of the solution. Compare the behavior here with the homogeneous wave equation.

b.) Repeat (a) using $a = 0.5$. How would you compare this situation with the one in (a) and the homogeneous wave equation solution?

c.) Repeat (b) using $a = 0.25, 2$, and 10. How is the forcing function affecting the long-term behavior of the solution?

d.) Repeat (a)-(c) using the exponential option for the initial velocity. Keep everything else the same. How is the initial velocity affecting the long-term behavior of the solution?

e.) Repeat (a)-(d) using the exponential initial position option and for the initial velocity, zero. Did you observe any behavior contradictory to your comparisons and conjectures made in (a)-(d)? If so, how?

f.) Repeat (a)-(d) using the step function initial position option and for the initial velocity, zero. Did you observe any behavior contradictory to your comparisons and conjectures made in (a)-(d)? If so, how?

■

EXPLORE!

11.16 Continuous Dependence on Parameters

MATLAB-Exercise 11.3.3. To explore continuous dependence with the GUI one must first solve the nonhomogeneous wave equation with a particular propagation constant, initial conditions, and forcing function. Once a solution has been constructed, click the *Perturb Parameters and Solve* button to specify the perturbed propogation constant and a perturbation size for the initial conditions. The GUI will also allow for a perturbed forcing function parameter. When prompted to select a norm choose *All* to investigate dependence using both norms at the same time. To begin, open **MATLAB** and in the command line, type:

MATLAB_Wave_Equation_Non_1D

Use the GUI to answer the following questions. Make certain to click on the HELP button for instructions on how to use the GUI. Screenshots of inputting parameters and GUI output can be found in Figures 11.52–11.56.

i.) Consider IBVP (11.42) with the sine initial position option, zero as the initial velocity option, and the forcing function option f1.

  a.) Investigate the continuous dependence of the solution on the propagation constant for IBVP (11.42) . Use $c^2 = 2$ and when prompted set A in the definition of $f_1(x,t)$ equal to 1. Plot solutions until $t = 0.5$. Once the movie has finished playing, click the *Perturb Parameters and Solve* button and choose a perturbed diffusivity constant $k^2 = 1.8$. Keep all other parameters the same as in the original run of the GUI. Describe the differences in the solutions. Compare and contrast the results using the different norms.

  b.) Choose $c^2 = 1.9, 2.1, 1.99$, and 2.01. Record your observations regarding the difference in the solutions. Make a conjecture regarding continuous dependence on the propogation constant.

  c.) Repeat (a) and (b) using $A = 5$ in the definition of $f_1(x,t)$.

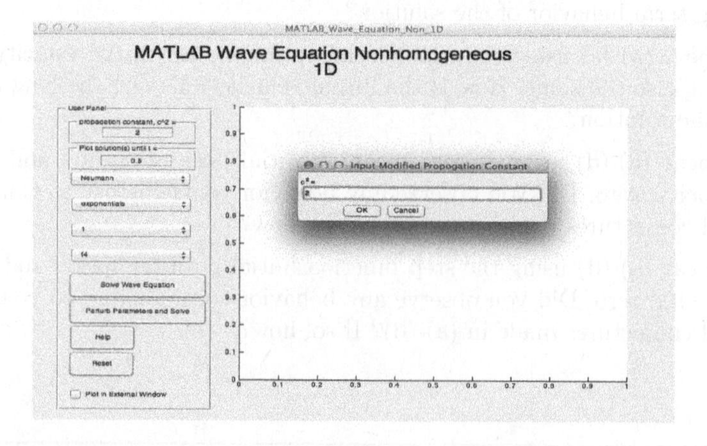

FIGURE 11.52: Perturbing parameters, initial conditions, and forcing function parameters with MATLAB Wave Equation Nonhomogeneous 1D GUI

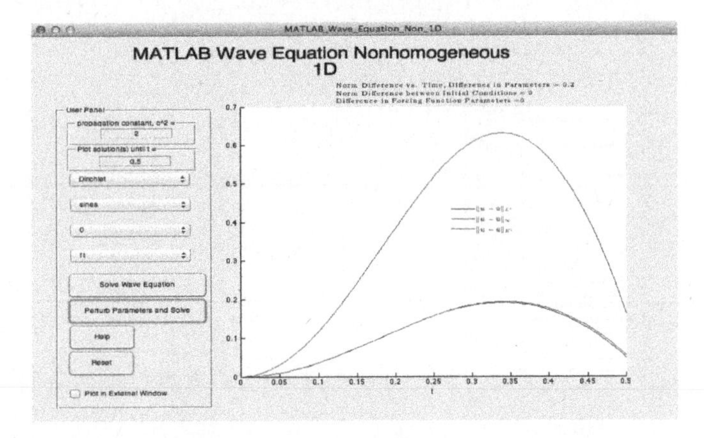

FIGURE 11.53: Perturbing the propagation constant in IBVP (11.42). Forcing function, $f_1(x,t) = 1$.

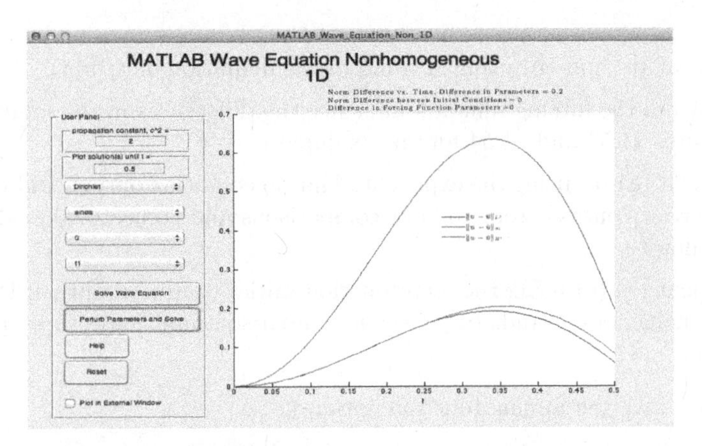

FIGURE 11.54: Perturbing the diffusivity constant in IBVP (11.42). Forcing function, $f_1(x,t) = 5$.

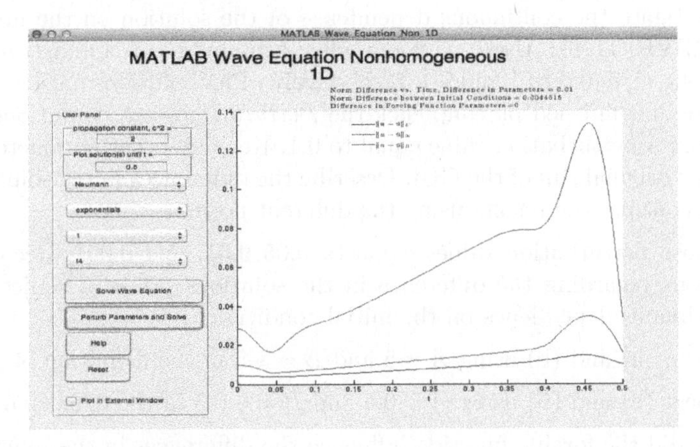

FIGURE 11.55: Perturbing the propagation constant and initial conditions in IBVP (11.43). Forcing function, $f_4(x,t) = e^{-5t}$.

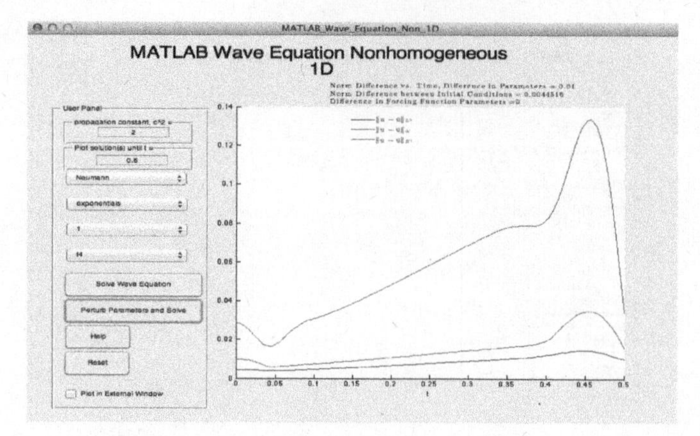

FIGURE 11.56: Perturbing the propagation constant and initial conditions in IBVP (11.43). Forcing function, $f_4(x,t) = e^{-0.5t}$.

d.) Repeat (a) and (b) using $A = 0.5$ in the definition of $f_1(x,t)$.

e.) How did the forcing function influence the differences in the solution? (Hint: See Figures 11.53 and 11.54 for one example.)

d.) Repeat (a)-(e) using the exponential initial condition option. Did you observe any behavior contradictory to your comparisons and conjectures made in (a)-(e)? If so, how?

e.) Repeat (a)-(e) using the step function initial condition option. Did you observe any behavior contradictory to your comparisons and conjectures made in (a)-(e)? If so, how?

ii.) Repeat (i) with the forcing function option f5.

iii.) Consider IBVP (11.43) with the cosine initial condition option, zero as the initial velocity, and the forcing function option f2.

a.) Investigate the continuous dependence of the solution on the initial conditions for IBVP (11.43). Use $c^2 = 2$ and when prompted, set A and B in the definition of $f_2(x,t)$, equal to 1 and -1, respectively . Plot solutions until $t = 0.5$. Once the movie has finished playing, click the *Perturb Parameters and Solve* button and choose a perturbation value equal to 0.1. Keep all other parameters the same as in the original run of the GUI. Describe the differences in the solutions. Compare and contrast the results using the different norms.

b.) Choose perturbation values equal to $0.05, 0.01$, and 0.001. Record your observations regarding the difference in the solutions. Make a conjecture regarding continuous dependence on the initial condition.

c.) Repeat (a) and (b) using $A = 5$ and $B = -5$ in the definition of $f_2(x,t)$.

d.) Repeat (a) and (b) using $A = 0.5$ and $B = -0.5$ in the definition of $f_2(x,t)$.

e.) How did the forcing function influence the differences in the solution?

d.) Repeat (a)-(e) using the exponential initial condition option and the initial velocity option 1. Did you observe any behavior contradictory to your comparisons and conjectures made in (a)-(e)? If so, how?

e.) Repeat (a)-(e) using the step function initial condition option. Did you observe any behavior contradictory to your comparisons and conjectures made in (a)-(e)? If so, how?

iv.) Consider IBVP (11.43) with the cosine initial condition option, zero as the initial velocity, and the forcing function option f4.

a.) Investigate the continuous dependence of the solution on the initial conditions and propagation constants for IBVP (11.43). Use $c^2 = 2$ and when prompted, set a in the definition of $f_4(x,t)$ equal to 1. Plot solutions until $t = 0.5$. Once the movie has finished playing, click the *Perturb Parameters and Solve* button and choose a perturbed propogation constant equal to 1.8 and perturbation value equal to 0.1. Keep all other parameters the same as in the original run of the GUI. Describe the differences in the solutions. Compare and contrast the results using the different norms.

b.) Choose perturbed propogation constants equal to $2.1, 1.9$, and 2.001 and perturbation values equal to $0.05, 0.01$, and 0.001. Record your observations regarding the difference in the solutions. Make a conjecture regarding continuous dependence on the initial conditions and the propagation constants.

c.) Repeat (a) and (b) using $a = 5$ in the definition of $f_4(x,t)$.

d.) Repeat (a) and (b) using $a = 0.5$ in the definition of $f_4(x,t)$.

e.) How did the forcing function influence the differences in the solution? (Hint: See Figures 11.55 and 11.56 for one example.)

e.) Repeat (a)-(e) using the exponential initial condition option. Did you observe any behavior contradictory to your comparisons and conjectures made in (a)-(e)? If so, how?

f.) Repeat (a)-(e) using the step function initial condition option. Did you observe any behavior contradictory to your comparisons and conjectures made in (a)-(e)? If so, how?

v.) Consider IBVP (11.43) with the cosine initial condition option, zero as the initial velocity, and the forcing function option f3.

a.) Investigate the IBVP continuous dependence on the initial condition and the forcing function parameter. Use $c^2 = 2$ and when prompted, set ω in the definition of $f_3(x,t)$ equal to 1. Plot solutions until $t = 0.5$. Once the movie has finished playing, click the *Perturb Parameters and Solve* button to choose a perturbation value equal to 0.1 and forcing function parameter equal to 1.1. Describe the differences in the solutions. Compare and contrast the results using the different norms.

b.) Choose perturbation values equal to $0.05, 0.01$, and 0.001, and forcing function parameters equal to $0.9, 0.95$, and 1.01. Record your observations regarding the difference in the solutions. Make a conjecture regarding continuous dependence on the initial condition and the forcing function parameter.

c.) Repeat (a)-(b) using the exponential initial condition option and any non-zero initial velocity. Did you observe any behavior contradictory to your comparisons and conjectures made in (a)-(b)? If so, how?

d.) Repeat (a)-(b) using the step function initial condition option. Did you observe any behavior contradictory to your comparisons and conjectures made in (a)-(b)? If so, how?

vi.) Summarize your observations regarding continuous dependence. Make sure you address the potential issue of using different norms and the influence the forcing function had on the differences in the solutions.

■

Next, suppose the string is embodied in a material that resists its movement, and we now account for source effects as a forcing term. Such resistance, or *viscous damping*, is interpreted as a force acting in the direction opposite to the velocity and is directly proportional to its magnitude. As such, the term $\alpha\frac{\partial}{\partial t}z(x,t)$, where $\alpha > 0$ is the damping coefficient, is added the left side of the PDE in the IBVPs (as we did for the spring-mass system model) leads to the following variant of (9.67) :

$$\begin{cases} \frac{\partial^2}{\partial t^2}z(x,t) + \alpha\frac{\partial}{\partial t}z(x,t) + c^2\frac{\partial^2}{\partial x^2}z(x,t) = f(x,t), \ 0 < x < L, \ t > 0, \\ z(x,0) = z_0(x), \ \frac{\partial z}{\partial t}(x,0) = z_1(x), \ 0 < x < L, \\ z(0,t) = z(L,t) = 0, \ t > 0, \end{cases} \tag{11.48}$$

Replacing the Dirichlet BCs with Neumann BCs in (11.48) yields

$$\begin{cases} \frac{\partial^2}{\partial t^2}z(x,t) + \alpha\frac{\partial}{\partial t}z(x,t) + c^2\frac{\partial^2}{\partial x^2}z(x,t) = f(x,t), \ 0 < x < L, \ t > 0, \\ z(x,0) = z_0(x), \ \frac{\partial z}{\partial t}(x,0) = z_1(x), \ 0 < x < L, \\ \frac{\partial z(0,t)}{\partial x} = \frac{\partial z(L,t)}{\partial x} = 0, \ t > 0, \end{cases} \tag{11.49}$$

or with periodic BCs yields

$$\begin{cases} \frac{\partial^2}{\partial t^2}z(x,t) + \alpha\frac{\partial}{\partial t}z(x,t) + c^2\frac{\partial^2}{\partial x^2}z(x,t) = f(x,t), \ 0 < x < L, \ t > 0, \\ z(x,0) = z_0(x), \ \frac{\partial z}{\partial t}(x,0) = z_1(x), \ 0 < x < L. \\ z(0,t) = z(L,t), \ t > 0. \end{cases} \tag{11.50}$$

EXPLORE!
11.17 Parameter Play!

In this **EXPLORE!** you will investigate the effects of adding dampening to (11.42)-(11.44). To do this, you will use the MATLAB Nonhomogeneous Wave Equation Damped 1D GUI to visualize the solutions for different choices of α and c^2.

MATLAB-Exercise 11.3.4. To begin, open **MATLAB** and in the command line, type:

MATLAB_Wave_Equation_Damp_Non_1D

Use the GUI to answer the following questions. Make certain to click on the HELP button for instructions on how to use the GUI. Screenshots of GUI output can be found in Figures 11.57, 11.58, and 11.59. After clicking the *Solve Wave Equation* button the GUI will automatically compare the damped nonhomogeneous solution to the undamped.

i.) Consider IBVP (11.48) with the sine initial position option, the zero initial velocity option, and the forcing function option f1.

a.) Use $c^2 = 1 = \alpha$, plot solutions until $t = 1$ and when prompted set A in the definition of $f_1(x,t)$ equal to 1. Describe the evolution of the solution. How did it differ from the corresponding undamped nonhomogeneous wave equation?

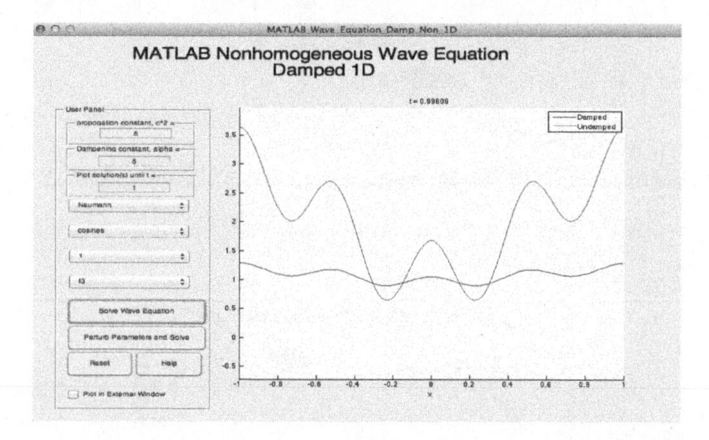

FIGURE 11.57: MATLAB Nonhomogeneous Wave Equation Damped 1D GUI

b.) Based on your intuition gained in Chapter 9 regarding the damped wave equation, how do you expect the solution of (11.42) to evolve when $\alpha = 5$, as compared to (a)? Check your intuition by using the GUI to visualize this situation.

c.) What do you expect to happen if $\alpha = 0.25$? Test your hypothesis with the GUI.

d.) Repeat (a) using $A = 0.5$. How would you compare this situation with the one in (a)?

e.) Repeat (d) using $A = 0.25, 0.1$, and 0.001. In addition, make a conjecture for whether or not these nonhomogeneous solutions converge to the damped homogeneous solution as $f_1(x,t)$ converges to the zero function.

f.) What if A gets larger? Explore this by choosing a sequence of increasingly larger values for A. Summarize your observations.

g.) Repeat (a)-(f) using the exponential initial position option. Did you observe any behavior contradictory to your conjectures in (b), (c), and (e)? If so, how?

h.) Repeat (a)-(f) using the step function initial position option. Did you observe any behavior contradictory to your conjectures in (b), (c), and (e)? If so, how?

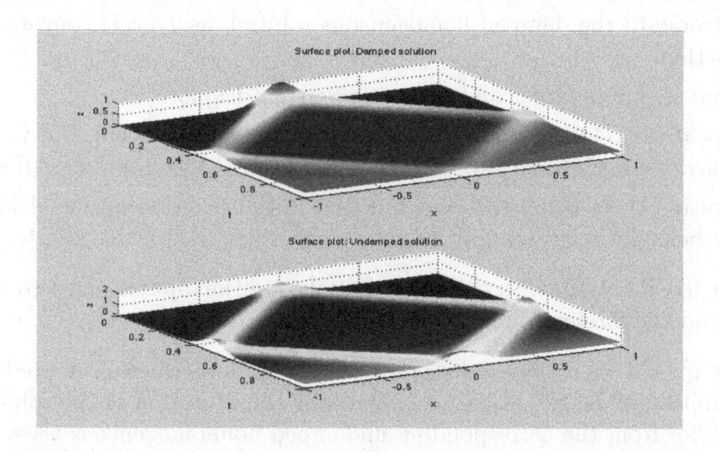

FIGURE 11.58: IBVP (11.49) with $c^2 = 2, \alpha = 5$, $f_4(x,t) = e^{-t}$, initial position: e^{-40x^2}, initial velocity: 0.

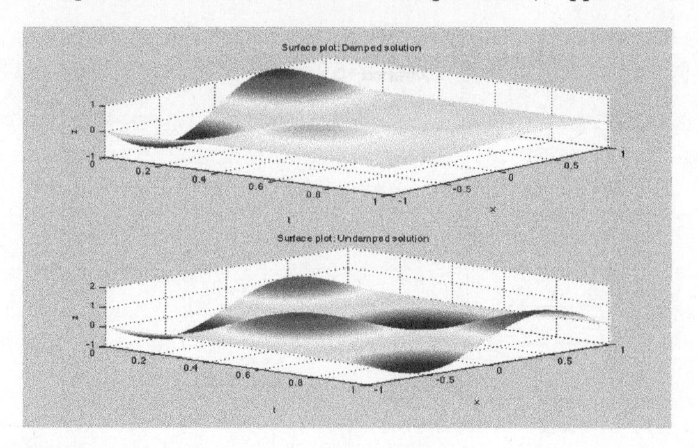

FIGURE 11.59: IBVP (11.48) with $c^2 = 2, \alpha = 5$, $f(x,t) = 1$, initial position: $\sin \pi x$, initial velocity: 0.

ii.) Repeat (i) with the forcing function option f5.

iii.) Consider IBVP (11.49) with the cosine initial position option, the cosine initial velocity option, and the forcing function option f2.

a.) Use $c^2 = 1 = \alpha$, plot solutions until $t = 1$, and when prompted, set A and B in the definition of $f_2(x,t)$, equal to 1 and -1, respectively . Describe the evolution of the solution. How did it differ from the corresponding undamped nonhomogeneous wave equation?

b.) Based on your intuition gained in Chapter 9 regarding the heat equation, how do you expect the solution of (11.49) to evolve when $\alpha = 5$, as compared to (a)? Check your intuition by using the GUI to visualize this situation.

c.) What do you expect to happen if $\alpha = 0.25$? Test your hypothesis with the GUI.

d.) Repeat (a) using $A = 0.5$ and $B = -0.5$. How would you compare this situation with the one in (a)?

e.) Repeat (d) using $A = 0.25, 0.1$, and 0.001 and $B = -0.25, -0.1$, and -0.001. In addition, make a conjecture for whether or not these nonhomogenous solutions converge to the damped homogeneous solution as $f_2(x,t)$ converges to the zero function.

f.) What if A gets larger and B more negative? Explore!

g.) Repeat (a)-(f) using the exponential initial position option. Did you observe any behavior contradictory to your conjectures in (b), (c), and (e)? If so, how?

h.) Repeat (a)-(f) using the step function initial position option. Did you observe any behavior contradictory to your conjectures in (b), (c), and (e)? If so, how?

iv.) Consider IBVP (11.49) with the cosine initial condition option, the zero initial velocity option, and the forcing function option f4.

a.) Use $c^2 = 1 = \alpha$, plot solutions until $t = 1$, and when prompted, set a in the definition of $f_4(x,t)$ equal to 1. Describe the evolution of the solution. How did it differ from the corresponding undamped nonhomogeneous wave equation?

b.) Based on your intuition gained in Chapter 9 regarding the heat equation, how do you expect the solution of (11.49) to evolve when $\alpha = 5$, as compared to (a)? Check your intuition by using the GUI to visualize this situation.

c.) What do you expect to happen if $\alpha = 0.25$? Test your hypothesis with the GUI.

d.) Repeat (a) using $a = 0.5$. How would you compare this situation with the one in (a) and the homogeneous wave equation solution?

e.) Repeat (d) using $a = 0.25, 0.1$, and 0.001. Does the solution appear to get closer to the damped homogeneous solution?

f.) What if a gets larger? Explore!

g.) Repeat (a)-(f) using the exponential initial position option. Did you observe any behavior contradictory to your conjectures in (b), (c), and (e)? If so, how?

h.) Repeat (a)-(f) using the step function initial position option. Did you observe any behavior contradictory to your conjectures in (b), (c), and (e)? If so, how?

v.) Consider IBVP (11.50) with the cosine initial condition option, any choice for the initial velocity, and the fourth forcing function option, f3.

a.) Use $c^2 = 1 = \alpha$, plot solutions until $t = 1$, and when prompted, set ω in the definition of $f_3(x,t)$ equal to 1. Describe the evolution of the solution. How did it differ from the corresponding homogeneous wave equation?

b.) Based on your intuition gained in Chapter 9 regarding the wave equation, how do you expect the solution of (11.50) to evolve when $\alpha = 5$, as compared to (a)? Check your intuition by using the GUI to visualize this situation.

c.) What do you expect to happen if $\alpha = 0.25$? Test your hypothesis with the GUI.

d.) Repeat (a) using $a = 0.5$. How would you compare this situation with the one in (a)?

e.) Repeat (d) using $a = 0.25, 0.1$, and 0.001. Explain why the solution should **not** tend to the homogeneous solution. Note: Unlike the previous two models, the periodic nature of the wave equation along with the periodic forcing function are making it difficult to compare with the homogeneous case. It's there...watch the boundaries!

f.) What if a gets larger? Explore!

g.) Repeat (a)-(f) using the exponential initial position option and any initial velocity. Did you observe any behavior contradictory to your conjectures in (b), (c), and (e)? If so, how?

h.) Repeat (a)-(f) using the step function initial position option. Did you observe any behavior contradictory to your conjectures in (b), (c), and (e)? If so, how?

■

EXPLORE!

11.18 Long Term Behavior?

MATLAB-Exercise 11.3.5. To begin, open **MATLAB** and in the command line, type:

MATLAB_Wave_Equation_Non_Damp_1D

Use the GUI to answer the following questions. Make certain to click on the HELP button for instructions on how to use the GUI.

i.) Consider IBVP (11.48) with the sine initial position option, zero as the initial velocity option, and the forcing function option f1.

a.) Investigate the behavior of the solution of IBVP (11.48) for large times. Use $c^2 = 3, \alpha = 1$, and when prompted set A in the definition of $f_1(x,t)$ equal to 1. Begin by plotting solutions until $t = 2$. If necessary, slowly increase the plotting time until you are able to describe the behavior of the solution. Remember, a computer cannot handle the infinite! Compare the behavior here with the homogeneous wave equation.

b.) Repeat (a) using $\alpha = 2$. How would you compare this situation with the one in (a) and the undamped nonhomogeneous wave equation solution?

c.) Repeat (b) using $\alpha = 0.25, 5$, and 10. How is the dampening constant affecting the long-term behavior of the solution?

d.) Repeat (a)-(c) using the exponential option for the initial velocity. Keep everything else the same. Does the non-zero initial velocity alter the pattern you observed in (a) ?

e.) Repeat (a)-(d) using the exponential initial position option and for the initial velocity, zero. Did you observe any behavior contradictory to your comparisons and conjectures made in (a)-(d)? If so, how?

f.) Repeat (a)-(d) using the step function initial position option and for the initial velocity, zero. Did you observe any behavior contradictory to your comparisons and conjectures made in (a)-(d)? If so, how?

ii.) Repeat (i) with the forcing function option f5.

iii.) Consider IBVP (11.49) with the cosine initial position option, zero as the initial velocity, and the forcing function option f2.

a.) Investigate the behavior of the solution of IBVP (11.49) for large times. Use $c^2 = 3, \alpha = 1$ and when prompted, set A and B in the definition of $f_2(x,t)$, equal to 1 and -1. Begin by plotting solutions until $t = 2$. If necessary, slowly increase the plotting time until you are able to describe the behavior of the solution. Compare the behavior here with the homogeneous wave equation.

b.) Repeat (a) using $\alpha = 2$. How would you compare this situation with the one in (a) and the homogeneous wave equation solution?

c.) Repeat (b) using $\alpha = 0.25, 5$, and 10. How is the dampening constant affecting the long-term behavior of the solution?

d.) Repeat (a)-(c) using the exponential option for the initial velocity. Keep everything else the same. Does the non-zero initial velocity alter the pattern you observed in (a) ?

e.) Repeat (a)-(d) using the exponential initial position option and for the initial velocity, zero. Did you observe any behavior contradictory to your comparisons and conjectures made in (a)-(d)? If so, how?

f.) Repeat (a)-(d) using the step function initial position option and for the initial velocity, zero. Did you observe any behavior contradictory to your comparisons and conjectures made in (a)-(d)? If so, how?

iv.) Consider IBVP (11.49) with the cosine initial condition option, zero as the initial velocity, and the forcing function option f4.

a.) Investigate the behavior of the solution of IBVP (11.49) for large times. Use $c^2 = 3, \alpha = 1$, and when prompted, set a in the definition of $f_4(x,t)$ equal to 1. Begin by plotting solutions until $t = 2$. If necessary, slowly increase the plotting time until you are able to describe the behavior of the solution. Compare the behavior here with the homogeneous wave equation.

b.) Repeat (a) using $\alpha = 2$. How would you compare this situation with the one in (a) and the homogeneous wave equation solution?

c.) Repeat (b) using $\alpha = 0.25, 5$, and 10. How is the dampening constant affecting the long-term behavior of the solution?

d.) Repeat (a)-(c) using the exponential option for the initial velocity. Keep everything else the same. Does the non-zero initial velocity alter the pattern you observed in (a) ?

e.) Repeat (a)-(d) using the exponential initial position option and for the initial velocity, zero. Did you observe any behavior contradictory to your comparisons and conjectures made in (a)-(d)? If so, how?

f.) Repeat (a)-(d) using the step function initial position option and for the initial velocity, zero. Did you observe any behavior contradictory to your comparisons and conjectures made in (a)-(d)? If so, how?

v.) Consider IBVP (11.49) with the cosine initial condition option, zero as the initial velocity, and the fourth forcing function option, f3.

a.) Investigate the IBVP solution's behavior for large times. Use $c^2 = 3, \alpha = 1$, and when prompted, set ω in the definition of $f_3(x,t)$ equal to 1. Begin by plotting solutions until $t = 2$. If necessary, slowly increase the plotting time until you are able to describe the behavior of the solution. Compare the behavior here with the homogeneous wave equation.

b.) Repeat (a) using $\alpha = 2$. How would you compare this situation with the one in (a) and the homogeneous wave equation solution?

c.) Repeat (b) using $\alpha = 0.25, 5$, and 10. How is the dampening constant affecting the long-term behavior of the solution?

d.) Repeat (a)-(c) using the exponential option for the initial velocity. Keep everything else the same. Does the non-zero initial velocity alter the pattern you observed in (a) ?

e.) Repeat (a)-(d) using the exponential initial position option and for the initial velocity, zero. Did you observe any behavior contradictory to your comparisons and conjectures made in (a)-(d)? If so, how?

f.) Repeat (a)-(d) using the step function initial position option and for the initial velocity, zero. Did you observe any behavior contradictory to your comparisons and conjectures made in (a)-(d)? If so, how?

■

EXPLORE!
11.19 Continuous Dependence on Parameters

MATLAB-Exercise 11.3.6. To explore continuous dependence with the GUI one must first solve the nonhomogeneous wave equation with a particular propagation constant, initial conditions, and forcing function. Once a solution has been constructed, click the *Perturb Parameters and Solve* button to specify the perturbed propagation constant and a perturbation size for the initial conditions. The GUI will also allow for a perturbed forcing function parameter. When prompted to select a norm choose *All* to investigate dependence using both norms at the same time. To begin, open **MATLAB** and in the command line, type:

MATLAB_Wave_Equation_Non_Damp_1D

Use the GUI to answer the following questions. Make certain to click on the HELP button for instructions on how to use the GUI. Screenshots of inputting parameters and of GUI output can be found in Figures 11.60–11.63.

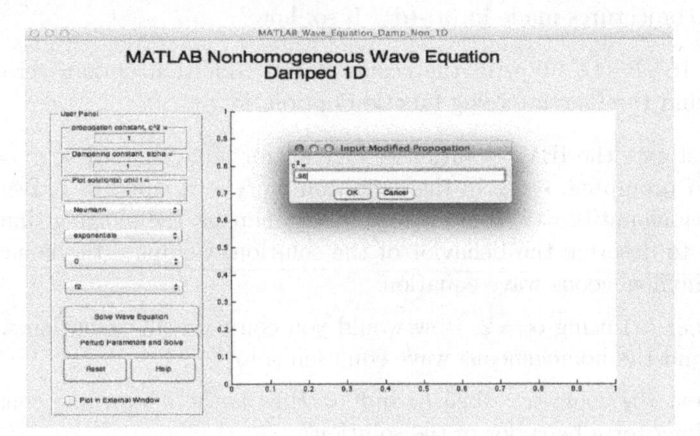

FIGURE 11.60: Perturbing parameters, initial conditions, and forcing function parameters with MATLAB Wave Equation Nonhomogeneous 1D GUI

i.) Consider IBVP (11.48) with the sine initial position option, zero as the initial velocity option, and the forcing function option f1.

a.) Investigate the continuous dependence of the solution on the propagation constant for IBVP (11.48). Use $c^2 = 2, \alpha = 1$ and when prompted set A in the definition of $f_1(x,t)$ equal to 1. Plot solutions until $t = 0.5$. Once the movie has finished playing, click the *Perturb Parameters and Solve* button and choose a perturbed propagation constant $c^2 = 1.8$. Keep all other parameters the same as in the original run of the GUI. Describe the differences in the solutions. Compare and contrast the results using the different norms.

b.) Choose $c^2 = 1.9, 2.1, 1.99,$ and 2.01. Record your observations regarding the difference in the solutions. Make a conjecture regarding continuous dependence on the propagation constant.

c.) Repeat (a) and (b) using $A = 5$ in the definition of $f_1(x,t)$.

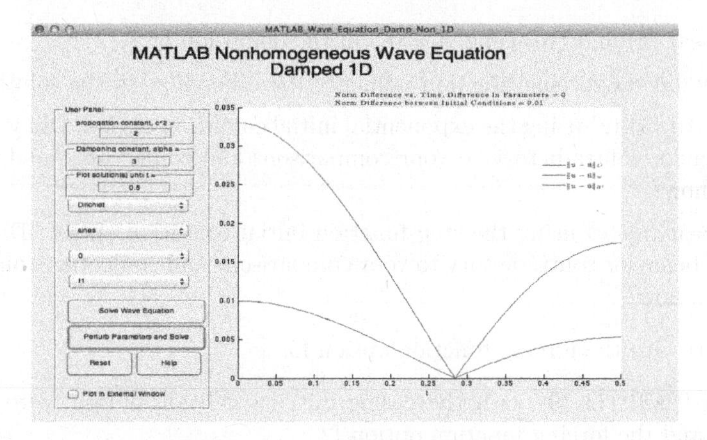

FIGURE 11.61: Perturbing the initial conditions in IBVP (11.48). Forcing function, $f_1(x,t) = 1$.

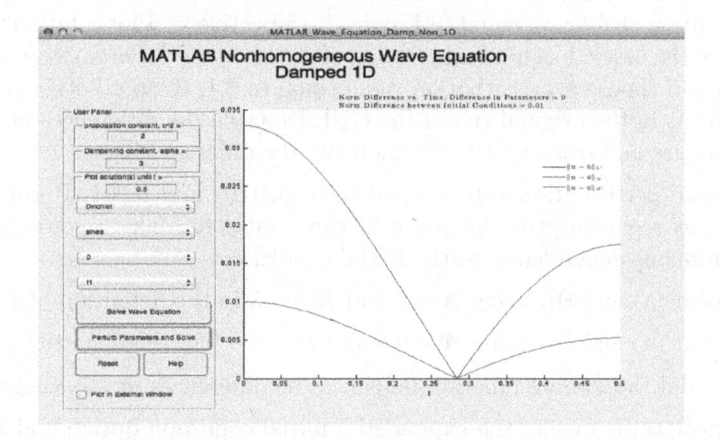

FIGURE 11.62: Perturbing the initial conditions in IBVP (11.48). Forcing function, $f_1(x,t) = 5$.

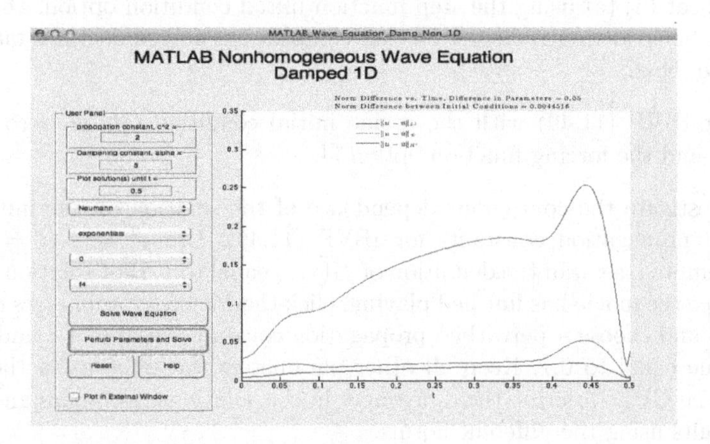

FIGURE 11.63: Perturbing the propagation constant and initial conditions in IBVP (11.49). Forcing function, $f_4(x,t) = e^{-5t}$.

d.) Repeat (a) and (b) using $A = 0.5$ in the definition of $f_1(x,t)$.

e.) How did the forcing function influence the differences in the solution?

d.) Repeat (a)-(e) using the exponential initial condition option. Did you observe any behavior contradictory to your comparisons and conjectures made in (a)-(e)? If so, how?

e.) Repeat (a)-(e) using the step function initial condition option. Did you observe any behavior contradictory to your comparisons and conjectures made in (a)-(e)? If so, how?

ii.) Repeat (i) with the forcing function option f5.

iii.) Consider IBVP (11.49) with the cosine initial condition option, zero as the initial velocity, and the forcing function option f2.

a.) Investigate the continuous dependence of the solution on the initial conditions for IBVP (11.49). Use $c^2 = 2, \alpha = 3$, and when prompted, set A and B in the definition of $f_2(x,t)$, equal to 1 and -1, respectively . Plot solutions until $t = 0.5$. Once the movie has finished playing, click the *Perturb Parameters and Solve* button and choose a perturbation value equal to 0.1. Keep all other parameters the same as in the original run of the GUI. Describe the differences in the solutions. Compare and contrast the results using the different norms.

b.) Choose perturbation values equal to $0.05, 0.01$, and 0.001. Record your observations regarding the difference in the solutions. Make a conjecture regarding continuous dependence on the initial condition.

c.) Repeat (a) and (b) using $A = 5$ and $B = -5$ in the definition of $f_2(x,t)$.

d.) Repeat (a) and (b) using $A = 0.5$ and $B = -0.5$ in the definition of $f_2(x,t)$.

e.) How did the forcing function influence the differences in the solution?

d.) Repeat (a)-(e) using the exponential initial condition option and the initial velocity option 1. Did you observe any behavior contradictory to your comparisons and conjectures made in (a)-(e)? If so, how?

e.) Repeat (a)-(e) using the step function initial condition option. Did you observe any behavior contradictory to your comparisons and conjectures made in (a)-(e)? If so, how?

iv.) Consider IBVP (11.49) with the cosine initial condition option, zero as the initial velocity, and the forcing function option f4.

a.) Investigate the continuous dependence of the solution on the initial conditions and propagation constants for IBVP (11.49). Use $c^2 = 2, \alpha = 5$, and when prompted, set a in the definition of $f_4(x,t)$ equal to 1. Plot solutions until $t = 0.5$. Once the movie has finished playing, click the *Perturb Parameters and Solve* button and choose a perturbed propagation constant equal to 1.8 and perturbation value equal to 0.1. Keep all other parameters the same as in the original run of the GUI. Describe the differences in the solutions. Compare and contrast the results using the different norms.

b.) Choose perturbed propagation constants equal to $2.1, 1.9$, and 2.001 and perturbation values equal to $0.05, 0.01$, and 0.001. Record your observations regarding the difference in the solutions. Make a conjecture regarding continuous dependence on the initial conditions and the propagation constants.

c.) Repeat (a) and (b) using $a = 5$ in the definition of $f_4(x,t)$.

d.) Repeat (a) and (b) using $a = 0.5$ in the definition of $f_4(x,t)$.

e.) How did the forcing function influence the differences in the solution? (Hint: See Figures 11.55 and 11.56 for one example.)

d.) Repeat (a)-(e) using the exponential initial condition option. Did you observe any behavior contradictory to your comparisons and conjectures made in (a)-(e)? If so, how?

e.) Repeat (a)-(e) using the step function initial condition option. Did you observe any behavior contradictory to your comparisons and conjectures made in (a)-(e)? If so, how?

v.) Consider IBVP (11.49) with the cosine initial condition option, zero as the initial velocity, and the forcing function option f3.

a.) Investigate the continuous dependence of the solution on the initial condition and the forcing function parameter for IBVP (11.49). Use $c^2 = 2, \alpha = 3$, and when prompted, set ω in the definition of $f_3(x,t)$ equal to 1. Plot solutions until $t = 0.5$. Once the movie has finished playing, click the *Perturb Parameters and Solve* button to choose a perturbation value equal to 0.1 and forcing function parameter equal to 1.1. Describe the differences in the solutions. Compare and contrast the results using the different norms.

b.) Choose perturbation values equal to $0.05, 0.01$, and 0.001, and forcing function parameters equal to $0.9, 0.95$, and 1.01. Record your observations regarding the difference in the solutions. Make a conjecture regarding continuous dependence on the initial condition and the forcing function parameter.

c.) Repeat (a)-(b) using the exponential initial condition option and any non-zero initial velocity. Did you observe any behavior contradictory to your comparisons and conjectures made in (a)-(b)? If so, how?

d.) Repeat (a)-(b) using the step function initial condition option. Did you observe any behavior contradictory to your comparisons and conjectures made in (a)-(b)? If so, how?

vi.) Summarize your observations regarding continuous dependence. Make sure you address the potential issue of using different norms and the influence the forcing function had on the differences in the solutions.

vii.) Investigate the continuous dependence of the solution on the dampening coefficient for IBVP (11.50). (Hint: follow the format of previous continuous dependence questions.)

■

11.3.2 Two-Dimensional Wave Equation with Source Effects

Generalizing the above discussion from a one-dimensional spatial domain to a bounded domain $\Omega \subset \mathbb{R}^N$ with smooth boundary $\partial\Omega$ is not difficult. Indeed, the resulting IBVP is given by

$$\begin{cases} \frac{\partial^2}{\partial t^2} z(\mathbf{x},t) - c^2 \Delta z(\mathbf{x},t) = f(\mathbf{x},t), \ \mathbf{x} \in \Omega, \, t > 0, \\ z(\mathbf{x},0) = z_0(\mathbf{x}), \ \frac{\partial z}{\partial t}(\mathbf{x},0) = z_1(\mathbf{x}), \ \mathbf{x} \in \Omega, \\ z(\mathbf{x},t) = 0, \ \mathbf{x} \in \partial\Omega, \, t > 0. \end{cases} \tag{11.51}$$

As was the case in Chapter 9, if Ω is a rectangle in $\mathbb{R}^2$ then (11.51) becomes

$$\begin{cases} \frac{\partial z}{\partial t^2}(x,y,t) = c^2\left(\frac{\partial^2 z}{\partial x^2}(x,y,t) + \frac{\partial^2 z}{\partial y^2}(x,y,t)\right) + f(x,y,t),\ 0<x<a,\ 0<y<b,\ t>0, \\ z(x,y,0) = z_0(x,y),\ 0<x<a,\ 0<y<b, \\ z(x,0,t) = 0 = z(x,b,t),\ 0<x<a,\ t>0, \\ z(0,y,t) = 0 = z(a,y,t),\ 0<y<b,\ t>0, \end{cases} \tag{11.52}$$

If we replace the Dirichlet BCs with Neumann then (11.52) transforms to

$$\begin{cases} \frac{\partial z}{\partial t^2}(x,y,t) = c^2\left(\frac{\partial^2 z}{\partial x^2}(x,y,t) + \frac{\partial^2 z}{\partial y^2}(x,y,t)\right) + f(x,y,t),\ 0<x<a,\ 0<y<b,\ t>0, \\ z(x,y,0) = z_0(x,y),\ 0<x<a,\ 0<y<b, \\ \frac{\partial z(x,0,t)}{\partial y} = 0 = \frac{\partial z(x,b,t)}{\partial y},\ 0<x<a,\ t>0, \\ \frac{\partial z(0,y,t)}{\partial x} = 0 = \frac{\partial z(a,y,t)}{\partial x},\ 0<y<b,\ t>0, \end{cases} \tag{11.53}$$

And equipping the wave equation with periodic BCs in two-dimensions with a forcing term yields the IBVP

$$\begin{cases} \frac{\partial z}{\partial t^2}(x,y,t) = c^2\left(\frac{\partial^2 z}{\partial x^2}(x,y,t) + \frac{\partial^2 z}{\partial y^2}(x,y,t)\right) + f(x,y,t),\ 0<x<a,\ 0<y<b,\ t>0, \\ z(x,y,0) = z_0(x,y),\ 0<x<a,\ 0<y<b, \\ z(x,0,t) = z(x,b,t),\ 0<x<a,\ t>0, \\ z(0,y,t) = z(a,y,t),\ 0<y<b,\ t>0, \end{cases} \tag{11.54}$$

Exercise 11.3.2. Mimic the procedure outlined in Section 12.1.2 to provide a similar procedure that can be used to construct the solution of (11.54), assuming appropriate smoothness conditions on all functions involved.

EXPLORE!

11.20 Effects of Parameters on Solutions

In this **EXPLORE!** you will investigate the effects of changing the parameter c^2 in IBVPs (11.52)-(11.54). To do this, you will use the MATLAB Wave Equation Nonhomogeneous 2D GUI to visualize the solutions for different choices of c^2.

MATLAB-Exercise 11.3.7. To begin, open **MATLAB** and in the command line, type:

MATLAB_Wave_Equation_Non_2D

Use the GUI to answer the following questions. Make certain to click on the HELP button for instructions on how to use the GUI. A screenshot from the GUI can found in Figure 11.64.

i.) Consider IBVP (11.52) with the sine initial position option, the zero initial velocity option, and the forcing function option f1.

a.) Use $c^2 = 1$, plot solutions until $t = 1$ and when prompted set A in the definition of $f_1(x,t)$ equal to 1. Describe the evolution of the solution. How did it differ from the corresponding homogeneous wave equation?

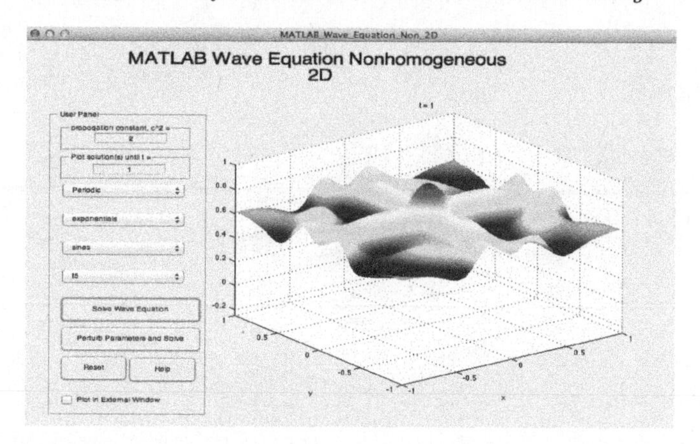

FIGURE 11.64: MATLAB Wave Equation Nonhomogeneous 2D GUI

b.) Based on your intuition gained in the one-dimensional case regarding the wave equation, how do you expect the solution of (11.52) to evolve when $c^2 = 2$, as compared to (a)? Check your intuition by using the GUI to visualize this situation.

c.) What do you expect to happen if $c^2 = 0.25$? Test your hypothesis with the GUI.

d.) Repeat (a) using $A = 0.5$. How would you compare this situation with the one in (a) and the homogeneous wave equation solution?

e.) Repeat (d) using $A = 0.25, 0.1$, and 0.001. In addition, make a conjecture for whether or not these nonhomogeneous solutions converge to the homogeneous solution as $f_1(x,t)$ converges to the zero function.

f.) What if A gets larger? Explore this by choosing a sequence of increasingly larger values for A. Summarize your observations.

g.) Repeat (a)-(f) using the exponential initial position option. Did you observe any behavior contradictory to your conjectures in (b), (c), and (e)? If so, how?

h.) Repeat (a)-(f) using the step function initial position option. Did you observe any behavior contradictory to your conjectures in (b), (c), and (e)? If so, how?

ii.) Repeat (i) with the forcing function option f5.

iii.) Consider IBVP (11.53) with the cosine initial position option, the cosine initial velocity option, and the forcing function option f2.

a.) Use $c^2 = 1$, plot solutions until $t = 1$, and when prompted, set A and B in the definition of $f_2(x,t)$, equal to 1 and -1, respectively . Describe the evolution of the solution. How did it differ from the corresponding homogeneous wave equation?

b.) Based on your intuition gained in the one-dimensional case regarding the wave equation, how do you expect the solution of (11.53) to evolve when $c^2 = 2$, as compared to (a)? Check your intuition by using the GUI to visualize this situation.

c.) What do you expect to happen if $c^2 = 0.25$? Test your hypothesis with the GUI.

d.) Repeat (a) using $A = 0.5$ and $B = -0.5$. How would you compare this situation with the one in (a) and the homogeneous wave equation solution?

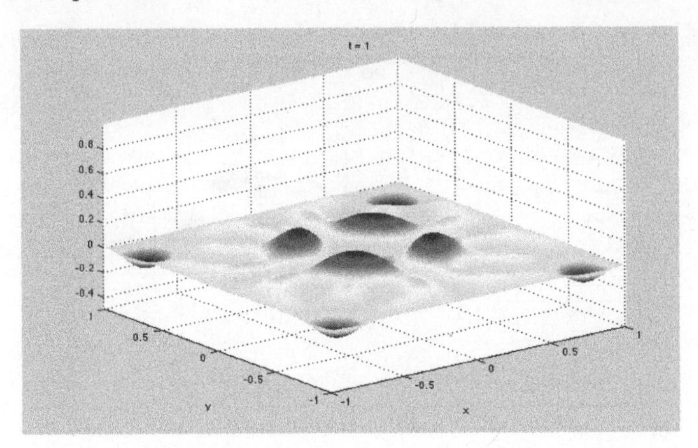

FIGURE 11.65: IBVP (11.52) with $c^2 = 3$, $f_1(x,t) = 1$, initial position: e^{-40x^2}, initial velocity: 0.

e.) Repeat (d) using $A = 0.25, 0.1$, and 0.001 and $B = -0.25, -0.1$, and -0.001. In addition, make a conjecture for whether or not these nonhomogenous solutions converge to the homogeneous solution as $f_2(x,t)$ converges to the zero function.

f.) What if A gets larger and B more negative? Explore!

g.) Repeat (a)-(f) using the exponential initial position option. Did you observe any behavior contradictory to your conjectures in (b), (c), and (e)? If so, how?

h.) Repeat (a)-(f) using the step function initial position option. Did you observe any behavior contradictory to your conjectures in (b), (c), and (e)? If so, how?

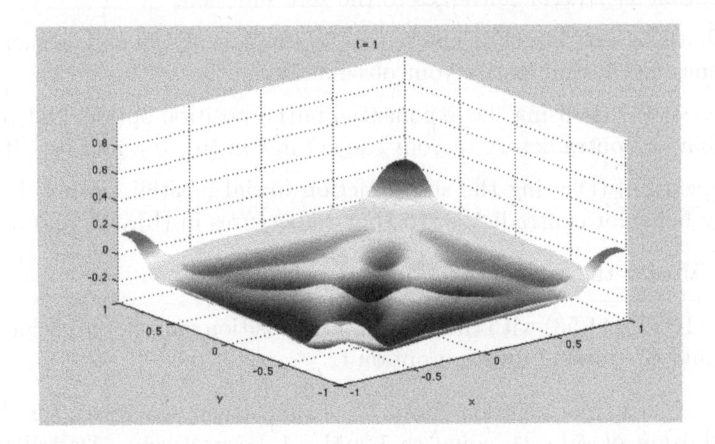

FIGURE 11.66: IBVP (11.53) with $c^2 = 3$, $f_2(x,t) = x - y$, initial position: e^{-40x^2}, initial velocity: 0.

iv.) Consider IBVP (11.53) with the cosine initial condition option, the zero initial velocity option, and the forcing function option f4.

a.) Use $c^2 = 1$, plot solutions until $t = 1$, and when prompted, set a in the definition of $f_4(x,t)$ equal to 1. Describe the evolution of the solution. How did it differ from the corresponding homogeneous wave equation?

b.) Based on your intuition gained in the one-dimensional case regarding the wave equation, how do you expect the solution of (11.53) to evolve when $c^2 = 2$, as compared to (a)? Check your intuition by using the GUI to visualize this situation.

c.) What do you expect to happen if $c^2 = 0.25$? Test your hypothesis with the GUI.

d.) Repeat (a) using $a = 0.5$. How would you compare this situation with the one in (a) and the homogeneous wave equation solution?

e.) Repeat (d) using $a = 0.25, 0.1$, and 0.001. Does the solution appear to get closer to the homogeneous solution?

f.) What if a gets larger? Explore!

g.) Repeat (a)-(f) using the exponential initial position option. Did you observe any behavior contradictory to your conjectures in (b), (c), and (e)? If so, how?

h.) Repeat (a)-(f) using the step function initial position option. Did you observe any behavior contradictory to your conjectures in (b), (c), and (e)? If so, how?

v.) Consider IBVP (11.54) with the cosine initial condition option, any choice for the initial velocity, and the fourth forcing function option, f3.

a.) Use $c^2 = 1$, plot solutions until $t = 1$, and when prompted, set ω in the definition of $f_3(x,t)$ equal to 1. Describe the evolution of the solution. How did it differ from the corresponding homogeneous wave equation?

b.) Based on your intuition gained in the one-dimensional case regarding the wave equation, how do you expect the solution of (11.54) to evolve when $c^2 = 2$, as compared to (a)? Check your intuition by using the GUI to visualize this situation.

c.) What do you expect to happen if $c^2 = 0.25$? Test your hypothesis with the GUI.

d.) Repeat (a) using $a = 0.5$. How would you compare this situation with the one in (a) and the homogeneous wave equation solution?

e.) Repeat (d) using $a = 0.25, 0.1$, and 0.001. Explain why the solution should **not** tend to the homogeneous solution. Note: Unlike the previous two models, the periodic nature of the wave equation along with the periodic forcing function are making it difficult to compare with the homogeneous case. It's there...watch the boundaries!

f.) What if a gets larger? Explore!

g.) Repeat (a)-(f) using the exponential initial position option. Did you observe any behavior contradictory to your conjectures in (b), (c), and (e)? If so, how?

h.) Repeat (a)-(f) using the step function initial position option. Did you observe any behavior contradictory to your conjectures in (b), (c), and (e)? If so, how?

■

EXPLORE!
11.21 Long Term Behavior?

MATLAB-Exercise 11.3.8. To begin, open **MATLAB** and in the command line, type:

MATLAB_Wave_Equation_Non_2D

Use the GUI to answer the following questions. Make certain to click on the HELP button for instructions on how to use the GUI.

i.) Consider IBVP (11.52) with the sine initial position option, zero as the initial velocity option, and the forcing function option f1.

  a.) Investigate the behavior of the solution of IBVP (11.52) for large times. Use $c^2 = 3$ and when prompted set A in the definition of $f_1(x,t)$ equal to 1. Begin by plotting solutions until $t = 2$. If necessary, slowly increase the plotting time until you are able to describe the behavior of the solution. Remember, a computer cannot handle the infinite! Compare the behavior here with the homogeneous wave equation.

  b.) Repeat (a) using $A = 0.5$. How would you compare this situation with the one in (a) and the homogeneous heat equation solution?

  c.) Repeat (b) using $A = 5, 0.25$, and 0.001. How is the forcing function affecting the long-term behavior of the solution?

  d.) Repeat (a)-(c) using the exponential option for the initial velocity. Keep everything else the same. How is the initial velocity affecting the long-term behavior of the solution?

  e.) Repeat (a)-(d) using the exponential initial position option and for the initial velocity, zero. Did you observe any behavior contradictory to your comparisons and conjectures made in (a)-(d)? If so, how?

  f.) Repeat (a)-(d) using the step function initial position option and for the initial velocity, zero. Did you observe any behavior contradictory to your comparisons and conjectures made in (a)-(d)? If so, how?

ii.) Repeat (i) with the forcing function option f5.

iii.) Consider IBVP (11.53) with the cosine initial position option, zero as the initial velocity, and the forcing function option f2.

  a.) Investigate the behavior of the solution of IBVP (11.53) for large times. Use $c^2 = 3$ and when prompted, set A and B in the definition of $f_2(x,t)$, equal to 1 and -1. Begin by plotting solutions until $t = 2$. If necessary, slowly increase the plotting time until you are able to describe the behavior of the solution. Compare the behavior here with the homogeneous wave equation.

  b.) Repeat (a) using $A = 0.5$ and $B = -0.5$. How would you compare this situation with the one in (a) and the homogeneous wave equation solution?

  c.) Repeat (b) using $A = 5, 0.25$, and 0.001 and $B = -5, -0.25$, and -0.001.

  d.) Repeat (a)-(c) using the exponential option for the initial velocity. Keep everything else the same. How is the initial velocity affecting the long-term behavior of the solution?

e.) Repeat (a)-(d) using the exponential initial position option and for the initial velocity, zero. Did you observe any behavior contradictory to your comparisons and conjectures made in (a)-(d)? If so, how?

f.) Repeat (a)-(d) using the step function initial position option and for the initial velocity, zero. Did you observe any behavior contradictory to your comparisons and conjectures made in (a)-(d)? If so, how?

iv.) Consider IBVP (11.53) with the cosine initial condition option, zero as the initial velocity, and the forcing function option f4.

a.) Investigate the behavior of the solution of IBVP (11.53) for large times. Use $c^2 = 3$ and when prompted, set a in the definition of $f_4(x,t)$ equal to 1. Begin by plotting solutions until $t = 2$. If necessary, slowly increase the plotting time until you are able to describe the behavior of the solution. Compare the behavior here with the homogeneous wave equation.

b.) Repeat (a) using $a = 0.5$. How would you compare this situation with the one in (a) and the homogeneous wave equation solution?

c.) Repeat (b) using $a = 0.25, 2$, and 10. How is the forcing function affecting the long-term behavior of the solution?

d.) Repeat (a)-(c) using the exponential option for the initial velocity. Keep everything else the same. How is the initial velocity affecting the long-term behavior of the solution?

e.) Repeat (a)-(d) using the exponential initial position option and for the initial velocity, zero. Did you observe any behavior contradictory to your comparisons and conjectures made in (a)-(d)? If so, how?

f.) Repeat (a)-(d) using the step function initial position option and for the initial velocity, zero. Did you observe any behavior contradictory to your comparisons and conjectures made in (a)-(d)? If so, how?

v.) Consider IBVP (11.53) with the cosine initial condition option, zero as the initial velocity, and the fourth forcing function option, f3.

a.) Investigate the IBVP solution's behavior for large times. Use $c^2 = 3$ and when prompted, set ω in the definition of $f_3(x,t)$ equal to 1. Begin by plotting solutions until $t = 2$. If necessary, slowly increase the plotting time until you are able to describe the behavior of the solution. Compare the behavior here with the homogeneous wave equation.

b.) Repeat (a) using $a = 0.5$. How would you compare this situation with the one in (a) and the homogeneous wave equation solution?

c.) Repeat (b) using $a = 0.25, 2$, and 10. How is the forcing function affecting the long-term behavior of the solution?

d.) Repeat (a)-(c) using the exponential option for the initial velocity. Keep everything else the same. How is the initial velocity affecting the long-term behavior of the solution?

e.) Repeat (a)-(d) using the exponential initial position option and for the initial velocity, zero. Did you observe any behavior contradictory to your comparisons and conjectures made in (a)-(d)? If so, how?

f.) Repeat (a)-(d) using the step function initial position option and for the initial velocity, zero. Did you observe any behavior contradictory to your comparisons and conjectures made in (a)-(d)? If so, how?

■

EXPLORE!

11.22 Continuous Dependence on Parameters

MATLAB-Exercise 11.3.9. To explore continuous dependence with the GUI one must first solve the nonhomogeneous wave equation with a particular propagation constant, initial conditions, and forcing function. Once a solution has been constructed, click the *Perturb Parameters and Solve* button to specify the perturbed propagation constant and a perturbation size for the initial conditions. The GUI will also allow for a perturbed forcing function parameter. When prompted to select a norm choose *All* to investigate dependence using both norms at the same time. To begin, open **MATLAB** and in the command line, type:

MATLAB_Wave_Equation_Non_2D

Use the GUI to answer the following questions. Make certain to click on the HELP button for instructions on how to use the GUI. Examples of inputting parameters and GUI outputs can be found in Figures 11.67–11.71.

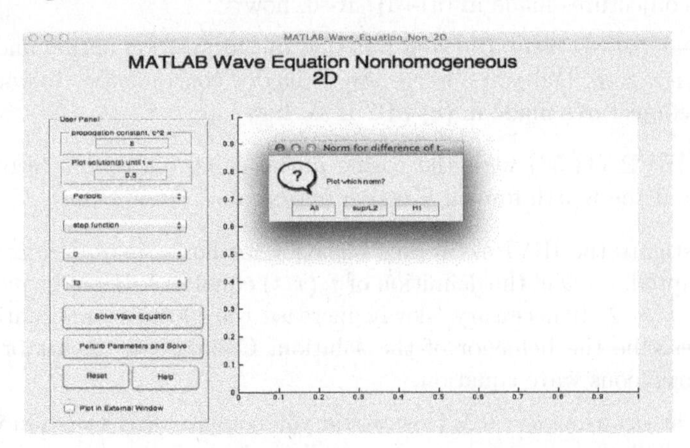

FIGURE 11.67: Perturbing parameters, initial conditions, and forcing function parameters with MATLAB Wave Equation Nonhomogeneous 2D GUI

i.) Consider IBVP (11.52) with the sine initial position option, zero as the initial velocity option, and the forcing function option f1.

a.) Investigate the continuous dependence of the solution on the propogation constant for IBVP (11.52). Use $c^2 = 2$ and when prompted set A in the definition of $f_1(x,t)$ equal to 1. Plot solutions until $t = 0.5$. Once the movie has finished playing, click the *Perturb Parameters and Solve* button and choose a perturbed diffusivity constant $k^2 = 1.8$. Keep all other parameters the same as in the

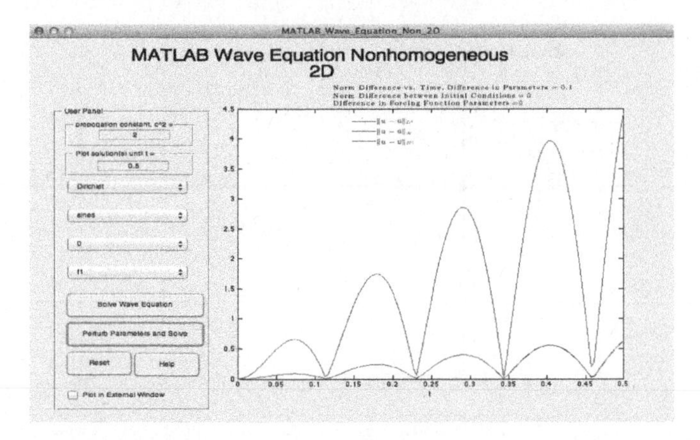

FIGURE 11.68: Perturbing the propagation constant in IBVP (11.52). Forcing function, $f_1(x,t) = 1$.

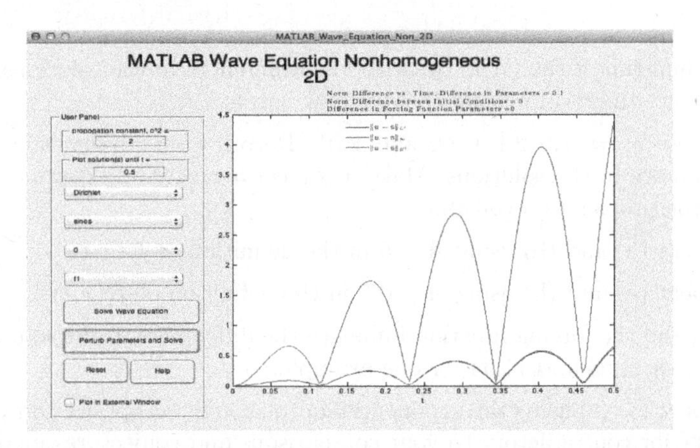

FIGURE 11.69: Perturbing the diffusivity constant in IBVP (11.52). Forcing function, $f_1(x,t) = 5$.

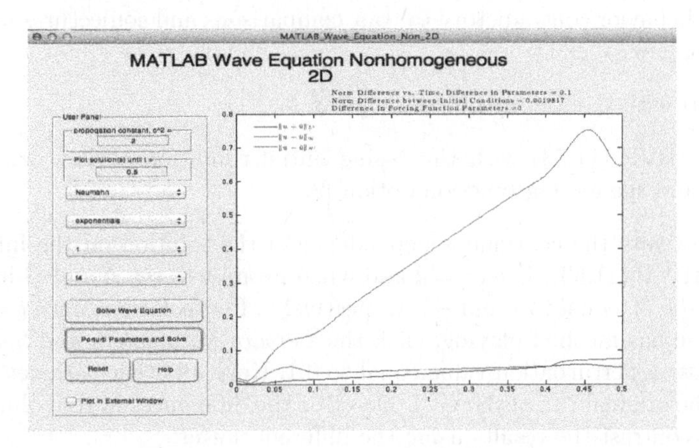

FIGURE 11.70: Perturbing the propagation constant and initial conditions in IBVP (11.53). Forcing function, $f_4(x,t) = e^{-5t}$.

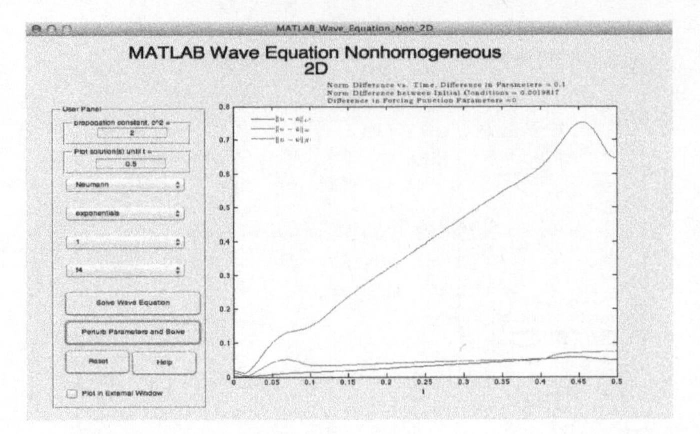

FIGURE 11.71: Perturbing the propagation constant and initial conditions in IBVP (11.53). Forcing function, $f_4(x,t) = e^{-0.5t}$.

original run of the GUI. Describe the differences in the solutions. Compare and contrast the results using the different norms.

b.) Choose $c^2 = 1.9, 2.1, 1.99$, and 2.01. Record your observations regarding the difference in the solutions. Make a conjecture regarding continuous dependence on the propagation constant.

c.) Repeat (a) and (b) using $A = 5$ in the definition of $f_1(x,t)$.

d.) Repeat (a) and (b) using $A = 0.5$ in the definition of $f_1(x,t)$.

e.) How did the forcing function influence the differences in the solution? (Hint: See Figures 11.53 and 11.54 for one example.)

d.) Repeat (a)-(e) using the exponential initial condition option. Did you observe any behavior contradictory to your comparisons and conjectures made in (a)-(e)? If so, how?

e.) Repeat (a)-(e) using the step function initial condition option. Did you observe any behavior contradictory to your comparisons and conjectures made in (a)-(e)? If so, how?

ii.) Repeat (i) with the forcing function option f5.

iii.) Consider IBVP (11.53) with the cosine initial condition option, zero as the initial velocity, and the forcing function option f2.

a.) Investigate the continuous dependence of the solution on the initial conditions for IBVP (11.53). Use $c^2 = 2$ and when prompted, set A and B in the definition of $f_2(x,t)$, equal to 1 and -1, respectively . Plot solutions until $t = 0.5$. Once the movie has finished playing, click the *Perturb Parameters and Solve* button and choose a perturbation value equal to 0.1. Keep all other parameters the same as in the original run of the GUI. Describe the differences in the solutions. Compare and contrast the results using the different norms.

b.) Choose perturbation values equal to $0.05, 0.01$, and 0.001. Record your observations regarding the difference in the solutions. Make a conjecture regarding continuous dependence on the initial condition.

c.) Repeat (a) and (b) using $A = 5$ and $B = -5$ in the definition of $f_2(x,t)$.

d.) Repeat (a) and (b) using $A = 0.5$ and $B = -0.5$ in the definition of $f_2(x, t)$.

e.) How did the forcing function influence the differences in the solution?

d.) Repeat (a)-(e) using the exponential initial condition option and the initial velocity option 1. Did you observe any behavior contradictory to your comparisons and conjectures made in (a)-(e)? If so, how?

e.) Repeat (a)-(e) using the step function initial condition option. Did you observe any behavior contradictory to your comparisons and conjectures made in (a)-(e)? If so, how?

iv.) Consider IBVP (11.53) with the cosine initial condition option, zero as the initial velocity, and the forcing function option f4.

a.) Investigate the continuous dependence of the solution on the initial conditions and propagation constants for IBVP (11.53). Use $c^2 = 2$ and when prompted, set a in the definition of $f_4(x, t)$ equal to 1. Plot solutions until $t = 0.5$. Once the movie has finished playing, click the *Perturb Parameters and Solve* button and choose a perturbed propagation constant equal to 1.8 and perturbation value equal to 0.1. Keep all other parameters the same as in the original run of the GUI. Describe the differences in the solutions. Compare and contrast the results using the different norms.

b.) Choose perturbed propagation constants equal to $2.1, 1.9$, and 2.001 and perturbation values equal to $0.05, 0.01$, and 0.001. Record your observations regarding the difference in the solutions. Make a conjecture regarding continuous dependence on the initial conditions and the propagation constants.

c.) Repeat (a) and (b) using $a = 5$ in the definition of $f_4(x, t)$.

d.) Repeat (a) and (b) using $a = 0.5$ in the definition of $f_4(x, t)$.

e.) How did the forcing function influence the differences in the solution? (Hint: See Figures 11.55 and 11.56 for one example.)

d.) Repeat (a)-(e) using the exponential initial condition option. Did you observe any behavior contradictory to your comparisons and conjectures made in (a)-(e)? If so, how?

e.) Repeat (a)-(e) using the step function initial condition option. Did you observe any behavior contradictory to your comparisons and conjectures made in (a)-(e)? If so, how?

v.) Consider IBVP (11.53) with the cosine initial condition option, zero as the initial velocity, and the forcing function option f3.

a.) Investigate the IBVP continuous dependence on the initial condition and the forcing function parameter. Use $c^2 = 2$ and when prompted, set ω in the definition of $f_3(x, t)$ equal to 1. Plot solutions until $t = 0.5$. Once the movie has finished playing, click the *Perturb Parameters and Solve* button to choose a perturbation value equal to 0.1 and forcing function parameter equal to 1.1. Describe the differences in the solutions. Compare and contrast the results using the different norms.

b.) Choose perturbation values equal to $0.05, 0.01$, and 0.001, and forcing function parameters equal to $0.9, 0.95$, and 1.01. Record your observations regarding the difference in the solutions. Make a conjecture regarding continuous dependence on the initial condition and the forcing function parameter.

c.) Repeat (a)-(b) using the exponential initial condition option and any non-zero initial velocity. Did you observe any behavior contradictory to your comparisons and conjectures made in (a)-(b)? If so, how?

d.) Repeat (a)-(b) using the step function initial condition option. Did you observe any behavior contradictory to your comparisons and conjectures made in (a)-(b)? If so, how?

vi.) Summarize your observations regarding continuous dependence. Make sure you address the potential issue of using different norms and the influence the forcing function had on the differences in the solutions.

■

11.3.3 Abstract Formulation

Exercise 11.3.3. Follow the discussion from Part I in which the nonhomogeneous spring-mass system was converted to a matrix equation to convert (11.52) and (11.53) to matrix equations on an appropriate Banach space. Look back at the development for the homogeneous wave equation in Chapter 9 and modify appropriately.

Chapter 12

Remaining Mathematical Models

12.1 Population Growth—Fisher's Equation

As seen in Chapters 1 and 3, the rate at which certain quantities change is dependent on the amount present at a given instant in time. This concept led to the model

$$\begin{cases} \frac{du(t)}{dt} &= au(t) \\ u(0) &= u_0 \end{cases} \tag{12.1}$$

where a is a non-zero real number and u_0 is a positive real number. The solution to (12.1) is an exponential function which was useful in some situations, but not realistic in others. For example, if we wanted to model the interaction of two chemical or biological species then an unbounded solution, like the exponential, is not meaningful. We should expect some type of limited growth or decline to account for resources, such as space and nutrients.

Consider a population model for the human race as an example. At some point our population will reach a level for which the natural resources of the planet will not be plentiful enough to sustain everyone. There will not be enough food, water, or space to facilitate population growth as in the past. The growth must slow down and the total population would be bounded above by some limiting value called the *carrying capacity.*

This suggests that we look for a new model where the proportionality constant in (12.1) depends on the population as well; that is, a is a function of the population. A launching point for our discussion is the ODE model

$$\begin{cases} \frac{du(t)}{dt} = a\left(u(t)\right)u(t) \\ u(0) = u_0. \end{cases} \tag{12.2}$$

This begs the question, "What should a($u(t)$) equal?" We know that if $u(t)$ nears the carrying capacity, denoted by u_∞, then $\frac{du(t)}{dt}$ should be close to zero. And, if $u(t)$ is much smaller than u_∞ we should expect exponential growth. For these two reasons let us assume

$$a\left(u(t)\right) = \alpha\left(1 - \frac{u(t)}{u_\infty}\right) \tag{12.3}$$

where α and u_∞ are positive constants.

STOP! Verify that (12.3) satisfies the limited growth scenario presented above.

Next, we incorporate the mechanism of diffusion into the mathematical model to allow for a spatial dependence. To do this, we add a diffusion term to (12.2) similar to the one in heat equation and impose periodic boundary conditions. The one-dimensional model is

$$\begin{cases} \frac{\partial u(x,t)}{\partial t} &= D\frac{\partial^2 u(x,t)}{\partial x^2}u(x,t) + \alpha u(x,t)\left(1 - \frac{u(x,t)}{u_\infty}\right),\ t > 0,\ x \in (-L, L) \\ u(x,0) &= u_0(x),\ x \in (-L, L) \\ u(-L,t) &= u(L,t),\ t > 0 \end{cases} \tag{12.4}$$

where L, D, α, u_∞ are positive constants and the population at locale x and time t is denoted by $u(x,t)$. IBVP (12.4) is called *Fisher's equation* and is viewed as a limited growth model in biology and chemistry [37, 13].

Up to this point in the text, we have skimmed over the importance of simplifying a model by the way of *nondimensionalization.* In fact, many models found in Chapters 3, 6, and 9 underwent a nondimensionalization process to simplify the discussion. Nondimensionalization is used to partially or fully remove the units of a model in order to simplify and rescale the system to a "unit" intrinsic to the problem. Removing units in order to produce a unit sounds like a paradox, but you already have experience with this concept. Probably the most familar example of this is the radian. You learned in trigonometry you could measure angles in either degrees or radians. The use of the word measure implies units are involved, but actually radians are unitless. Defining "radians" involved the ratio of arc length to circumference of the circle. Since both the arc length and the circle are measured using the same unit of length, their ratio is unitless. But, since the radian is relative to the circumference of the circle we retain the sense that we are dealing with units of measure.

To nondimensionalize a model one must do the following:

1. Identify all independent and dependent variables;
2. Take each variable and create a new variable by dividing it by a combination of parameters that have the same dimension in order to create a dimensionless variable (There is not always a unique way to do this.);
3. Use chain rule to derive a new differential equation using the new variables.

The nondimensionalization process is more of an art than a science and requires experimentation. Step 2 can be quite difficult especially if your goal is to reduce as many model parameters as you can to one.

If we denote the spatial variable unit as l and time variable unit as T, then the diffusivity parameter, D, has units $\frac{l^2}{T}$, the growth rate, α, has units $\frac{1}{T}$, and the carrying capacity has the units of number of individuals. To form a nondimensional version of (12.4) we introduce the variables

$$\tau = \alpha t, \quad z = x\sqrt{\frac{\alpha}{D}}, \quad v = \frac{u}{u_\infty}$$

which enables us to rewrite (12.4) as

$$\begin{cases} \frac{\partial v(z,\tau)}{\partial \tau} = \frac{\partial^2 v(z,\tau)}{\partial z^2} + v(z,\tau)\left(1 - v(z,\tau)\right),\ \tau > 0, z \in (-L', L'), \\ v(0,z) = v_0(z),\ z \in (-L', L'), \\ z(-L',\tau) = z(L',\tau), t > 0. \end{cases} \tag{12.5}$$

where $L' = L\sqrt{\frac{\alpha}{D}}$.

STOP! Verify that τ and z are dimensionless.

Exercise 12.1.1. Use the change of variable provided and change rule to verify (12.5) is equivalent to (12.4).

12.1.1 Two-Dimensional Fisher's Equation

Now, we account for a population diffusing in two spatial dimensions. To account for this feature in the model we can replace the second spatial derivative in IBVP (12.5) with the two-dimensional laplacian. Doing so yields

$$\begin{cases} \frac{\partial \bar{u}(\bar{x},\bar{y},\bar{t})}{\partial \bar{t}} = \Delta \bar{u}(\bar{x},\bar{y},\bar{t}) + \bar{u}(\bar{x},\bar{y},\bar{t})\left(1-\bar{u}(\bar{x},\bar{y},\bar{t})\right), \bar{t}>0,\ \bar{x}\in(-\bar{L},\bar{L}),\ \bar{y}\in(-\bar{L},\bar{L}) \\ \bar{u}(\bar{x},\bar{y},0) = \bar{u}_0(\bar{x},\bar{y}),\ \bar{x}\in[-\bar{L},\bar{L}],\ \bar{y}\in[-\bar{L},\bar{L}] \\ \bar{u}(\bar{x},-\bar{L},\bar{t}) = \bar{u}(\bar{x},\bar{L},\bar{t}),\ \bar{t}>0,\ \bar{x}\in[-\bar{L},\bar{L}] \\ \bar{u}(-\bar{L},\bar{y},\bar{t}) = \bar{u}(\bar{L},\bar{y},\bar{t}),\ \bar{t}>0,\ \bar{y}\in[-\bar{L},\bar{L}] \end{cases} \tag{12.6}$$

where $\bar{L}$ is a positive real number and $\bar{u}_0(x,y)$ represents the normalized initial population.

Exercise 12.1.2. Use a change of variable similar to the one-dimensional case to transform

$$\begin{cases} \frac{\partial u(x,y,t)}{\partial t} = D\Delta u(x,y,t) + \alpha u(x,y,t)\left(1-\frac{u(x,y,t)}{u_\infty}\right), t>0,\ x\in(-L,L),\ y\in(-L,L) \\ u(x,y,0) = u_0(x,y),\ x\in[-L,L],\ y\in[-L,L] \\ u(x,-L,t) = u(x,L,t),\ t>0,\ x\in[-L,L] \end{cases}$$

12.2 Zombie Apocalypse! Epidemiological Models

If you have seen a zombie television program, or movie, or read a novel or comic book involving these undead creatures, you know that zombies strike fear in the human population by spreading a plague that turns humans into zombies. Depending on the genre, a human can be infected by the zombie virus through bites, scratches, etc., but in every case *contact* must take place between a human and zombie before transmission can occur. In that moment of contact, the human may become infected by the virus and turn into a zombie *or* retain their humanity.

The underlying dynamics in the humans vs. zombies scenario applies to many real viruses that affect particular populations [37]. It is in the interest of public health to model the evolution or spread of the virus throughout a population.

Suppose a population consists of two groups: infectives (zombies), denoted by $I(x,t)$, who carry the virus and susceptibles (humans), denoted by $S(x,t)$. The two groups can interact at different points in space x and at different times t. For simplicity, we have assumed one spatial component.

When an infective comes in contact with a susceptible, the virus *may* transmit to the individual, tranforming him into an infective. The *transmission rate* depends on the virus and the person, but for simplicity we assume the rate is constant and is denoted by r. Momementarily, let us leave the zombie apocalypse scenario and consider a less severe virus. An individual may leave the infective group when the virus works its way out of the infective's system or through death. Thus, we assume that an infective leaves the group by recovering and producing an immunity or dying. We assume this *removal rate* is constant and is denoted by a.

Following the Rate-in minus Rate-out approach to modeling introduced in Part I yields

$$\begin{cases} \frac{dS}{dt} &= -rIS, \; t > 0 \\ \frac{dI}{dt} &= rIS - aI, \; t > 0 \end{cases} \tag{12.7}$$

The model (12.7) is slightly misleading because the Rate-in minus Rate-out approach only accounts for changes in one-variable (usually time). We want to allow for interactions in both time and space, so we need to allow the infectives and susceptibles to diffuse in the spatial dimension. To accomplish this effect, we incorporate diffusion terms into (12.7) resulting in

$$\begin{cases} \frac{\partial S(x,t)}{\partial t} = D_s \frac{\partial^2 S(x,t)}{\partial x^2} - rI(x,t)S(x,t), \; x \in (-L, L), \; t > 0, \\ \frac{\partial I(x,t)}{\partial t} = D_I \frac{\partial^2 I(x,t)}{\partial x^2} + rI(x,t)S(x,t) - aI(x,t), \; x \in (-L, L), \; t > 0, \\ S(x,0) = S_0, \; x \in [-L, L], \\ I(x,0) = I_0(x), \; x \in [-L, L], \\ S(-L,t) = S(L,t), \; t > 0, \\ I(-L,t) = I(L,t), \; t > 0, \end{cases} \tag{12.8}$$

where L and S_0 are positive real numbers and $I_0(x)$ is nonnegative function. The susceptible initial condition is assumed to be homogeneous, meaning that initially, the group's population is constant in space.

To simplify (12.8) we nondimensionalize the model by introducing the variables

$$x^* = \left(\frac{rS_0}{D_I}\right)^{\frac{1}{2}} x$$

$$t^* = rS_0 t$$

$$u(x^*, t^*) = \frac{I(\frac{D_I x^*}{rS_0}, \frac{t^*}{rS_0})}{S_0} = \frac{I(x,t)}{S_0}$$

$$v(x^*, t^*) = \frac{S(\frac{D_I x^*}{rS_0}, \frac{t^*}{rS_0})}{S_0} = \frac{S(x,t)}{S_0}$$

which yields

$$\frac{\partial^2 S(x,t)}{\partial x^2} = rS_0 \frac{1}{D_I} \frac{\partial^2 S(x,t)}{(\partial x^*)^2} \tag{12.9}$$

$$\frac{\partial S(x,t)}{\partial t} = rS_0 \frac{\partial S(x,t)}{\partial t^*} \tag{12.10}$$

$$\frac{\partial^2 I(x,t)}{\partial x^2} = rS_0 \frac{1}{D_I} \frac{\partial^2 I(x,t)}{(\partial x^*)^2} \tag{12.11}$$

$$\frac{\partial I(x,t)}{\partial t} = rS_0 \frac{\partial I(x,t)}{\partial t^*}. \tag{12.12}$$

Exercise 12.2.1. Verify (12.9)-(12.12).

Substituting (12.9)-(12.12) into the differential equations in (12.8) and dividing the initial conditions by S_0 yields

$$\begin{cases} rS_0 \frac{\partial S(x,t)}{\partial t^*} = rS_0 \frac{D_S}{D_I} \frac{\partial^2 S(x,t)}{(\partial x^*)^2} - rI(x,t)S(x,t),\ x \in (-L,L),\ t > 0 \\ rS_0 \frac{\partial I(x,t)}{\partial t^*} = rS_0 \frac{\partial^2 I(x,t)}{(\partial x^*)^2} + rI(x,t)S(x,t) - aI(x,t),\ x \in (-L,L),\ t > 0 \\ \frac{S(x,0)}{S_0} = 1, x \in [-L,L] \\ \frac{I(x,0)}{S_0} = \frac{I_0(x)}{S_0}, x \in [-L,L] \end{cases} \tag{12.13}$$

Dividing the differential equations in (12.13) by rS_0, and letting $\lambda = \frac{a}{rS_0}$ and $\epsilon = \frac{D_S}{D_I}$ enables us to rewrite (12.13) as

$$\begin{cases} \frac{\partial v(x^*,t^*)}{\partial t^*} = \frac{\partial^2 v(x^*,t^*)}{(\partial x^*)^2} - u(x^*,t^*),\ x \in (-L,L),\ t^* > 0 \\ \frac{\partial u(x^*,t^*)}{\partial t^*} = \epsilon \frac{\partial^2 u(x^*,t^*)}{(\partial x^*)^2} + u(x^*,t^*)(v(x^*,t^*) - \lambda),\ x^* \in (-L^*,L^*),\ t^* > 0 \\ v(x^*,0) = 1,\ x^* \in [-L^*,L^*] \\ u(x^*,0) = u_0(x^*),\ x^* \in [-L^*,L^*] \end{cases} \tag{12.14}$$

where $L^* = \sqrt{\frac{rS_0}{D_I}}L$ and $u_0(x^*) = \frac{I_0(x)}{S_0}$.

Exercise 12.2.2. Verify that (12.14) follows from (12.9)-(12.12) and that (12.15) follows from (12.14).

It is common to revert back to the original variables in notation after nondimensionalizing a model. For example, IBVP (12.14) would be written as

$$\begin{cases} \frac{\partial v(x,t)}{\partial t} = \epsilon \frac{\partial^2 v(x,t)}{(\partial x)^2} - u(x,t),\ x \in (-L,L),\ t > 0 \\ \frac{\partial u(x,t)}{\partial t} = \frac{\partial^2 u(x,t)}{(\partial x)^2} + u(x,t)(v(x,t) - \lambda),\ x \in (-L,L),\ t > 0 \\ v(x,0) = 1,\ x \in [-L,L] \\ u(x,0) = u_0(x),\ x \in [-L,L] \end{cases} \tag{12.15}$$

even though the x and t variables in (12.15) are not the physical variables described in the epidemic scenario. Along those lines, we will refer to ϵ as the diffusivity constant and λ the removal rate throughout the MATLAB GUIs and **EXPLORE!** projects, even though they are neither the physical diffusion constants nor the removal rate. However, we can recover one from the other using the relationships

$$\epsilon = \frac{D_S}{D_I} \quad \lambda = \frac{a}{rS_0}.$$

12.3 How Did That Zebra Gets Its Stripes? A First Look at Spatial Pattern Formation

Pattern formation can be viewed as a *reaction-diffusion* system which involves two chemical species reacting with each other and diffusing through a medium [37, 41, 9] . The population model developed in Section 12.1 and the epidemic model developed in Section 12.2 each individually touched upon the features of pattern formation. We can expect the pattern formation model to involve similar equations. The interesting features of pattern formation are *pulses* in one spatial dimension and *spots* in two spatial dimensions. Pulses and spots are defined by a region where the concentration of the two chemical species greatly differ.

Let U and V represent the two chemical species. The irreversible Gray-Scott model governs the pattern formation reaction

$$\begin{aligned} \mathrm{U} + \mathrm{V} &\longrightarrow 3\,\mathrm{V} \\ \mathrm{V} &\longrightarrow \mathrm{P} \end{aligned} \tag{12.16}$$

where P is an inert product. Letting $u(x,t) = [U](x,t)$ and $v(x,t) = [V](x,t)$, both normalized concentrations, the reaction-diffusion system governing (12.16) is

$$\begin{cases} \frac{\partial u}{\partial t} &= D_u \frac{\partial^2 u}{\partial x^2} - uv^2 + A(1-u) \\ \frac{\partial v}{\partial t} &= D_v \frac{\partial^2 v}{\partial x^2} + uv^2 - Bv \end{cases} \tag{12.17}$$

The first equation in (12.17) describes how the quantity u changes in time. The first term is the diffusion term with diffusivity constant D_u. The second term in the first equation is the reaction due to the law of mass action. The third term, $A(1-u)$, is called the *replenishment term.* According to the reaction, all of U will be used up in forming V unless it is replenished. The replenishment term accounts for adding U back to the system with a *feed rate* A. The second equation in (12.17) is essentially the same as the first except for the third term called the *diminishment term.* The diminishment term is added to limit the growth of V.

Here we consider a slight variant of (12.17) to "encourage" pulse and spot formation. We set D_u equal 1 and relabel D_v as ϵ, which will be a small positive diffusivity constant. To complete the modeling process we specify initial conditions and assume periodic BCs to yield the Gray-Scott reaction-diffusion IBVP given by

$$\begin{cases} \frac{\partial u(x,t)}{\partial t} = \frac{\partial^2 u(x,t)}{\partial x^2} - u(x,t)v(x,t)^2 + A\left(1 - u(x,t)\right), \, t > 0, \, x \in [-L, L] \\ \frac{\partial v(x,t)}{\partial t} = \epsilon \frac{\partial^2 v(x,t)}{\partial x^2} + u(x,t)v(x,t)^2 - Bv(x,t), \, t > 0, \, x \in [-L, L] \\ u(x,0) = u_0(x), \, x \in [-L, L] \\ v(x,0) = v_0(x), \, x \in [-L, L] \\ u(-L,t) = u(L,t), \, t > 0 \\ v(-L,t) = v(L,t), \, t > 0 \end{cases} \tag{12.18}$$

Exercise 12.3.1. Propose a pattern formation model that generalizes IBVP (12.18) to two spatial dimensions. (Hint: See Section 12.1.1.)

12.4 Autocatalysis Combustion!

An *autocatalyst* is a molecular species which, in the presence of suitable reagents, catalyzes a reaction in which one or more of the catalyst species can be found in the product. The catalyst molecule from the reaction and at least one more remains in the product. These types of molecules can be viewed as self-replicating. Examples of autocatalytic reactions can be found in biological and chemical systems. These examples include protein misfoldings, mad cow disease, and Parkinson's disease. In fact, we saw an example of this in Section 12.3 where V was an autocatalyst in the first step of the reaction before becoming an inert product in the second step.

Here we focus on a reaction and model from combustion theory. Let U and V be two chemical species that react according to

$$mU + V \longrightarrow (m+1)U \tag{12.19}$$

where m is a positive real parameter. We further assume both species may diffuse with different diffusivities. The nondimensional model we consider for (12.19) is given by

$$\begin{cases} \frac{\partial u(x,t)}{\partial t} = \frac{\partial^2 u(x,t)}{\partial x^2} + v(x,t)f(u(x,t)),\, t>0,\, x\in(-L,L) \\ \frac{\partial v(x,t)}{\partial t} = \epsilon\frac{\partial^2 v(x,t)}{\partial x^2} - v(x,t)f(u(x,t)),\, t>0,\, x\in(-L,L) \\ u(x,0) = u_0(x),\, x\in[-L,L] \\ v(x,0) = v_0(x),\, x\in[-L,L] \\ u(-L,t) = u(L,t),\, t>0 \\ v(-L,t) = v(L,t),\, t>0 \end{cases} \tag{12.20}$$

where u and v are nondimensionalized concentrations of U and V, respectively, and

$$f(u) = \begin{cases} u^m, & u\geq 0 \\ 0, & u<0 \end{cases}$$

The parameter ϵ represents a ratio of the diffusion rates, $u_0(x)$ and $v_0(x)$ the initial nondimensional concentrations of the reagants, and m a positive real number. The derivation of IBVP (12.20) is similar to the one for IBVP (12.18) and is omitted. An interested reader may refer to [2, 34, 37].

12.5 Money, Money, Money—A Simple Financial Model

Imagine that you own a shipping business where fuel costs affect your profits and ultimately the livelihood of your business and its employees. Typically, the cost of gas rises in the summer, so you would like to offset that premium in some way. Let us assume it is January 1st and you enter a contract with another company that owns an excess of gasoline. The contract states that you pay an upfront cost of \$0.05 per gallon to earn the *right*, but not obligation, to purchase gas on July 1st for \$3.65 per gallon.

Let us suppose on July 1st gas costs \$4.00 per gallon. In this case you are a happy business owner because you will exercise your right to purchase the gas at \$3.65 per gallon and realize a savings of \$(0.35 − 0.05) per gallon. But what if the price of gas falls to \$3.50 per gallon? In this case, you would <u>not</u> choose to exercise your contract and pay \$3.65 per gallon. Rather, you would go into the market and pay only \$3.50 per gallon. In some sense, you are happy gas prices fell, but on the other hand, you have realized a loss of \$0.05 per gallon from purchasing the contract on January 1st.

The situation described above illustrates what is commonly referred to as a *European call Option*. A European call option is a contract in which at a prescribed time in the future called the *expiration date* the holder of the option may purchase a prescribed asset, known as the *underyling asset*, for a prescribed amount called the *strike price*. The cost to purchase such a contract is called the *contract price*.

Exercise 12.5.1. Use the example described in this section to answer the following:

i.) What was the contract price?

ii.) What was the underlying asset?

iii.) What was the strike price?

iv.) When is the expiration date?

The question of interest is, "How much would one pay for such a contract"? In other words, what is the value of the call option? Before answering this question, let us try to answer the more basic question, "On what should the value of the option depend?" It seems plausible that the value of a European call option on gas will differ from one where the underlying asset is grain, coffee, or a stock of a company. This suggests the value depends at least on the underlying asset.

The expiration date is integral in determining the value since a shorter time period lends itself to less variablity or *risk*. Or, in other words, if the expiration was set to one *day* you would feel confident about the value of the option. **(Why?)** However, if it was set to one *year* the value would not be as clear. Another important feature of the contract is the strike price. The right to buy an underlying asset for $1.00 must be more expensive than one for which the strike price is one million dollars.

Another common option is the *European put option* in which the holder purchases the right to *sell* an underlying asset at a specified strike price on the expiration date.

STOP! What should the value of a European put option depend on? Do the same arguments for the call option apply?

Let us begin the process of modeling the value of a European option V which is a function of the price of the underlying asset S and time t. When pricing a call option we set $V = C(S,t)$ and when pricing a put option we set $V = P(S,t)$. Some essential model parameters are the volatility of the asset, denoted by σ; the expiration date T; and r the interest rate of a risk-free investment like a savings account or goverment issued bond.

According to the definition of a call option we can determine the value of the option at the expiration date T. In the case, we know

$$C(T,S) = \max\left(S(T) - E, 0\right)$$

where E is the strike price. **(Why?)** If we consider the put option at time T we need to think through two scenarios. If the price of the underlying asset falls below the strike price, then the contract holder will exercise his right to sell because he is able to sell something for more than it's worth in the market. In this case he realizes a gain of $E - S(T)$. However, if the value of the underlying asset is above the strike price he will not exercise his right to sell. **(Why?)** From this argument we have

$$P(T,S) = \max\left(E - S(T), 0\right)$$

Remark 12.5.1. In both the European call and put option scenarios, we know a final condition as opposed to an initial condition.

Before providing the model to value an option there is one more relationship we can use to our advantage when modeling. Suppose you own one stock in the fictitious company named *Operator Exponential* whose stock symbol is `eAt`. Since you own an asset, you purchase a put option for one stock of `eAt` and sell a call option for one stock of `eAt` where both have the same expiration date T. On the expiration date you will realize

$$S(T) + \max(E - S(T), 0) - \max(S(T) - E, 0)$$

Notice that if $S(T) \leq E$ then you will realize a gain of

$$S(T) - (E - S(T)) - 0 = E$$

whereas if $S(T) \geq E$ the gain is

$$S(T) + 0 - (S(T) - E)) = E.$$

In both cases the payoff is the same! In other words the investment has a guaranteed pay out of E.

A *riskless investment* is an investment strategy that yields guaranteed payoffs. Examples include savings accounts and goverment issued bonds. If $\$(t)$ represents the amount of money in a riskless investment with interest rate r then $\$(t)$ is modeled by

$$\begin{cases} \frac{d\$(t)}{dt} & = r\$(t) \\ \$(0) & = \$_0 \end{cases} \tag{12.21}$$

In Part I we derived the solution of IBVP (12.21) and found it to be $\$(t) = \$_0 e^{rt}$. If the riskless investment is to equal E at time T then solving for S_0 yields

$$\$(t) = Ee^{-r(T-t)}$$

Since the put-call strategy involving the stock `eAt` is riskless at time T it is equivalent (12.21) where we initially invest Ee^{-rT} dollars into a riskless investment. Based on this argument, it seems reasonable that

$$S(t) + P(t) - C(t) = E^{-r(T-t)} \tag{12.22}$$

In actuality, in order for (12.22) to be valid we must assume there are no *arbitrage* opportunities. Arbitrage is essentially the ability to realize a riskless profit at no cost. This is not the same as buying low and selling high at a future date because there was a chance the price could have dropped resulting in a loss. However, if you could instantaneously buy low and sell high then that would be an example of arbitrage. For more details on arbitrage and modeling options refer to [45, 20].

We assume the market is arbitrage-free so that (12.22) called the *put-call parity* is valid. The put-call parity allows you to know the value of a put if we know the value of a call or vice-versa. **(Tell How.)** Furthermore, it makes clear that riskless investment strategies should play a role in modeling the value of an option.

The derivation of the equation that models the value of an option is beyond the scope of this text. It states that the value of the option, $V(S, t)$, satisfies

$$\frac{\partial V}{\partial t} + \frac{1}{2}\sigma^2 S^2 \frac{\partial^2 V}{\partial S^2} + rS\frac{\partial V}{\partial S} - rV = 0 \tag{12.23}$$

where boundary conditions and the "initial condition" depend on the type of option. Equation (12.23) is called the *Black and Scholes equation* [5]. Let us focus on the value of a call option and derive the appropriate boundary and "initial" conditions. As discussed above, the call does not have an initial conditon, but rather a final condition at $t = T$. Namely,

$$C(T, S(T)) = \max(S(T) - E, 0)$$

This implies that we begin at $t = T$ and travel back in time to the date in which the contract was purchased.

Since the underlying asset is playing the role as the spatial variable, the boundary conditions will depend on the value of the underlying asset. The value of underlying asset can never become negative, so $S \geq 0$. To determine the boundary condition at $S = 0$ we must determine the value of call option in this case. What is the value of a contract that gives you the right to buy an asset that is worthless? Nothing! Thus, $C(t, 0) = 0$ for all t.

Theoretically, the value of the underlying asset can be arbitrarily large or "priceless," so mathematically $S \in [0, \infty)$. Therefore, there is no right-boundary condition. However, in practical implementations of the model one needs to know the value of the option when S is large.

STOP! What is the value of a contract that gives the right to purchase an asset that is extremely valuable?

As S becomes large, the value of the call becomes that of the asset, so we expect

$$C(S, t) \sim S, \text{ for large } S \text{ and for all } t.$$

Putting everything together, the value of a European call option satisfies the IBVP

$$\begin{cases} \frac{\partial C(S,t)}{\partial t} + \frac{1}{2}\sigma^2 S^2 \frac{\partial^2 C(S,t)}{\partial S^2} + rS\frac{\partial C(S,t)}{\partial S} - rV = 0, \ 0 < t < T, \ 0 \leq S < \infty \\ C(T, S) = \max(S(T) - E, 0), \ 0 \leq S < \infty \\ C(t, 0) = 0, \ 0 \leq t < T. \end{cases} \tag{12.24}$$

Exercise 12.5.2. Derive the boundary conditions for the European put using a similar financial argument used for the call along with the put-call parity.

12.5.1 Black and Scholes Equation and the Heat Equation

Unlike previously considered IBVPs, (12.24) has a final time condition as opposed to an initial condition. Also, the structure of the differential equation in (12.24) is unlike any we have encountered so far. Next, we go through a series of simplifications to change (12.24) into a proper *initial* boundary value problem. Along the way, we will transform (12.24) into the heat equation.

IBVP (12.24) has $\frac{\partial}{\partial t}$ and $\frac{\partial^2}{\partial S^2}$ terms like the heat equation, but unlike the heat equation, it also has non-constant coefficients, a first-order term, and a final condition. In order to rewrite the European call model as a heat equation we introduce a change of variable.

Let x and τ be two variables such that

$$S = Ee^x \text{ and } t = T - \frac{\tau}{\frac{1}{2}\sigma^2}. \tag{12.25}$$

Then x becomes dimensionless and takes on values between $(-\infty, \infty)$ and $\tau = 0$ represents

$t = T$. Thus, the final condition in IBVP (12.24) becomes an initial condition when using the τ variable.

Next, we introduce the function, v, in two variables x and τ such that

$$C = Ev(x, \tau).$$

STOP! What are the units of $v(x, \tau)$ for all x and τ?

The following **EXPLORE!** activity completes the nondimensionlization and simplification process for IBVP (12.24).

EXPLORE!

12.1 Are Option Values The Same As Heat Diffusion?

Complete the following exercises to transform IBVP (12.24) into a heat equation. If you are unable to complete one part move on to the next.

i.) Use the change of variables described in (12.25) to verify

$$\frac{\partial C}{\partial t} = -\frac{1}{2}\sigma^2 E \frac{\partial v}{\partial \tau} \tag{12.26}$$

$$\frac{\partial C}{\partial S} = \frac{E}{S}\frac{\partial v}{\partial x} \tag{12.27}$$

$$\frac{\partial^2 C}{\partial S^2} = \frac{E}{S^2}\left[\frac{\partial^2 v}{\partial x^2} + \frac{\partial v}{\partial x}\right] \tag{12.28}$$

ii.) Use (12.26)-(12.28) to find a k such that (12.23) is equivalent to

$$\frac{\partial v}{\partial \tau} - \frac{\partial^2 v}{\partial x^2} + (1-k)\frac{\partial v}{\partial x} + kv = 0 \tag{12.29}$$

Progress has been made. Notice that in (12.29) the non-constant coefficients have been removed. Next, we remove the $\frac{\partial v}{\partial x}$ and kv terms by introducing some change of variable. However, we do not want to change the independent variables since they allowed us to get rid of the non-constant coefficients.

iii.) Let u be a function in x and τ such that

$$v(x, \tau) = e^{ax+b\tau}u(x, \tau) \tag{12.30}$$

where a, b are constants to be determined shortly. Verify

$$\frac{\partial v}{\partial \tau} = e^{ax+b\tau}\left[\frac{\partial u}{\partial t} + bu\right] \tag{12.31}$$

$$\frac{\partial v}{\partial x} = e^{ax+b\tau}\left[\frac{\partial u}{\partial x} + au\right] \tag{12.32}$$

$$\frac{\partial^2 v}{\partial x^2} = e^{ax+b\tau}\left[\frac{\partial u}{\partial x^2} + 2a\frac{\partial u}{\partial x}a^2 u\right] \tag{12.33}$$

iv.) Verify that (12.34) is equivalent to

$$\frac{\partial u}{\partial \tau} - \frac{\partial^2 u}{\partial x^2} + [1-k-2a]\frac{\partial u}{\partial x} + \left[b - a^2 + (1-k)a + k\right]u = 0 \tag{12.34}$$

In order for (12.34) to be a heat equation the coefficients in front of the $\frac{\partial u}{\partial x}$ and u terms need to be zero, which motivates the next **EXPLORE!** activity.

v.) Show $a = \frac{1-k}{2}$ and $b = -\frac{1}{4}(k+1)^2$ solve

$$\begin{cases} 1 - k - 2a & = 0 \\ b - a^2 + (1-k)a + k & = 0 \end{cases}$$

The transformation is almost complete. All we need to take care of the initial condition in the new variables.

vi.) Show the corresponding initial condition to (12.34) is

$$u(x,0) = \max\left(e^{\frac{1}{2}(k+1)x} - e^{\frac{1}{2}(k-1)x}, 0\right)$$

and IBVP (12.24) is equivalent to

$$\begin{cases} \frac{\partial u(x,\tau)}{\partial \tau} & = \frac{\partial^2 u(x,\tau)}{\partial x^2}, \; 0 < \tau < \frac{1}{2}\sigma^2 T, \; -\infty < x < \infty \\ u(x,0) & = \max\left(e^{\frac{1}{2}(k+1)x} - e^{\frac{1}{2}(k-1)x}, 0\right), \; -\infty < x < \infty \end{cases}. \tag{12.35}$$

Carefully describe the step-by-step process that transforms the solution of IBVP (12.35) into the solution of IBVP (12.24).

Chapter 13

Formulating a Theory for (A-NonCP)

Overview

The presence of a forcing term in the PDEs of the mathematical models leads to a more accurate explanation of the phenomena, as it did in Part I. We would like to extend the theory for (Non-CP) in Part I to the present setting, taking advantage of the theory outlined in Chapter 5. This is the goal of this chapter.

13.1 Introducing (A-NonCP)

The following abstract evolution equation emerged when reformulating IBVPs in Chapter 11 as abstract IVPs in a Hilbert space.

Definition 13.1.1. An abstract evolution equation of the form

$$(\mathbf{A-NonCP}) \begin{cases} \frac{d}{dt}\left(u(t)\right) = A\left(u(t)\right) + f(t), \; t_0 < t < t_0 + T, \\ u(t_0) = u_0, \end{cases} \tag{13.1}$$

in a Hilbert space $\mathcal{H}$, where $u : [t_0, t_0 + T] \to \mathcal{H}$ is the unknown function, $A : \text{dom}(A) \subset \mathcal{H} \to \mathcal{H}$ is an appropriately-defined linear operator (like $\frac{d^2}{dx^2}$, $\triangle + I$, etc.), $u_0 \in \mathcal{H}$ is a function describing the initial profile, and $f : [t_0, t_0 + T] \to \mathcal{H}$ is a forcing term is called an *abstract nonhomogeneous Cauchy problem*, denoted (A-NonCP).

STOP! Before moving on, go back through Chapter 11 and review the steps involved when reformulating the IBVPs equipped with forcing terms abstractly in the above form. Make certain to understand all of the identifications along the way.

The fact that forming the theoretical framework for the homogeneous version of (A-NonCP) (that is, when $f = 0$) followed naturally from the development of Chapter 4 gives reason for optimism in our current situation. Assuming that $\left\{e^{At} : \mathcal{H} \to \mathcal{H} \,|t \geq 0\right\}$ is a semigroup of bounded linear operators on $\mathcal{H}$ and that $A : \text{dom}(A) \subset \mathcal{H} \to \mathcal{H}$ is a sufficiently nice operator, we begin with the variation of parameters formula. Given that each term of this formula is now a member of a general Hilbert space $\mathcal{H}$, the integral is $\mathcal{H}$−valued. Upon tracing through the steps of the derivation of (13.18) given in Chapter 5, we find that we simply need to reinterpret the notion of "multiplying both sides of an equality by the term $e^{-\mathbf{A}t}$" as "applying the operator e^{-At} to both sides." All steps of the derivation then remain valid since all of the usual calculus operations have been extended to the present setting. As such, assuming that f is integrable, all calculations are justified *provided that* $u(t) \in \text{dom}(A)$, $\forall t > t_0$. In such case,

$$u(t) = e^{At}u_0 + \int_0^t e^{A(t-s)} f(s)ds. \tag{13.2}$$

Stop! Why is the condition "$u(t) \in \text{dom}(A)$, $\forall t > t_0$" needed?

As in Part I, this formula is called the *variation of parameters formula* for (A-NonCP).

We saw various examples of constructing a solution to certain IBVPs in Chapter 5, but there the motivation was to show the emergence of the variation of parameters formula from elementary techniques. Now, we are *starting* with the variation of parameters formula and ask the question, "How do we interpret this formula in the Hilbert space setting?" As in Part I (see Example 5.3.1), we first need to have in hand the formula for e^{At}, which is often easier said than done. If we are lucky enough to have it, we can then apply it to the forcing term $f(\cdot)$ and integrate. But remember, $f(\cdot)$ is now $\mathcal{H}$-valued in this abstract form. So, to produce a useable practical function, we must revisit the dependence on the spatial variable that we worked so hard to suppress in the first place. In effect, this amounts to working backward through the derivation illustrated in Chapter 11.

STOP! Look back at the heat equation discussion in the previous section to see what this means.

A more practical approach is to consider a numerical approximation scheme. We extend the argument provided in Section (5.2.2) to the variation of parameters formula for (A-NonCP).

Let $t > 0$, $n \in \mathbb{N}$, and define $\bar{u}$ to be the solution of the approximate IVP

$$\begin{cases} \frac{\bar{u}(s+\frac{t}{n})-\bar{u}(s)}{\frac{t}{n}} & = A\bar{u}(s+\frac{t}{n}) + f(s), \; s \geq 0 \\ \bar{u}(0) & = u_0 \end{cases} \tag{13.3}$$

Solving the first equation in (13.3) for $\bar{u}(s+\frac{t}{n})$ yields

$$\bar{u}\left(s+\frac{t}{n}\right) = \left(1-\frac{t}{n}A\right)^{-1}\bar{u}(s) + \left(1-\frac{t}{n}A\right)^{-1} f(s)\frac{t}{n} \tag{13.4}$$

Taking the first step in the march towards t amounts to evaluating (13.4) at $s = 0$ and realizing that

$$\bar{u}\left(\frac{t}{n}\right) = \left(1-\frac{t}{n}A\right)^{-1}\bar{u}_0 + \left(1-\frac{t}{n}A\right)^{-1} f(0)\frac{t}{n}$$

As was the case in Section (5.2.2),

$$\begin{aligned}\bar{u}\left(\frac{2t}{n}\right) &= \left(1-\frac{t}{n}A\right)^{-1}\bar{u}\left(\frac{t}{n}\right) + \left(1-\frac{t}{n}A\right)^{-1} f\left(\frac{t}{n}\right)\frac{t}{n} \\ &= \left(1-\frac{t}{n}A\right)^{-2}\bar{u}_0 + \left(\left(1-\frac{t}{n}A\right)^{-2} f(0) + \left(1-\frac{t}{n}A\right)^{-1} f\left(\frac{t}{n}\right)\right)\frac{t}{n}\end{aligned}$$

where the first term is exactly the same as in the homogeneous case and the second term involves a sum for which each term is multiplied by $\frac{t}{n}$. Following the same argument as the one provided in Section (5.2.2), we have

$$\bar{u}(t) = \left(1 - \frac{t}{n}A\right)^{-n} u_0 + \sum_{j=0}^{n-1} \left(1 - \frac{t}{n}A\right)^{-(n-j)} f\left(\frac{t}{n}j\right)\frac{t}{n} \tag{13.5}$$

If A and u_0 satisfy Pseudo-Thereom 2 and $\bar{u}$ converges to u, then taking the limit of (13.5) as n goes to infinity produces

$$u(t) = e^{At}u_0 + \lim_{n\to\infty} \sum_{j=0}^{n-1} \left(1 - \frac{t}{n}A\right)^{-(n-j)} f\left(\frac{t}{n}j\right)\frac{t}{n}.$$

STOP! Go back and look at the variation of parameters formula for (A-NonCP). What is different? The same?

We know an integral must play a role in the formula for $u(t)$. In fact, if we apply the exact same argument as in Section 5.2.2 we obtain

$$\lim_{n\to\infty} \sum_{j=0}^{n-1} \left(1 - \frac{t - s_j}{n-j}A\right)^{-(n-j)} f(s_j)\Delta s_j = \int_0^t e^{A(t-s)} f(s)ds$$

Thus, we have obtained the variations of parameters formula via an iterative approach that makes sense of (13.2).

Despite the complexity of simplifying the variation of parameters formula, it is definitely preferable to not having a representation formula for a solution. Many results (for instance, continuous dependence and stability) benefit greatly from having it. Recall from Part I that there is a bit of a hitch when deciding whether or not this variation of parameters formula actually constituted a solution of (Non-CP). Overcoming this hurdle depended solely on the nature of the forcing term.

STOP! Review the development of the variation of parameters formula again. What is required in order to prove this formula actually satisfies the equation in (A-NonCP)?

This prompts us to introduce two different notions of a solution as we did in Part I.

Definition 13.1.2. A function $u : [t_0, t_0 + T] \to \mathcal{H}$ is a

i.) *classical solution* of (A-NonCP) on $[t_0, t_0 + T]$ if

a.) $u(t) \in \mathrm{dom}(A)$, $\forall t \in [t_0, t_0 + T]$,
b.) $u(\cdot)$ is continuous on $[t_0, t_0 + T]$,
c.) $u(\cdot)$ is differentiable on $(t_0, t_0 + T)$,
d.) (A-NonCP) is satisfied.

ii.) *mild solution* of (A-NonCP) on $[t_0, t_0 + T]$ if

a.) $u(\cdot)$ is continuous on $[t_0, t_0 + T]$,
b.) $u(\cdot)$ is given by (13.2), $\forall t \in [t_0, t_0 + T]$.

We explore the existence of both types of solutions in the next section.

13.2 Existence and Uniqueness of Solutions of (A-NonCP)

The fact that the derivation of the variation of parameters formula goes through as it did in Part I, assuming only continuity of the forcing term f, enables us to easily prove that a mild solution of (A-NonCP) exists under this assumption, and the variation of parameters formula provides us with a representation for this solution. Uniqueness of a solution is argued as in the proof of Theorem 5.3.1 due to the fact that e^{At} satisfies the same properties in this more general setting. But, when is this solution also a classical solution? Interestingly, the theory from Part I breaks down a bit here in the sense that continuity of f alone does not imply the variation of parameters formula is differentiable and so, it need not satisfy the equation in (A-NonCP).

This is a very subtle point. The main difference from the $\mathbb{R}^N$ setting is that for a general Hilbert space $\mathcal{H}$, it need not be the case that $\mathrm{dom}(A) = \mathcal{H}$, whereas equality *must* occur when $\mathcal{H} = \mathbb{R}^N$. This difference is precisely what makes it possible to construct an example for which a mild solution exists that is not a classical solution.

This raises the question, "What is sufficient to guarantee the differentiability of (A-NonCP)?" Certainly, we need $u_0 \in \mathrm{dom}(A)$ in order for $t \mapsto e^{At}u_0$ to be differentiable **(Why?)**, but what about the second term in (A-NonCP)? Since we ultimately need u to be differentiable, a natural sufficient condition would be to require that the forcing term $f : [t_0, t_0 + T] \to \mathrm{dom}(A)$ be a continuous function for which the mapping $t \mapsto \int_0^t e^{A(t-s)} f(s)ds$ is differentiable. This suggests the following result.

Proposition 13.2.1. If $u_0 \in \mathrm{dom}(A)$ and $f : [t_0, t_0 + T] \to \mathrm{dom}(A)$ is such that $t \mapsto \int_0^t e^{A(t-s)} f(s)ds$ is differentiable, then (A-NonCP) has a unique classical solution given by (13.1).

Proof. Uniqueness was established earlier. We must argue that u given by (13.1) satisfies the above definition.

The continuity of u is clear. **(Why?)** We begin by showing that $u(t) \in \mathrm{dom}(A)$, $\forall t \in [0, T]$. This requires that we verify that $\lim\limits_{h\to 0} \frac{e^{Ah}u(t)-u(t)}{h}$ exists. Since $e^{At}u_0 \in \mathrm{dom}(A)$, $\forall t \in [0, T]$, **(Why?)** we need only to argue the existence of

$$\lim_{h\to 0} \frac{e^{Ah} - I}{h} \left(\int_0^t e^{A(t-s)} f(s)ds \right).$$

(Tell why.) Observe that

$$\begin{aligned}
A \int_0^t e^{A(t-s)} f(s)ds &= \lim_{h\to 0} \frac{e^{Ah} - I}{h} \left(\int_0^t e^{A(t-s)} f(s)ds \right) \\
&= \lim_{h\to 0} \int_0^t \frac{e^{Ah} - I}{h} \left(e^{A(t-s)} f(s) \right) ds \\
&= \lim_{h\to 0} \int_0^t \frac{e^{Ah}e^{A(t-s)} f(s) - e^{A(t-s)} f(s)}{h} ds \qquad (13.6) \\
&= \lim_{h\to 0} \Bigg[\frac{\int_0^t e^{A(t+h-s)} f(s)ds - \int_0^t e^{A(t-s)} f(s)ds}{h} \\
&\quad - \frac{1}{h} \int_t^{t+h} e^{A(t+h-s)} f(s)ds \Bigg] \\
&= \frac{d}{dt} \left(\int_0^t e^{A(t-s)} f(s)ds \right) - f(t).
\end{aligned}$$

The right-side of the last equality in (13.6) is a well-defined member of dom(A) (by assumption) and $f(t) \in \text{dom}(A)$. Thus, $u(t) \in \text{dom}(A)$, $\forall t \in [0,T]$.

Finally, we show that u given by (13.1) satisfies (13.2). Observe that (13.6) implies

$$\frac{d}{dt}\left(\int_0^t e^{A(t-s)} f(s)ds\right) = A\int_0^t e^{A(t-s)} f(s)ds + f(t). \tag{13.7}$$

Also,

$$\frac{d}{dt}\left(e^{At}u_0\right) = Ae^{At}u_0. \tag{13.8}$$

Using the linearity of A, adding (13.7) and (13.8) yields

$$\begin{aligned} u'(t) &= \frac{d}{dt}\left(e^{At}u_0 + \int_0^t e^{A(t-s)} f(s)ds\right) = A\left(e^{At}u_0 + \int_0^t e^{A(t-s)} f(s)ds\right) + f(t) \\ &= Au(t) + f(t), \end{aligned}$$

and $u(0) = u_0$. This completes the proof. □

STOP! Check which of the IBVPs in Chapter 11 satisfy the hypotheses of this theorem.

We can build on the perturbation result from Chapter 10 as well. Consider the following exercise.

Exercise 13.2.1. Consider the following perturbed heat equation

$$\begin{cases} \frac{\partial}{\partial t}z(x,t) = \frac{\partial^2}{\partial x^2}z(x,t) + f(x), \ 0 < x < a, \, t > 0, \\ z(x,0) = 0, \ 0 < x < a, \\ z(0,t) = z(a,t) = 0, \ t > 0, \end{cases} \tag{13.9}$$

i.) Explain carefully why (13.9) has a unique classical solution on $0, T$.
ii.) Does the same approach apply to the following IBVP? Explain carefully.

$$\begin{cases} \frac{\partial}{\partial t}z(x,y,t) &= k\left(\triangle z(x,y,t) + z(x,y,t)\right) + t^2 + x + 2y, \ 0 < x < a, \ 0 < y < b, \ t > 0, \\ z(x,y,0) &= \sin 2x + \cos 2y, \ 0 < x < a, \ 0 < y < b, \\ z(x,0,t) &= 0 = z(x,b,t), \ 0 < x < a, \ t > 0, \\ z(0,y,t) &= 0 = z(a,y,t), \ 0 < y < b, \ t > 0, \end{cases}$$

13.3 Dealing with a Perturbed (A-NonCP)

The variation of parameters formula is useful when establishing a continuous dependence result formally.

STOP! Review Section 5.4 before moving onward.

We begin by considering a perturbation result in which only the initial condition is tweaked. To this end, let $u_0 \in \text{dom}(A)$ and consider a sequence $(u_0)_n \subset \text{dom}(A)$ such that $\|(u_0)_n - u_0\|_{\mathcal{H}} \to 0$. Now, consider (A-NonCP) and for every $n \in \mathbb{N}$, consider the following IVP in which we replace u_0 by $(u_0)_n$:

$$\begin{cases} \frac{d}{dt}(u_n(t)) = A(u_n(t)) + f(t), \; t_0 < t < t_0 + T, \\ u_n(t_0) = (u_0)_n, \end{cases} \tag{13.10}$$

We know that as long as f satisfies the assumptions of the existence-uniqueness result of the previous section, then (13.10) has a unique classical solution given by the variation of parameters formula

$$u_n(t) = e^{At}(u_0)_n + \int_0^t e^{A(t-s)} f(s)ds. \tag{13.11}$$

Denoting the solution of (A-NonCP) by $u(t)$ and using its representation formula, observe that

$$\begin{aligned} \lim_{n\to\infty} \max_{t_0 \le t \le t_0+T} \{\|u_n(t) - u(t)\|_{\mathcal{H}}\} &= \lim_{n\to\infty} \max_{t_0 \le t \le t_0+T} \left[\left\| e^{A(t-t_0)}(u_0)_n - e^{A(t-t_0)} u_0 \right\|_{\mathcal{H}} \right] \\ &\le \underbrace{\max_{t_0 \le t \le t_0+T} \{\|e^{At}\|_{\mathcal{H}}\}}_{\text{Bounded}} \cdot \lim_{n\to\infty} \|(u_0)_n - u_0\|_{\mathcal{H}} = 0 \end{aligned} \tag{13.12}$$

The right-side of equation (13.12) comes from substituting (13.11) for $u_n(t)$ and the corresponding variation of parameters representation of $u(t)$. The integral terms

$$\int_0^t e^{A(t-s)} f(s)ds$$

in each are the same and cancel due to the subtraction.

APPLY IT!!

To put the continuous dependence result above in context for the models listed below complete the following:

I.) List the physcial parameters, in context of the model, for which the theory guarantees a continuous dependence result.

II.) Compare and contrast your answer in (I) with your observations from the continuous dependence **EXPLORE!** activities in Chapter 11.

i.) The Neumann BC nonhomogeneous heat equation

$$\begin{cases} \frac{\partial}{\partial t} z(x,t) = k^2 \frac{\partial^2}{\partial x^2} z(x,t) + f(x,t), \; 0 < x < a, \; t > 0, \\ z(x,0) = z_0(x), \; 0 < x < a, \\ \frac{\partial z}{\partial x}(0,t) = \frac{\partial z}{\partial x}(a,t) = 0, \; t > 0. \end{cases} \tag{13.13}$$

ii.) The two-dimensional Dirichlet BC nonhomogeneous heat equation

$$\begin{cases} \frac{\partial}{\partial t} z(x,y,t) &= k^2 \Delta z(x,y,t) + f(x,y,t), \; 0 < x < a, \; 0 < y < b, \; t > 0, \\ z(x,y,0) &= z_0(x,y), \; 0 < x < a, \; 0 < y < b, \\ z(x,0,t) &= 0 = z(x,b,t), \; 0 < x < a, \; t > 0, \\ z(0,y,t) &= 0 = z(a,y,t), \; 0 < y < b, \; t > 0, \end{cases} \tag{13.14}$$

iii.) The Dirichlet BC nonhomogeneous Fluid Seepage model

$$\begin{cases} \frac{\partial}{\partial t}\left(p(x,t) - \alpha\frac{\partial^2}{\partial x^2}p(x,t)\right) = \beta\frac{\partial^2}{\partial x^2}p(x,t) + f(x,t),\ 0 < x < \pi,\ t > 0, \\ p(x,0) = p_0(x),\ 0 < x < \pi, \\ p(0,t) = p(\pi,t) = 0,\ t > 0. \end{cases} \tag{13.15}$$

iv.) The Dirichlet BC nonhomogeneous wave equation

$$\begin{cases} \frac{\partial^2}{\partial t^2}z(x,t) - c^2\frac{\partial^2}{\partial x^2}z(x,t) &= f(x,t),\ 0 < x < L, t > 0, \\ z(x,0) = z_0(x),\ \frac{\partial z}{\partial t}(x,0) &= z_1(x),\ 0 < x < L, \\ z(0,t) = z(L,t) &= 0,\ t > 0. \end{cases} \tag{13.16}$$

v.) The periodic BC nonhomogeneous damped wave equation

$$\begin{cases} \frac{\partial^2}{\partial t^2}z(x,t) + \alpha\frac{\partial}{\partial t}z(x,t) + c^2\frac{\partial^2}{\partial x^2}z(x,t) = f(x,t),\ 0 < x < L, t > 0, \\ z(x,0) = z_0(x),\ \frac{\partial z}{\partial t}(x,0) = z_1(x),\ 0 < x < L. \\ z(0,t) = z(L,t),\ t > 0, \end{cases} \tag{13.17}$$

What about the other parameters and forcing functions in the "continuous dependence" **EXPLORE!** activities?

STOP! Look back at the **EXPLORE!** activities in Chapter 11 that discussed this issue. Pay particular attention to all of the different model parameters and forcing terms that were perturbed.

Precisely, consider the following two related IVPs:

$$\begin{cases} u'(t) = au(t) + f(t),\ t > t_0 \\ u(t_0) = u_0, \end{cases} \tag{13.18}$$

$$\begin{cases} \overline{u}'(t) = a\overline{u}(t) + \overline{f}(t),\ t > t_0 \\ \overline{u}(t_0) = \overline{u}_0, \end{cases} \tag{13.19}$$

where we assume there exist $\delta_1, \delta_2 > 0$ for which

$$\|u_0 - \overline{u}_0\|_{\mathcal{H}} < \delta_1 \text{ and } \max_{t_0 \le t \le t_0+T} \left\|f(t) - \overline{f}(t)\right\|_{\mathcal{H}} < \delta_2. \tag{13.20}$$

The solutions of (13.18) and (13.19) are, respectively, given by

$$u(t) = e^{A(t-t_0)}u_0 + \int_{t_0}^{t} e^{A(t-s)}f(s)ds, \tag{13.21}$$

$$\overline{u}(t) = e^{A(t-t_0)}\overline{u}_0 + \int_{t_0}^{t} e^{A(t-s)}\overline{f}(s)ds. \tag{13.22}$$

The question of interest is, "How far apart are these solutions at a given time t?" In Part I, when the outputs were real numbers, we visualized this by computing the vertical distance between the graphs of $y = u(t)$ and $y = \overline{u}(t)$ at all times t in an interval on which we are studying the solutions. Now, we must replace the norm by the $\|\cdot\|_{\mathcal{H}}$-norm. With the exception of this change, and interpreting all quantities in a Hilbert space, the steps involved in estimating the difference between the solutions is the same as in Part I.

Proceeding as before, we subtract the two solution formulas, and estimate using the properties of absolute value. This yields the following for all $t \in [t_0, T]$.

$$
\begin{aligned}
&\|u(t) - \overline{u}(t)\|_{\mathcal{H}} = \\
&\left\| \left(e^{A(t-t_0)} u_0 - e^{A(t-t_0)} \overline{u}_0 \right) + \left(\int_{t_0}^{t} e^{A(t-s)} f(s) ds - \int_{t_0}^{t} e^{A(t-s)} \overline{f}(s) ds \right) \right\|_{\mathcal{H}} = \\
&\left\| e^{A(t-t_0)} \left(u_0 - \overline{u}_0 \right) + \int_{t_0}^{t} e^{A(t-s)} \left(f(s) - \overline{f}(s) \right) ds \right\|_{\mathcal{H}} \leq \\
&\left\| e^{A(t-t_0)} \left(u_0 - \overline{u}_0 \right) \right\|_{\mathcal{H}} + \left\| \int_{t_0}^{t} e^{A(t-s)} \left(f(s) - \overline{f}(s) \right) ds \right\|_{\mathcal{H}} \leq \\
&\left(\max_{t_0 \leq t \leq t_0 + T} \left\| e^{A(t-t_0)} \right\|_{\mathcal{H}} \right) \cdot \|u_0 - \overline{u}_0\|_{\mathcal{H}} + \left\| \int_{t_0}^{t} e^{A(t-s)} \left(f(s) - \overline{f}(s) \right) ds \right\|_{\mathcal{H}} \leq \\
&\left(\max_{t_0 \leq t \leq t_0 + T} \left\| e^{A(t-t_0)} \right\|_{\mathcal{H}} \right) \delta_1 + \left\| \int_{t_0}^{t} e^{a(t-s)} \left(f(s) - \overline{f}(s) \right) ds \right\|_{\mathcal{H}}
\end{aligned}
\tag{13.23}
$$

The lingering problem is the second term in the last line in (13.23). How do we handle the norm of an integral term?

STOP! Assuming similar properties hold when the absolute value is replaced by the Hilbert space norm, work line by line through the derivation of the same result in Section 5.4 to show that

$$
\|u(t) - \overline{u}(t)\|_{\mathcal{H}} \leq \left(\max_{t_0 \leq t \leq t_0 + T} \left\| e^{A(t-t_0)} \right\|_{\mathcal{H}} \right) \left(\delta_1 + \delta_2 (T - t_0) \right). \tag{13.24}
$$

APPLY IT!!

According to (13.24), solutions of IBVPs (10.12)-(10.16) depend continuously on the initial conditions and forcing functions. Compare this theoretical result with your observations from the "continuous dependence" **EXPLORE!** activities corresponding to (10.12)-(10.16). If any of your observations stated that a solution *did not* depend continuously on the initial condition and forcing function, try to determine small enough δ_1 and δ_2 such that GUI confirms (13.24).

13.4 Long-Term Behavior

Regarding the long-term behavior (as $t \to \infty$) of a mild solution, we saw in Part I that the boundedness of the forcing term $\mathbf{f}$ was sufficient to guarantee that $\int_0^{\infty} e^{\mathbf{A}(t-s)} \mathbf{f}(s) ds < \infty$, $\forall t > 0$, provided that $\lim_{t \to \infty} \left\| e^{\mathbf{A}t} \right\|_{\mathbb{M}^N} = 0$. How does this translate into the present setting? Observe that if

$$
\left\| e^{At} \right\| \leq M_A e^{\omega t}, \text{ for some } \omega < 0, \tag{13.25}
$$

and

$$\sup_{s \geq 0} \|f(s)\|_{\mathcal{X}} \leq M_f, \tag{13.26}$$

then $\forall t, s \geq 0$,

$$\left\| e^{A(t-s)} f(s) \right\|_{\mathcal{H}} \leq \left\| e^{A(t-s)} \right\| \|f(s)\|_H \leq M_A M_f e^{\omega(t-s)}. \tag{13.27}$$

If, in addition, f is continuous and

$$\lim_{t \to \infty} \|f(t) - L\|_{\mathcal{H}} = 0, \tag{13.28}$$

then (13.26) holds, and we have more information to aid us in calculating $\lim_{t\to\infty} \|u(t)\|_{\mathcal{H}}$. Precisely, we have the following result.

Proposition 13.4.1. Let $u_0 \in \text{dom}(A)$, $f : [0, \infty) \to \mathcal{H}$, and u be the mild solution of (A-NonCP).

i.) If (13.25) and (13.26) hold, then $\lim_{t\to\infty} \|u(t)\|_{\mathcal{H}} = 0$.

ii.) If (13.25) and (13.28), then $\lim_{t\to\infty} \left\| u(t) - A^{-1}L \right\|_{\mathcal{H}} = 0$.

APPLY IT!!

Use the "long-term behavior" **EXPLORE!** activities in Chapter 11 to make a conjecture regarding the stability properties of the operators A found in IBVPs (13.13)-(13.17).

13.5 Looking Ahead

External forces acting on a system are often state-dependent. For instance, if the forcing term represents temperature regulation of a material, then it necessarily takes into account the temperature of the material at various times t and makes appropriate adjustments. This is easily illustrated by the following adaption of the forced heat equation:

$$\begin{cases} \frac{\partial}{\partial t} z(x,y,t) &= k\Delta z(x,y,t) + z(x,y,t) + \alpha e^{-\frac{\beta}{z(t,x,y)}}, \; 0 < x < a, \; 0 < y < b, \; t > 0, \\ z(x,y,0) &= \sin 2x + \cos 2y, \; 0 < x < a, \; 0 < y < b, \\ z(x,0,t) &= 0 = z(x,b,t), \; 0 < x < a, \; t > 0, \\ z(0,y,t) &= 0 = z(a,y,t), \; 0 < y < b, \; t > 0, \end{cases} \tag{13.29}$$

where $z(x, y, t)$ represents the temperature at the point (x, y) on the plate at time $t > 0$. In Chapter (12) we encountered several models with state-dependent forcing functions. To see how and the added complexity that arises when compared to (A-NonCP), complete the next exercise.

Exercise 13.5.1. For each of the listed models complete the following:

I.) Write the model as an abstract evolution equation in the form

$$\begin{cases} \frac{d}{dt}(u(t)) = A(u(t)) + f(t, u(t)), \; t_0 < t < t_0 + T, \\ u(0) = u_0, \end{cases} \tag{13.30}$$

II.) Using the ideas and theorems in Chapter 10, argue that the corresponding (A-HCP) of (13.30) has a unique solution and e^{At} exists. (Disregard the forcing term in (13.30).)

III. Propose a solution to (13.30) using a variation of parameters formula similar (13.2), but which accounts for the state-dependent term like the variation of parameters formula (7.24) found in Part I.

IV. Compare and contrast (13.2) and your answer in III. Indicate any new complications that you expect to arise in any solution process.

i.) Fisher's Equation

$$\begin{cases} \frac{\partial v(z,\tau)}{\partial \tau} = \frac{\partial^2 v(z,\tau)}{\partial z^2} + v(z,\tau)\left(1 - v(z,\tau)\right), \tau > 0, z \in (-L', L') \\ v(0,z) = v_0(z), \; z \in (-L', L') \\ z(-L',\tau) = z(L',\tau), t > 0. \end{cases}$$

ii.) The epidemiological model

$$\begin{cases} \frac{\partial v(x,t)}{\partial t} = \epsilon \frac{\partial^2 v(x,t)}{(\partial x)^2} - u(x,t), \; x \in (-L, L), \; t > 0 \\ \frac{\partial u(x,t)}{\partial t} = \frac{\partial^2 u(x,t)}{(\partial x)^2} + u(x,t)(v(x,t) - \lambda), \; x \in (-L, L), \; t > 0 \\ v(x,0) = 1, \; x \in [-L, L] \\ u(x,0) = u_0(x), \; x \in [-L, L] \end{cases}$$

iii.) The Gray-Scott model

$$\begin{cases} \frac{\partial u(x,t)}{\partial t} = \frac{\partial^2 u(x,t)}{\partial x^2} - u(x,t)v(x,t)^2 + A\left(1 - u(x,t)\right), \; t > 0, \; x \in [-L, L] \\ \frac{\partial v(x,t)}{\partial t} = \epsilon \frac{\partial^2 v(x,t)}{\partial x^2} + u(x,t)v(x,t)^2 - Bv(x,t), \; t > 0, \; x \in [-L, L] \\ u(x,0) = u_0(x), \; x \in [-L, L] \\ v(x,0) = v_0(x), \; x \in [-L, L] \\ u(-L,t) = u(L,t), \; t > 0 \\ v(-L,t) = v(L,t), \; t > 0 \end{cases}$$

iv.) The autocatalysis model

$$\begin{cases} \frac{\partial u(x,t)}{\partial t} = \frac{\partial^2 u(x,t)}{\partial x^2} + v(x,t)f(u(x,t)), \; t > 0, \; x \in (-L, L) \\ \frac{\partial v(x,t)}{\partial t} = \epsilon \frac{\partial^2 v(x,t)}{\partial x^2} - v(x,t)f(u(x,t)), \; t > 0, \; x \in (-L, L) \\ u(x,0) = u_0(x), \; x \in [-L, L] \\ v(x,0) = v_0(x), \; x \in [-L, L] \\ u(-L,t) = u(L,t), \; t > 0 \\ v(-L,t) = v(L,t), \; t > 0 \end{cases}$$

where

$$f(u) = \begin{cases} u^m, & u \geq 0 \\ 0, & u < 0 \end{cases}$$

v.) The forced heat equation

$$\begin{cases} \frac{\partial}{\partial t} z(x,y,t) & = k\triangle z(x,y,t) + z(x,y,t) + \alpha e^{-\frac{\beta}{z(t,x,y)}},\ 0 < x < a,\ 0 < y < b,\ t > 0, \\ z(x,y,0) & = \sin 2x + \cos 2y,\ 0 < x < a,\ 0 < y < b, \\ z(x,0,t) & = 0 = z(x,b,t),\ 0 < x < a,\ t > 0, \\ z(0,y,t) & = 0 = z(a,y,t),\ 0 < y < b,\ t > 0, \end{cases}$$

Chapter 14

A Final Wave of Models—Accounting for Semilinear Effects

Overview

As in the ODE models explored in Part I, external forces are often more complicated than just functions depending on the independent variables. In fact, they often depend on not only time and position, but also on the solution itself and its partial derivatives. More precisely, we shall consider forcing terms of the form $f(x, t, z(x, t))$. The forcing term can also depend on the partial derivatives of the unknown function (of order equal to at most one less than the order of the PDE). We shall just consider some examples in this chapter, primarily for visualization purposes.

14.1 Turning Up the Heat—Semi-Linear Variants of the Heat Equation

The heat production source can change with time and the temperature of the material being heated. For instance, it is reasonable for certain chemical reactions to exhibit an exponentially decaying heat source. The following IBVP for heat conduction through a metal sheet takes this into account :

$$\begin{cases} \frac{\partial}{\partial t} z(x,y,t) = k^2 \triangle z(x,y,t) + Ae^{-\frac{B}{|z(x,y,t)|}},\ 0 < x < a,\ 0 < y < b,\ t > 0, \\ z(x,y,0) = z_0(x,y),\ 0 < x < a,\ 0 < y < b, \\ \frac{\partial z}{\partial y}(x,0,t) = 0 = \frac{\partial z}{\partial y}(x,b,t),\ 0 < x < a,\ t > 0, \\ \frac{\partial z}{\partial x}(0,y,t) = 0 = \frac{\partial z}{\partial x}(a,y,t),\ 0 < y < b,\ t > 0. \end{cases} \tag{14.1}$$

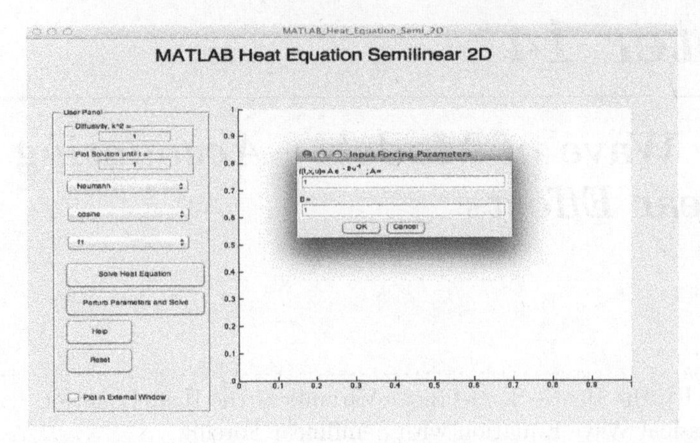

FIGURE 14.1: MATLAB Heat Equation Semilinear 2D GUI

Exercise 14.1.1. Try to reformulate (14.1) as an abstract evolution equation.

MATLAB-Exercise 14.1.1. To begin, open **MATLAB** and in the command line, type:

MATLAB_Heat_Equation_Semi_2D

Use the GUI to answer the following questions. Make certain to click on the HELP button for instructions on how to use the GUI. Screenshots of inputting parameters and GUI output can be found in Figure 14.1 and Figure 14.2.

i.) Consider IBVP (14.1) with the cosine initial position option and the forcing function option f1.

a.) Use $k^2 = 1$, plot solutions until $t = 1$ and when prompted set A in the definition of $f_1(x, t)$ equal to 1 and $B = 0.5$. Describe the evolution of the solution. How does it differ from the homogeneous heat equation?

b.) Investigate how the solution of IBVP (14.1) depends on the forcing parameters. For example, does increasing A or B cause the diffusion to speed up? slow down?

c.) Investigate the long-term behavior of solutions of IBVP (14.1). Do solutions tend to a steady-state temperature as time goes on and on? How do the model parameters play a role in the long-term behavior of solutions?

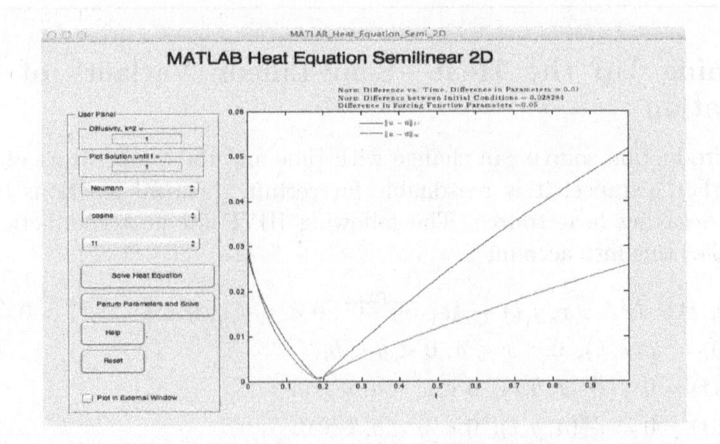

FIGURE 14.2: Investigating Continuous Dependence for IBVP (14.1)

d.) Investigate the continuous dependence of the solution on the model parameters and initial conditions for IBVP (14.1). Make sure you carefully describe the effects of using different norms.

e.) Repeat (a)-(d) using the exponential initial condition option.

f.) Repeat (a)-(d) using the step function initial condition option.

ii.) Reformulate (14.1) to account for Dirichlet BCs.

iii.) Repeat (i) using Dirichlet BCs and sine for the initial condition.

iv.) Reformulate (14.1) to account for periodic BCs.

v.) Repeat (i) using periodic BCs.

■

The exponentially decaying heat source in (14.1) is a particular choice for a forcing function that may depend on the solution itself. Generalizing IBVP (14.1) yields a class of IBVPs of the form

$$\begin{cases} \frac{\partial}{\partial t}z(x,y,t) = k^2\Delta z(x,y,t) + f(x,y,t,z), \ 0 < x < a, \ 0 < y < b, \ t > 0, \\ z(x,y,0) = z_0(x,y), \ \ 0 < x < a, \ 0 < y < b, \\ \frac{\partial z}{\partial y}(x,0,t) = 0 = \frac{\partial z}{\partial y}(x,b,t), \ \ 0 < x < a, \ t > 0, \\ \frac{\partial z}{\partial x}(0,y,t) = 0 = \frac{\partial z}{\partial x}(a,y,t), \ \ 0 < y < b, \ t > 0. \end{cases} \tag{14.2}$$

where $f(x,y,t,z)$ may depend on t,x,y and the solution z.

Exercise 14.1.2. Try to reformulate (14.2) as an abstract evolution equation.

Such generalitity in an IBVP allows for fairly complicated forcing terms in the heat equation. For example, consider the forcing function

$$f(t,x,y,z) = \begin{cases} A\left|\cos(t)\cos^3(x)\cos^5(y)\right|\sin(z), & t > 0, \, 0 < x < a, \, 0 < y < b, \, z > 0, \\ 0, & t > 0, \, 0 < x < a, \, 0 < y < b, \, z \leq 0. \end{cases} \tag{14.3}$$

where $A \in \mathbb{R}$. Although this function may appear to be quite complicated, it is continuous in each of the variables.

STOP! Explain why (14.3) is continuous in t,x,y, and z. In the non-homogeneous, continuity of the forcing function guaranteed what with regard to the variation of parameters formula?

The next exercise will illustrate solutions, if any, to (14.2) with (14.3) as the forcing function.

MATLAB-Exercise 14.1.2. To begin, open **MATLAB** and in the command line, type:

MATLAB_Heat_Equation_Semi_2D

Use the GUI to answer the following questions. A typical screenshot of the GUI can be found in Figure 14.3.

i.) Consider IBVP (14.2) with the cosine initial position option and the forcing function option f2 which MATLAB recognizes as (14.3).

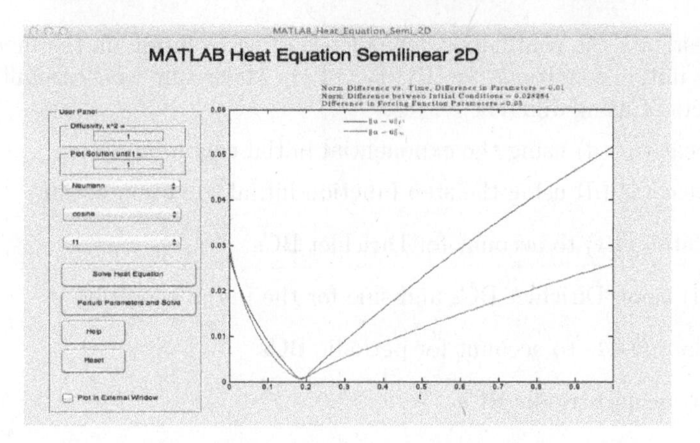

FIGURE 14.3: Investigating Continuous Dependence for IBVP (14.1)

a.) Use $k^2 = 1$, plot solutions until $t = 1$ and when prompted set A in the definition of $f_1(x,t)$ equal to 1. Describe the evolution of the solution. How does it differ from the homogeneous heat equation?

b.) Investigate how the solution depends on the forcing parameters. For example, does increasing A cause the diffusion to speed up? Slow down?

c.) Investigate the long-term behavior of solutions. How do the model parameters play a role?

d.) Investigate the continuous dependence of the solution on the model parameters and initial conditions for IBVP (14.2). Carefully articulate the results of using different norms.

e.) Repeat (a)-(d) using the exponential initial condition option.

f.) Repeat (a)-(d) using the step function initial condition option.

ii.) Reformulate (14.2) to account for Dirichlet BCs.

iii.) Repeat (i) using Dirichlet BCs and sine for the initial condition.

iv.) Reformulate (14.2) to account for periodic BCs.

v.) Repeat (i) using periodic BCs.

vi.) Investigate how this semilinear heat equation differs from the homogeneous version by running the GUI for $A = 1$, then clicking *Perturb Parameters and Solve* button. When prompted, choose the new function parameter to equal $A = 0$.

■

14.2 The Classical Wave Equation with Semilinear Forcing

In Section 11.3, you encountered the classical wave equation and its variants for which the forcing term had the form $f(t,x)$ or in two dimensions $f(t,x,y)$. Here we allow the forcing function to depend on the solution as well. For the MATLAB exercises to follow regarding the wave equation, you shall use each of the following forcing terms:

$$f_1(t,x,y,z) = Az(x,y,t), \text{ where } A \in \mathbb{R}\setminus\{0\} \tag{14.4}$$
$$f_2(t,x,y,z) = A(x+y) + Bz(x,y,t)^2, \text{ where } A, B \in \mathbb{R}\setminus\{0\} \tag{14.5}$$
$$f_3(t,x,y,z) = \cos\left(\omega z(x,y,t)t\right), \text{ where } \omega > 0 \tag{14.6}$$
$$f_4(t,x,y,z) = e^{-a|z(x,y,t)|}, \text{ where } a > 0 \tag{14.7}$$

The natural extension of (11.52) to account for position dependent forcing is

$$\begin{cases} \frac{\partial z}{\partial t^2}(x,y,t) = c^2\left(\frac{\partial^2 z}{\partial x^2}(x,y,t) + \frac{\partial^2 z}{\partial y^2}(x,y,t)\right) + f(t,x,y,z),\ 0 < x < L,\ 0 < y < L,\ t > 0, \\ z(x,y,0) = z_0(x,y),\ 0 < x < L,\ 0 < y < L, \\ z(x,0,t) = 0 = z(x,b,t),\ 0 < x < L,\ t > 0, \\ z(0,y,t) = 0 = z(a,y,t),\ 0 < y < L,\ t > 0 \end{cases} \tag{14.8}$$

where L is a positive real number.

Exercise 14.2.1. Try to reformulate IBVP (14.8) when the forcing term equals (14.4). Then again using (14.5). Which case cannot be represented in the form we have already discussed?

MATLAB-Exercise 14.2.1. To begin, open **MATLAB** and in the command line, type:

MATLAB_Wave_Equation_Semi_2D

Make certain to click on the HELP button for instructions on how to use the GUI. Screenshots of inputting parameters and GUI output can be found in Figures 14.4–14.7. Use the

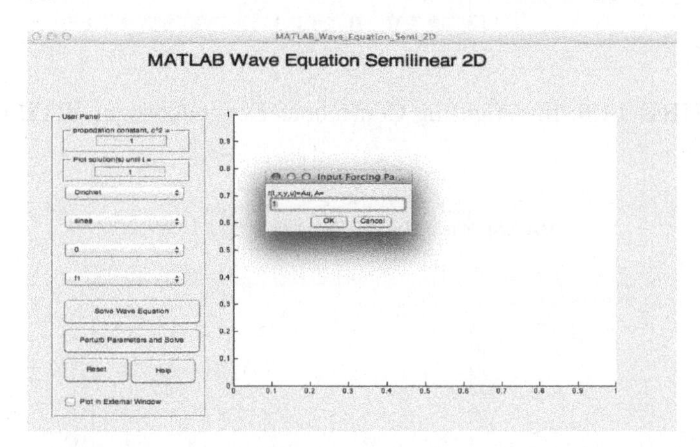

FIGURE 14.4: MATLAB Wave Equation Semilinear 2D GUI

GUI to answer the following questions.

i.) Consider IBVP (14.8) with the sine initial position option, the zero initial velocity option, and the forcing function option f1.

   a.) Use $c^2 = 1$, plot solutions until $t = 1$ and when prompted set A in the definition of $f_1(x,t)$ equal to 1. Describe the evolution of the solution. How does it differ from the homogeneous wave equation?

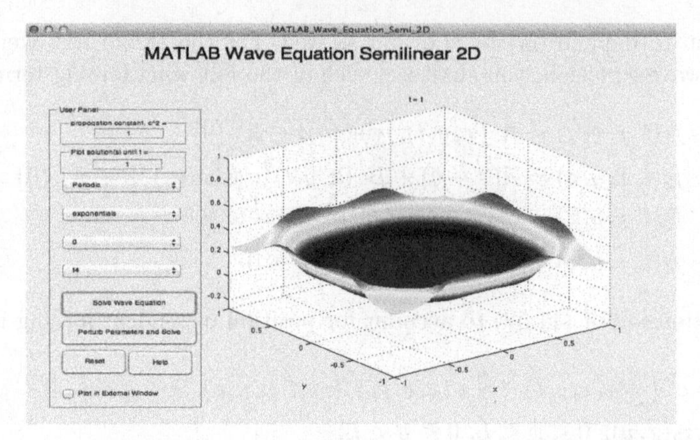

FIGURE 14.5: Investigating Effect of Parameters in IBVP (14.8)

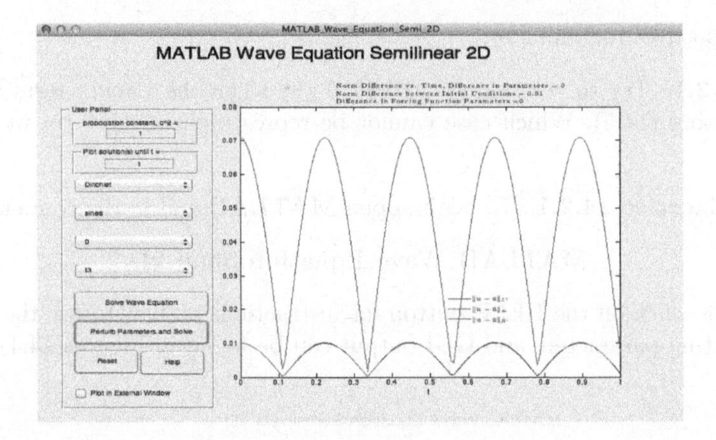

FIGURE 14.6: Investigating Continuous Dependence for IBVP (14.8)

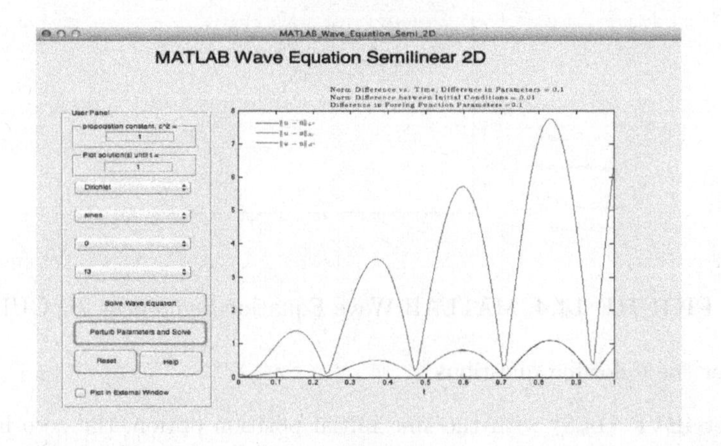

FIGURE 14.7: Investigating Continuous Dependence for IBVP (14.8)

b.) Investigate how the solution depends on the forcing parameters. For example, does increasing A cause wave propagation to speed up? Slow down?

c.) Investigate the long-term behavior of solutions. How do the model parameters play a role?

d.) Investigate the continuous dependence of the solution on the model parameters and initial conditions for IBVP (14.8). Carefully articulate the results of using different norms.

e.) Repeat (a)-(d) using the exponential initial position option.

f.) Repeat (a)-(d) using the step function initial position option.

ii.) Reformulate (14.8) to account for Neumann BCs.

iii.) Consider IBVP (14.8) equipped with Neumann BCs, the cosine initial position option, the zero initial velocity option, and the forcing function option f2.

a.) Use $k^2 = 1$, plot solutions until $t = 1$ and when prompted set A in the definition of $f_2(t, x, y, z)$ equal to 1 and $B = 1$. Describe the evolution of the solution.

b.) Based on the numerical results produced in MATLAB, make a conjecture regarding the existence of a solution for IBVP (14.8) when the forcing term equals (14.5).

i.) Consider IBVP (14.8) equipped with Neumann BCs, the cosine initial position option, the zero initial velocity option, and the forcing function option f3.

a.) Use $c^2 = 1$, plot solutions until $t = 1$ and when prompted set ω in the definition of $f_3(t, x, y, z)$ equal to 5. Describe the evolution of the solution. How does it differ from the homogeneous heat equation?

b.) Investigate how the solution depends on the forcing parameters. For example, does increasing ω cause wave propagation to speed up? Slow down?

c.) Investigate the long-term behavior of solutions. How do the model parameters play a role?

d.) Investigate the continuous dependence of the solution on the model parameters and initial conditions for IBVP (14.8) equipped with Neumann BCs. Carefully articulate the results of using different norms.

e.) Repeat (a)-(d) using the exponential initial position option.

f.) Repeat (a)-(d) using the step function initial position option.

iv.) Reformulate (14.8) to account for periodic BCs.

v.) Repeat (iii) using periodic BCs and the forcing function option f4.

vi.) Figure 14.6 illustrates the situation when only the initial position and velocity were perturbed whereas in Figure 14.7 all model parameters including the initial conditions were perturbed. Explain how the figures suggest the solutions for IBVP (14.8) do not depend continuously on on the model parameters.

■

14.3 Population Growth—Fisher's Equation

Consider the nondimensionalized Fisher's equation

$$\begin{cases} \frac{\partial u(x,t)}{\partial t} = \frac{\partial^2 u(x,t)}{\partial x^2} + u(x,t)(1-u(x,t)),\ t>0,\ x\in(-L,L) \\ u(0,x) = u_0(x),\ x\in[-L,L] \\ u(t,-L) = u(t,L),\ t>0 \end{cases} \tag{14.9}$$

For the **EXPLORE!** projects to follow, you shall use each of the following initial conditions:

$$u_0^1(x) = 0,\ x\in[-L,L] \tag{14.10}$$

$$u_0^2(x) = \frac{1}{2\cosh(\delta x)},\ x\in[-L,L] \tag{14.11}$$

$$u_0^3(x) = e^{-\delta x^2},\ x\in[-L,L] \tag{14.12}$$

where δ is a nonnegative real parameter that will be used to perturb the initial condition. In the case of the two-dimensional model, (14.11) and (14.12) are replaced by

$$u_0^2(x,y) = \frac{1}{2\cosh(\delta(x^2+y^2))},\ x\in[-L,L],\ y\in[-L,L]$$

$$u_0^3(x,y) = e^{-\delta(x^2+y^2)},\ x\in[-L,L],\ y\in[-L,L]$$

EXPLORE!

14.1 Parameter Play!

In this **EXPLORE!** you will investigate the effects of changing the initial condition along with model parameter δ used in defining the initial condition. To do this, you will use the MATLAB Fisher's Equation GUI to visualize the solutions to IBVP (14.9) for different choices of δ and $u_0(x)$.

MATLAB-Exercise 14.3.1. To begin, open **MATLAB** and in the command line, type:

MATLAB_Fishers

Use the GUI to answer the following questions. Make certain to click on the HELP button for instructions on how to use the GUI. Screenshots of GUI output can be found in Figures 14.8–14.11.

i.) Consider IBVP (14.9) with the zero initial condition option.

  a.) How would describe the physical situation in (i)? How do you expect the solution to behave under these conditions?

  b.) Use the GUI to visualize the solution, and comment on any differences between the solution presented in MATLAB and the situation you described in (a).

ii.) Consider IBVP (14.9) with the exponential initial condition option.

  a.) Use $\delta = 0.5$ and plot solutions until $t = 2$. Describe the evolution of the solution. How did it differ from the solution of the heat equation? (Recall, Fisher's equation contains a diffusion term just like the one in the heat equation.)

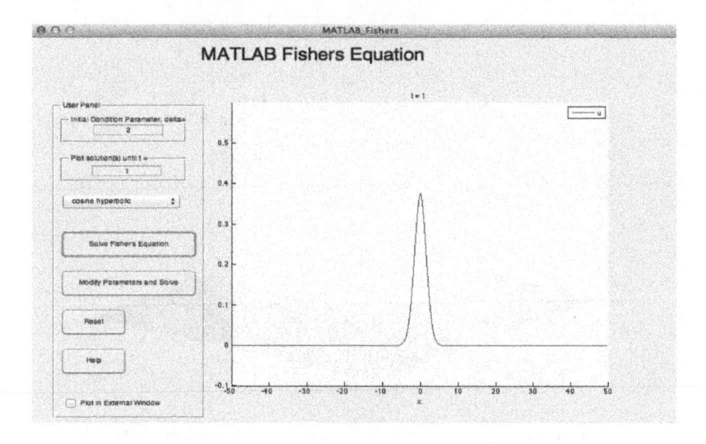

FIGURE 14.8: MATLAB Fisher's Equation GUI

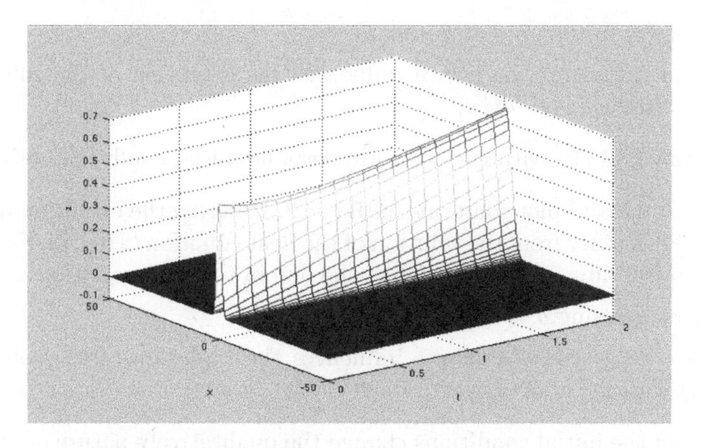

FIGURE 14.9: IBVP (14.9), initial profile: $e^{-0.5x^2}$.

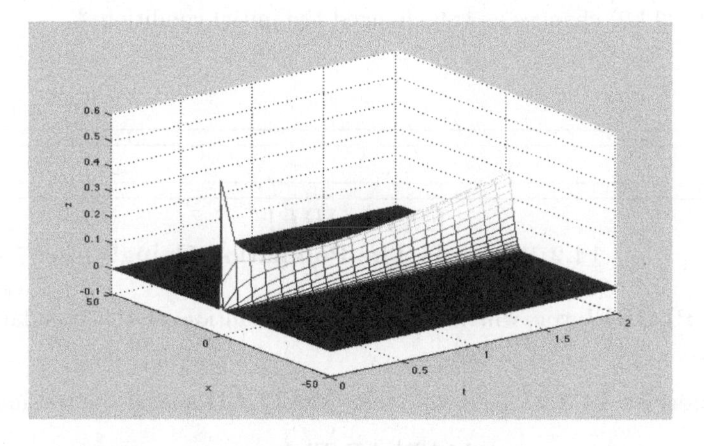

FIGURE 14.10: IBVP (14.9), initial profile: e^{-10x^2}.

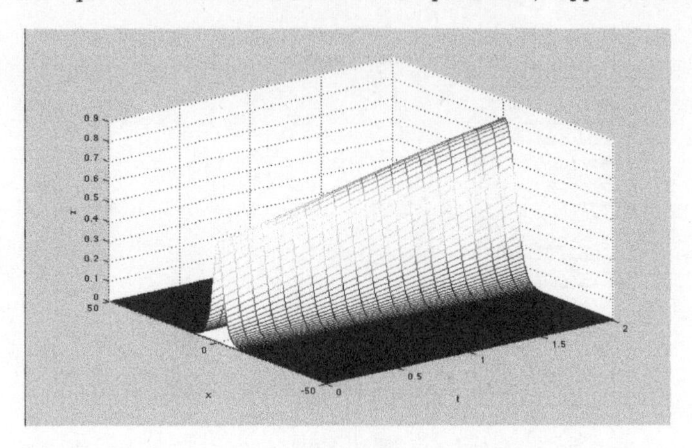

FIGURE 14.11: IBVP (14.9), initial profile: $\frac{1}{2\cosh(0.5x)}$.

b.) Repeat (a) using $\delta = 1, 2, 5$ and 10.

c.) Make a conjecture for how δ influences the evolution of the solution to IBVP (14.9) with $u_0(x) = u_0^2(x)$.

iii.) Consider IBVP (14.9) with the cosine hyperbolic initial condition option.

a.) Use $\delta = 0.5$ and plot solutions until $t = 2$. Describe the evolution of the solution. How did it differ from the solution of the heat equation? (Recall, Fisher's equation contains a diffusion term just like the one in the heat equation.)

b.) Repeat (a) using $\delta = 1, 2, 5$ and 10.

c.) Make a conjecture for how δ influences the evolution of the solution to IBVP (14.9) with $u_0(x) = u_0^3(x)$.

iv.) Did changing the initial conditions change the qualitatively nature of the evolution of the solution to (14.9)? Explain.

v.) Based on your intuition gained in this **EXPLORE!**, How did the evolution of the solution to (14.9) changes as you changed the initial conditions?

■

EXPLORE!

14.2 Where Is The Population Going!

In this **EXPLORE!** you will investigate the evolution of the population over long periods of time.

MATLAB-Exercise 14.3.2. To begin, open **MATLAB** and in the command line, type:

MATLAB_Fishers

Use the GUI to answer the following questions. Make certain to click on the HELP button for instructions on how to use the GUI. Screenshots of GUI output can be found in Figures 14.12, 14.13, and 14.14.

i.) Consider IBVP (14.9) with the zero initial condition option.

a.) Based on the physical situation and your findings in **EXPLORE!** 14.1, where is the population going in the long-run?

b.) Use the GUI to visualize the solution, and comment on any differences between the solution presented in MATLAB and the situation you described in (a).

ii.) Consider IBVP (14.9) with the exponential initial condition option.

a.) Investigate the behavior of the solution for large times. Use $\delta = 1$. Begin by plotting solutions to $t = 10$. If necessary, slowly increase the plotting time until you are able to describe the behavior of the solution. Remember, a computer cannot handle infinity!

b.) Repeat (a) using $\delta = 2, 5$, and 10.

c.) Make a conjecture for the long term behavior of the solution to (14.9) with $u_0(x) = u_0^2(x)$.

d.) Compare your conjecture in (c) to the long-term behavior of the heat equation.

iii.) Consider IBVP (14.9) with the cosine hyperbolic initial condition option.

a.) Investigate the behavior of the solution for large times. Use $\delta = 1$. Begin by plotting solutions to $t = 10$. If necessary, slowly increase the plotting time until you are able to describe the behavior of the solution. Remember, a computer cannot handle infinity!

b.) Repeat (a) using $\delta = 2, 5$, and 10.

c.) Make a conjecture for the long term behavior of the solution to (14.9) with $u_0(x) = u_0^3(x)$.

d.) Compare your conjecture in (c) to the long-term behavior of the heat equation.

■

EXPLORE!

14.3 Continuous Dependence on Parameters

In this **EXPLORE!** we investigate the continuous dependence of the solution to (14.9) on the initial condition. Here we perturb the initial condition by perturbing the parameter δ in $u_0^2(x)$ and $u_0^1(x)$.

MATLAB-Exercise 14.3.3. Open **MATLAB** and in the command line, type:

MATLAB_Fishers

To explore continuous dependence using the GUI we must first solve Fisher's equation with a particular initial condition and δ. Once a solution has been constructed, click the *Modify Parameters and Solve* button to specify a modified value of δ. The GUI will automatically output norm differences in the solutions. Use the GUI to answer the following questions. Make sure to click the Help button to find more information about the GUI. Screenshots of inputting parameters and GUI output can be found in Figures 14.15, 14.16, and 14.17.

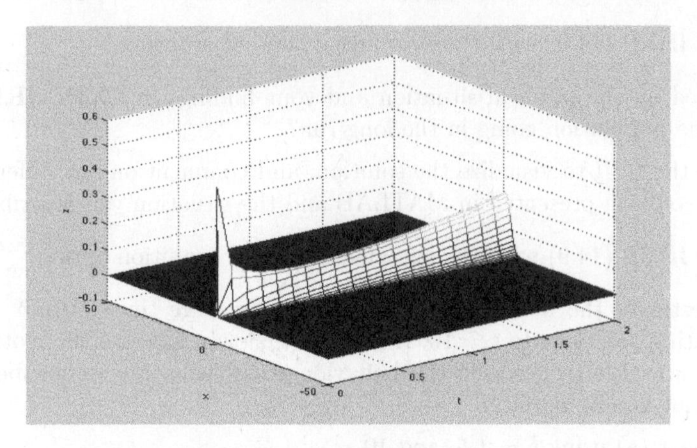

FIGURE 14.12: IBVP (14.9), initial profile: $\frac{1}{2\cosh(10x)}$.

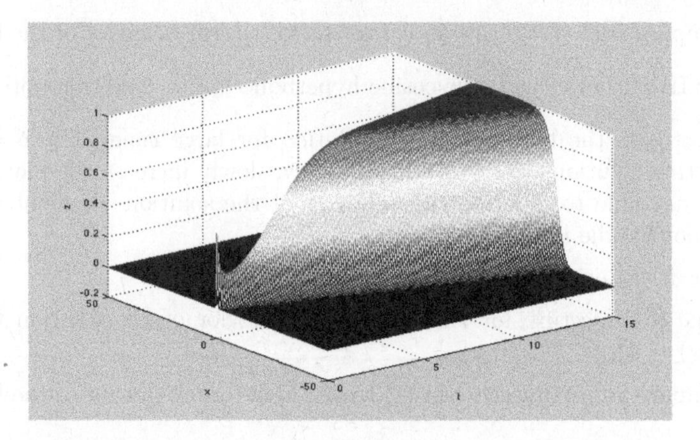

FIGURE 14.13: IBVP (14.9), initial profile: e^{-5x^2}.

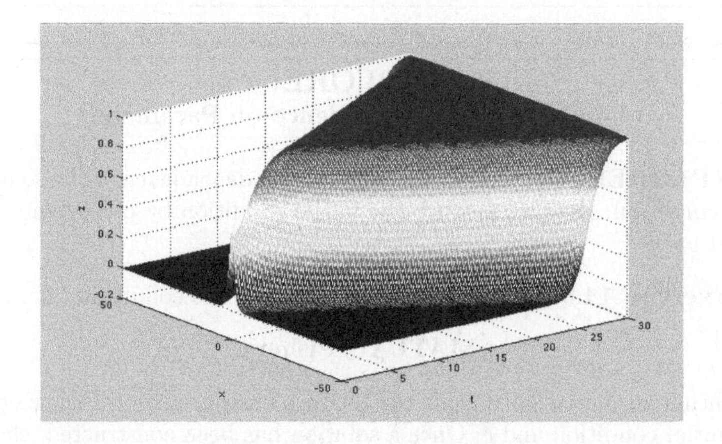

FIGURE 14.14: IBVP (14.9), initial profile: $\frac{1}{2\cosh(x)}$.

i.) Consider IBVP (14.9). Explore whether or not the solution of this IBVP depends continuously on the exponential initial condition option.

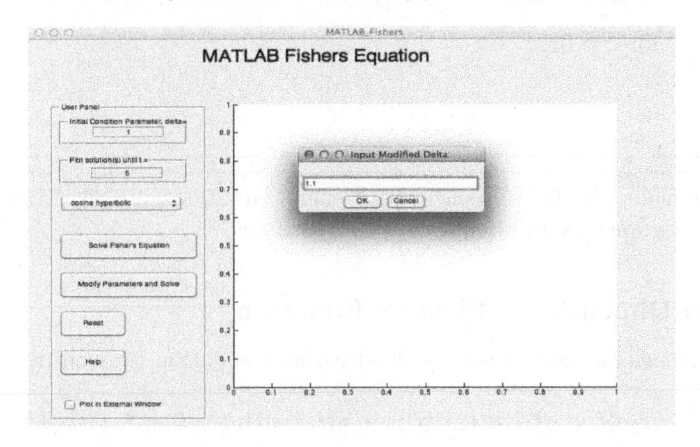

FIGURE 14.15: Perturbing δ in MATLAB Fisher's Equation GUI

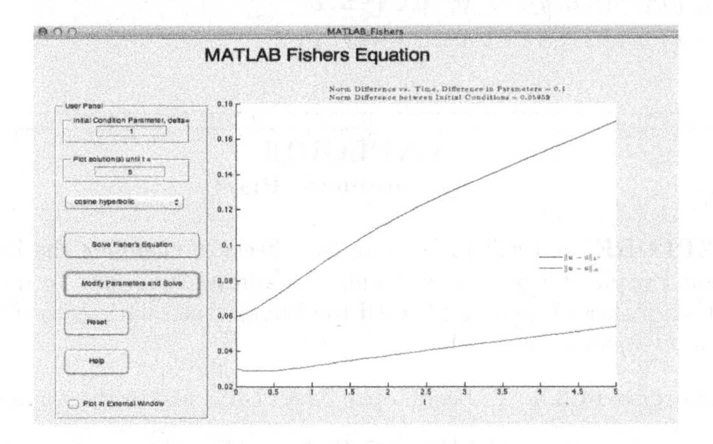

FIGURE 14.16: Perturbing the initial condition in (14.9)

a.) Use $\delta = 1$ and plot solutions until $t = 3$. Choose a modified δ to equal 1.1. How would you describe the differences in the solutions?

b.) Repeat (a) with modified δ values equal to $0.9, 0.95, 1.05$ and 1.01.

c.) Make a conjecture regarding IBVP (14.9) continuous dependence on the initial condition for $\delta = 1$.

ii.) Repeat (i) for $\delta = 5$.

iii.) Repeat (i) for $\delta = 0.5$.

iv.) Consider IBVP (14.9). Explore whether or not the solution of this IBVP depends continuously on the cosine hyperbolic initial condition option.

a.) Use $\delta = 1$ and plot solutions until $t = 3$. Choose a modified δ to equal 1.1. How would you describe the differences in the solutions?

b.) Repeat (a) with modified δ values equal to $0.9, 0.95, 1.05$ and 1.01.

c.) Make a conjecture regarding IBVP (14.9) continuous dependence on the initial condition for $\delta = 1$.

v.) Repeat (iv) for $\delta = 5$.

vi.) Repeat (iv) for $\delta = 0.5$.

■

Next, we consider the two-dimensional Fisher's equation and investigate its solution's dependence on parameters and behavior for large times.

14.3.1 Two-Dimensional Fisher's Equation

The two-dimensional nondimensionalized Fisher's equation is given by

$$\begin{cases} \frac{\partial u(t,x,y)}{\partial t} = \Delta u(t,x,y) + u(t,x,y)(1 - u(t,x,y)),\ t > 0,\ x \in (-L, L),\ y \in (-L, L) \\ u(0,x,y) = u_0(x,y),\ x \in [-L, L],\ y \in [-L, L] \\ u(t,-L,y) = u(t,L,y),\ t > 0,,\ y \in [-L, L] \\ u(t,x,-L) = u(t,x,L),\ t > 0,,\ x \in [-L, L] \end{cases} \cdot \tag{14.13}$$

EXPLORE!

14.4 Parameter Play!

In this **EXPLORE!** you will investigate the effects of changing the initial condition along with model parameter δ used in defining the initial condtion. To do this, you will use the MATLAB Fisher's Equation 2D GUI to visualize the solutions to IBVP (14.13) for different choices of δ and $u_0(x)$.

MATLAB-Exercise 14.3.4. To begin, open **MATLAB** and in the command line, type:

MATLAB_Fishers_2D

Use the GUI to answer the following questions. Make certain to click on the HELP button for instructions on how to use the GUI. Screenshots of GUI output can be found in Figures 14.18–14.22.

i.) Consider IBVP (14.13) with the zero initial condition option.

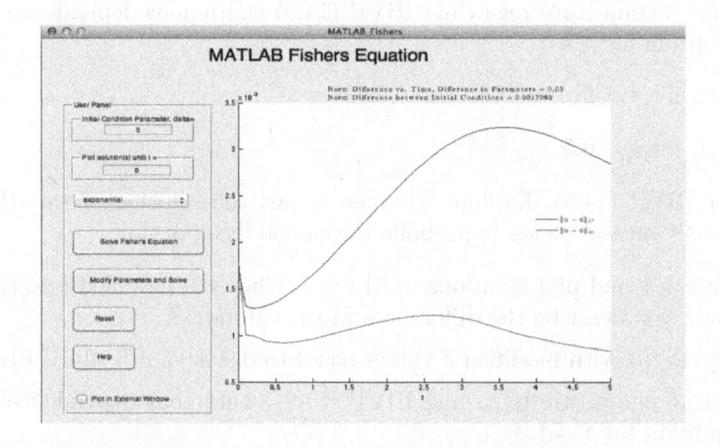

FIGURE 14.17: Perturbing the initial condition in (14.9)

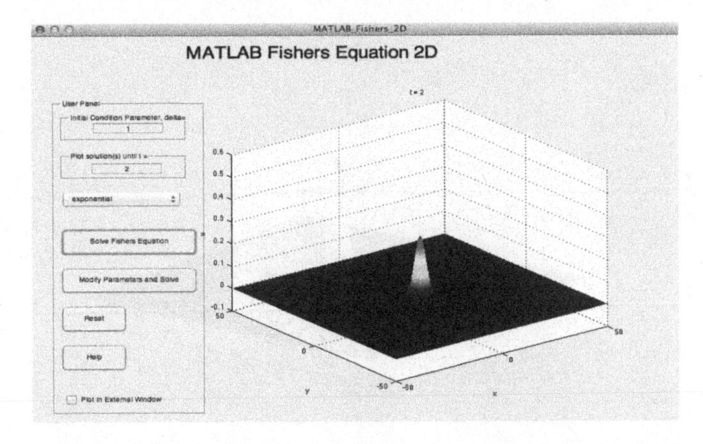

FIGURE 14.18: MATLAB Fisher's Equation 2D GUI

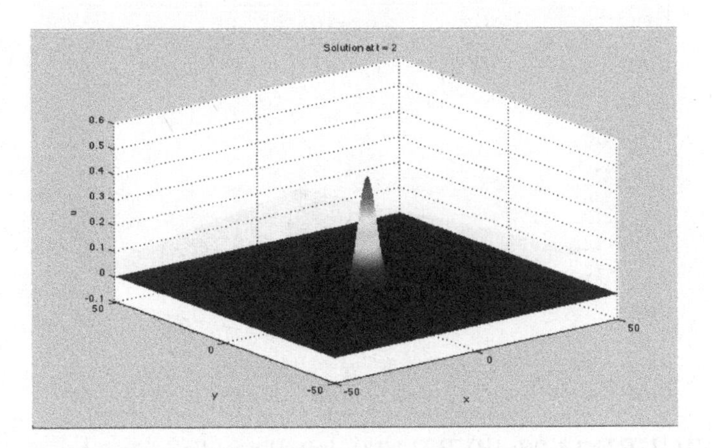

FIGURE 14.19: IBVP (14.13), initial profile: $e^{-0.5(x^2+y^2)}$.

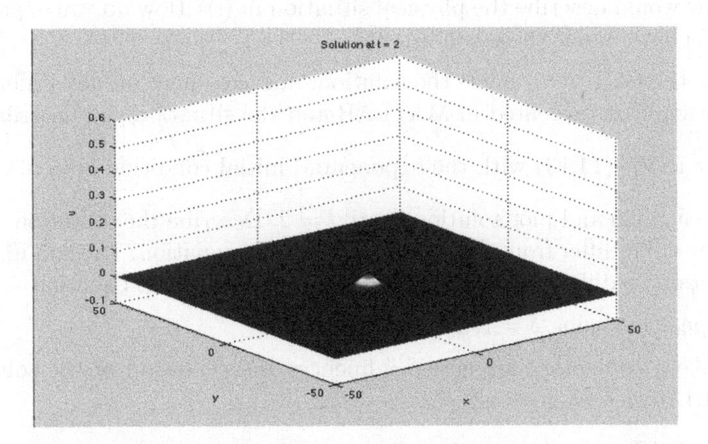

FIGURE 14.20: IBVP (14.13), initial profile: $e^{-10(x^2+y^2)}$.

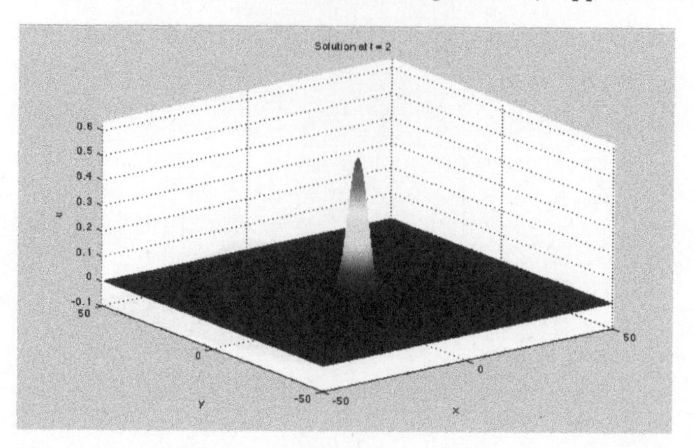

FIGURE 14.21: IBVP (14.13), initial profile: $\frac{1}{2\cosh\left(0.5(x^2+y^2)\right)}$.

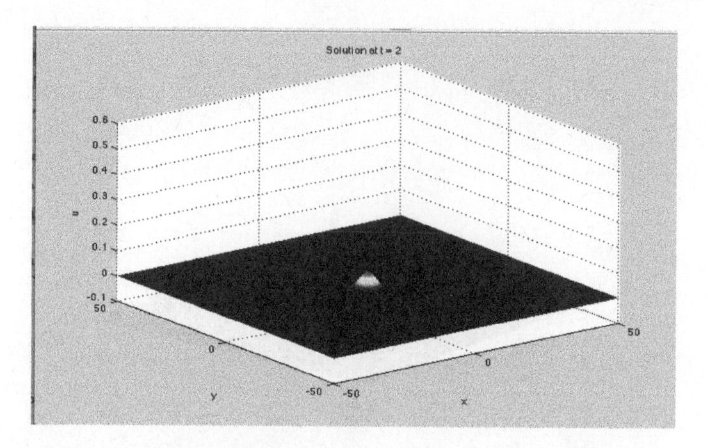

FIGURE 14.22: IBVP (14.13), initial profile: $\frac{1}{2\cosh\left(10(x^2+y^2)\right)}$.

a.) How would describe the physical situation in (i)? How do you expect the solution to behave under these conditions?

b.) Use the GUI to visualize the solution, and comment on any differences between the solution presented in MATLAB and the situation you described in (a).

ii.) Consider IBVP (14.13) with the exponential initial condition option.

a.) Use $\delta = 0.5$ and plot solutions until $t = 2$. Describe the evolution of the solution. How did it differ from the solution of the heat equation? (Recall, Fisher's equation contains a diffusion term just like the one in the heat equation.)

b.) Repeat (a) using $\delta = 1, 2, 5$ and 10.

c.) Make a conjecture for how δ influences the evolution of the solution to IBVP (14.13) with $u_0(x) = u_0^2(x)$.

iii.) Consider IBVP (14.13) with the cosine hyperbolic initial condition option.

a.) Use $\delta = 0.5$ and plot solutions until $t = 2$. Describe the evolution of the solution. How did it differ from the solution of the heat equation? (Recall, Fisher's equation contains a diffusion term just like the one in the heat equation.)

b.) Repeat (a) using $\delta = 1, 2, 5$ and 10.

c.) Make a conjecture for how δ influences the evolution of the solution to IBVP (14.13) with $u_0(x) = u_0^3(x)$.

iv.) How did the evolution of the solution to (14.13) change as you chnaged the initial conditions?.

v.) Based on your intuition gained in this **EXPLORE!**, how do you expect the population to evolve over long periods of time?

■

EXPLORE!

14.5 Where Is The Population Going!

In this **EXPLORE!** we will investigate the evolution of the population over long periods of time.

MATLAB-Exercise 14.3.5. To begin, open **MATLAB** and in the command line, type:

MATLAB_Fishers_2D

Use the GUI to answer the following questions. Make certain to click on the HELP button for instructions on how to use the GUI. Screenshots of GUI output can be found in Figure 14.23 and Figure 14.24.

i.) Consider IBVP (14.13) with the zero initial condition option.

a.) Based on the physical situation and your findings in **EXPLORE!** 14.4, where is the population going in the long-run?

b.) Use the GUI to visualize the solution, and comment on any differences between the solution presented in MATLAB and the situation you described in (a).

ii.) Consider IBVP (14.13) with the exponential initial condition option.

a.) Investigate the behavior of the solution for large times. Use $\delta = 1$. Begin by plotting solutions to $t = 10$. If necessary, slowly increase the plotting time until you are able to describe the behavior of the solution. Remember, a computer cannot handle infinity!

b.) Repeat (a) using $\delta = 2, 5$, and 10.

c.) Make a conjecture for the long–term behavior of the solution to (14.13) with $u_0(x) = u_0^2(x)$.

d.) Compare your conjecture in (c) to the long-term behavior of the heat equation.

iii.) Consider IBVP (14.13) with the cosine hyperbolic initial condition option.

a.) Investigate the behavior of the solution for large times. Use $\delta = 1$. Begin by plotting solutions to $t = 10$. If necessary, slowly increase the plotting time until you are able to describe the behavior of the solution. Remember, a computer cannot handle infinity!

b.) Repeat (a) using $\delta = 2, 5$, and 10.

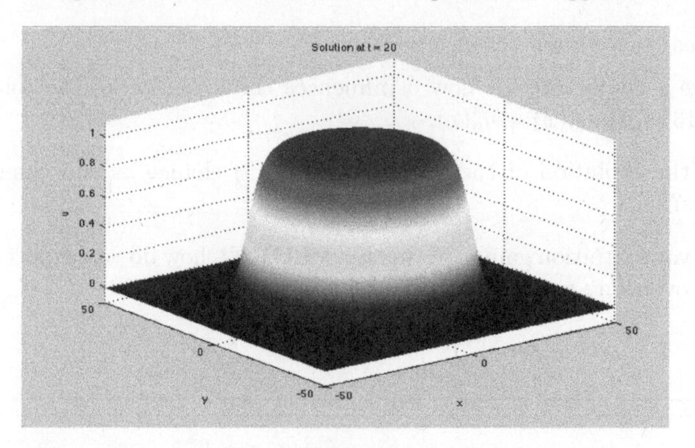

FIGURE 14.23: IBVP (14.13), initial profile: $e^{-5(x^2+y^2)}$.

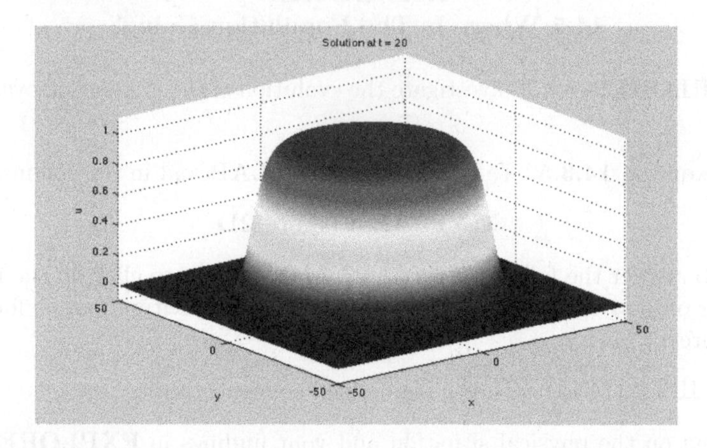

FIGURE 14.24: IBVP (14.13), initial profile: $\frac{1}{2\cosh((x^2+y^2))}$.

c.) Make a conjecture for the long–term behavior of the solution to (14.13) with $u_0(x) = u_0^3(x)$.

d.) Compare your conjecture in (c) to the long-term behavior of the heat equation.

■

EXPLORE!

14.6 Continuous Dependence on Parameters

In this **EXPLORE!** we investigate the continuous dependence of the solution to (14.13) on the initial condition. Here we perturb the initial condition by perturbing the parameter δ in $u_0^2(x)$ and $u_0^1(x)$.

MATLAB-Exercise 14.3.6. Open **MATLAB** and in the command line, type:

MATLAB_Fishers_2D

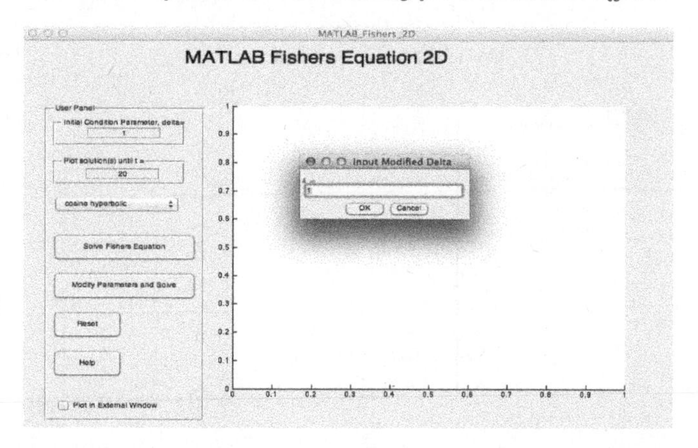

FIGURE 14.25: Perturbing δ in MATLAB Fisher's Equation GUI

To explore continuous dependence using the GUI we must first solve Fisher's equation with a particular initial condition and δ. Once a solution has been constructed, click the *Modify Parameters and Solve* button to specify a modified value of δ. The GUI will automatically output norm differences in the solutions. Use the GUI to answer the following questions. Make certain to click on the HELP button for more information about the GUI. Screenshots on inputting parameters and GUI output can be found in Figures 14.25, 14.26, and 14.27.

i.) Consider IBVP (14.13). Explore whether or not the solution of this IBVP depends continuously on the exponential initial condition option.

  a.) Use $\delta = 1$ and plot solutions until $t = 3$. Choose a modified δ to equal 1.1. How would you describe the differences in the solutions?

  b.) Repeat (a) with modified δ values equal to $0.9, 0.95, 1.05$ and 1.01.

  c.) Make a conjecture regarding IBVP (14.13) continuous dependence on the initial condition for $\delta = 1$.

ii.) Repeat (i) for $\delta = 5$.

iii.) Repeat (i) for $\delta = 0.5$.

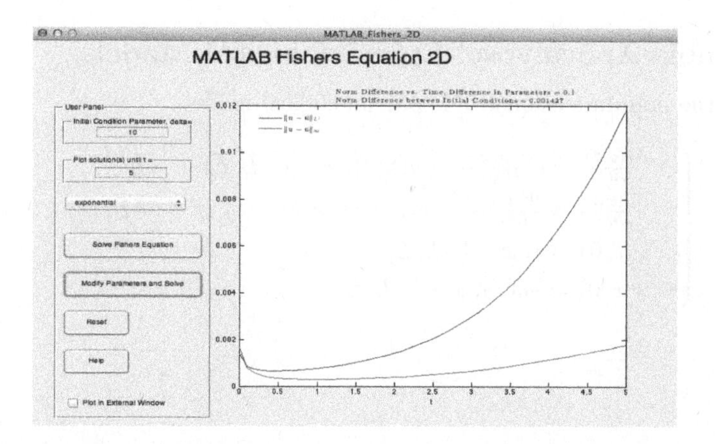

FIGURE 14.26: Perturbing the initial condition in (14.13)

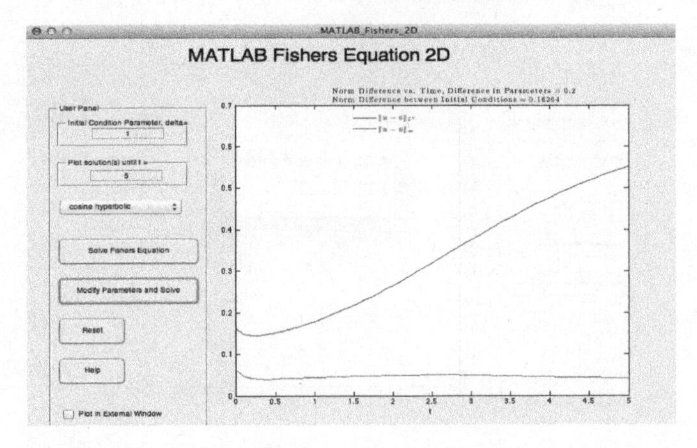

FIGURE 14.27: Perturbing the initial condition in (14.13)

iv.) Consider IBVP (14.13). Explore whether or not the solution of this IBVP depends continuously on the cosine hyperbolic initial condition option.

  a.) Use $\delta = 1$ and plot solutions until $t = 3$. Choose a modified δ to equal 1.1. How would you describe the differences in the solutions?

  b.) Repeat (a) with modified δ values equal to $0.9, 0.95, 1.05$ and 1.01.

  c.) Make a conjecture regarding IBVP (14.13) continuous dependence on the initial condition for $\delta = 1$.

v.) Repeat (iv) for $\delta = 5$.

vi.) Repeat (iv) for $\delta = 0.5$.

■

14.4 Zombie Apocalypse! Epidemiological Models

Consider the non-dimensionalized epidemiological model

$$\begin{cases} \frac{\partial v(x,t)}{\partial t} = \epsilon \frac{\partial^2 v(x,t)}{(\partial x)^2} - u(x,t),\ x \in (-L,L),\ t > 0 \\ \frac{\partial u(x,t)}{\partial t} = \frac{\partial^2 u(x,t)}{(\partial x)^2} + u(x,t)(v(x,t) - \lambda),\ x \in (-L,L),\ t > 0 \\ v(x,0) = 1,\ x \in [-L,L] \\ u(x,0) = u_0(x),\ x \in [-L,L] \end{cases} \tag{14.14}$$

where ϵ and λ are the diffusivity and removal rate model parameters. For the **EXPLORE!** activities to follow, we shall use each of the following initial conditions for u,

$$u_0^1(x) = 0, x \in [-L, L] \tag{14.15}$$

$$u_0^2(x) = e^{-0.1x^2}, x \in [-L, L] \tag{14.16}$$

$$u_0^3(x) = \begin{cases} 1, & x \in [-\frac{L}{10}, \frac{L}{10}] \\ 0, & \text{otherwise} \end{cases} \tag{14.17}$$

EXPLORE!
14.7 Parameter Play!

In this **EXPLORE!** you will investigate the effects of changing the initial condition along with model parameters. To do this, you will use the MATLAB Epidemiology GUI to visualize the solutions of IBVP (14.14) for different choices of $u_0(x)$, ϵ, and λ.

MATLAB-Exercise 14.4.1. To begin, open **MATLAB** and in the command line, type:

MATLAB_Epidemiology

Use the GUI to answer the following questions. Make certain to click on the HELP button for instructions on how to use the GUI. Screenshots of GUI output can be found in Figures 14.28, 14.29, and 14.30.

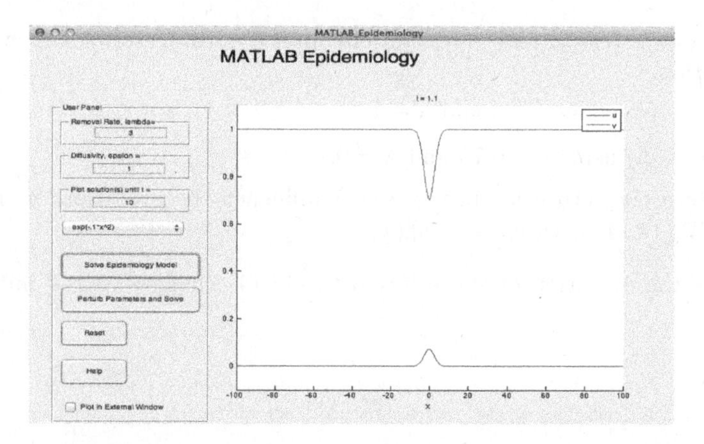

FIGURE 14.28: MATLAB Epidemiology GUI

i.) Consider IBVP (14.14) with the zero initial condition option.

a.) How would you describe the physical situation in (i)? How do you expect the solution to behave under these conditions? Will the model parameters play any role in the evolution of the solution?

b.) Use the GUI to visualize the solution, and comment on any differences between the solution presented in MATLAB and the situation you described in (a).

ii.) Consider IBVP (14.14) with initial condition (14.16).

a.) Use $\epsilon = 1$, $\lambda = 3$, and plot solutions until $t = 10$. Describe the evolution of the solution.

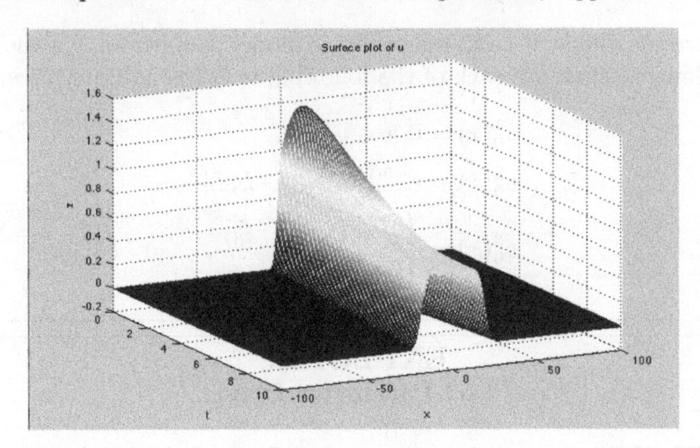

FIGURE 14.29: Surface plot of *Infectives* in IBVP (14.14) with initial condition (14.16).

b.) Repeat (a) using $\epsilon = 3$ and $\lambda = 1$.

c.) Repeat (a) using $\epsilon = 1.5$ and $\lambda = 0.1$.

d.) Make a conjecture for how ϵ and λ influence the evolution of the solution to IBVP (14.14) with $u_0(x) = u_0^2(x)$.

iii.) Consider IBVP (14.14) with initial condition (14.17).

a.) Use $\epsilon = 1$, $\lambda = 3$, and plot solutions until $t = 10$. Describe the evolution of the solution.

b.) Repeat (a) using $\epsilon = 3$ and $\lambda = 1$.

c.) Repeat (a) using $\epsilon = 0.75$ and $\lambda = 0.1$.

d.) Make a conjecture for how ϵ and λ influence the evolution of the solution to IBVP (14.14) with $u_0(x) = u_0^2(x)$.

iv.) How does the evolution of the solution to (14.14) changes as the initial conditions change?

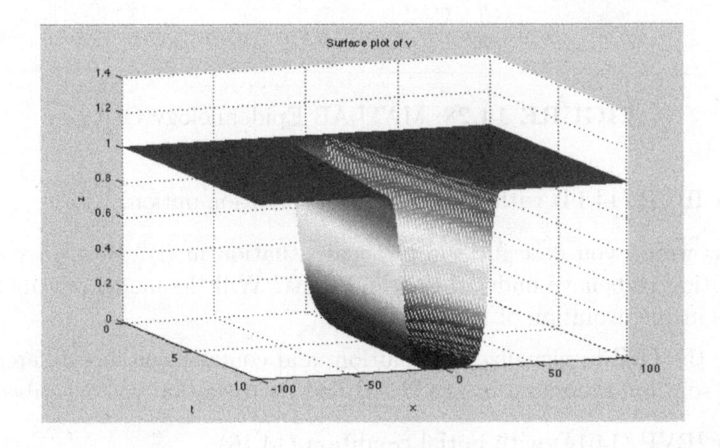

FIGURE 14.30: Surface plot of *Susceptibles* in IBVP (14.14) with initial condition (14.17).

v.) Based on your intuition gained in this **EXPLORE!**, how do you expect the populations to evolve over long periods of time? Do you believe the model allows for an apocalypse for which the susceptible population becomes extinct? Explain.

■

EXPLORE!
14.8 Where Is The Virus Going?

In this **EXPLORE!** you will investigate the evolution of the Susceptibles and Infectives over long periods of time.

MATLAB-Exercise 14.4.2. To begin, open **MATLAB** and in the command line, type:

MATLAB_Epidemiology

Use the GUI to answer the following questions. Make certain to click on the HELP button for instructions on how to use the GUI. Examples of GUI output can be found in Figures 14.31, 14.32, and 14.33.

i.) Consider IBVP (14.14) with the zero initial condition option.

a.) Based on the physical situation and your findings in **EXPLORE!** 14.7, what will happen to the two groups in the long-run?

b.) Use the GUI to visualize the solution, and comment on any differences between the solution presented in MATLAB and the situation you described in (a).

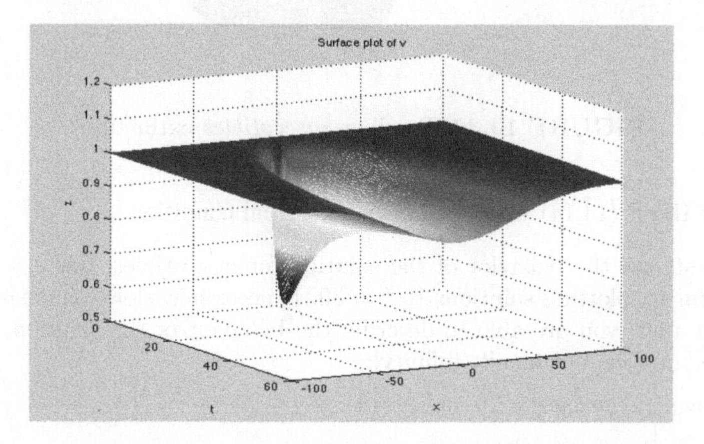

FIGURE 14.31: Surface plot of *Susceptibles* in IBVP (14.14) with (14.16).

ii.) Consider IBVP (14.14) with (14.16) as the initial condition.

a.) Investigate the behavior of the solution for large times. Use $\epsilon = 3$ and $\lambda = 5$. Begin by plotting solutions to $t = 30$. If necessary, slowly increase the plotting time until you are able to describe the behavior of the solution. Remember, a computer cannot handle infinity!

b.) Repeat (a) using $\epsilon = 10$ and $\lambda = 1$.

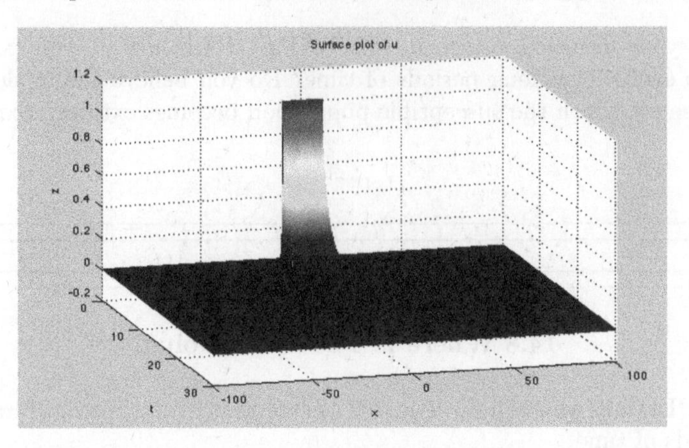

FIGURE 14.32: Surface plot of *Infectives* in IBVP (14.14) with (14.16).

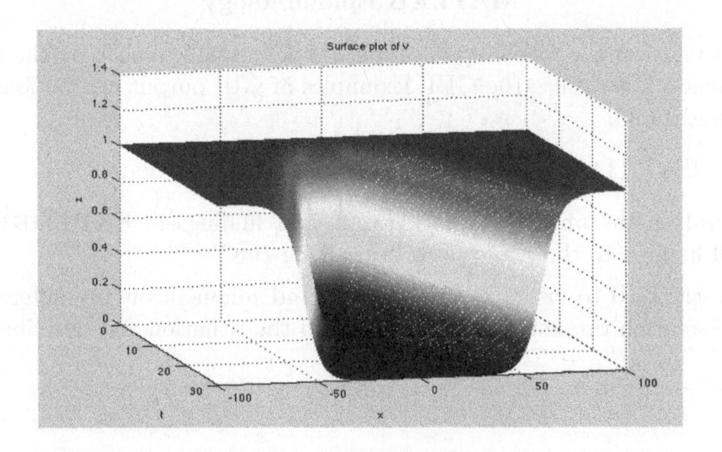

FIGURE 14.33: Possible *Susceptibles* extinction?

iii.) Consider IBVP (14.14) with (14.17) as the initial condition.

a.) Investigate the behavior of the solution for large times. Use $\epsilon = 3$ and $\lambda = 5$. Begin by plotting solutions to $t = 30$. If necessary, slowly increase the plotting time until you are able to describe the behavior of the solution. Remember, a computer cannot handle infinity!

b.) Repeat (a) using $\epsilon = 10$ and $\lambda = 1$.

iv.) Is extinction possible?

a.) Use $\epsilon = 10$, $\lambda = 0.01$, initial condition (14.17), and plot solutions until $t = 30$. Carefully watch the movie to get a sense of how the Infectives are interacting with (chasing) the Susceptibles.

b.) Based on (a) do you believe the Susceptible group will go extinct?

c.) It turns out that if you plot solutions a little further in time the model breaks down due to the periodic boundary conditions. Try it! Plot until $t = 36$.

■

EXPLORE!

14.9 Continuous Dependence on Parameters

In this **EXPLORE!** we investigate the continuous dependence of the solution to (14.14) on the Infectives' initial condition, the diffusivity constant, and the removal rate.

MATLAB-Exercise 14.4.3. Open **MATLAB** and in the command line, type:

MATLAB_Epidemiology

To explore continuous dependence using the GUI we must first solve (14.14) with a particular initial condition, ϵ, and λ. Once a solution has been constructed, click the *Perturb Parameters and Solve* button to specify a modified value of ϵ, λ and perturbation size for the initial condition. The GUI will automatically output norm differences in the solutions. Use the GUI to answer the following questions. Make certain to click the HELP button for more information about the GUI. Screenshots of inputting parameters and GUI output can be found in Figures 14.34–14.37.

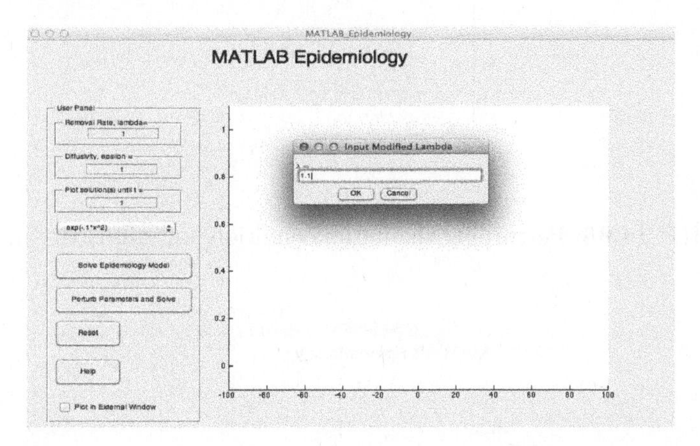

FIGURE 14.34: Perturbing parameters in MATLAB Epidemiology GUI

i.) Consider IBVP (14.14). Explore whether or not the solution of this IBVP depends continuously on the initial condition (14.16).

a.) Use $\epsilon = 1$, $\lambda = 1$, and plot solutions until $t = 10$. Choose a perturbation size of 0.1 for the initial condition and keep everything else the same. How would you describe the differences in the solutions?

b.) Repeat (a) with perturbation sizes equal to 0.05 and 0.01.

c.) Use $\epsilon = 1$, $\lambda = 0.1$, and plot solutions until $t = 10$. Choose a perturbation size of 0.1 for the initial condition and keep everything else the same. How would you describe the differences in the solutions?

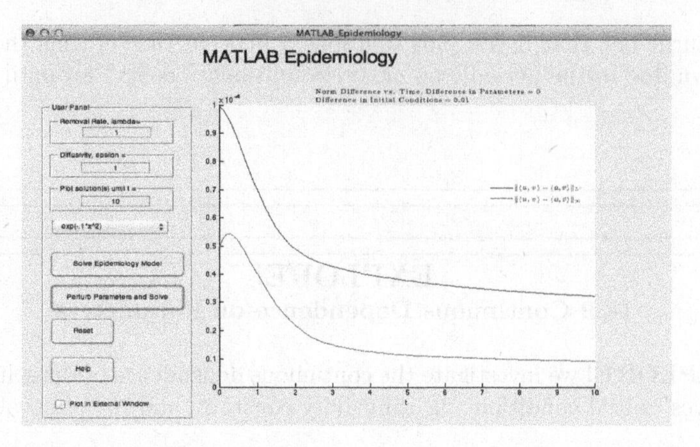

FIGURE 14.35: Perturbing the initial condition in (14.14)

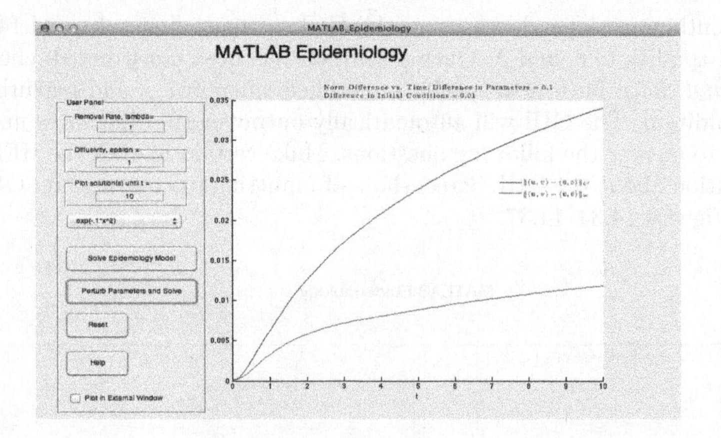

FIGURE 14.36: Perturbing the initial condition and diffusivity in (14.14)

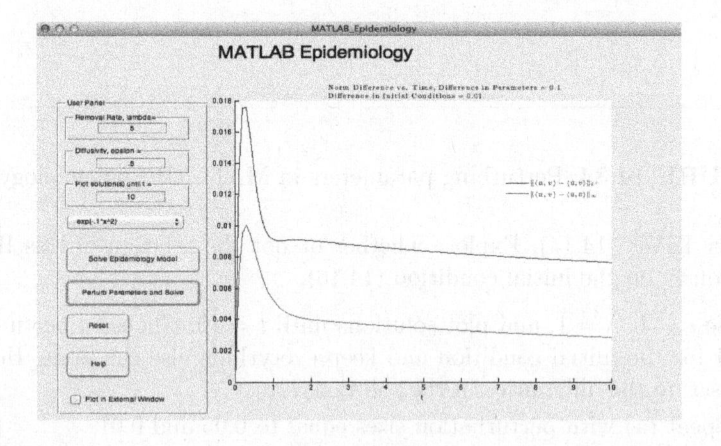

FIGURE 14.37: Perturbing the initial condition and removal rate in (14.14)

d.) Repeat (c) with perturbation sizes equal to 0.05 and 0.01.

e.) Make a conjecture regarding IBVP (14.14) continuous dependence on the initial condition (14.15).

ii.) Consider IBVP (14.14). Explore whether or not the solution of this IBVP depends continuously on the initial condition (14.16) and the diffusivity constant.

a.) Use $\epsilon = 1$, $\lambda = 1$, and plot solutions until $t = 10$. Choose 1.1 as the perturbed diffusivity constant and a perturbation size of 0.1 for the initial condition. How would you describe the differences in the solutions?

b.) Repeat (a) with perturbed diffusivites equal to 0.95 and 1.01 and perturbation sizes equal to 0.05 and 0.01.

c.) Use $\epsilon = 0.5$, $\lambda = 5$, and plot solutions until $t = 10$. Choose 0.4 as the perturbed diffusivity constant and a perturbation size of 0.1 for the initial condition. How would you describe the differences in the solutions?

d.) Repeat (a) with perturbed diffusivites equal to 0.55 and 0.49 and perturbation sizes equal to 0.05 and 0.01.

e.) Make a conjecture regarding IBVP (14.14) continuous dependence on the initial condition and diffusivity constant (14.15).

iii.) Consider IBVP (14.14). Explore whether or not the solution of this IBVP depends continuously on the initial condition (14.16) and the removal rate.

a.) Use $\epsilon = 1$, $\lambda = 1$, and plot solutions until $t = 10$. Choose 1.1 as the perturbed removal rate and a perturbation size of 0.1 for the initial condition. How would you describe the differences in the solutions?

b.) Repeat (a) with perturbed rates equal to 0.95 and 1.01 and perturbation sizes equal to 0.05 and 0.01.

c.) Use $\epsilon = 0.5$, $\lambda = 5$, and plot solutions until $t = 10$. Choose 4.9 as the perturbed removal rate and a perturbation size of 0.1 for the initial condition. How would you describe the differences in the solutions?

d.) Repeat (a) with perturbed removal rates equal to 5.05 and 4.99 and perturbation sizes equal to 0.05 and 0.01.

e.) Make a conjecture regarding IBVP (14.14) continuous dependence on the initial condition and the removal rate (14.15).

iv.) Repeat (i) using (14.17) as the initial condition.

v.) Repeat (ii) using (14.17) as the initial condition.

vi.) Repeat (iii) using (14.17) as the initial condition.

■

14.5 How Did That Zebra Get Its Stripes? A First Look at Spatial Pattern Formation

Consider the one-dimensional Gray-Scott reaction-diffusion IBVP,

$$\begin{cases} \frac{\partial u(x,t)}{\partial t} = \frac{\partial^2 u(x,t)}{\partial x^2} - u(x,t)v(x,t)^2 + A\left(1 - u(x,t)\right), \, t > 0, \, x \in [-L, L] \\ \frac{\partial v(x,t)}{\partial t} = \epsilon \frac{\partial^2 v(x,t)}{\partial x^2} + u(x,t)v(x,t)^2 - Bv(x,t), \, t > 0, \, x \in [-L, L] \\ u(x,0) = u_0(x), \, x \in [-L, L] \\ v(x,0) = v_0(x), \, x \in [-L, L] \\ u(-L,t) = u(L,t), \, t > 0 \\ v(-L,t) = v(L,t), \, t > 0 \end{cases} \tag{14.18}$$

where $\epsilon > 0$ is the diffusivity constant, A the replenishment constant, and B the diminishment constant. For the **EXPLORE!** activities to follow, we use the following initial condition for IBVP (14.18),

$$\begin{cases} u_0(x) &= 1 - \frac{1}{2}\sin^{100}\left(\frac{\pi(x-L)}{2L}\right) \\ v_0(x) &= \frac{1}{4}\sin^{100}\left(\frac{\pi(x-L)}{2L}\right) \end{cases} \tag{14.19}$$

EXPLORE!
14.10 Parameter Play!

In this **EXPLORE!** you will investigate the effects of changing the diffusivity, replenishment, and diminishment constants. To do this, you will use the MATLAB Gray-Scott Model GUI to visualize the solutions to IBVP (14.18) for different choices of ϵ, A, and B.

MATLAB-Exercise 14.5.1. To begin, open **MATLAB** and in the command line, type:

MATLAB_Gray_Scott_Model

Use the GUI to answer the following questions. Make certain to click on the HELP button for instructions on how to use the GUI. Solutions of GUI output can be found in Figures 14.38 and 14.39.

i.) Consider IBVP (14.18) with $\epsilon = 0.01$, $A = 0.5$, and $B = 0.5$.

   a.) Plot solutions until $t = 10$. How would you describe the evolution of the two concentrations?

   b.) Repeat (a) using $\epsilon = 1$ keeping everything else fixed. How did increasing ϵ affect the evolution?

ii.) Consider IBVP (14.18) with $\epsilon = 0.01$, $A = 0$, and $B = 1$.

   a.) With this particular set of model parameters, what should we expect to happen to the concentrations?

   b.) Plot solutions until $t = 10$. How would you describe the evolution of the two concentrations? Did the evolution match your expectation?

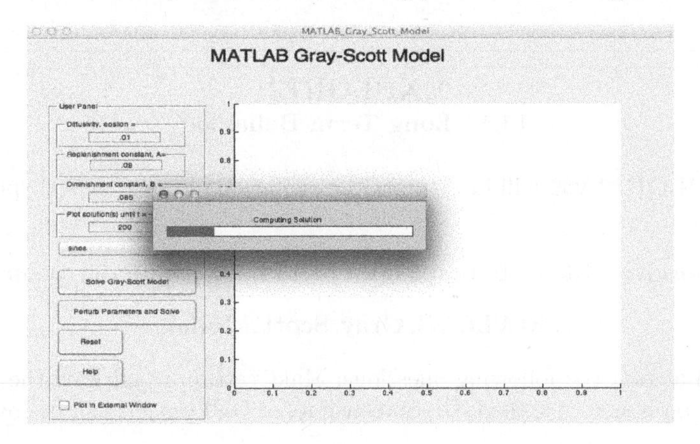

FIGURE 14.38: MATLAB Gray-Scott Model GUI

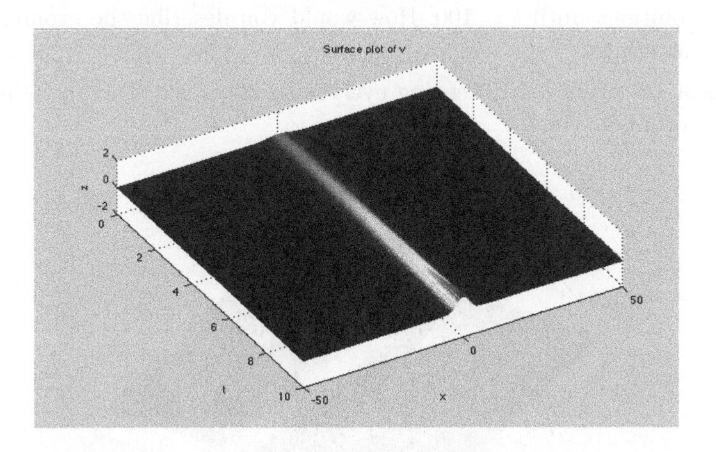

FIGURE 14.39: Surface plot of v in IBVP (14.18)

c.) Do the solutions remind you of another model's solutions? If so, which one?

iii.) Consider IBVP (14.18) with $\epsilon = 0.01$, $A = 1$, and $B = 0$.

a.) With this particular set of model parameters, what should we expect to happen to the concentrations?

b.) Plot solutions until $t = 10$. How would you describe the evolution of the two concentrations? Did the evolution match your expectation?

c.) It is worth pointing out that the chaotic behavior seen in the plot from the GUI *could* conceivably be due to the way the computer is solving for the answer. However, reaction (12.16) coupled with replenishment and no diminishment should lead to non-physical solutions.

■

EXPLORE!

14.11 Long Term Behavior

In this **EXPLORE!** you will investigate the evolution of the chemical species over long periods of time.

MATLAB-Exercise 14.5.2. To begin, open **MATLAB** and in the command line, type:

MATLAB_Gray_Scott_Model

Use the GUI to answer the following questions. Make certain to click on the HELP button for instructions on how to use the GUI. Screenshots of GUI output can be found in Figures 14.40 and 14.41.

i.) Consider IBVP (14.18) with $\epsilon = 0.01$, $A = 0.5$, and $B = 0.5$.

  a.) Plot solutions until $t = 100$. How would you describe the evolution of the two concentrations?

  b.) Repeat (a) using $\epsilon = 1$ keeping everything else fixed. Did changing ϵ affect the long-term behavior?

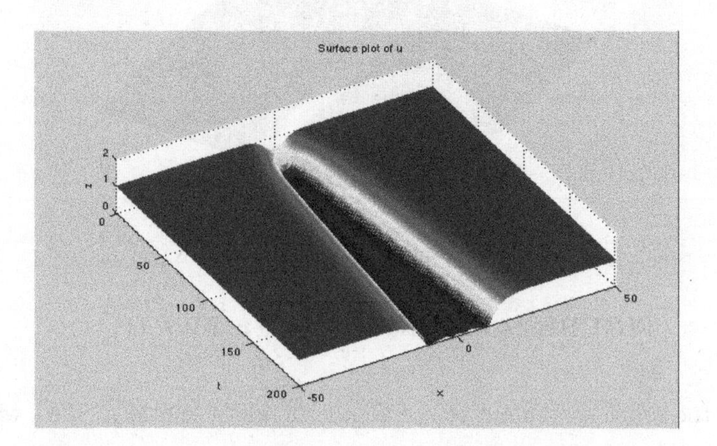

FIGURE 14.40: Surface Plot of u in IBVP (14.18).

ii.) Consider IBVP (14.18) with $\epsilon = 0.01$, $A = 0$, and $B = 1$.

  a.) With this particular set of model parameters, what should we expect to happen to the concentrations?

  b.) Plot solutions until $t = 200$. How would you describe the long-term behavior of the two concentrations?

  c.) How does the long-term behavior in (b) compare to that in the epidemiology model (14.14)?

ii.) Consider IBVP (14.18) with $\epsilon = 0.01$, $A = 0.09$, and $B = 0.085$.

  a.) Plot solutions until $t = 200$. How would you describe the long-term behavior of the two concentrations?

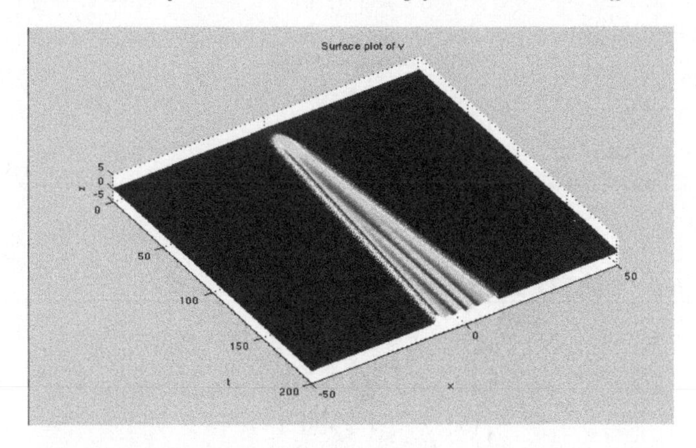

FIGURE 14.41: Surface Plot of v in IBVP (14.18).

c.) Based on your observations in (b), do you believe u and v will tend to constants as was the case in (i) and (ii)? Explain.

■

EXPLORE!

14.12 Continuous Dependence on Parameters

In this **EXPLORE!** we investigate the continuous dependence of the solution to (14.18) the models' parameters and the initial condition.

MATLAB-Exercise 14.5.3. Open **MATLAB** and in the command line, type:

MATLAB_Gray_Scott_Model

To explore continuous dependence using the GUI we must first solve IBVP (14.18) with a particular set of parameters. Once a solution has been constructed, click the *Perturb Parameters and Solve* button to specify modified parameters and a perturbation size for the initial condition. The GUI will automatically output norm differences in the solutions. Use the GUI to answer the following questions. Make certain to click on the HELP button to find more information about the GUI. Screenshots of GUI output can be found in Figures 14.42, 14.43, and 14.44.

i.) Explore whether or not the solution of this IBVP (14.18) depends continuously on the initial condition.

a.) Use $\epsilon = 0.01$, $A = 0.5, B = 0.5$, and plot solutions until $t = 10$. Choose a perturbation size of 0.1 for the initial condition and keep everything else the same. How would you describe the differences in the solutions?

b.) Repeat (a) with perturbation sizes equal to 0.05 and 0.01.

ii.) Repeat (i) for $A = 0$ and $B = 1$ keeping everything else the same and then again for $A = 0.09$ and $B = 0.085$.

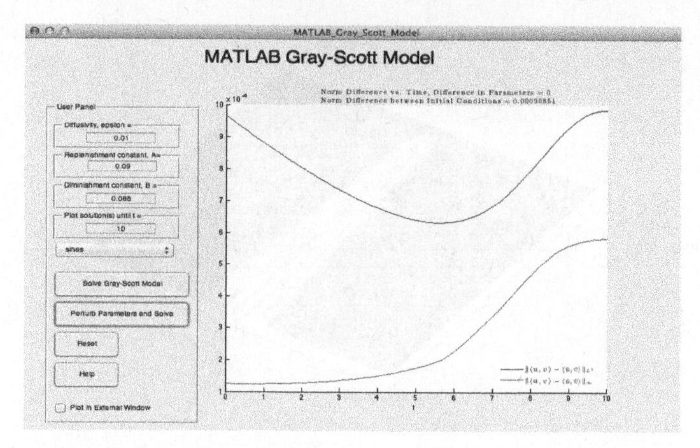

FIGURE 14.42: Perturbing the initial condition in (14.18)

iii.) Make a conjecture regarding IBVP (14.18) continuous dependence on the initial condition.

iv.) Explore whether or not the solution of this IBVP (14.18) depends continuously on the replinishment constant.

   a.) Use $\epsilon = 0.01$, $A = 0.5, B = 0.5$, and plot solutions until $t = 10$. Choose a modified replinishment constant equal to 0.6 and keep everything else the same. How would you describe the differences in the solutions?

   b.) Repeat (a) with modified replinishment constants equal to $0.55, 0.45$ and 0.049.

vi.) Repeat (i) for $A = 0$ and $B = 1$. Choose modified replinishment terms equal to $0.1, 0.05$, and 0.01.

vii.) Repeat (i) for $A = 0.09$ and $B = 0.085$. Choose modified replinishment terms equal to $0.15, 0.07$, and 0.08.

viii.) Make a conjecture regarding IBVP (14.18) continuous dependence on the replinishment term.

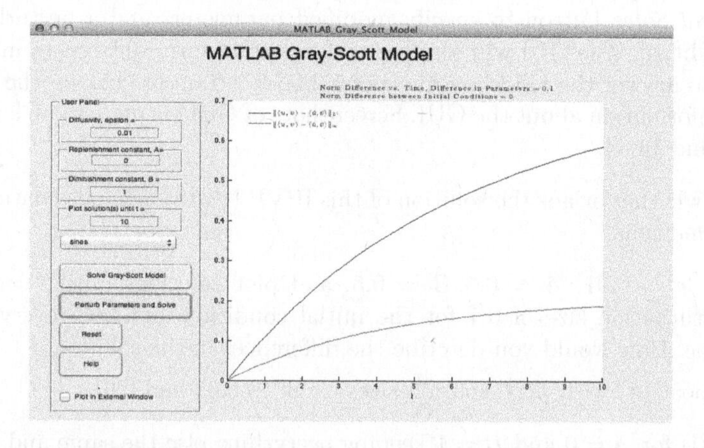

FIGURE 14.43: Perturbing the replinishment term in (14.18)

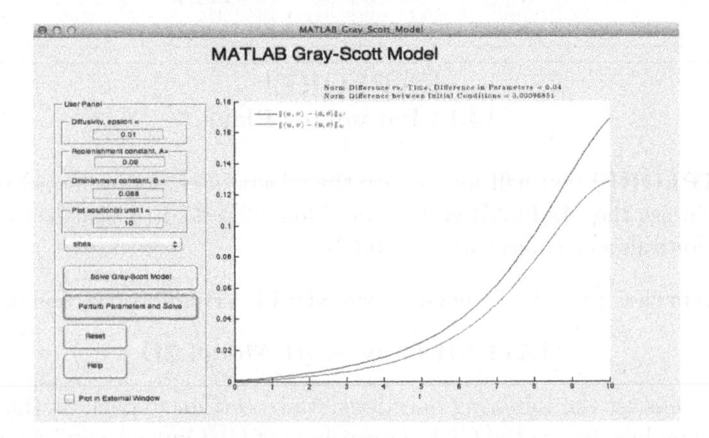

FIGURE 14.44: Perturbing the diffusivity constant and initial condition in (14.18)

ix.) Explore whether or not the solution of this IBVP (14.18) depends continuously on the diffusivity constant and the initial conditions.

a.) Use $\epsilon = 0.01$, $A = 0.5, B = 0.5$, and plot solutions until $t = 10$. Choose a modified diffusivity constant equal to 0.1 and a perturbation size of 0.1 for the initial condition and keep everything else the same. How would you describe the differences in the solutions?

b.) Repeat (a) with modified diffusivity constants equal to 0.07 and 0.03 and perturbation sizes equal to 0.05 and 0.01.

c.) Repeat (a) and (b) using $A = 0$ and $B = 1$.

d.) Repeat (a) and (b) using $A = 0.09$ and $B = 0.085$.

■

Next, we consider the two-dimensional Gray-Scott equation and investigate its solution's dependence on parameters and behavior for large times.

14.5.1 Two-Dimensional Gray-Scott Equation

The two-dimensional nondimensionalized Gray-Scott equation is given by

$$\begin{cases} \frac{\partial u(x,y,t)}{\partial t} = \Delta u(x,y,t) - u(x,y,t)v(x,y,t)^2 + A\left(1 - u(x,y,t)\right), \, t > 0, \, x \in [-L,L], \, y \in [-L,L] \\ \frac{\partial v(x,y,t)}{\partial t} = \epsilon \Delta v(x,y,t) + u(x,y,t)v(x,y,t)^2 - Bv(x,y,t), \, t > 0, \, x \in [-L,L], \, y \in [-L,L] \\ u(x,y,0) = u_0(x,y), \, x \in [-L,L], \, y \in [-L,L] \\ v(x,y,0) = v_0(x,y), \, x \in [-L,L], \, y \in [-L,L] \\ u(-L,y,t) = u(L,y,t), \, t > 0, \, y \in [-L,L] \\ u(x,-L,t) = u(x,L,t), \, t > 0, \, x \in [-L,L] \\ v(-L,y,t) = v(L,y,t), \, t > 0, \, y \in [-L,L] \\ v(x,-L,t) = v(x,L,t), \, t > 0, \, x \in [-L,L] \end{cases} \tag{14.20}$$

EXPLORE!
14.13 Parameter Play!

In this **EXPLORE!** you will investigate the effects of changing model parameters. To do this, you will use the MATLAB Gray-Scott Model 2D GUI to visualize the solutions to IBVP (14.20) for different choices of ϵ, A and B.

MATLAB-Exercise 14.5.4. To begin, open **MATLAB** and in the command line, type:

MATLAB_Gray_Scott_Model_2D

Use the GUI to answer the following questions. Make certain to click on the HELP button for instructions on how to use the GUI. Screenshots of GUI output can be found in Figures 14.45 and 14.46..

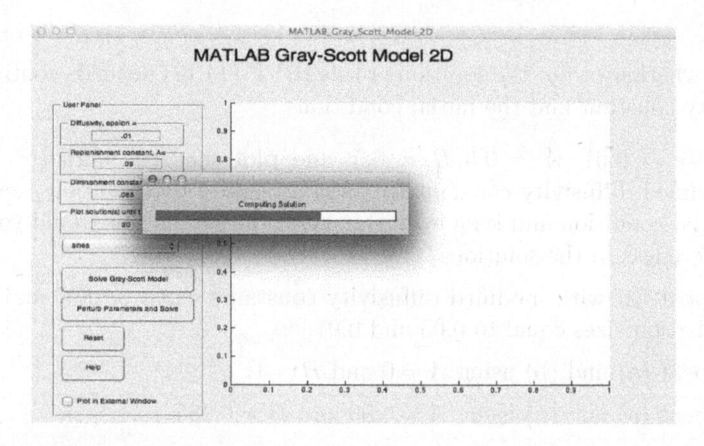

FIGURE 14.45: MATLAB Gray-Scott Equation 2D GUI

i.) Consider IBVP (14.20) with $\epsilon = 0.01$, $A = 0.5$, and $B = 0.5$.

a.) Plot solutions until $t = 10$. How would you describe the evolution of the two concentrations?

b.) Repeat (a) using $\epsilon = 1$ keeping everything else fixed. How did increasing ϵ affect the evolution?

ii.) Consider IBVP (14.20) with $\epsilon = 0.01$, $A = 0$, and $B = 1$.

a.) With this particular set of model parameters, what should we expect to happen to the concentrations?

b.) Plot solutions until $t = 10$. How would you describe the evolution of the two concentrations? Did the evolution match your expectation?

c.) Do the solutions remind you of another model's solutions? If so, which one?

iii.) Consider IBVP (14.20) with $\epsilon = 0.01$, $A = 1$, and $B = 0$.

a.) With this particular set of model parameters, what should we expect to happen to the concentrations?

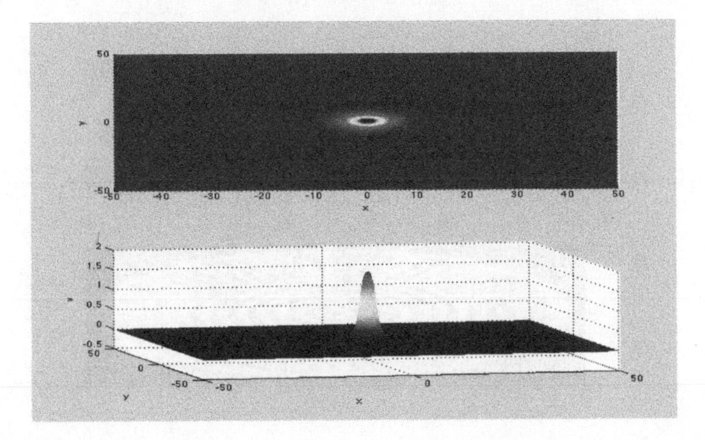

FIGURE 14.46: Aerial view of u and side view of v in (14.20) at $t = 10$

b.) Plot solutions until $t = 10$. How would you describe the evolution of the two concentrations? Did the evolution match your expectation?

c.) It is worth pointing out that the chaotic behavior seen in the plot from the GUI *could* conceivably be due to the way the computer is solving for the answer. However, reaction (12.16) coupled with replinishment and no diminishment should lead to non-physical solutions.

iv.) Compare your two-dimensional model parameters observations with those made in the one-dimesional case.

■

EXPLORE!
14.11 Long–Term Behavior

In this **EXPLORE!** you will investigate the evolution of the chemical species over long periods of time.

MATLAB-Exercise 14.5.5. To begin, open **MATLAB** and in the command line, type:

MATLAB_Gray_Scott_Model_2D

Use the GUI to answer the following questions. Make certain to click on the HELP button for instructions on how to use the GUI. A typical screenshot of GUI output can be found in Figure 14.47.

i.) Consider IBVP (14.20) with $\epsilon = 0.01$, $A = 0.5$, and $B = 0.5$.

a.) Plot solutions until $t = 75$. How would you describe the evolution of the two concentrations?

b.) Repeat (a) using $\epsilon = 1$ keeping everything else fixed. Did changing ϵ affect the long-term behavior?

c.) Do you believe the concentrations will converge to constants? Explain.

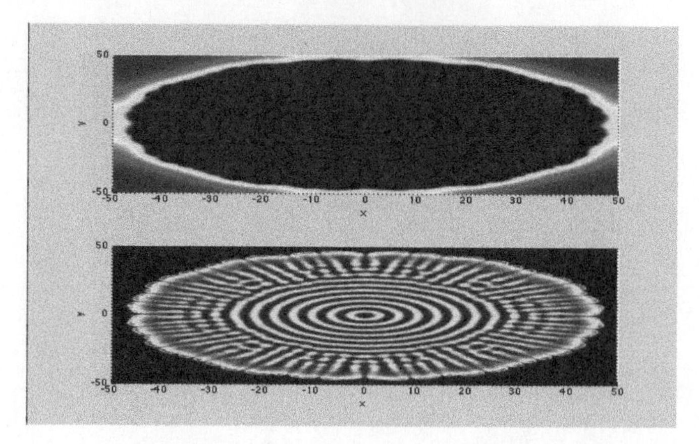

FIGURE 14.47: Aerial view of solutions to IBVP (14.20) at $t = 1000$.

ii.) Consider IBVP (14.20) with $\epsilon = 0.01$, $A = 0$, and $B = 1$.

a.) With this particular set of model parameters, what should we expect to happen to the concentrations?

b.) Plot solutions until $t = 100$. How would you describe the long-term behavior of the two concentrations?

ii.) Consider IBVP (14.20) with $\epsilon = 0.01$, $A = 0.09$, and $B = 0.085$.

a.) Plot solutions until $t = 500$. How would you describe the long-term behavior of the two concentrations?

c.) Based on your observations in (b), do you believe u and v will tend to constants as was the case in (i) and (ii)? Explain.

■

EXPLORE!

14.12 Continuous Dependence on Parameters

In this **EXPLORE!** we investigate the continuous dependence of the solution to (14.20) the models' parameters and the initial condition.

MATLAB-Exercise 14.5.6. Open **MATLAB** and in the command line, type: Use the GUI to answer the following questions. Make certain to click the HELP button for more information about the GUI. Screenshots of GUI output can be found in Figures 14.48, 14.49, and 14.50.

MATLAB_Gray_Scott_Model_2D

i.) Explore whether or not the solution of this IBVP (14.20) depends continuously on the initial condition.

a.) Use $\epsilon = 0.01$, $A = 0.5, B = 0.5$, and plot solutions until $t = 10$. Choose a perturbation size of 0.1 for the initial condition and keep everything else the same. How would you describe the differences in the solutions?

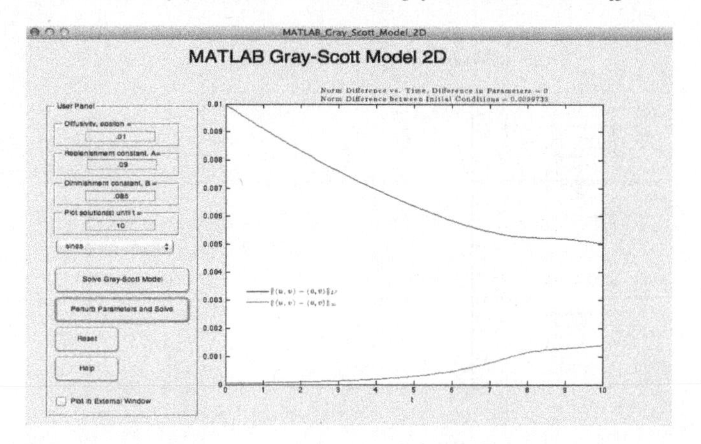

FIGURE 14.48: Perturbing the initial condition in (14.20)

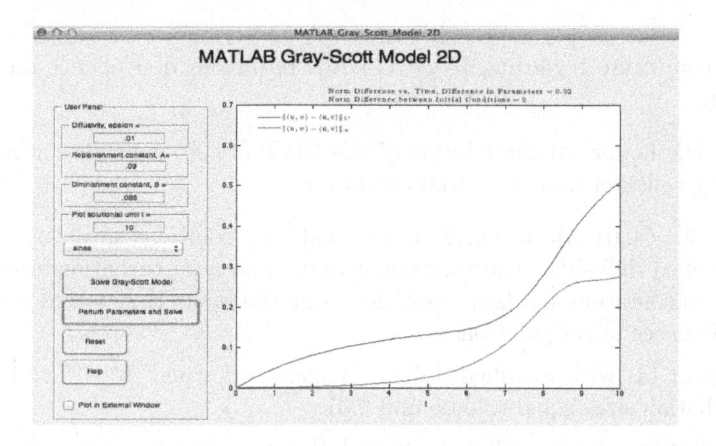

FIGURE 14.49: Perturbing the replinishment term in (14.20)

b.) Repeat (a) with perturbation sizes equal to 0.05 and 0.01.

ii.) Repeat (i) for $A = 0$ and $B = 1$ keeping everything else the same and then again for $A = 0.09$ and $B = 0.085$.

iii.) Make a conjecture regarding IBVP (14.20) continuous dependence on the initial condition.

iv.) Explore whether or not the solution of this IBVP (14.20) depends continuously on the replinishment constant.

a.) Use $\epsilon = 0.01$, $A = 0.5, B = 0.5$, and plot solutions until $t = 10$. Choose a modified replinishment constant equal to 0.6 and keep everything else the same. How would you describe the differences in the solutions?

b.) Repeat (a) with modified replinishment constants equal to $0.55, 0.45$ and 0.049.

vi.) Repeat (i) for $A = 0$ and $B = 1$. Choose modified replinishment terms equal to $0.1, 0.05$, and 0.01.

vii.) Repeat (i) for $A = 0.09$ and $B = 0.085$. Choose modified replinishment terms equal to $0.15, 0.07$, and 0.08.

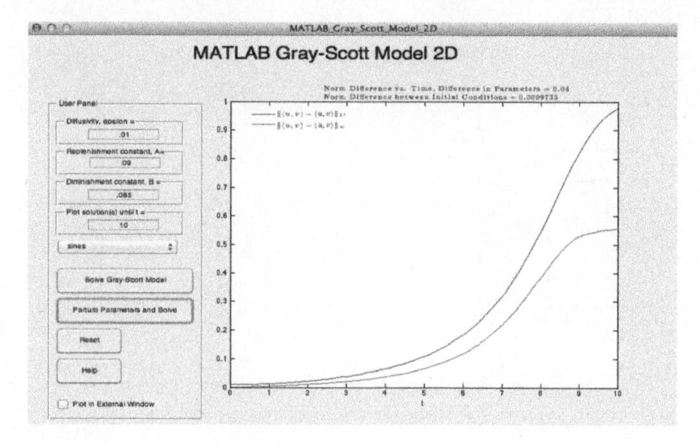

FIGURE 14.50: Perturbing the diffusivity constant and initial condition in (14.20)

viii.) Make a conjecture regarding IBVP (14.20) continuous dependence on the replinishment term.

ix.) Explore whether or not the solution of this IBVP (14.20) depends continuously on the diffusivity constant and the initial conditions.

a.) Use $\epsilon = 0.01$, $A = 0.5, B = 0.5$, and plot solutions until $t = 10$. Choose a modified diffusivity constant equal to 0.1 and a perturbation size of 0.1 for the initial condition and keep everything else the same. How would you describe the differences in the solutions?

b.) Repeat (a) with modified diffusivity constants equal to 0.07 and 0.03 and perturbation sizes equal to 0.05 and 0.01.

c.) Repeat (a) and (b) using $A = 0$ and $B = 1$.

d.) Repeat (a) and (b) using $A = 0.09$ and $B = 0.085$.

■

14.6 Autocatalysis—Combustion!

Consider the non-dimensionalized autocatalysis model

$$\begin{cases} \frac{\partial u(x,t)}{\partial t} = \frac{\partial^2 u(x,t)}{\partial x^2} + v(x,t)f(u(x,t)),\ t > 0,\ x \in (-L, L) \\ \frac{\partial v(x,t)}{\partial t} = \epsilon\frac{\partial^2 v(x,t)}{\partial x^2} + v(x,t)f(u(x,t)),\ t > 0,\ x \in (-L, L) \\ u(x,0) = u_0(x),\ x \in [-L, L] \\ v(x,0) = v_0(x),\ x \in [-L, L] \\ u(-L,t) = u(L,t),\ t > 0 \\ v(-L,t) = v(L,t),\ t > 0 \end{cases} \tag{14.21}$$

for

$$f(u) = \begin{cases} u^m, & u \geq 0 \\ 0, & u < 0 \end{cases}.$$

The model parameters are ϵ and m representing the diffusivity and reaction coefficient, respectively. For the **EXPLORE!** activities to follow, we use the following initial condition for IBVP (14.21):

$$\begin{cases} u_0(x) & = \frac{1}{2}\left(1 + \tanh\left(\frac{L}{10}\left(\frac{L}{10} - |x|\right)\right)\right) \\ v_0(x) & = 1 - \frac{1}{4}\left(1 + \tanh\left(\frac{L}{10}\left(\frac{L}{10} - |x|\right)\right)\right) \end{cases} \tag{14.22}$$

EXPLORE!
14.16 Parameter Play!

In this **EXPLORE!** you will investigate the effects of changing the model parameters. To do this, you will use the MATLAB Autocatalysis Reaction GUI to visualize the solutions of IBVP (14.21) for different choices of ϵ and m.

MATLAB-Exercise 14.6.1. To begin, open **MATLAB** and in the command line, type:

MATLAB_Autocatalysis_Reaction

Use the GUI to answer the following questions. Make certain to click on the HELP button for instructions on how to use the GUI. A typical screenshot of the GUI can be found in Figure 14.51.

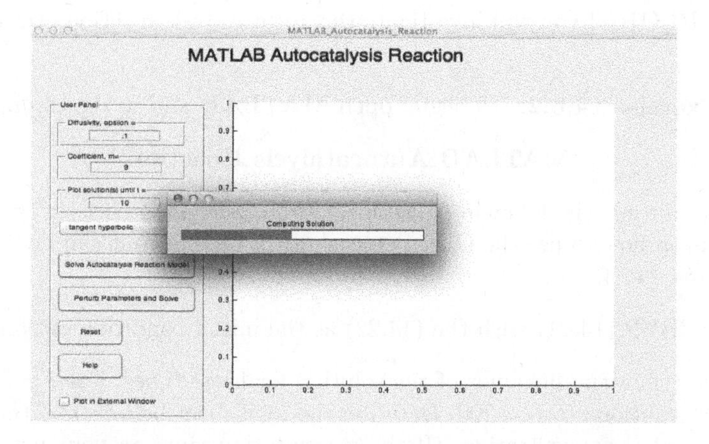

FIGURE 14.51: MATLAB Autocatalysis GUI

i.) Consider IBVP (14.21) with (14.22) as the initial condition option.

a.) Use $\epsilon = 0.1$, $m = 1$, and plot solutions until $t = 10$. Describe the evolution of the solution. Retain the surface plot outputs from the GUI for the next question.

b.) Repeat (a) using $\epsilon = 0.1$ and $m = 11$. How did changing m affect the solution to IBVP (14.21)?

c.) Repeat (a) using $\epsilon = 5$ and $m = 1$.

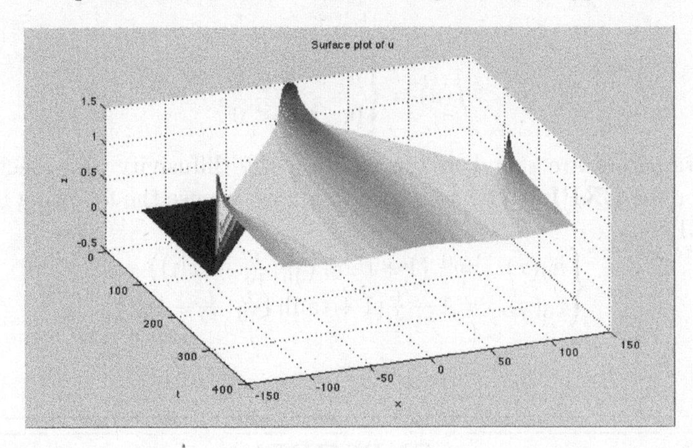

FIGURE 14.52: Surface plot of u in IBVP (14.21) with $m = 4$

d.) Repeat (a) using $\epsilon = 5$ and $m = 11$. How did changing m affect the solution in this case?

■

EXPLORE!
14.17 Long-Term Behavior

In this **EXPLORE!** you will investigate the autocatalysis reaction over long periods of time.

MATLAB-Exercise 14.6.2. To begin, open **MATLAB** and in the command line, type:

MATLAB_Autocatalysis_Reaction

Use the GUI to answer the following questions. Make certain to click on the HELP button for instructions on how to use the GUI. Screenshots of GUI output can be found in Figures 14.52, 14.53, and 14.54.

i.) Consider IBVP (14.21) with the (14.22) as the initial condition option.

a.) Investigate the behavior of the solution for large times. Use $\epsilon = 0.1$, $m = 2$, and plot solutions to $t = 200$. Describe the long term behavior of the autocatalytic molecules' concentration. (If the computation time on your personal computer was not too great for $t = 200$, use $t = 400$ in the next two questions.)

b.) Repeat (a) using $m = 9$.

c.) Repeat (a) using $m = 11$.

ii.) Repeat (i) using $\epsilon = 1$. Comment on the differences between the long-time evolution of the autocatalytic molecule's concentration for $\epsilon = 1$ and for $\epsilon = 0.1$.

■

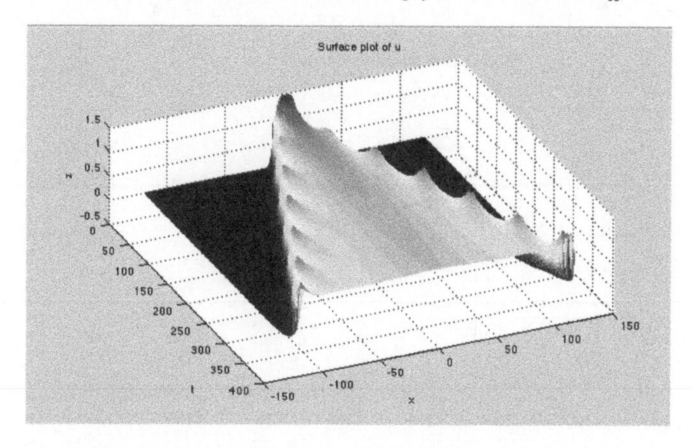

FIGURE 14.53: Surface plot of u in IBVP (14.21) with $m = 9$

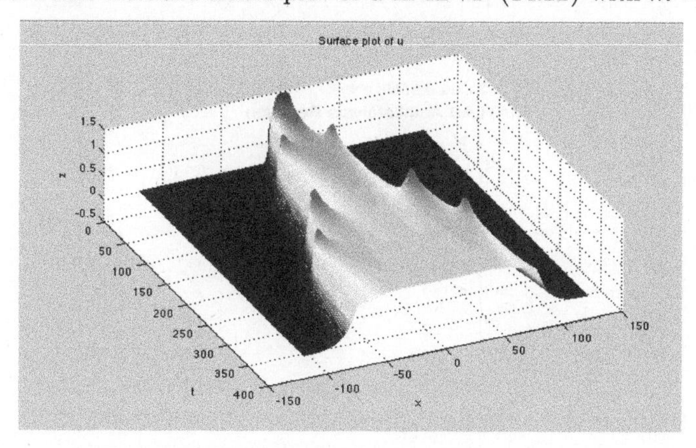

FIGURE 14.54: Surface plot of u in IBVP (14.21) with $m = 11$

EXPLORE!
14.18 Continuous Dependence on Parameters

In this **EXPLORE!** we investigate the continuous dependence of the solution to (14.21) on the initial condition, the diffusivity constant, and the reaction coefficient.

MATLAB-Exercise 14.6.3. Open **MATLAB** and in the command line, type:

MATLAB_Autocatalysis_Reaction

To explore continuous dependence using the GUI we must first solve (14.21) with a particular ϵ and λ. Once a solution has been constructed, click the *Perturb Parameters and Solve* button to specify a modified value of ϵ, m and perturbation size of for the initial condition. The GUI will automatically output norm differences in the solutions. Use the GUI to answer the following questions. Make certain to click the HELP button for more information about the GUI. Screenshots of inputting parameters and GUI output can be found in Figures 14.55–14.58.

i.) Consider IBVP (14.21) with initial condition (14.22). Explore whether or not the solution of this IBVP depends continuously on the initial condition.

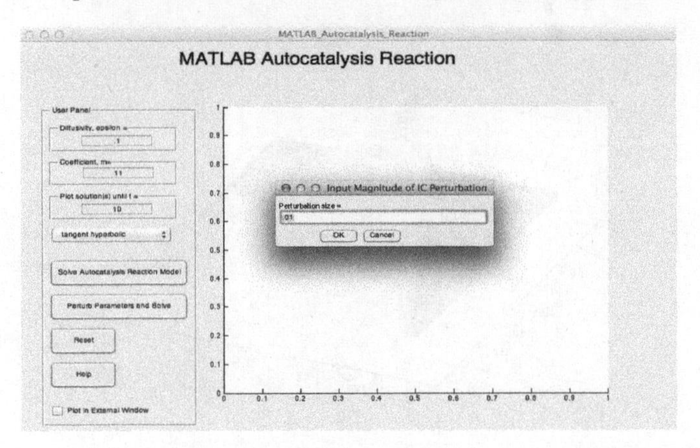

FIGURE 14.55: Perturbing parameters in MATLAB Autocatalysis Reaction GUI

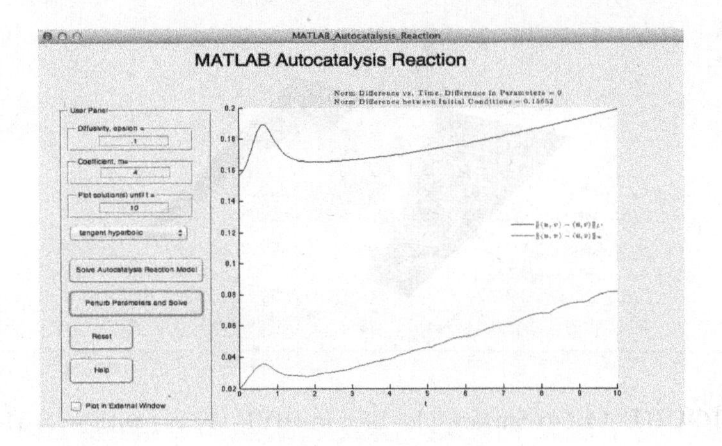

FIGURE 14.56: Perturbing the initial condition in (14.21)

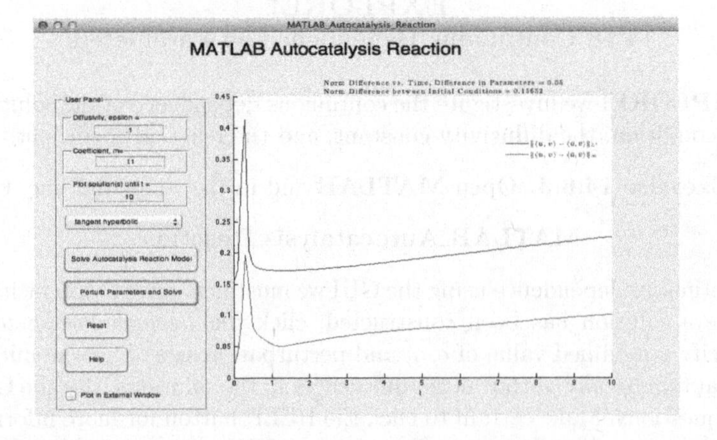

FIGURE 14.57: Perturbing the initial condition and diffusivity in (14.21)

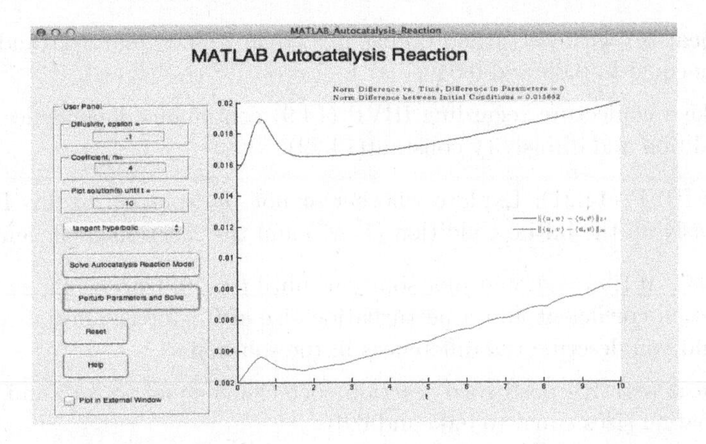

FIGURE 14.58: Perturbing the initial condition and reaction coefficient in (14.21)

a.) Use $\epsilon = 0.1$, $m = 4$, and plot solutions until $t = 10$. Choose a perturbation size of 0.1 for the initial condition and keep everything else the same. How would you describe the differences in the solutions?

b.) Repeat (a) with perturbation sizes equal to 0.05 and 0.01.

c.) Use $\epsilon = 0.1$, $m = 9$, and plot solutions until $t = 10$. Choose a perturbation size of 0.1 for the initial condition and keep everything else the same. How would you describe the differences in the solutions?

d.) Repeat (c) with perturbation sizes equal to 0.05 and 0.01.

e.) Use $\epsilon = 0.1$, $m = 11$, and plot solutions until $t = 10$. Choose a perturbation size of 0.1 for the initial condition and keep everything else the same. How would you describe the differences in the solutions?

f.) Repeat (e) with perturbation sizes equal to 0.05 and 0.01.

g.) Make a conjecture regarding IBVP (14.21) continuous dependence on the initial condition (14.22).

ii.) Consider IBVP (14.21). Explore whether or not the solution of this IBVP depends continuously on the initial condition (14.22) and the diffusivity constant.

a.) Use $\epsilon = 0.1$, $m = 4$, and plot solutions until $t = 10$. Choose 0.2 as the perturbed diffusivity constant and a perturbation size of 0.1 for the initial condition. How would you describe the differences in the solutions?

b.) Repeat (a) with perturbed diffusivites equal to 0.05 and .011 and perturbation sizes equal to 0.05 and 0.01.

c.) Use $\epsilon = 0.1$, $m = 9$, and plot solutions until $t = 10$. Choose 0.2 as the perturbed diffusivity constant and a perturbation size of 0.1 for the initial condition. How would you describe the differences in the solutions?

d.) Repeat (c) with perturbed diffusivites equal to 0.05 and .011 and perturbation sizes equal to 0.05 and 0.01.

e.) Use $\epsilon = 0.1$, $m = 11$, and plot solutions until $t = 10$. Choose 0.2 as the perturbed diffusivity constant and a perturbation size of 0.1 for the initial condition. How would you describe the differences in the solutions?

f.) Repeat (c) with perturbed diffusivites equal to 0.05 and .011 and perturbation sizes equal to 0.05 and 0.01.

e.) Make a conjecture regarding IBVP (14.9) continuous dependence on the initial condition and diffusivity constant(14.22).

iii.) Consider IBVP (14.21). Explore whether or not the solution of this IBVP depends continuously on the initial condition (14.22) and the reaction coefficient.

a.) Use $\epsilon = 0.1$, $m = 4$, and plot solutions until $t = 10$. Choose 4.5 as the perturbed reaction coefficient and a perturbation size of 0.1 for the initial condition. How would you describe the differences in the solutions?

b.) Repeat (a) with perturbed reaction coefficients equal to 3.75 and 4.01 and perturbation sizes equal to 0.05 and 0.01.

c.) Use $\epsilon = 0.1$, $m = 9$, and plot solutions until $t = 10$. Choose 8.5 as the perturbed reaction coefficient and a perturbation size of 0.1 for the initial condition. How would you describe the differences in the solutions?

d.) Repeat (c) with perturbed reaction coefficient to 9.15 and 8.9 and perturbation sizes equal to 0.05 and 0.01.

e.) Use $\epsilon = 0.1$, $m = 11$, and plot solutions until $t = 10$. Choose 11.2 as the perturbed reaction coefficient and a perturbation size of 0.1 for the initial condition. How would you describe the differences in the solutions?

f.) Repeat (c) with perturbed reaction coefficients equal to 100.95 and 11.01 and perturbation sizes equal to 0.05 and 0.01.

e.) Make a conjecture regarding IBVP (14.21) continuous dependence on the initial condition and diffusivity constant(14.22).

iv.) The most interesting features of the autocatalytic molecule's concentration are the periodic fluctuations that present themselves in the long run. Determine if perturbing any of the model parameters affects the occurrence of these fluctuations. To do this, only consider model parameters and types of perturbations that you *believe* preserve continuous dependence in the short run. Use (i)-(iii) to determine such conditions and plot solutions only until $t = 100$.

■

Epilogue

We have come to the end of our journey. It is instructive to look back and carefully examine the outcome of your plodding through the past thirteen chapters. Although we started our discussion with finite-dimensional systems of linear ordinary differential equations with frequent reference to the even more special one-dimensional case and developed the material for such three progressively more general versions of such systems (homogeneous, nonhomogeneous, and semi-linear), we could very easily have simply recovered those results as corollaries of the more abstract theory established in Part I. However, we would have lost the opportunity to instill important intuition about what ought to happen in the more abstract case. In effect, this could have arguably made understanding the development of the theory of abstract linear differential equations a much more difficult task.

Looking back, it would be instructive to convince yourself that each of the main results established in Part I is indeed a corollary of a result in Part II, and so all of the models developed in Part I can be subsumed under the Part II paradigm. That said, the development of the material in Part II was more intuitive than rigorous, and often the results and definitions formulated were stated more simply than precisely in order to make them accessible. If you are interested in a more rigorous account of the development of this theory, and you have taken a course in real analysis, then allow us to encourage you to continue your studies with the book Discovering Evolution Equations with Applications [33].

A natural question to ask at the end of such a journey is, "What comes next?" This is a loaded question when it comes to the world of differential equations. Arguably, the next step should be to carefully study real analysis so that your toolbox is strengthened prior to embarking on the next journey in differential equations. The reason for this is that the next natural steps involve considering fully nonlinear ordinary and partial differential equations and stochastic differential equations (that is, those for which environmental noise is accounted). Both directions involve fairly substantial amounts of real and functional analysis and the latter also involves the use of probability theory. Suffice it to say that we have only cracked open a portal to the world of differential equations and provided you with a quick peek inside. What lies beyond the portal is an immensely rich body of material, full of new surprises! We both hope that you have enjoyed your journey through this textbook and that you have found it as fulfilling as we have in guiding you through it.

Mark A. McKibben
Micah Webster

Appendix

Proof of Theorem 7.6.1

Proof. Consider the following sequence:

$$\mathbf{U}_n(t) = e^{\mathbf{A}(t-t_0)}\mathbf{U}_0 + \int_{t_0}^{t} e^{\mathbf{A}(t-s)}\mathbf{F}(s.\mathbf{U}_{n-1}(s))ds \tag{14.23}$$

$$\mathbf{U}_0(t) = e^{\mathbf{A}(t-t_0)}\mathbf{U}_0.$$

We need to prove the existence of $\mathbf{U} \in \mathbb{C}\left([t_0, T]; \mathbb{R}^N\right)$ such that

(a) $\lim_{n\to\infty} \|\mathbf{U}_n - \mathbf{U}\|_{\mathbb{C}([t_0,T];\mathbb{R}^N)} = 0$, and
(b) $\mathbf{U}$ is a mild solution of (Semi-CP) on $[t_0, T]$.

Claim 1: $\{\mathbf{U}_n : n \in \mathbb{N}\}$ is a uniformly bounded subset of $\mathbb{C}\left([t_0, T]; \mathbb{R}^N\right)$.
Let $t_0 \leq t \leq T$ be fixed. For any $n \in \mathbb{N}$,

$$\begin{aligned}
\|\mathbf{U}_n(t)\|_{\mathbb{R}^N} &= \left\|e^{\mathbf{A}(t-t_0)}\mathbf{U}_0\right\|_{\mathbb{R}^N} + \int_0^t \left\|e^{\mathbf{A}(t-s)}\left[\mathbf{F}(s.\mathbf{U}_{n-1}(s)) - \mathbf{F}(s,0) + \mathbf{F}(s,0)\right]\right\|_{\mathbb{R}^N} ds \\
&\leq \overline{M_{\mathbf{A}}}\left[\|\mathbf{U}_0\|_{\mathbb{R}^N} + \int_0^t \left[\|\mathbf{F}(s.\mathbf{U}_{n-1}(s)) - \mathbf{F}(s,0)\|_{\mathbb{R}^N} + \|\mathbf{F}(s,0)\|_{\mathbb{R}^N}\right] ds\right] \\
&\leq \overline{M_{\mathbf{A}}}\left[\|\mathbf{U}_0\|_{\mathbb{R}^N} + \|\mathbf{F}(\cdot,0)\|_{\mathbb{L}^1(0,T;\mathbb{R}^N)}\right] + \overline{M_{\mathbf{A}}}M_{\mathbf{F}}\int_0^t \|\mathbf{U}_{n-1}(s)\|_{\mathbb{R}^N} ds.
\end{aligned} \tag{14.24}$$

Let $C_1 = \overline{M_{\mathbf{A}}}\left[\|\mathbf{U}_0\|_{\mathbb{R}^N} + \|\mathbf{F}(\cdot,0)\|_{\mathbb{L}^1(0,T;\mathbb{R}^N)}\right]$. For any $K \in \mathbb{N}$, taking supremums on both sides of (14.24) yields

$$\begin{aligned}
\sup_{1\leq n\leq K} \|\mathbf{U}_n(t)\|_{\mathbb{R}^N} &\leq C_1 + \overline{M_{\mathbf{A}}}M_{\mathbf{F}}\int_0^t \sup_{1\leq n\leq K} \|\mathbf{U}_{n-1}(s)\|_{\mathbb{R}^N} ds \\
&\leq C_1 + \overline{M_{\mathbf{A}}}M_{\mathbf{F}}\int_0^t \left[\|\mathbf{U}_0\|_{\mathbb{R}^N} + \sup_{1\leq n\leq K} \|\mathbf{U}_n(s)\|_{\mathbb{R}^N}\right] ds \\
&\leq \left[C_1 + \overline{M_{\mathbf{A}}}M_{\mathbf{F}}\|\mathbf{U}_0\|_{\mathbb{R}^N} T\right] + \overline{M_{\mathbf{A}}}M_{\mathbf{F}}\int_0^t \sup_{1\leq n\leq K} \|\mathbf{U}_n(s)\|_{\mathbb{R}^N} ds.
\end{aligned} \tag{14.25}$$

Applying Gronwall's lemma to (14.25) yields

$$\sup_{1\leq n\leq K} \|\mathbf{U}_n(t)\|_{\mathbb{R}^N} \leq \left[C_1 + \overline{M_{\mathbf{A}}}M_{\mathbf{F}}\|\mathbf{U}_0\|_{\mathbb{R}^N} T\right] e^{\left(\overline{M_{\mathbf{A}}}M_{\mathbf{F}}\right)t}$$

so that

$$\sup_{1\leq n\leq K} \|\mathbf{U}_n\|_{\mathbb{C}([0,T];\mathbb{R}^N)} \leq \left[C_1 + \overline{M_{\mathbf{A}}}M_{\mathbf{F}}\|\mathbf{U}_0\|_{\mathbb{R}^N} T\right] e^{\left(\overline{M_{\mathbf{A}}}M_{\mathbf{F}}\right)T}. \tag{14.26}$$

Since the right-side of (14.26) is independent of K, we conclude that $\{\mathbf{U}_n n \in \mathbb{N}\}$ is a uniformly bounded subset of $\mathbb{C}\left([0,T];\mathbb{R}^N\right)$.$\diamondsuit$

Claim 2: There exists $u \in \mathbb{C}\left([0,T];\mathbb{R}^N\right)$ such that $\lim_{n\to\infty} \|\mathbf{U}_n - \mathbf{U}\|_{\mathbb{C}} = 0$.

For every $m \in \mathbb{N}$ and $0 \le t \le T$, observe that

$$\mathbf{U}_m(t) = u_0(t) + \sum_{k=0}^{m-1} \left[\mathbf{U}_{k+1}(t) - \mathbf{U}_k(t)\right]. \tag{14.27}$$

It can be shown that $\sum_{k=0}^{\infty} \|\mathbf{U}_{k+1}(t) - \mathbf{U}_k(t)\|_{\mathbb{R}^N}$ is uniformly convergent on $[0,T]$. (Why?) Arguing iteratively, we see that $\forall k \in \mathbb{N}$,

$$\begin{aligned}
\|\mathbf{U}_{k+1}(t) - \mathbf{U}_k(t)\|_{\mathbb{R}^N} &= \left\| \int_0^t e^{\mathbf{A}(t-s)} \left[\mathbf{F}(s, \mathbf{U}_k(s)) - \mathbf{F}(s, \mathbf{U}_{k-1}(s))\right] ds \right\|_{\mathbb{R}^N} \\
&\le \overline{M_{\mathbf{A}}} M_{\mathbf{F}} \int_0^t \|\mathbf{U}_k(s_1) - \mathbf{U}_{k-1}(s_1)\|_{\mathcal{X}}\, ds_1 \\
&\le \left(\overline{M_{\mathbf{A}}} M_{\mathbf{F}}\right)^2 \int_0^t \int_0^{s_1} \|\mathbf{U}_{k-1}(s_2) - \mathbf{U}_{k-2}(s_2)\|_{\mathcal{X}}\, ds_2 ds_1 \\
&\;\;\vdots \\
&\le \left(\overline{M_{\mathbf{A}}} M_{\mathbf{F}}\right)^k \int_0^t \cdots \int_0^{s_{k-1}} \|\mathbf{U}_1(s_k) - \mathbf{U}_0(s_k)\|_{\mathcal{X}}\, ds_k \ldots ds_1 \\
&\le \left(\overline{M_{\mathbf{A}}} M_{\mathbf{F}}\right)^k \int_0^t \cdots \int_0^{s_{k-1}} \left\| \int_0^t e^{\mathbf{A}(s_k-\tau)} \mathbf{F}(\tau, e^{\mathbf{A}\tau}\mathbf{U}_0) d\tau \right\|_{\mathcal{X}} ds_k \ldots ds_1 \\
&\le M_{\mathbf{F}}^k \left(\overline{M_{\mathbf{A}}}\right)^{k+1} \overline{M_{\mathbf{F}}} \frac{t^{k+1}}{(k+1)!}.
\end{aligned} \tag{14.28}$$

Therefore, $\forall 0 \le t \le T$,

$$\sum_{k=0}^{\infty} \|\mathbf{U}_{k+1}(t) - \mathbf{U}_k(t)\|_{\mathbb{R}^N} \le \left(\frac{\overline{M_{\mathbf{A}}}\,\overline{M_{\mathbf{F}}}}{M_{\mathbf{F}}}\right) \sum_{k=0}^{\infty} \frac{(tM_{\mathbf{F}})^{k+1}}{(k+1)!} \le \left(\frac{\overline{M_{\mathbf{A}}}\,\overline{M_{\mathbf{F}}}}{M_{\mathbf{F}}}\right) e^{M_{\mathbf{F}} T} \tag{14.29}$$

We conclude from (14.29) that

$$\lim_{k\to\infty} \|\mathbf{U}_{k+1}(t) - \mathbf{U}_k(t)\|_{\mathbb{R}^N} = 0 \text{ uniformly } \forall 0 \le t \le T. \tag{14.30}$$

Hence, we can conclude that

$$\mathbf{U_m} \longrightarrow \mathbf{U} \text{ uniformly on } [0,T] \text{ as } m \to \infty. \tag{14.31}$$

(Tell how carefully.) $\diamondsuit$

Claim 3: Show that u is a mild solution of (Semi-CP).

Let $\epsilon > 0$. There exists $M \in \mathbb{N}$ such that

$$m \ge M \Longrightarrow \|\mathbf{U_m} - \mathbf{U}\|_{\mathbb{C}} < \frac{\epsilon}{\overline{M_{\mathbf{A}}} M_{\mathbf{F}} T + 1}. \tag{14.32}$$

For such m, it follows that

$$\begin{aligned}
\left\| \int_0^t e^{A(t-s)} \left[f(s, u_{m-1}(s)) - f(s, u(s))\right] ds \right\|_{\mathcal{X}} &\le \overline{M_{\mathbf{A}}} M_{\mathbf{F}} \int_0^t \|u_{m-1}(s) - u(s)\|_{\mathcal{X}}\, ds \\
&\le \overline{M_{\mathbf{A}}} M_{\mathbf{F}} T \|\mathbf{U_{m-1}} - \mathbf{U}\|_{\mathbb{C}} \\
&< \epsilon.
\end{aligned}$$

Thus, $\forall t \in [0, T]$,

$$\lim_{m\to\infty} \int_0^t e^{\mathbf{A}(t-s)} \mathbf{F}(s, \mathbf{U}_{m-1}(s))ds = \int_0^t e^{\mathbf{A}(t-s)} \mathbf{F}(s, \mathbf{U}(s))ds. \tag{14.33}$$

Using (14.33), we now take a limit as $n \to \infty$ to conclude that u satisfies the variation of parameters formula and hence is a mild solution of (Semi-CP).◇

Claim 4: Verify that the mild solution of (Semi-CP) is unique.

Let v_1 and v_2 be mild solutions of (Semi-CP). Then, they both satisfy the variation of parameters formula. Subtracting these two expressions yields, $\forall 0 \le t \le T$,

$$\|v_1(t) - v_2(t)\|_{\mathcal{X}} \le \overline{M_{\mathbf{A}}} M_{\mathbf{F}} \int_0^t \|v_1(s) - v_2(s)\|_{\mathbb{R}^N} ds. \tag{14.34}$$

An application of Gronwall's lemma in (14.34) leads us to conclude that $v_1(t) = v_2(t)$, $\forall 0 \le t \le T$. (Why?) Thus, uniqueness follows. ◇

This completes the proof. □

Bibliography

[1] L.F. Abbott and P. Dayan. *Theoretical Neuroscience.* MIT Press, 2001.

[2] N.J. Balmforth, R.V. Craster, and S.J.A Malham. Unsteady fronts in an autocatalytic system. *Proc. R. Soc. Land. A*, 455:1401–1433, 1999.

[3] G.I. Barenblatt, Y.P. Zheltov, and Kochina. Basic concepts in the theory of seepage of homogeneous liquids in fissured rocks (strata). *PMM, J. Appl. Math. Mech.*, 24:1286–1303, 1961.

[4] H.C. Berg. Chemotaxis in bacteria. *Annual Review of Biophysics and Bioengineering*, 4(1):119–136, 1975.

[5] F. Black and M. Scholes. The pricing of options and corporate liabilities. *J. Pol. Econ.*, 81:637–659, 1973.

[6] J.R. Brannan and W.E. Boyce. *Differential Equations with Boundary Value Problems.* John Wiley & Sons, 2011.

[7] M. Braun. *Differential Equations and Their Applications.* Springer-Verlag, 1978.

[8] R.V. Craster and R. Sassi. A pseudospectral procedure for the solution of nonlinear wave equations with examples from free-surface flows. Technical Report 99, Universita degli Studi di Milano, Polo Didattico e di Ricerca di Crema, Dec 2006.

[9] A. Doelman, T.J. Kaper, and P.A. Zegeling. Pattern formation in the one-dimensional gray-scott model. *Nonlinearity*, 10:523–563, 1997.

[10] L. Edelstein-Keschet. *Mathematical Models in Biology.* Society for Industrial and Applied Mathematics, 2005.

[11] L. Edelstein-Keschet. *Mathematical Models in Biology4.* Society for Industrial and Applied Mathematics, 2005.

[12] R.E. Edwards. *Fourier Series: A Modern Introduction.* Holt, Rinehart and Winston, 1967.

[13] R.A. Fisher. The wave of advance of advantageous genes. *Ann. Eugenics*, 7:19–26, 1937.

[14] W.H. Fleming. Diffusion processes in population biology. *Advances in Applied Probability*, 7:100–105, 1975.

[15] G.R. Fowles and G.L. Cassiday. *Analytical Mechanics.* Brookes Cole, 2005.

[16] J. Goldberg and M.C. Potter. *Differential Equations A Systems Approach.* Prentice-Hall, 1998.

[17] M. Gorunescu. Prognostication of hospital beds occupancy using the kinetic modeling. In *6th International Conference on Artificial Intelligence and Digital Communications (AIDC 2006)*, volume 106, pages 19–25. Research Notes in Artificial Intelligence and Data Communications, Reprograph Press, 2006.

[18] J. Herod, R. Shonkwiler, and E. Yeargers. *An Introduction to the Mathematics of Biology.* Birkhäuser, 1996.

[19] F.C. Hoppensteadt and C.S. Peskin. *Mathematics in Medicine and the Life Sciences.* Springer-Verlag, 1992.

[20] J.C. Hull. *Options, Futures, and Other Derivatives.* Prentice-Hall, 2006.

[21] A. Ida, S. Oharu, and Y. Oharu. A mathematical approach to HIV infection dynamics. *Journal of Compuational and Applied Mathematics*, 204(1):172–186, 2007.

[22] E. Izhikevich. *Dynamical Systems in Neuroscience.* MIT Press, 2007.

[23] A.K. Kassam and L.N. Trefethen. Fourth-order time-stepping for stiff PDEs. *SIAM J. Sci. Comput.*, 26(4):1214–1233 (electronic), 2005.

[24] J. Keener and J. Sneyd. *Mathematical Physiology I: Cellular Physiology.* Springer, 2009.

[25] J. Keener and J. Sneyd. *Mathematical Physiology II: Systems Physiology.* Springer, 2009.

[26] R. Knobel. *An Introduction to the Mathematical Theory of Waves.* American Mathematical Society, 2000.

[27] E. Kreyszig. *Introductory Functional Analysis with Applications.* John Wiley and Sons, 1978.

[28] S. Krogstad. Generalized integrating factor methods for stiff PDEs. *J. Comput. Phys.*, 203(1):72–88, 2005.

[29] M.A. Lahiji, Z.A. Aziz, M. Ghanbari, and H.P. Mini. A note on fourth-order time stepping for stiff PDE via spectral method. *Appl. Math. Sci. (Ruse)*, 7(37-40):1881–1889, 2013.

[30] K.J. Laidler and J.H. Meiser. *Physical Chemistry.* Houghton Mifflin Company, 1995.

[31] J.R. Ledwell and A.J. Watson. Evidence for slow mixing across the pycnocline from an open-ocean tracer-release experiment. *Nature*, 364(6439):701–703, 1993.

[32] C.R. MacCluer. *Boundary Value Problems and Orthogonal Expansions: Physical Problems from a Sobolev Viewpoint.* Dover Publications, 1994.

[33] M.A. McKibben. *Discovering Evolution Equations with Application Volume I-Deterministic Equations.* CRC Press, 2011.

[34] M.J. Metcalf, J.H. Merkin, and S.K. Scott. Oscillating wave fronts in isothermal chemical systems with arbitrary powers of autocatalysis. *Proc. R. Soc. Land. A*, 447:155–174, 1994.

[35] P.A. Milewski and E.G. Tabak. A pseudospectral procedure for the solution of nonlinear wave equations with examples from free-surface flows. *SIAM J. Sci. Comput.*, 21(3):1102–1114, 1999.

[36] J.W. Moore and R.G. Pearson. *Kinetics and Mechanism.* John Wiley & Sons, 1981.

[37] J.D. Murray. *Mathematical Biology.* Springer-Verlag, 2003.

[38] G.W. Parker. Projectile motion with air resistance quadratic in the speed. *American Journal of Physics*, 45(7):606–610, 1977.

[39] D. Powers. *Boundary Value Problems.* Academic Press, 1979.

[40] J.D. Rawn. *Biochemistry.* Neil Patterson Publishers, 1989.

[41] W.N. Reynolds, S. Ponce-Dawson, and J.E. Pearson. Self-replicating spots in reaction-diffusion systems. *Phys. Rev E*, 56:185–198, 1997.

[42] R. Rosloff, M.A. Ratner, and W.B. Davis. Dynamics and relaxation in interacting systems: Semigroup methods. *Journal of Chemical Physics*, 106(17):7036–7043, 1997.

[43] J.R. Taylor. *Classical Mechanics.* University Science Books, 2005.

[44] L.N. Trefethen. *Spectral Methods in MATLAB.* Society for Industrial and Applied Mathematics, 2000.

[45] P. Wilmott, S. Howison, and J. Dewynne. *The Mathematics of Financial Derivatives.* Cambridge, 1995.

[46] G. Wu. Use of a five-compartment closed model to describe the effects of ethanol inhalation on the transport and elimination of injected pyruvate in the rat. *Alcohol & Alcoholism*, 32(5):555–561, 1997.

[47] J. Wu and H. Xia. Self-sustained oscillations in a ring array of coupled lossless transmission lines. *Journal of Differential Equations*, 124(1):209–238, 1996.

[48] P. Yan and S. Liu. SEIR epidemic model with delay. *Anziam Journal*, 48(1):119–134, 2006.

[49] D.G. Zill. *Differential Equations with Boundary Value Problems.* Brooks Cole, 1992.

[50] D.G. Zill. *Differential Equations with Computer Lab Experiments.* Brooks Cole, 1998.

Index